PLATO'S GHOST

PLATO'S GHOST

THE MODERNIST TRANSFORMATION OF MATHEMATICS

JEREMY GRAY

PRINCETON UNIVERSITY PRESS

PRINCETON AND OXFORD

Copyright © 2008 by Princeton University Press
Published by Princeton University Press, 41 William Street, Princeton,
New Jersey 08540

In the United Kingdom: Princeton University Press, 6 Oxford Street,
Woodstock, Oxfordshire OX20 1TW

Library of Congress Cataloging-in-Publication Data
Gray, Jeremy, 1947–
    Plato's ghost : the modernist transformation of mathematics / Jeremy Gray.
        p.   cm.
    Includes bibliographical references and index.
    ISBN 978-0-691-13610-3 (alk. paper)
    1. Mathematics–History–19th century.   2. Mathematics–Philosophy.
    3. Aesthetics, Modern–19th century.   I. Title.
    QA26.G73 2008
    510.9'034–dc22   2007061027

This book has been composed in

Printed on acid-free paper.∞

press.princeton.edu

Printed in the United States of America

1 3 5 7 9 10 8 6 4 2

WHAT THEN?

His chosen comrades thought at school
He must grow a famous man;
He thought the same and lived by rule,
All his twenties crammed with toil;
"What then?" sang Plato's ghost. "What then?"

Everything he wrote was read,
After certain years he won
Sufficient money for his need;
Friends that have been friends indeed;
"What then?" sang Plato's ghost. "What then?"

All his happier dreams came true—
A small old house, wife, daughter, son,
Grounds where plum and cabbage grew,
Poets and Wits about him drew;
"What then?" sang Plato's ghost. "What then?"

"The work is done," grown old he thought,
"According to my boyish plan;
Let the fools rage, I swerved in naught,
Something to perfection brought";
But louder sang that ghost, "What then?"

—William Butler Yeats

CONTENTS

PLATO'S GHOST

INTRODUCTION

I.1 Opening Remarks

I.1.1. Mathematical Modernism

In this book I argue that the period from 1890 to 1930 saw mathematics go through a modernist transformation. Here, modernism is defined as an autonomous body of ideas, having little or no outward reference, placing considerable emphasis on formal aspects of the work and maintaining a complicated—indeed, anxious—rather than a naïve relationship with the day-to-day world, which is the de facto view of a coherent group of people, such as a professional or discipline-based group that has a high sense of the seriousness and value of what it is trying to achieve.

This brisk definition is certainly compatible with what many creators of the artistic modernisms thought they were doing. Consider, for example, these remarks by Guillaume Apollinaire from his *The Beginnings of Cubism*, published in 1912, where, speaking of many young painters, he said:

> These painters, while they still look at nature, no longer imitate it, and carefully avoid any representation of natural scenes which they may have observed, and then reconstructed from preliminary studies. Real resemblance no longer has any importance, since everything is sacrificed by the artist to truth, to the necessities of a higher nature whose existence he assumes, but does not lay bare. The subject has little or no importance any more. . . . Thus we are moving towards an entirely new art which will stand, with respect to painting as envisaged heretofore, as music stands to literature. It will be pure painting, just as music is pure literature. . . . This art of pure painting, if it succeeds in freeing itself from the art of the past, will not necessarily cause the latter to disappear; the development of music has not brought in its train the abandonment of the various genres of literature, nor has the acridity of tobacco replaced the savoriness of food.[1]

---

[1] In Chipp 1968, 222–223.

Or the American art critic Arthur Dow, writing in 1917, who characterized modernism in these seven points (not all of which met with his approval):

1. Freedom from the constraint of juries, critics or any law making art-body, involving
2. The rejection of most of the traditional ideas of art, even to the denial that beauty is worth seeking. As this seems opposed to the principle of evolution, and is only negative, I do not see how it can be maintained.
3. Interest in the expression of each individual, whether it conforms to a school or not, whether it be agreeable or the reverse.
4. Less attention to subject, more to form. Line, mass and color have pure aesthetic value whether they represent anything or not. Ceasing to make representation a standard but comparing the visual arts with music. Finding a common basis for all the visual arts.
5. Convincing us that there are limitless fields yet unrevealed by art. C. Lewis Hind says that "Matisse flashes upon canvas the unexplored three-fourths of life."
6. New expression by color, not by the colors of things, or color in historic art. Seeking hitherto unexpressed relations of color.
7. Approaching, through non-applied design and in other methods the creation of new types of design, decoration and craft work.[2]

The modernization of mathematics is most apparent at the foundational level, but it had important ramifications throughout the subject. Indeed, its roots are in explicit mathematical practice—problem solving and theory building—and modernism emerged and was successful in mathematics because it connected fruitfully with what mathematicians were doing and with the image they were creating for themselves as an autonomous body of professionals within, or alongside, the disciplines of philosophy and science, more specifically physics.[3] It is important to stress this point, in order to obviate misconceptions about the role of foundational inquiries in mathematics. I argue that the explosion of interest in the foundation of mathematics around 1900 has quite specific historical features that separate it from earlier and later phases. When German mathematicians spoke of a new conceptual mathematics (*begriffliche Mathematik*) they were directing attention not merely to the increasingly abstract nature of their subject—mathematics has always struck some people as rarefied—but to its newly self-contained aspect and the way it was being built up independently of references to the outside world and even the world of science.

This book argues the case for a decisive transformation of mathematical ontology around 1900, not only geometry and the well-known but perhaps misunderstood domain of analysis, where I also draw out, after Epple (1999), the way in which

---

[2] See Dow 1917, quote on 115–116. Dow goes on to say that not all of these demands were new, and in particular the Japanese artist, Keisai Yeisen, was more expressive in his use of line than Kandinsky. I am indebted to Linda Henderson for bringing this to my attention.

[3] Science on any definition is not synonymous with or reducible to physics. But physics was the paradigmatic science for mathematicians throughout the period covered by this book, the one with the most significant overlap and the greatest use of advanced mathematics, so for present purposes I have sometimes allowed the terms "science" and "physics" to elide, in keeping with the usage of the time.

traditional ideas persisted for a time behind a modern façade of Cauchy-Weierstrass style definition, but in algebra with its shift toward set theory.[4] I give my reasons below (see §2.2) for not finding Cauchy's rigorous real analysis modernist; they derive from my view of Cauchy's traditional mathematical ontology. There was, in addition, a radical transformation in mathematical epistemology. Mathematicians have almost always been concerned with proof, but the autonomous ontology raised new standards and eventually led to a decisive break with previous formulations. Thus, even in analysis, the Weierstrassian impulse led more and more to a reliance on algebra and arithmetic, and an interesting polemic between Karl Weierstrass and Leopold Kronecker, the leaders of the Berlin mathematics department, the largest in the world during the later nineteenth century. The formal language of those in Peano's school in Turin a little later marks a further step away from geometric intuition. Then in geometry itself the idea of relative consistency proofs, and the analysis of geometries in Hilbert's *Grundlagen der Geometrie* (1899), went even further in its analysis of proof. Finally, there was Hilbert's remarkable idea of Proof Theory.

In this context there was a profound reawakening of interest in the philosophy of mathematics, which was matched by many coming from a philosophical background (notably but not exclusively Bertrand Russell). The overlap of logic, mathematics, and language was particularly stimulating at this time, and is worked through here in a more inclusive way than hitherto; much still remains to be done. Inasmuch as this activity also overlaps with contemporary debates in psychology, such an investigation has been greatly eased by the work of Martin Kusch (1997). A significant methodological feature of the present work is its systematic attempt to reopen doors to history of science and the philosophy of mathematics.

I.1.1.1 WHAT IS IN THIS BOOK

Francis Bacon compared the activity of the historian to raising a gigantic wreck from the ocean floor.[5] All one can hope for is that fragments can be brought to the surface, that they are not too damaged in the process, and that their relative positions are not too disturbed. I have tried to bring more to the surface than has been attempted before, and in so doing I have been assisted by the work of many active historians of mathematics. I believe the main novelties of this book are these:

It gives a new picture of the shift into modern mathematics, one that sees it as a characteristically modernist development.

It uncovers a rich interconnected web of mathematical ideas joining the foundations of mathematics with questions at the frontiers of research.

It does this across every branch of mathematics.

It shows that these questions, patently internal as they are, forged links in the minds of contemporary mathematicians between mathematics, logic, philosophy, and language.

---

[4] As discussed in Avigad 2006, Corry 1996, Ferreirós 1999, Gray 1992a, and other writers.

[5] Bacon, *The Advancement of learning*, Book II; the comparison is made in the Everyman edition, ed. G. W. Kitchin, 73. The origins of the metaphor are eloquently discussed by Anthony Grafton in the *New York Review of Books*, October 5, 2006.

Accordingly, it brings to light aspects of the life of mathematics not previously discussed: notably, interactions with ideas of international and artificial languages, and with issues in psychology.

It integrates a number of issues too often treated separately: philosophy of mathematics in the hands of Husserl and Frege, Russell and Peirce; interactions between mathematics and physics; theories of measurement.

It opens a door to the history of the philosophy of mathematics.

In particular, it brings in the important figure of Wilhelm Wundt, who is almost always omitted.

It also reminds the historian of mathematics of a number of forgotten figures: Josiah Royce, W. B. Kempe (for once, not for his fallacious proof of the four-color theorem).

It brings in a number of philosophers hitherto marginalized in the history of mathematics (Herbart, Fries, Erdmann, and Lotze) or even completely forgotten, such as W. Tobias, G. F. Lipps, and S. Santerre.

It documents how the very names of Leibniz and Kant were profitably taken as marking major, rival, and evolving positions in logic and the philosophy of mathematics.

It takes the popularization of mathematics around 1900 seriously, as indicative of the fundamental changes mathematics was undergoing, and being seen to undergo.

It looks at the importance of the history of mathematics, which had a resurgence in the period, and considers the ways in which it was written.[6]

It raises the issue of modernism in theology, which coexisted with the situation in mathematics but incurred outright opposition and repression.

The book shows that major changes occurred almost everywhere where mathematics was done at a high level. It therefore argues that taken together all the changes in mathematics here described and the connections to other intellectual disciplines that were then animated constitute a development that cannot be described adequately as progress in this or that branch of mathematics (logic and philosophy) but must be seen as a single cultural shift, which I call mathematical modernism. Such a cultural shift was enabled by the growing autonomy of the mathematicians within the academic profession. The suggestion that the changes in mathematics around 1900 constitute a modernism, much as contemporary changes in art, literature, or music do, is intended to be provocative, because the literature on the artistic modernisms is vast and full of disagreements. I give my grounds for establishing the case on a core definition of modernism, and I believe that mathematical modernism provides a handle with which to grasp otherwise sprawling developments, as well as a store of analogies. Readers familiar with the literature on cultural modernisms are invited to bring its various approaches to the history of mathematics, and see which prove fruitful. One analogy that is pursued here is anxiety, a well-established theme in writing about modernism; I find that in mathematics, too, anxiety was a growing presence. Another theme I have adopted is that of the deep modernist interest in the history of its subject, which was often used as a way of legitimizing the new style, at least in the eyes of its adherents.

---

[6] Discussed briefly in Gray 1998.

I.1.1.2 THE SPREAD OF MATHEMATICAL MODERNISM

The arrival of modernism in mathematics cannot be understood as a simple event with a definite arrival time and a clear before and after. As with any modernism, there are a few important figures who are early enough to disturb the chronology: two notable examples are the German mathematician Bernhard Riemann and the American philosopher and polymath Charles Sanders Peirce. They illustrate how questions of its success and its spread, especially of a causal kind, have both an institutional and a national aspect. Riemann was centrally placed in Göttingen and managed with no more than a dozen papers to shape the work of two generations of mathematicians, first in Germany and then further afield. Peirce marginalized himself in the already mathematically marginal American scene, and as a result his impact has been greater on historians than on his contemporaries.

Mathematical modernism was strongest in settings where there already was a professional separation between mathematics and physics, and where there was an existing commitment to research for its own sake. This was the case in Germany, especially in Göttingen and Berlin, the dominant places for mathematics in the world in the late nineteenth century. It was also true in the much thinner atmosphere of the United States, whose fledgling mathematics departments took their lead from Germany where many of the staff had received their training. In France, modernism arrived with what I shall argue, following Gispert (1991), was the emergence of the first coherent group of researchers, as opposed to gifted individuals, and at a time when the cherished link between mathematics and the sciences in France was at its most attenuated. In Italy, too, it was the school around Giuseppe Peano that most thoroughly identified with modernism. In Britain, however, the movement first needed to establish a recognizable group of pure mathematicians, something that had been long delayed by the successes of applied mathematics as a research subject in Cambridge and its visible technological triumphs.

If there is a single exemplary work that ushered in modernism, it is perhaps Hilbert's *Grundlagen der Geometrie* (or, *The Foundations of Geometry*), although Dedekind's *Was sind und was sollen die Zahlen* (or, *What Are Numbers and What Are They For*) would be another strong candidate.[7] But to say this is to make clear that it is not individual works that change the world, but the messages they convey. Those messages go from people to people and, to succeed, must articulate genuine concerns that are also expressed elsewhere. The intellectual concerns therefore must be those of people able to advance them, and so those of significant groups of people with the right opportunities. The professional situation of mathematicians, in particular their relative autonomy from scientists, did not cause modernism to happen, but it enabled it and it promoted it. The social and institutional settings are therefore also necessarily considered in this account.

It is important to my argument to describe an international change in mathematics. One way I have tried to manage the large amount of material this generates is to put mathematicians in the lead. I have been most interested in tracing how mathematicians saw their subject, and the images of it that they tried to present.

---

[7] It is interesting to note that this was one of the books Einstein and his friends read in 1902 in their Olympia Academy, along with Poincaré's *La science et l'hypothèse*, and various other works. See Howard 2005, 36. For Dedekind's influence on Hilbert and Bernays, see Sieg and Schlimm 2005.

There are other books that could profitably reverse the focus; I hope they will be written and find this one useful. But it has been a deliberate decision to give the mathematicians more space, and if that seems to you, the reader, to distort matters unduly, then I look forward to your reply. Distortion, inevitably, there has been; not everything could be brought to the surface and choices had to be made.

Almost everyone who writes about mathematics for an audience larger than the professional mathematicians apologizes for the difficulty of the subject. I do not. Rather, I think that the issues the mathematicians confronted have a directness and an importance which can be brought out with a minimum of technicalities and a judicious use of analogy. There is no reason why mathematics should be more alarming than many a topic in the history and philosophy of science, which covers alchemy, ancient cosmology, and everything down to zoology with a wealth of detail, many implicit references to the history not only of Europe but all the world, draws in many languages, and so forth. I have tried to show that the mathematicians' questions were generally reasonable, asked for good reasons, and answered in ways that make a certain amount of sense, and because I am concerned to restore voices to the original debates, I have often let the mathematicians speak at length. I invite the reader to meet them halfway. A few passages where I judged that some readers will be helped by explicit mathematics, while others will not, are set off from the rest with an asterisk and the use of an italic font. There is also a glossary of mathematical terms at the end of the book.

I.1.1.3 A FIRST OVERVIEW

The first chapter of this book is an introductory one. It sketches out the themes of modernism in mathematics, then indicates the relation of mathematics to logic, philosophy, and language at some length, and concludes with some remarks about mathematicians and their audiences. This sets out the principal themes of the book. In the chapters that follow, I describe three broad phases in the construction of modernist mathematics: a period before modernism; the piecemeal arrival of modernism in various disciplines; and then modernism's open avowal as a new, and improved, way of doing mathematics. Within each of these chapters, the order of mathematical topics is broadly the same: geometry, mathematical analysis, algebra, philosophy, and logic. This is so that readers may navigate the material in a number of ways and be somewhat selective if they wish. I then widen the picture and look at the changing relationship between mathematics and physics, at the theory of measurement, and at mathematics as a subject fit for energetic popularization. I then widen the picture still further and consider the connections between mathematics and languages both natural and artificial, and the vexed subject of the psychology of mathematics. A strong theme of the period was investigations of how we can know mathematics, and many answers were given.

The final chapter looks at the way some of these themes worked out after the First World War. The war changed the intellectual landscape in many ways, and would require a book of its own to describe, but I feel strongly that it would be wrong to stop the stories I have to tell in 1914 or 1918. I have not attempted a full account, but I hope to have seen the debates of the earlier chapters safely home. At the end I raise, but by no means entirely answer, the question of what establishing that there was a modernist movement in mathematics enables us to say. I discuss the extent to which modernism succeeded in mathematics, and how platonism became the default philosophy of most mathematicians.

I.1.1.4 MODERNISMS

How, if at all, were the forces promoting modernism in mathematics related to the better-known modernisms of twentieth-century cultural life?[8] It is clear that some artists and writers picked up on these changes in mathematics. Henderson's treatment (1983) of modern painters' ideas of non-Euclidean geometry and the fourth dimension is exemplary. The use of quasi-mathematical forms in the sculptures of Brancusi, Gabo, and Pevsner is another indication worth analyzing. It is also possible to trace literary implications, although much remains to be done. In the substantial case of Musil, who trained as an engineer, there are many mathematical allusions in *Der Mann ohne Eigenschaften* (*The Man without Qualities*). Considering that Hermann Hesse left school with little formal education, his brief remarks about mathematics in his novel *Das Glasperlenspiel* (*The Glass Bead Game*) are strikingly accurate.

Reference to the best-established cultural modernisms (art and literature, to which one must add music) lead to the ultimately insoluble question of how the modernism in mathematics relates to them. There are certainly occasional overlaps, in the person of Hausdorff, for example (see Mehrtens's book *Moderne Sprache Mathematik*— Modern Language Mathematics). But there are significant departures, most notably in the political dispositions of the actors. Forman's and Ringer's analyses of the decline of the German mandarinate after World War I are the very opposite of an account of cultural modernizers turning in hostility on the bourgeoisie.[9] That said, the political dispositions of musicians and authors was by no means simply, or uniformly, left wing; painters seem to have taken up the most provocative political positions.

There seems to have been little direct influence of the broad cultural shifts into modernism on the practice of mathematics. It is indeed hard to see how a mathematician, drawing whatever inspiration from a cubist painting or James Joyce's *Ulysses*, could do different mathematics, although Hausdorff seems to have been open to such influences. If anything, the reverse seems to have been the case. It therefore seems more profitable to see what historians' discussions of modernism in other spheres can helpfully bring to the mathematical case.

A proper comparative historiographic analysis would be dauntingly long (although instructive), and steps have been taken in this direction by a number of writers. A few tentative, and cautionary, points can be made. The paradigm case of literature offers the same sense of a collective change adopted by a professional group, away from naturalism toward an autonomous practice. It offers the same qualitative mixture of continuities and changes, shifts in critical attitudes and in popular opinion. There is a central focus (for literature, in France; for mathematics, Germany) and the same absence of a leadership. Perhaps even more interestingly, literature offers the same range of awkward cases that obstruct tidy definitions and snappy chronologies. For Hermann Melville, read C. S. Peirce.

---

[8] In this connection, see Everdell's richly informative cultural history, *The First Moderns* (1997), which covers innovation across many of the arts and sciences and gives mathematics pride of place.

[9] See Forman 1971 and Ringer 1990.

Much the same can be said of painting and sculpture, and also of architecture, as Geoffrey Scott demonstrated in his *The Architecture of Humanism* (1914), but the case of music is more challenging. Here it can be argued that there was not a modernist shift at all. The standard argument that there was one is, one might say, Viennese. It has as its central figures such composers as Brahms, Bruckner, Mahler, then shifts us into Schoenberg, Berg, Webern. It is not at all obvious that the French passage centered on Debussy or that the eastern Europeans Bartok and Stravinsky fit into a modernist transition. The absence of an agreed alternative to the classical model is worrying, all the more so when the twelve-tone movement itself seems to have run out of energy. Careful writers on musical modernism take one of two positions: either they stay close to Vienna, or, as a few do, they argue that indeed modernism can be consistently defined in music.

The awkward case of music can be used to make the analysis of modernism in mathematics more precise, and has indeed been to some extent taken on board in formulating the present thesis. It should certainly counsel against any easy argument to the effect that there was a widespread shift into modernism and mathematics joined in. Rather, it reminds us that modernism is a twofold category. At one level it belongs to the actors themselves. On my definition of the terms the original protagonists must adopt recognizably modernist positions and do recognizably modernist things in a more-or-less coherent way, or my whole analysis collapses. At another level, however, the ascription of the term is the historian's. Historians tidy up the past, they identify trends that were at least partly hidden, and when they do so they have to argue for the legitimacy and the cogency of their actions. The music example is therefore a spur to an analysis of the extent to which the present account of modernism in mathematics conforms to the paradigm cases of literature and art, and how much it is problematic.

I shall argue that the mathematical case conforms rather strongly. In view of the high degree of independence of mathematics from the cases of literature and art, and the close agreement of chronology, this looks like a case of what biologists call convergent evolution. Some features of bats and birds, or ichthyosaurs and dolphins, are alike because the requirements of efficient flying or swimming promote them, but one species does not inherit them from the other. The common features in the present case are hard to discern, but the sheer size of society, its extensive diversification, the existence of cultural activities remote from immediate practical needs, and their high degree of cultural hegemony are certainly present in each.

If the case for modernism in mathematics is established, it is possible to examine a number of related historiographic issues. I have argued (in Gray 1998) that mathematicians have written the history of non-Euclidean geometry in ways that reflect the modernist perspective, and indeed to some extent distort the history accordingly. I shall also argue here that views on mathematics, and especially on the proper place of pure mathematics, affected the creation of the Greek canon. In each case, the simultaneous emergence of the modernist movement with the first modern historians of ancient mathematics invites analysis. A case treated at length by Peckhaus (1997) is particularly worth considering: the Leibniz revival of the turn of the century.[10]

---

[10] Here is the place to mention another book that deserves to be better known, Volkert's *Die Krise der Anschauung* (The Crisis of Intuition), 1986. It has a Lakatosian interest in monsters, and it covers a longer

I.1.2 Mehrtens's Moderne Sprache Mathematik

The first historian of mathematics to address the theme of modernism specifically was Herbert Mehrtens, in his book *Moderne Sprache Mathematik* of 1990. This book not only drew together a number of investigations into the history of recent mathematics, it generated several subsequent studies that included a discussion of the "modern" in their accounts of the history of the mathematics of the twentieth century. My book is close to his, and it seems appropriate to discuss it explicitly now.

Mehrtens came to his book after writing a number of social-historical accounts of mathematics, and, shamefully, when he wrote it, sources that would have enabled him to write a book on mathematics in the Nazi time were kept closed to him.[11] Only in the last two or three years has a younger generation of German historians begun to gain access to the relevant archives.[12]

The idea that mathematics underwent a profound transformation around 1900 is not new. It was the view of many of the protagonists at the time, and it was argued for intelligently by Ernst Cassirer (1910). But the first historian of mathematics to associate this view with modernism and to articulate it in a way that brought fresh insights was Mehrtens, and he did this partly through the richness of his scholarship and partly through the way he situated the transformation of mathematical practice in a wider social context. Mehrtens posited a conflict between moderns (*die Moderne*) and countermoderns (*die Gegenmoderne*) that began in the nineteenth century and which he pursued to its end in the 1960s. This is not only a characterization of two opposing tendencies that Mehrtens perceived to have been at work, it reflects an ambiguity he took to be at the heart of the very idea of the modern and which he saw in particular as essential to any adequate account of the Nazi time. He innovated further in the history of mathematics by following Foucault into discourse analysis and stressing the difference between the language of mathematics (its written texts) and the words of the mathematicians. "Modern" mathematicians claimed to work with words, not with (idealized) objects or transcendentally defined objects, and "countermoderns" objected that the very fundament of mathematics had been removed. But, since the moderns won, it is their language that dominates the discourse about mathematics, and so it was only possible to write this history, said Mehrtens, either from within the modern perspective or impossibly and ahistorically from without. This position is mandated, for Mehrtens, by his sense that modernism in mathematics is inseparable from the arrival of modernity, the social condition of the modern world, which he discusses at various points throughout his book.

The conflict opened, or rather, Mehrtens's account of it opens, with Cantor's paradise of set theory and transfinite numbers. Dedekind and Peano are discussed for their radical reformulation of the number concept. Once these dramatic novelties

---

chronological sweep than this book, but where the periods overlap so too do many of the examples. It does not specifically address the topic of modernism, but by reaching to the middle of the twentieth century it shows more of the philosophical consequences of what I argue was a modernist transformation of mathematics.

[11] This also explains why Mehrtens, who speaks and writes excellent English, wrote the book in German. It remains a minor scandal that no one has translated the book, which it would still be worthwhile to do.

[12] See, among several of Volker Remmert's essays, Remmert 1999 and Remmert 2004.

have been explained, Mehrtens turned to the more vexed question of space and introduces his second modern, Bernhard Riemann, and the first of his counter-moderns, Felix Klein. More precisely, Riemann opened up a separation in mathematics that grew into the distinction between the modern and the countermodern. Klein, with his deep attachment to meanings in mathematics and to the uses of mathematics in science, went one way. His insistence on the importance of intuition made him the opposite of Riemann in certain respects and, later, of David Hilbert in others, for Hilbert, of course, is Mehrtens's general director of the modernism project. The conflict grows ever more heated: what on one view brings monstrous functions and loss of control over the concepts and the purpose of the mathematical enterprise is seen on the other as scope for freedom and creativity.

Hilbert in particular has a major role, and his work is presented ironclad as a program to make all of mathematics abstract, axiomatic, and internally self-consistent. He took up the cause of Georg Cantor, and had the support of Zermelo and, albeit on his own more radical terms, Felix Hausdorff. The opposition was first represented by Klein, then Henri Poincaré, and after his death by Hermann Weyl and above all by Luitzen Brouwer. And so the scene is set for the *Grundlagen Krise* (Foundational crisis) of the 1920s, portrayed as a clash not over the foundations of mathematics, but as an upheaval in the ideas of truth, sense, object, and existence in mathematics. An upheaval, moreover, with an intense social and political context. Victory, Mehrtens suggests, went to the formalists and created in Hilbertian proof theory and Turing's analysis of calculation a paradise for machines. In the next generation the victors were exemplified by Bourbaki, and they strove for rationalization, objectivity, effectiveness, and economy of thought. The losers made a fetish of intuition, turned it into a racial category, and some, led by Ludwig Bieberbach and Oswald Teichmüller, became Nazis.

Mehrtens did not depict the clash of modern and countermodern as a simple battle between two sides. Certainly he saw it as a deep division between two views of mathematics, and one that was of a piece with the other ways in which the modern world was created. But it is only if anyone descends into Nazi ideology that Mehrtens labels them "antimodern"; modern and countermodern are yoked together. Mehrtens's critique was written in a post-Marxist spirit, influenced by such writers as Foucault. Modernization, for him, is not progress, it is also part of the catastrophe of Nazism. If he is less clear that the search for meaning and a place in the world has its good side, it is only because he sees more clearly the ways in which late capitalism is antithetical to all of that.[13]

It will be clear that I believe Mehrtens identified all the major German players in the modernization of mathematics, and a number of its leading opponents. They were mostly known to have been in place before—only the discussion of Hausdorff was strikingly new in 1990—but they are richly described and held together in a particularly interesting matrix. The polarity of modern and countermodern is a genuine dialectic: both sides need each other for their full expression. They express themselves in a way that reflects the wider cultural and social changes of modernization, and which, moreover, Mehrtens sees as thoroughly ambiguous. Given his natural wish to see modernization as part of the story of the Nazi catastrophe, this is

---

[13] Two further avowedly speculative chapters close the book, which go further into cultural criticism than I need to follow here.

only reasonable. Nonetheless, my disagreements with his treatment are most noticeable at this, as it might be called, outermost level.

Where Mehrtens succeeds most impressively is in his chapter on the modernization of mathematics. Here his focus on the German community and in particular on Göttingen pays off with a subtlety of analysis that is necessitated by the awkward fact that, while it is Hilbert and Göttingen that are the hub of modernity, it is Göttingen and Klein who are the very center of a powerful and diverse identification of mathematics. Klein pushed with extraordinary effect for the institutionalization of mathematics within school teaching, technology, engineering, and science, as well as a subject in its own right. A victory then for the moderns, or the countermoderns? The answer Mehrtens gives is "both." The moderns and countermoderns are inseparable parts of the mathematics profession: they offer views that are both contradictory and complementary. They came to cohabit in Göttingen and in Germany generally to their mutual advantage because the institutional base of mathematics was broad and secure and because the moderns exercised a hegemonic role within the profession.

That said, explanations of major social movements do not sit comfortably with individuals and their actions, and while Mehrtens does well to give his central figures the autonomy they have, ultimately they are small parts of a machine that seems to have a logic of its own, giving his whole account an oddly Hegelian ring. I am not sure there is more than a resonance that links mathematical modernism to the arrival of modernity, modern capitalism, and its horrific opponent, the Nazi state. We are far from knowing much about the societies we inhabit. We do not really know, I believe, which economic and social features will prove to be essential to any dynamic account of a culture, country or community, and even if we did, why should we prefer generalities of wide application to a series of more localized but richer accounts? The new capitalisms of China and India will offer endless material for studies comparing them with the American, European, and Japanese versions, but we may yet manage better without grand overarching accounts (whether or not they also serve immediate political agendas).

It is unfortunate for Mehrtens that the Nazis shared the widespread dislike of mathematics and found mathematicians irrelevant to their purposes, just as liberals, conservatives, and communists did. They tore apart the great mathematical community in Göttingen and later in all of Germany because of their hatred of Jews, but finding no racial component to mathematics had no real interest in trashing the subject itself.[14] It so happens that Bieberbach, who had few scruples about striking poses that would advance his career, chose when the time came to be a Nazi, that Teichmüller was a passionate Nazi, and that some other figures, such as Helmut Hasse, were right-wing conservatives who had no great problems with the Nazis. It so happens that Emil Artin, one of the leading modern mathematicians and very much part of Emmy Noether's group, was a cultural modernist in many other ways, just as it does that Weyl, deeply rooted as he was in German philosophy, emigrated when the Nazis came to power. But Hasse also worked closely with Emmy Noether, as did the Dutchman Bartel van der Waerden, who chose to stay in Germany from 1933 right through the war, and Oscar Perron, who also stayed but was a courageous

[14] See Segal 2003 and also the review by Remmert, http://www.h-net.org/reviews/showpdf.cgi?path=220081064463863, H-Net Reviews in Humanities and the Social Sciences.

anti-Nazi throughout, was strongly opposed to abstract modern mathematics. It ceases to be obvious that the dance of modern and countermodern lines up very well with the grander cultural clash of modern and traditional. And indeed, with the turn back to applications in mathematics in the last twenty years, which was scarcely visible in 1990, it is less clear that objects, sense, and meaning have been so thoroughly driven out of mathematics. For all these reasons, the book's tight focus on Germany is unfortunate. Its core story is at the mercy, to some extent, of the people one looks at, and its outermost level, the account of modernity, for all its sources in contemporary writing, is to an extent a priori.

If we eschew such lofty considerations, Mehrtens's book can, however, be seen to point the way to a number of interesting questions. What, for example, was the contemporary situation outside Germany? Were many mathematicians caught up in the transition to modernism, or was it the business of only a few major figures? As some of its critics have pointed out, the book says very little about the presumably changing relation between mathematics and physics at the time. It says almost nothing about relationships with philosophers and philosophy, and despite its interests in *Sprache* (language) where discourses about mathematics are concerned, it says almost nothing about the mathematical language used in research or about the contemporary situation in linguistics. All of these topics are investigated here.

In this book, I chose to investigate how broadly an account of mathematical modernism could be drawn, going beyond its heartland in Germany, and my conclusion, evidently, is that it can be. This led me to explore the ways mathematics was thought about, and the ways it was used, in neighboring disciplines, and it took me away from the sociological dimension of Mehrtens's work. So I have not used his juxtaposition of modern and countermodern, even though I appreciate the insights it brought him and the complexity it enabled him to handle. The tensions and divides Mehrtens found in Germany and I find broadly repeated elsewhere are complicated and shifting, but they function in his book because they bridge the gap between society and the individual. I see that gap differently, and his terms would be mere labels in my setting. I have also had the advantage of more than fifteen years of research in the history of mathematics and other subjects, years in which post-modernist certainties have swelled, older Marxist certainties have diminished, and now too postmodern approaches seem to be losing what charm they had. The result is a broader picture of mathematics and its connections than Mehrtens attempted, which is perhaps more suggestive of a modernist sweep but is shorn of deliberate political overtones. I might conjecture that his list of modernists and mine agree, but I must admit that his vision of modernism is not mine.

I.1.3 Disclaimers

I.1.3.1 MODERNITY

It remains to discuss the relation of the historical developments described in this book to "modernity." This is the most nebulous of the "m" words. I have attempted to make clear what I mean by modernism; modernization is the process by which something is made modern, but modernity is something like a condition. It might be the condition of a state or society, it might be people's perception of that condition. Authors of books and articles, museum and gallery curators presume on occasion that we inhabit it, or have inhabited it. It is associated with novelty, rapid change, loss of control, social disturbance, the automobile and the plane, the telephone, films, social planning, all manner of aspects of twentieth-century living in Europe and America especially. It is in danger, like the concept of time in the well-known aphorism, of being something we understand until we think about it. It is certainly unclear what aspects of modern life particularly qualify for discussion under the heading of modernity. Unfortunately, this has not stopped a stream of authors writing about it until one can have the impression that it is a well-understood term— or at least that it is well understood by everyone else, and so may be written about without further thought.[15]

I do not doubt that people's sense of themselves and their society matters. There are times in this book when I allow a glimpse of the wider world and its concerns, because that glimpse amplifies the story I wish to tell. But modernity, and Modern with a capital "m," are vast and misty generalizations I have not tried to bring to the page.

I.1.3.2 WHAT THIS BOOK IS NOT

Of the many things this book is not, some are worth mentioning. It is not a whole history, or even a package history, of mathematics in the period around 1900. It presents an argument that mathematics underwent a modernist transformation, and to that end it deals with what I believe are enough good examples to make a persuasive case. Topics only qualified for inclusion insofar as they engage with the case. Among the topics that are missing are a number which would push the case even further, but I judged excessive. For example, and with some regrets, I left out the work of Noether and her school on structural algebra in the 1920s and '30s, although it is impeccably modernist, because it would have placed even more demands on the reader.[16] For the same reason I have not pushed the theme of modern topology, so well argued by Moritz Epple (1999). It has been persuasively argued that there was a revolution in probability theory around 1900, culminating in the work of the Russian mathematician Andrei Kolmogorov, but I chose not to discuss it for the same reasons.[17] And I decided not to say much about responses to Einstein's theories of special and general relativity, not just from a feeling that this was pushing on a (fast-moving?) open door, but because it was a dramatic change in our ideas about physics rather than one in mathematics.[18] The same goes for nineteenth-century studies in the psychology of perception that questioned whether we see Euclidean or non-Euclidean space, investigations of the horopter curve, discussions of whether space or time might be discrete, and such topics.[19]

Nor do I deal in any detail with developments that might imperil a thesis I have not argued for (which would claim that modernism deeply affected every branch of mathematics). There are indeed some branches of mathematics that did

---

[15] Two good books that can be mentioned here are Berman 1982, and Kern 1983.

[16] For accounts of this, see Corry 1997, 2007, and McLarty 2006 and forthcoming papers.

[17] See Krüger, Daston, and Heidelberger 1989 and on Kolmogorov see Plato 1994 and Hochkirchen 1999.

[18] On the responses by philosophers, one can consult Hentschel 1990. Otherwise, the Einstein literature is huge and growing. Many interesting papers were given at the rolling conferences in the Einstein year 2005 and will doubtless find their ways into print; the series Einstein Studies is always interesting, and as good a place to start as any are Stachel 1995 and, for different reasons, Pais 1982.

[19] On these matters, see Heelan 1981.

not modernize. I give my assessment of the importance of them at the end of the book. I do not believe either that it is necessary to argue that modernism won completely, or that the survival of more traditional forms of mathematics weakens my case. Indeed, there were major mathematicians who were only partly persuaded of the modernist cause, Poincaré and Weyl among them. But I do believe that by far the major part of mathematics was transformed in a modernist way. That is for the reader to judge, and while I am offering a view it cannot be the only view, even if it is found to be convincing—the historical record is far too large for that. Instead, I have tried to make it clear where I stand, and to survey enough of mathematics to establish how sweeping the modernist transformation was. Modernism is many things, and if taking some interpretation of it in, say, the history of music proves a productive spur to asking worthwhile questions in the history of mathematics, then that will be some justification for emphasising it so strongly here. I have argued for the existence of substantial links between mathematics and physics, philosophy, logic, psychology, and even the study of language in the period, in the belief that it will be valuable if these links survive further analysis and generate further research.

However, I claim no expertise in those areas. The relevant parts of this book offer a view, as it were, outward from the history of mathematics. I hope experts in those domains will be enabled to offer views outward from where they stand and toward mathematics, thus bringing about a dialogue between history of mathematics and other parts of the history of science and intellectual history.

Finally, I do not claim that the modernization of mathematics was part of a broader cultural push, animated by concurrent changes in the arts. I do claim that the changes were similar in kind and were helped along by a growing diversity and specialization in all walks of cultural and intellectual life. But the mathematical ones described here and the better-known artistic ones happened independently.

I.1.3.3 PLATO'S GHOST

I surely do not need to say that I am far from believing that I have brought this book to perfection. Yeats's poem, with its evocation of hard work leading to worldly success, culminating in a claim for perfection that Plato's ghost only too plausibly denies, stands at the front of this book because it fits the mathematical scene I have tried to describe here remarkably well. Not only is Plato the presiding ancestor of mathematicians, and Platonism commonly supposed to be their default philosophy, philosophers too work in his shadow. Hard work, for many of those discussed here, did lead to worldly success and to lasting fame. More pertinently, the mathematicians often did believe that they had brought something to perfection, and written, at last, the definitive account of this or that aspect of mathematics, or of its philosophy, or its relation to logic or the natural world. Each time, they were mistaken. They may have penetrated deeper than anyone before them, but only to make deeper mistakes. Who better than the ghost of Plato to ask them if indeed it was all over, if nothing more needed to be said?

I.1.4 Acknowledgments

It is a pleasure the thank some people for their particular help. Moritz Epple, José Ferreirós, and Erhard Scholz read the entire first draft of the book and made a large and generous set of suggestions for how it could be improved as well as rooting out errors. Many conversations with them over the years as well as their impressive publications have been a constant stimulus to me. A number of authors wrote essays I was able to read before they were published; for that and for numerous helpful conversations I wish to acknowledge their help here: Jeremy Avigad, Leo Corry, Philip Ehrlich, Colin McLarty, and Jamie Tappenden. It should be clear how much I owe to them at specific points in this book. Jed Buchwald's invitation for me to visit the Dibner Institute at MIT was the spur to start work on the book back in 1997, and his continuing support and that of the Dibner Institute and the people there has been a great help. Many conversations over the years with Martin Kusch have been a great source of inspiration as well as information. I also wish to thank Tom Archibald, June Barrow-Green, Mike Beaney, Umberto Bottazzini, Paola Cantù, Christina Chimisso, Menachem Fisch, Jean Mahwin, Philippe Nabonnand, Peter Neumann, Helmut Pulte, David Rowe, Steve Russ, Norbert Schappacher, Joachim Schwermer, Klaus Volkert, Scott Walter, and my colleagues at the Open University and the University of Warwick for their continuing interest in the history of mathematics. Thanks go too to the Department of the History and Philosophy of Science at the University of Cambridge, where being an Associated Research Scholar has not only been stimulating but opened up the resources of the University Library and other collections.

I also wish to thank everyone at Princeton University Press who helped with this book: Alice Calaprice, Kathleen Cioffi, Anna Pierrehumbert, Dimitri Karetnikov, and Vickie Kearn.

To Sue, Martha, and Eleanor go thanks of another order, mostly private, for the life and hope I have shared with them so far.

I.1.4.1 PERMISSIONS

Some material in this book is taken from earlier essays I have written and reworked to a lesser or greater extent for publication here. Copyright permissions have been sought for extracts from the following:

Material from Jeremy J. Gray, *Janos Bolyai, Non-Euclidian Geometry, and the Nature of Space*, ©2003 Burndy Library, is reproduced by kind permission of The MIT Press, pp. 49–55.

Material from Michael Friedman and Alfred Nordmann, eds., *The Kantian Legacy in Nineteenth-Century Science*, ©2006 Massachusetts Institute of Technology, is reproduced by kind permission of The MIT Press, pp. 295–314.

Material from Jeremy J. Gray, Languages for mathematics and the language of mathematics in a world of nations, in *Mathematics Unbound: The Evolution of an International Mathematical Community, 1800–1945*, pp. 201–228, K. H. Parshall, A. C. Rice (eds.), American and London Mathematical Societies, **23**, Providence, Rhode Island (2002), is reproduced by kind permission of the American Mathematical Society and the editors, K. H. Parshall and A. C. Rice.

Material from Jeremy Gray, Anxiety and abstraction in nineteenth-century mathematics, *Science in Context*, 17.2:23–48 (2004), Cambridge University Press, is reproduced by kind permission of Cambridge University Press.

Material from Jeremy Gray, The foundations of geometry and the history of geometry, *The Mathematical Intelligencer*, 20.2:54–59, Springer, The

1.2 Some Mathematical Concepts

The nineteenth and early twentieth centuries saw the introduction of a number of significant novel ideas that were to be of major importance in the years after 1880. For convenience, a number of these that will occupy us later in the book are noted here, with an indication of their value. For brevity the list is merely indicative, loosely thematic, and not chronological.

In *Geometry*

Non-Euclidean geometry. A consistent metrical geometry different from Euclidean geometry, its discovery showed that physical space is not necessarily described by Euclidean geometry.

Projective (non-metrical) geometry. A geometry without metrical concepts (angles, lengths, areas, etc.) only those of incidence; more fundamental than Euclidean or non-Euclidean geometry.

The Kleinian view of geometry (Erlangen Program). A hierarchy of geometries within projective geometry; the first clear association of groups of transformations with a geometry, elucidating the idea of different types of geometry.

Axiomatic geometry. A key place for the implicit definition of mathematical terms that are otherwise meaningless; axiomatization became a creative method for producing novel mathematical structures.

n-dimensional geometry. Geometry (Euclidean, non-Euclidean, projective, or of any other kind, such as differential geometry) in any number of dimensions, not merely two or three.

In *Analysis*

The natural numbers. Defined mathematically by Dedekind, Peano, and Frege.

The real numbers. Defined mathematically by Dedekind and Cantor (and others) in terms of infinite sets.

Continuous but nowhere differentiable functions. Highly counterintuitive, they show that mathematical objects, and therefore physical processes, need not be analytic or obey the naive theory of the calculus.

Continuous curve. A naive concept that admits different, more precise definitions, and allows for counterintuitive examples, such as curves with no length and curves with finite area. The challenge was to match intuitions with proofs: even the Jordan curve theorem (a closed curve which does not cross itself has an inside and an outside) is surprisingly hard to prove.

Measure theory. A fundamental reformulation of the theory of integration independent of the concept of area.

Topology. A branch of mathematics underlying analysis (especially but not exclusively subsets of the real line), complex function theory, and geometry; the abstract setting for the study of continuity and properties invariant under continuous change.

In *Algebra*

Group theory. A group is a set with a composition law; group theory is an abstract setting for generalizing addition (of numbers), composition (of functions), and successive transformations in geometry. Finite groups were shown to have various distinguishing properties, opening the way to a structural classification of groups.

Algebraic number theory. The study of novel types and classes of integers and their properties, including prime and ideal factors.

Galois theory. A theory of groups and fields that arose from earlier studies of the solution of polynomial equations and became the paradigm structural theory in modern algebra.

In *Set Theory*

Set theory. As a foundation of analysis and algebraic number theory, it was introduced by Dedekind. The discovery that there are both countable and uncountable sets, opening the way to a theory of infinite sets, is due to Cantor.

Ordinal and cardinal numbers. The generalization of familiar concepts to infinite sets of arbitrary sizes. Their properties are described by the theory of *transfinite arithmetic.*

In *Logic*

Intuitionistic logic. A logic characterized by the claim that the law of the excluded middle (a statement is either true or false) does not necessarily apply when infinite sets are under consideration.

Formal languages. Introduced to make logic more precise and to handle distinctions almost inaccessible in natural languages.

Proof theory. A mathematical analysis of mathematical arguments designed to provide a rigorous account of all mathematics.

1st- and 2nd- order logics. A logic is first order if its quantifiers (all, some, none) refer only to elements of a set of objects, but second order if they refer to sets of objects.

1

MODERNISM AND MATHEMATICS

1.1 Modernism in Branches of Mathematics

The origins of modern mathematics can be found in the mathematical practices of the nineteenth century. It has become a commonplace that the nineteenth century saw the rigorization of analysis under the slogan, coined by Felix Klein in a public lecture in 1895, of the "arithmetization of analysis."[1] Klein was then making his bid to be the leading mathematician in Germany, with a vision of the subject as a whole, and, as he was eager to point out, the arithmetization he was criticizing underestimated the flourishing nineteenth-century line in applied mathematics, but it is true nonetheless that analysis was rigorized, and indeed based on arithmetic. However, this was only intermittently the aim of mathematicians, and their motives were various (correct reasoning, proving theorems, resolving contradictory answers, obtaining good applications, and pedagogy belong among them).

The late nineteenth century saw many overlapping kinds of mathematics being done, as well as a growing awareness of the possibility of error in mathematical reasoning. Attempts to give some "established truths" the security of a decent proof can seem obscure or pedantic—as they did to many physicists—but once the point is successfully put across that a proof is lacking, the search for one can simply become part of accepted mathematical practice. Developments in geometry displaced old "certainties" about the nature of mathematics and mathematical objects. The discovery of non-Euclidean geometry by János Bolyai and Nicolai Lobachevskii (and, if one is charitable, Carl Friedrich Gauss) resolved ancient difficulties about the parallel postulate in the most contentious way. Mathematicians finally accepted their work after it was given foundations in differential geometry by Bernhard Riemann, who was widely recognized as the most original mathematician of the mid-nineteenth century, and Eugenio Beltrami. It was then speedily incorporated into projective geometry by Klein. This, the most famous "revolution" in mathematics, put an end to

---

[1] Klein 1895.

any idea that mathematics was, as it were, distilled science. If there are two distinct geometries, then neither can be necessarily true.

The situation was, however, even more complicated and difficult for mathematicians. A chain of writers from Gauss and August Crelle at the start of the nineteenth century to Moritz Pasch and Bertrand Russell at the end found problems with the very apparatus of Euclid's geometry: terms once obvious seemed harder and harder to define, gaps in reasoning harder and harder to fill. There were discussions of what counts as a plane, for example. Hitherto unnoticed gaps in reasoning (or missing axioms) were picked out and filled, most famously by Pasch as part of wholesale rewrite of elementary (projective) geometry. It was noticed, for example, that the opening definitions in Euclid's *Elements* do not actually succeed in defining anything. Consider the first: "A point is that which has no part." This is very evocative if you know what a point is, but it is of no use if you do not.[2] Deliberate errors (such as the famous purported proof that all triangles are isosceles) were put forward to highlight the degree to which Euclidean arguments presupposed diagrams and might for that reason be erroneous. These, and other criticisms I shall look at, all cumulatively undermined previous confidence that Euclid's *Elements* were indeed the epitome of reasoning.

Projective geometry, by any standards the most remarkable success story of nineteenth-century geometry, in both its synthetic and analytic modes deployed points, lines, and ultimately hyperplanes at infinity in a manner unintelligible to classical geometry. Moreover, as the philosopher and historian Ernest Nagel argued many years ago,[3] the use of duality in projective geometry plays havoc with intuition and, he argued, opens the door to purely logical reasoning. This claim has recently been contested in a nineteenth-century English setting by Richards,[4] but I think it is on the mark and the English are best seen here and in algebra as a case unto themselves (and a good test case for the modernist thesis).

Modernism in geometry arrives with Hilbert, and the difference between his *Grundlagen der Geometrie (Foundations of Geometry)* and the earlier work of Pasch. As recent scholars have emphasized, from Hans Freudenthal to Elena Marchisotto and Michael Hallett and Ullrich Majer, the Italian contribution is if anything more abstract in its axiomatic approach, and it is currently being integrated into a full historical account in the work of Avellone et al. and Umberto Bottazzini.[5] Several threads need to be disentangled. The absorption of non-Euclidean geometry into differential geometry need not have provoked a crisis in the foundations of mathematics (that it did is another matter, connected to the global character of Euclidean definitions). The break with any kind of geometric intuition in defining geometric terms is characteristically modernist, however. A similar shift took place in the presentation of projective geometry. Then there are the implications for other branches of mathematics—and so for mathematics as a whole—of the introduction

---

[2] This problem is particularly acute with the first seven definitions, and it should be noted that they have recently been claimed to be interpolations; see Russo 1998.

[3] Nagel 1939.

[4] Richards 1986.

[5] Freudenthal 1962, Marchisotto 1995, 2006, Bottazzini 2001, Avellone et al. 2002, and Hallett and Majer 2004. I am indebted to Avellone et al. for the chance to read their paper, which I found very useful in writing this book, before it was published.

of the abstract axiomatic method. A significant group here is the American postulation theorists,[6] who also influenced the logician Alfred Tarski in the 1920s. Last, but indeed by no means least, there are the philosophers' responses and debates about the import of non-Euclidean geometry for philosophy in general and the philosophies of mathematics and science in particular. Poincaré's famous conventionalism arose in this setting, but the passions the topic aroused (contagious even among modern painters, as Henderson has described in her *The Fourth Dimension and Non-Euclidean Geometry in Modern Art*, 1983) are worthy of independent attention.

A long-running topic of relevance to the development of proof is the convergence of series, especially Fourier series. Another is potential theory: the persistent failure of mathematicians to prove the existence of harmonic functions satisfying arbitrary boundary conditions, although physicists found strong heuristic grounds for accepting them. These often, and for once well-told, tales bear reexamination for two reasons. They exemplify the origins of rigorous analysis in broad lines of mathematical inquiry, and they culminate in what might be called the crisis of continuity. This is the widespread feeling among mathematicians around 1900, documented in many sources, that the basic topic of analysis, continuity, was profoundly counterintuitive. This realization marks a break with all philosophies of mathematics that present mathematical objects as idealizations from natural ones; it is characteristic, I argue, of modernism.

1.1.1 Ontology and Epistemology

There are indeed two foundational aspects at work—one largely ontological, the other largely epistemological. The first is the one usually called the arithmetization of analysis. It traces a path from Cauchy's novel theory of functions in the 1820s that was based on certain limiting processes defining continuity, differentiability, and integrability through its unclear notion of the real numbers to an eventual resolution somewhere in the use of limiting processes to define the real numbers. Such a destination presupposed a satisfactory theory of the integers, whence the familiar slogan. At least for mathematicians, ontological clarity was shed on the nature of the real numbers by the ways in which they were defined from the rational numbers (which were themselves constructed from the integers), and then, in a second phase, light was shed on the integers themselves. In the first phase it was the difficult and sometimes counterintuitive behavior of limit processes that forced the nature of irrational numbers onto the attention of mathematicians. In the second phase, formulations of arithmetic led inexorably to set theory and then to deep problems with set theory, one of the many instances where it was sometimes felt that problems had only been traded in for deeper problems.

The epistemological aspect stays with the notion of continuity: its separation from differentiability, the emergence of a class of phenomena sometimes called pathological and which required special techniques to handle that must (it was suggested)

---

[6] See Scanlon 1991.

demarcate mathematics from science or philosophy. These too provoked deep problems that set theory was created to handle.

The third major domain of mathematics sees a similar transformation; indeed, the term "modern algebra" has greater resonance than modern analysis or modern geometry. The latter two sound like marketing terms today, but the first carries real weight: the arrival of structural algebra with the work of Emmy Noether and her school in the 1920s and then Bourbaki in the next generation. The historian Leo Corry has recently analyzed the history in considerable detail, laying great emphasis on the structural aspects.[7] As with applied mathematics, there was also a strong sense that these developments grew out of real questions in mathematics and were not pursued merely for their own sake. It is well known to historians of mathematics that many of the key elementary terms of modern structural algebra can be found in nineteenth-century algebraic number theory and algebraic geometry, along with proofs of many of the key results. In view of this tradition, it is necessary to characterize the modernist aspect of the matter with greater precision, and again this can be done in ontological and epistemological terms. As with the analysis of continuity, the abstraction and generality of the new ontology promoted the new epistemology.

Ontologically, the erosion of the concept of number is crucial. The most famous example is that of the quaternions, a noncommutative system of numberlike objects discovered by William Rowan Hamilton in 1843.[8] I shall argue that the asymmetrical responses to non-Euclidean geometry and quaternions are interesting and need fresh thought: geometry was put with mechanics and empirical matters, while numbers remained with logic. But the high road to structural algebra leads from the algebraic integers used even by Leonhard Euler to Ernst Kummer's more mysterious ideal numbers to Richard Dedekind's ideals, and then joins a complicated and previously not-well-understood topic concerned with Kronecker's vision of algebra and its reception. This last stands at the junction of algebraic number theory and algebraic geometry. The outcome is a concept of algebraic integer vastly more general than that of the usual integers, and an array of theorems that show just how clearly the usual integers form a special case. This aspect is the epistemological one. The two together made it possible for mathematicians to take the decisive modernist step of setting up the new theory in opposition to all previous ones, as the new autonomous domain that includes, in reworked form, all "that matters" of earlier approaches.

Unlike the developments in classical mathematics, contemporary changes in logic grew out of a moribund field. It is generally agreed that investigations into logic grew quietly to a halt after Leibniz, his followers like Christian Wolff, and finally Johann Heinrich Lambert at the end of the eighteenth century. They were revived by a remarkable British school, prominent among whom were de Morgan and Boole. After them, but independently, Gottlob Frege recast the traditional theory of logic insofar as it was an analysis of mathematical language, and the period of modern logic began. The nature of Frege's contribution and its significance (then or since) remains controversial, and so too does the work of his contemporaries. Some (see van

---

[7] Corry 1996, 2007.
[8] See Hamilton 1843 and, among other accounts by historians, the essays in Flament 1997.

Heijenoort 1980) see a bifurcation into algebraic logic, associated with George Boole, Charles Sanders Peirce, and Ernst Schröder, and the alternative offered by Frege and Peano. An early exponent of this view was Russell, who, it is often said, was influential in marginalizing the algebraic viewpoint. Others, such as Anellis (1995), find the algebraic school more active in its day, and more substantial in their points, than Russell appreciated, and consequently differ in their historical assessment of the period.

The first significance of the explosion of interest in mathematical logic in the second half of the nineteenth century is that it took place at all, given the hitherto quiet state of affairs. Gradually during the first half of the nineteenth century it had come to be appreciated that classical logic was inappropriate to classical mathematics: Euclidean geometry was simply not written in syllogisms. What then emerged as an acknowledged unsolved problem in the subject was posed by quantifiers. This problem would not in itself have caught the attention of mathematicians, who were in the process of giving up Euclidean geometry anyway. What caught their attention was both the algebraic side of the new logic, growing as it did out of the formal approach to problems in analysis, and the possibility that logic was a possible resting place for mathematics, deprived as it was increasingly of bedrock in the natural sciences.

The second significance of the explosion is that it promoted successive waves of investigation into logic itself. Logic, it transpired, was not a simple matter of thinking clearly, which, to be sure, might not be a simple matter to describe fully and accurately. There were levels of logic, requiring assumptions about modes of reasoning, which seemed arbitrary and did not at all have the force of the elementary logical laws (which could quite naturally be thought of as laws of thought). This recognition was slowly and painfully won, and consequently many of the earliest analyses of the relation between logic and mathematics are bedeviled by obscurities. In particular, the existence of models, and the nature of various kinds of models for various kinds of axiom systems, was obscure for a long time, an obscurity that can be measured by the distance from Hilbert to Thoralf Skolem and Tarski.

The third significance is that the enterprise was to have no satisfactory resolution. Many, some might say all, of the major problems that were thrown up have been solved; their answers are theorems in modern mathematical logic. But the answers collectively show both that elementary logic is inadequate as a foundation for any kind of sophisticated mathematics, and that no adequate foundation has enough intuitive feel or force to command assent. In short, the modernist transformation of logic itself meant that even logic no longer has a straightforward connection with simple clear thinking.

Taken together, this plunge into the depths of the nature of truth and proof in mathematics epitomizes the introspective, even anxious character of modernist mathematics. The initial possibility that logic could give naive but acceptable answers to problems in the foundations of mathematics, upon which mathematicians could then erect sophisticated theories of an entirely mathematical kind, was found to be barren. Once this was discovered, there was no alternative but to seek sophisticated foundations, adequate to the sophisticated mathematics of the day. Issues in the foundations of mathematics became issues in the very philosophy of mathematics, and because of their intimate connection to other branches of mathematics caught the attention of mathematicians.

1.1.2 Psychology and Language

Others moved in the same waters: philosophers, analysts of language, and psychologists. As Kusch has shown (1995), the German-speaking world was drawn into heated debates about the nature of the so-called laws of thought. Were they descriptive or normative? What had priority: abstract logic or the workings of the mind? The debate had many sides, each holding slightly different positions and trenchant in their criticisms of others. While arguments ranged over the whole domain of the philosophy of mind, there were significant interactions with the philosophy of mathematics. Frege's name has already been mentioned, and notoriously he drew (even sought) criticism from mathematicians; but Edmund Husserl, a lapsed mathematician who emerged as a powerful philosopher with his *Logische Untersuchungen* (or, *Logical Investigations*) of 1900 and 1901, was a greater influence in his day. At the same time Wilhelm Wundt, one of the founding fathers of psychology who also wrote widely on such topics as philosophy and linguistics, and his followers discussed logic in a way that explicitly allowed for contemporary mathematics, notably the non-Euclidean geometry that Kantians found difficult to accommodate. But Wundt also wanted to analyze thought psychologically, and his approach was endorsed by Federigo Enriques in 1906 in his well-received accounts of the nature of mathematics (his *Problemi delle scienze*, or *Problems of Science*).

Among mathematicians the most famous psychologistic theory of mathematics was Poincaré's, endorsed by Enriques even though he rejected Poincaré's conventionalism. The question here is one of understanding: What is it for someone to understand a piece of mathematics? Poincaré distinguished this from the exercise of giving a satisfactory proof, and he sought to set bounds of a finitistic kind on what the mind can do. For this reason, and because of his emphasis on intuition, he is sometimes placed with Kronecker among the forerunners of Brouwer. The psychologizing was conscious and deliberate. Both Poincaré and Enriques felt that mathematics was possible because of the evolution of the human species and the training of each individual mind. Their psychological theories were of a secondhand, armchair kind, of course, and for this they were sometimes dismissed by professional psychologists, but it can be asked if their speculations differed much from their more professional peers, and if indeed that distinction matters.

The issue of mathematics and language is the trickiest to elucidate. It was a commonplace of Enlightenment thought that mathematics was the most powerful human intellectual activity because it employed the most eloquent language; indeed, mathematics was held up as the paradigm of a language, and the nearest people had reached a *characteristica universalis* of the kind advocated by Leibniz. A number of issues in the philosophy of language, such as the nature of names and the structure of sentences, formed part of the discipline of modern logic. Well before a sharp modern distinction was drawn between syntax and semantics, it was possible to feel that mathematics was a language that obeyed merely syntactical rules, just as it was possible to hold the quite different view that mathematics succeeded because it had a particularly clear view of its objects. But if it is possible to trace shifts in the latter view quite precisely, the view of mathematics as a language lacks a single well-defended position. I argue below that one profitable way to discuss this view is through the connections—of which there were many in the years around 1900—made between mathematical formalism and ideas in mathematics.

1.2 Changes in Philosophy

1.2.1 The Path Out of Kant

As noted above, the prevailing philosophy of mathematics at the end of the eighteenth century was that mathematics was the study of quantity. This was the view of Euler, Jean le Rond d'Alembert in the *Encyclopédie*, and of Immanuel Kant. The specifically Kantian twist concerns the nature of intuition, or *Anschauung*, by which knowledge (of any kind, and of every specific kind) is made possible. One way to understand nineteenth-century philosophy of mathematics is as a series of responses to, and nuanced rejections of, Kantianism. This prompts a difficult question that can only be discussed briefly here, concerning what Kantianism was taken to be through the nineteenth century. Not only is no simple answer possible, it became the lament of later philosophers such as Reichenbach that Kantianism had been defended to the point of incoherence (which is after all the fate of all major philosophies and religions).[9]

Idealism was the dominant German response to Kant between 1800 and 1850. It is marked by a dissatisfaction with the idea of a "thing in itself" (*Ding an sich*) and an enthusiasm for the role of the knowing subject. In the context of modernism in mathematics, the idealists (Hegel, Fichte, and Schelling) must, perforce, play a marginal role, but their influence was clearly conducive to the introduction of psychology. It is also their interpretations that constituted, by 1840, the orthodoxy against which a "back to Kant" movement rebelled, finding that idealism had led to metaphysical excess.

The phrase "back to Kant" is something of a refrain in a book that did much to help the movement along: Otto Liebmann's *Kant und die Epigonen* of 1865. Liebmann noted, by quoting Kuno Fischer, that "Kant dominates the nineteenth century as Leibniz did the eighteenth," and his refreshingly sharp book assesses Kant's contribution and those of his followers in four distinct directions. He saw much to approve of in Kant's original philosophy, but rejected the *Ding an sich* as an answer to an unanswerable question. He then considered whether the epigones followed Kant when he was correct and when he was in error, if they introduced new errors or corrected Kant's, and if they made advances. The idealist direction (Hegel, Fichte, and Schelling) and the transcendental direction (Schopenhauer) need not concern us here, except to note that for different reasons they did not represent an advance on Kant "and so one must go back to Kant." The realist direction had been taken, said Liebmann, by Herbart, but this self-proclaimed Kantian had failed to see the inconsistency in the impossible idea of the *Ding an sich*: "and so one must go back to Kant." The empiricist direction had been taken by Fries, but he too had adopted Kant's fundamental error—"and so one must go back to Kant."

It is impossible to give a fully adequate account of German philosophy in the nineteenth century, but fortunately that is not necessary.[10] The philosopher of

---

[9] See Reichenbach 1958, 31: "The philosophy of Kant has been subject to so many interpretations by his disciples that it can no longer serve as a sharply defined basis for present day epistemological analysis."

[10] The best account is Köhnke 1986, English translation 1991. Unfortunately, much of the most relevant scholarship can only be found in the footnotes in the German original.

greatest significance for the topic of modernism is undoubtedly Johann Friedrich Herbart. Before him, chronologically, comes Jakob Friedrich Fries, and after him Hermann Lotze. It is their responses to Kant that set the scene for Helmholtz's modernism, which was to be of crucial significance because of his dominant role in German scientific and intellectual life.

Although Kant intended his transcendental philosophy to proceed independently of any study of human minds, by making a sophisticated analysis of the operation of the mind in his epistemology Kant opened the door to the study of real human minds in a philosophical context. This path was the distinctive interpretation of Fries[11] followed by Helmholtz and Wilhelm Wundt. Fries proposed a theory close to Kant's but explicitly anthropological (his term) in that it sought to explain the existence of human knowledge in terms of the structure of the human mind. Through the work of Leonard Nelson at Göttingen, who regarded himself as a follower of Fries and founded the neo-Friesian school, his ideas connect to Hilbert (see Peckhaus 1990).

Johann Friedrich Herbart, who had studied under Fichte, lectured first in Göttingen, then occupied Kant's chair in Königsberg from 1809 to 1833 before returning to Göttingen, where he remained until his death in 1841 at the age of sixty-five. He disagreed with Fries about the role of psychology in philosophy. For Herbart it could not play a foundational role—that was the task of metaphysics—and Herbart's metaphysics was much more realist in its attitude to "things in themselves." In Herbart's view the world is understood by us through concepts we devise, and which philosophy disentangles and refines. Thinking is an activity of the mind, but what is thought—the concept—is abstracted from mental activity. (This move, made repeatedly in philosophy, might be called the abstractionist defense.) The concepts that Herbart himself distilled out are famously strange: monadic, like Leibniz's, yet also in contact with one another so that real causal changes can occur. This is perhaps a hint that we should expect scientists and mathematicians attracted to Herbart's realist interpretation of Kant nonetheless to pick and choose among his conclusions.

Fries and Herbart may be taken as signposts pointing toward different anti-metaphysical and positivistic readings of Kant. They also highlight two attitudes to what came to be seen as Kant's Achilles' heel: the allegedly Euclidean nature of space. Despite the views of later apologists, I hold that Kant's position was simply that space is (or rather must be perceived as) Euclidean. A possible non-Euclidean geometry of space is therefore a grievous blow to his system—although not necessarily fatal. Fries accepted Kant's views on geometry completely. Herbart, on the other hand, had a radically different idea. For him, every concept belonged to a space of its own (all these spaces, curiously, were never more than three dimensional) in which it varied (take color, for example). All these spaces, including physical space, were made up of discrete monads, and the continuum was a concept we employed to help us along because it captured the sense of nearness between the monads. As a result, mathematics could be used to study space which, note, is not the space we actually experience but an intellectual construction—Herbart rejected the idea that some concepts are known a priori. For Herbart, then, space was not, or at least not necessarily, Euclidean. Such matters as the dimension of space are therefore open to question. As Scholz (1982) has shown, this view directly influenced Riemann,

---

[11] See Hatfield 1990, especially 111–115.

although not in the way that Russell (1897) suggested. Specifically, it is Herbart's epistemology but not his ontology that Riemann was attracted to. Herbart was also an influence on Helmholtz, as Lenoir has described.[12]

Helmholtz is a central figure in nineteenth-century attitudes to Kant and Kantianism, just as he is in so many questions of science, and the originality and fairness to Kant of this powerful thinker can be better understood by teasing out what he took Kantian philosophy to be.[13] He trained in medicine and became a leading physicist, yet the bulk of his writings are on physiology and the psychology of visual perception. All the while he wrote extensively on philosophy, concentrating on epistemology in what he regarded as an essentially Kantian spirit. He was also among the first to appreciate the topological work of Riemann and to attempt to extend it, in the context of hydrodynamics.[14] We shall see (§3.1.4) that Helmholtz had provocative views on the nature of Space and spatial perception, which did much to put non-Euclidean geometry on a philosophical par with Euclidean geometry. As we have seen, the space question is a key one for the nature of Kantian orthodoxy, and the one point at which all orthodox Kantians believed that Helmholtz was no Kantian at all.

Another protagonist with a Kantian turn of mind is the often-misunderstood figure of Kronecker. He is usually presented as Mehrtens (1990) described him, as an arch finitist and forerunner of the intuitionists like Brouwer; Mehrtens calls him a *Gegenmoderne*. His position, and that of his few followers, is much more complicated than that and is authentically Kantian in its emphasis on construction.[15] In his *Critique of Pure Reason*, Kant put the distinction between mathematics and philosophy this way:[16]

> Philosophical knowledge is the knowledge gained by reason from concepts; mathematical knowledge is the knowledge gained by reason from the construction of concepts. To construct a concept means to exhibit a priori the intuition which corresponds to the concept. For the construction of the concept we therefore need a non-empirical intuition.[17]

Kant explained that this construction could proceed by imagination alone, in pure intuition, as when the mathematician constructs a triangle and draws a suitable parallel line. Exactly this capacity of the mathematician was what Kronecker insisted upon, and his finitism is a limitation, in his view inevitable, on what the human mind can construct. This informs both his philosophical views on numbers and his entire and ambitious program to unite all mathematics under the aegis of arithmetic and algebra.

Anyone thinking their way through the tumult of new ideas in late nineteenth century mathematics and science had necessarily to work out where they stood in relation to the ideas of Kant. Some, who may be said to have inscribed Leibniz on their banners, turned to logic and a position later known as logicism; Frege is the major figure here. Many located their ideas about mathematics in relation to science.

---

[12] Lenoir 2006.
[13] See the essays in Cahan 1993.
[14] See Epple 1998.
[15] See Gray 1997, Gray 1999.
[16] All quotations from Kant's *Kritik der reinen Vernunft* are taken from Norman Kemp Smith's translation, *Immanuel Kant's critique of pure reason* (1929), 2nd ed., reprint 1970.
[17] *Critique* A713, B741.

Yet others turned to a revived philosophy of mathematics that had been a strong but minority view of the subject in the eighteenth century: that mathematics was a particularly successful language. Formalism, as this view became known (first by its opponents), is a major part of Hilbert's philosophy of mathematics in the 1920s, and it turned out to raise intriguing questions about the nature of human reasoning.

1.2.2 The Path to Logic and Logicism

Changes in the foundations of mathematics have to be analyzed from a philosophical standpoint, specifically, in a historical work, as they were perceived, advocated, and opposed by contemporaries. Philosophical explanations of the power of mathematics changed during the nineteenth century as novel discoveries displaced old ideas, and by the end of the century consensus had given way to vigorous debate.

Two intertwined themes governed thinking about mathematics in the period. One concerned its truth, the other (to be discussed below) the nature of mathematical reasoning. Put simply, the prevailing view around 1800 among mathematicians was that mathematics was true. At its best it gave knowledge of the world, arguably because of the nature of its objects, which the mind could grasp with particular clarity. The validity of these truth claims was contested by the discovery and gradual acceptance of non-Euclidean geometry. The nature of its objects was thereafter found to be less and less clear, until the traditional hold on them, articulated by d'Alembert, was lost. That mathematics in the second half of the nineteenth century offered a plurality of meanings did not, of course, rule out that some of its conclusions were true. But the claim that the truth was built in because of the exceptional nature of the starting points was no longer plausible.

Among the allegedly clear objects were discrete quantities that one counted, and continuous magnitudes that one measured. Continuous quantities (geometrical magnitudes) were more doubtful than discrete ones, and the "God-given" nature of the integers was accepted by many—Gauss, Helmholtz, and Kronecker among them—who separated them off from magnitudes and accordingly placed geometry with mechanics. For this to be a demotion it must be the case that mechanics was understood to be inescapably empirical, and although the truth status of mechanics has been less well studied, Pulte (1998) has shown that the claims for the apodictic quality of mechanics à la Lagrange were initially accepted by Jacobi but later called into question by him, interestingly enough in lectures that Riemann probably heard. This offers an instructive three-way comparison with the truth claims of elementary mathematics, geometry, and physics.

The idea that at least elementary number theory lay among the eternal verities offered two possibilities. One might hope to give geometry and even mechanics the certainty of arithmetic by founding them impeccably on arithmetic. Or one might hope to ground all of mathematics, numbers included, on some absolutely secure foundation. The one that commended itself was logic, and its attendant philosophy, to be discussed below, became known as logicism (see §3.4.2). This move led directly to the Kantian question about the possibility of knowledge: Were there necessary limitations on human thought? Were there a priori dispositions of the mind that ensured that things could be known? Once it was established to the satisfaction of many, but not all, that some Kantian accounts of what we know and how we know it were in fact false, two avenues of retreat were possible. One led through anthropology to psychology, and to the idea of reasoning as a specifically human activity. The other led to the elementary laws of logic, and the hope that mathematics could be reerected upon those slender but sure foundations. The emphasis now was not on the caliber of mathematical objects, but upon mathematical reasoning. The neo-Kantians of the Marburg school at the start of the twentieth century also set much store by logic, precisely because they had lost confidence in the Kantian concept of intuition, which struck them as unduly psychological. But they had quite a different vision of logic in mind, as we shall see when Natorp's and Cassirer's work is discussed in §4.9.1.

Much interesting scholarship has surfaced in recent years on the history of logic.[18] The German-speaking world was caught up in a debate about the proper relation of philosophy to psychology, which grew up as a subdiscipline of the older subject and seemed capable of taking it over. This has been analyzed by Kusch. The logicians of the nineteenth century, of whom there were many, have also been considered afresh, the Germans among them not least for the way their work related to contemporary work in algebra, by Ferreirós (1996) and Peckhaus (1997). The place of Schröder and Peirce has been reassessed (positively), as has that of Russell (many times). Many technical matters have been examined, and our understanding of nineteenth-century logic has deepened, enabling us to examine the implications that were drawn, from Logicist and other standpoints, about the nature of mathematics.

1.2.3 Formalism

Modern mathematics does seem to offer a set of off-the-peg theories for scientists to interpret and use, as formalists such as Carnap were to suggest. What might be called uninterpreted but interpretable mathematics has its own philosophy, the minority view by the nineteenth century, going back to Leibniz and preferred by Lambert and Condorcet, that mathematics was a language—indeed the best language, because of the simplicity of its terms and the formality of its arguments. The nature of the allowable rules for an otherwise free play of symbols is a vexed one, as my arguably paradoxical formulation is intended to suggest. The view of the English mathematician George Peacock in 1830 was that arbitrary rules could be studied, but the only useful ones would be modeled on arithmetic.[19] This British view sustained a major blow with the advent of Hamilton's quaternions. Later German mathematicians, such as Hankel, held a interestingly semiformal view of real numbers, arguing that intuition was needed to go beyond the rational numbers (which could be defined formally).[20] Among the Germans, Thomae was signaled out for a furious and celebrated polemic by his colleague at Jena, Gottlob Frege. Frege is an interesting case for

---

[18] See, for example, Sluga 1980, Beaney 1996, the now five volumes of the *Collected works* of Gödel, Dawson 1997, A. B. and S. Feferman 2004, Ferreirós 1996, 1999, Ferreirós and Gray 2006, and Grattan-Guinness 2000 and the references cited therein.

[19] The British algebraists of the first half of the nineteenth century can be seen as quite formal in their study of systems of meaningless symbols governed by arbitrary rules, but they failed to pursue this insight to any depth, nor did they promote it successfully across other branches of mathematics. The result is a flimsy structure that subsequent generations abandoned, and did not contribute to the rise of modernist mathematics. For more about these matters, see, among a large literature, Fisch 1994, 1999.

[20] See Epple 2003.

rightly criticizing such mathematicians for assuming that their symbolic manipulations were free of contradiction, while he continued to hold a realist ontology about numbers and geometry that was increasingly unsustainable. Nonetheless, the view that mathematics is about arbitrary systems and effectively requires no ontology became a fiercely modernist one in the hands of Paul Mongré around 1898, although not with Hausdorff a few years later.[21] Less radically, many mathematicians were willing to put their trust not only in consistent formal axiom systems, but also in systems for which they had only a naive consistency argument.

Formal analyses of mathematics often tended to reduce it to the application of logic to creations of the mind; such was Dedekind's view, for example, when he formalized the real numbers and the natural numbers.[22] This, of course, leads us back to the question of what logic was taken to be.[23] The modernist period examined logic with intense concern, and the surprising answer was that there are many logics, which can be divided into first-order, second-order, and infinitary, with some other cases in between, as well as classical and intutionistic logics (see §7.1.3). A second-order logic permits quantifiers to range over sets and relations, and infinitary logics include such things as theories with formulas of infinite length. It was Skolem's discovery that first-order theories permit highly unexpected models of the integers and the real numbers, whereas second-order theories are often categorical in these respects (meaning that any two models are essentially the same). On the other hand, first-order logic permits the development of proof theory, which was Hilbert's enthusiasm in the 1920s. To make this explicit, another of the achievements of mathematical modernism is that it entirely reworked the concept of logic and shows, in particular, that mathematics is not a branch of logic (and so not all mathematical statements are tautologies). A further consequence is that many of the older usages of terms were redefined, so that even Frege's ideas about epistemology are not faithfully captured by modern terminology.[24]

Only by 1930 did a consensus emerge that there were many kinds of logic, of which first-order was taken to be the correct basis for set theory, as Skolem had argued (in opposition to Hilbert) but with the support of Bernays and Gödel. On the other hand, Gödel's incompleteness theorem established that first-order logic could not adequately support all of mathematics. The theorem also sharpened a growing awareness of the distinction between syntax and semantics in mathematics. This distinction illuminates unresolved disagreements between protagonists, and between historians, for example, the algebraic style of modern logic associated with Peirce and Schröder, and the opposing view due to Russell. Discussions of mathematics as a kind of language are all the more interesting in view of the emergence of Italian attempts to remove natural language from the discourse of mathematics, and Brouwer's brief involvement with the Significs movement led by Mannoury.[25]

The leading formalist (although apparently he never used the word "formalism") was David Hilbert, and Hilbert's gradual and uncertain path to the foundations of

---

[21] Let me unpack this below.

[22] Dedekind's friend Heinrich Weber later recalled that on a walk with him, Dedekind remarked, "We are a divine genus, so our mental activity is creative"; see Weber 1906, 173.

[23] Our guide here will be G. E. Moore's pathbreaking book *Zermelo's axiom of choice* (1982).

[24] The key moment to look for here is the elimination of the term "definit"; see §4.7.6.

[25] See the essays in Heijermann and Schmidt 1991.

mathematics forms an oft-told tale that is currently the object of much new research.[26] An early interest in the axiomatization of geometry led to a general interest in foundations that was only sharpened by the dispute with Brouwer and the apparent defection of Hilbert's star pupil, Hermann Weyl. The finitism of Hilbert's position in the 1920s can be seen as an attempt to derive as theorems, from a restricted logic, all the truths that Brouwer denied we could grasp intuitively. The inevitable failure of such attempts was established by Gödel and Turing, and the subsequent failure to find a starting point as impeccable as first-order logic led to the whole enterprise losing its force. Perhaps the strongest reason for revisiting this topic is that the philosophy of mathematics was simply left in disarray. No philosophical programme for mathematics won; the historical events led to the collapse of every original protagonist. The implications of this for the perception of mathematics, whether by mathematicians or the broader intellectual public, have not been fully analyzed.

A further reason is the subsequent perception of Hilbert as an arch formalist (the role he plays ubiquitously, but most visibly perhaps in Mehrtens's book). This is coupled to the tradition in philosophy of mathematics that grew up which treated mathematics as indeed Hilbert had ultimately presented it as strings of symbols. This extreme of formalism cuts mathematics off from almost every aspect of it that mathematicians recognize, and has led to a perception of modern mathematics quite distinct from what I want to argue was created. So it is important to disentangle mathematics, as it was practiced and modernized in the period, from what some historians and philosophers of mathematics have taken it to be.

It is also important to note that interest in how mathematics can be known did not cease with the failure of philosophies of mathematics.[27] Gödel's work initiated a flood of work (by Church, Post, Turing, and others) on the limitations of mechanical thought, elaborating the nature of recursive and computable objects. Negative resolutions of Hilbert's halting problem and his decision problem (to be defined in chapter 7) not only formed major results in the new theory of mathematical logic, they brought to prominence Turing's idea of a universal machine. This apparently very abstract work, usually noted for its philosophical implications, also connected to the burgeoning science of the mind. The Turing test is Alan Turing's proposal for how one could discriminate between human understanding and merely mechanical reasoning, and it carries the latent question that eventually one might indeed not be able to tell the two apart: What then? Another key connection of a different kind was made by the neurologists McCulloch and Pitts, in their paper of 1943 on how assemblies of neurons may do arithmetic. Their analysis of neural nets drew on the work of Carnap and Hilbert and Ackermann, advanced our understanding of Turing machines, and yet was explicitly aimed at understanding how the human brain can conduct simple arithmetical operations. This analogy captivated John von Neumann and was incorporated into his description of the EDVAC machine in 1945. The use of psychologistic ideas in computer design was not accepted by others in the field,

---

[26] See, for example, Hallett 1994, Sieg 2002, and Rowe 1994. I have only seen the first of the what will be six volumes collectively entitled *David Hilbert's Lectures on the Foundations of Mathematics and Physics, 1891 to 1933*, ed. William Ewald, Michael Hallett, Wilfried Sieg, and Ulrich Majer. This volume is Hallett and Majer 2004. Nor have I seen Hallett's provisionally entitled *Hilbert on the Foundations of Mathematics*, to be published by the Clarendon Press, Oxford.

[27] See, for example, the essays in Ferreirós and Gray 2006.

such as Mauchly and Eckhart, but it offers an example of the coming together of mathematics, logic, and psychology that is highly significant.

1.2.4 Science, Mathematics, and Philosophy

Science in the late nineteenth century had its own complicated and shifting relation with philosophy, as the example of Helmholtz makes clear. One of the driving features in the history of philosophy in the nineteenth century is the progress of science. It runs directly into Kant's position that there are limitations on what can be thought. The modernist position emphasizes the freedom of the creative mind, in contrast to the constraints envisioned by Kant. Science enters the discussion by seeming to render intelligible ideas beyond the limits laid down by philosophers.

Science is also what might be called object driven, a phenomenon I wish to keep deliberately vague but which is captured by the well-known joke that scientists go regularly to the laboratory to find that their ideas do not hold up, but the philosopher has no laboratory. Scientists do not indulge in arbitrary mental constructions for their own sake. Instead, they have a vivid sense of the universe as full of objects of unexpected kinds indulging in obscure but rule-governed behavior. They are therefore disposed to be realists, although the focus of their realism (specific objects, specific theories) is something they have learned to be hesitant about. But their sense of what there is, if it conflicts with a philosophical position, carries the force of the stone kicked by Dr. Johnson.

Arbitrariness in the mathematical domain is another matter. It is possible to imagine that the discovery of non-Euclidean geometry by Bolyai and Lobachevskii in the late 1820s could have been a mere intellectual curiosity. Its philosophical significance was, however, after much delay, that it dents (some were to argue, shatters) confidence in the Kantian theory of knowledge. But it is one thing if non-Euclidean geometry offends because it is a mere logical possibility, another if it is the actual geometry of space. It would be quite reasonable for humans to perceive space as Euclidean if that is what it is, although the force of necessity Kant sought to invoke might require further analysis. But it would be downright odd if space had to be thought of as Euclidean when it actually isn't.

But while science is highly intellectually constrained—indeed, that active sense of constraint is what gives science its claim to be true—once liberated, mathematics soon found that it was barely constrained at all. As the range of mathematical objects piled up, the idea grew that mathematicians could freely choose what they studied, and study it subject only to the laws of logic. Mathematics moved from being the favorite pastime of philosophers to challenge the whole idea that there were constraints to human thought.

The logicist program can be seen as an attempt to build an adequate philosophy entirely on the meager list of the laws of logic, whether, with Frege, it is coupled to a strongly realist ontology, or, as with Dedekind, it grants existence to any collection of objects given by rules that are free from self-contradiction. But others doubted that the bare minimum of logical reasoning could be enough to construct a rich ontology for all of mathematics.

The matter of ontology raises the question of Truth. Mathematicians, and philosophers of mathematics of all stripes, like to say that the conclusions of mathematical arguments are proved. Are they also true (as mathematicians habitually assert)? This question breaks into two further questions, one resolved by Gödel in 1931: true and provable are distinct concepts. The other is whether, according to any particular theory of Truth, mathematical statements can be true at all. Are true mathematical statements true in virtue of what they say about mathematical objects, however such objects are to be regarded? One orthodox view today is that the failure of mathematical objects to be perceptible makes it hard for mathematics to be true, and that any resolution of that difficulty makes it hard for mathematics to be provable.[28] But mathematicians often sound like Platonists nonetheless, and ascribe the correctness of their theorems to a correspondence with some mentally accessible non-spatio-temporal world of objects whose properties the theorems describe. This mathematical platonism is a twentieth-century phenomenon, and is very arguably the mathematicians' response to their own modernism (as is discussed in §7.3).

The question of the truth of mathematics cannot be ducked by opposing fanciful mathematics to truthful science. This facile move will not do, because mathematics is not just games. Many of the leading modernists, for example David Hilbert and Emmy Noether, explicitly disdained that view, and almost every mathematician regards the connection between mathematics and physics as intimate and vital, if also mysterious. Part of the force of the weird and the wonderful in mathematics is that it might be true (in some sense). Mathematical modernism respects the connection even as it finds it hard to articulate. One way it does this is by offering a view of mathematics as rule governed but open to interpretation.

1.3 The Modernization of Mathematics

1.3.1 Experts and Audiences

In discussing such questions, it is not sufficient to relay what was said. One must also ask: Who spoke, who listened, and to what effect? There were at least three types of speaker: professional mathematicians drawn into these debates (as more and more were, which is interesting in itself), experts in philosophy and logic, and psychologists of various kinds. Whether they listened to one another can be hard to determine and must be analyzed on a case-by-case basis. Behind them stood a wider public: teachers, scientists, engineers, and buyers of books and magazines who may have been no more than concerned citizens and taxpayers in the emerging, technological, modern world.

Among these are two groups that a thesis about modernism in mathematics must concern itself with chiefly. One is the professional mathematicians, those who became the modernists themselves and were accepted as such by their fellow professionals. The other is the educated public, if it is claimed, as here, that the new self-image of mathematics was widely accepted. Both because the question of public acceptance is not one of a philosophical character and for chronological reasons, I have separated the discussion of the professionals and their audiences into two chapters.

The activities of the mathematicians themselves went through two phases. The first is the technical work of many authors: Peirce in America, Schröder and Frege in

---

[28] Benacerraf 1973.

Germany, Peano and his followers in Italy, MacColl in Scotland. Historians who write on this subject seem agreed that these authors more-or-less wrote for each other, and seldom engaged the attention of mainstream mathematicians. All that changed around 1900, when Hilbert in Germany, Poincaré in France, and Enriques in Italy enter the debate, many other writers are drawn in, such as Couturat in France and Josiah Royce in America, and the issues were argued out in high-quality popular journals.

What brought about this change, I argue, is the emergence of a modernist consensus. Mathematicians have always been able to manage for long periods of time without a philosophy of mathematics, but by 1900 there was a new set of foundations that seemed to secure a wide range of mathematical activity and resolve a number of outstanding difficulties. Now that the profession could see the new consensus, experts among it could take stands for or against, and the public could be confronted with the radically new vision of what the oldest science was now taken to "be." What images of mathematics did this modernist philosophy of mathematics present? What were it supposed connections, for example, to contemporary science? How widely was it accepted, and why? I shall argue that the shift toward modernism has considerable ramifications, and cannot be understood without looking at the wider cultural context.

As science grew in the nineteenth century, there was a growing divorce between mathematics and science, which I shall analyze chiefly from the standpoint of the mathematicians. Mathematicians came to accept that there was increasingly a separate discipline, which we call physics, which had more and more of a monopoly on the study of the natural world (or that part of nature which could be captured in the laboratory). It followed that the nature of the mathematician's contribution to the scientific enterprise changed, from discovery to explanation, and an intermediate field grew up, applied mathematics, where the results of science and mathematics could be traded back and forth. It does not follow that a mathematician of the nineteenth century fitted tidily into one of the present-day categories, or that their acquaintance with physics was only secondhand. Riemann worked in the laboratory of Gauss's friend Wilhelm Weber, and Minkowski also studied physics firsthand, although he did feel, jokingly, that he had to apologize for doing so. But there is nonetheless a distinction between doing physics and doing mathematics that emerged in the nineteenth century, and it can be profitably traced in the work of a number of mathematicians, of whom the leading figures of Riemann, Klein, Poincaré, Minkowski, and Hilbert will be considered in detail. But it should be noted that there is a rich and largely ignored literature consisting of inaugural addresses and popular writings of French, German, and American professors in the period, for example Hölder and Pringsheim, that further amplifies the picture.

Each figure offers a distinctive perspective on the mathematics and the physics of their day. The chronologically extreme cases demonstrate the extent of the phenomena I am trying to delineate. While Riemann is a leading figure in modernizing mathematics (indeed to a degree that poses an awkward challenge to my chronology of modernism), his physics is more conventional. Two papers stand out, one on shock waves and one on the propagation of heat, and in each of these there is a mix of original mathematics and novel insights into traditional physics. Hilbert's contribution, on the other hand (notably analyzed by Corry 1999 and 2004), is very much that of the mathematician, seeking to clarify—and if possible axiomatize—the mathematical aspects of this or that physical theory. Even when he engaged directly with physics, as in the celebrated "race" with Einstein for the field equations of general relativity, his thinking was that of the mathematician, and on those grounds he was shrewdly criticised for naiveté by Einstein. Hilbert's attitudes have been profitably contrasted in (Rowe 1994) with those of Klein, who sought to promote what he regarded as a Gaussian unity of mathematics and physics by an energetic program of fund-raising and hirings at Göttingen.[29]

Poincaré offers an instructive intermediate position. He was accepted by French physicists as the ultimate authority in disputes (for example the Crémieu controversy) and ranged from adjudicating on the analysis of experiments to the high-level theory of his *Electricité et optique*. As I have argued (1996, 2006b) he explicitly adopted three standards or modes of work: one for physics, one for applied mathematics, and one for pure mathematics. His lengthy discussions of the role of mathematical rigor in scientific work document in themselves the separation of mathematics and physics and his conventionalism can be read as a modernist position of its own.

On the other side of this divide, British and German physicists have been most analyzed and should be briefly considered for their views on the role of mathematics in science. The accounts of numerous authors have built up a rich picture of different approaches to physics.[30] The impression is of mathematics, often high-powered mathematics, employed at the service of science. The question of arbitrariness in the use of mathematics is presently not well analyzed historically. There is interesting work by Alberts (2000) on the transition from applied mathematics to mathematical modeling, which he defines as a transition from interpreted mathematics to less interpreted, more phenomenological theories. The use of Hilbert space methods on quantum mechanics offers another example one could discuss, but I shall stay close to the mathematical coastline and not enter the deeper waters of the history of modern science.

An interesting topic that has been studied recently is that of measurement. Michell has distinguished between classical or quantity-based and modern, representational theories, and discusses the transition from the one to the other by considering the ideas of Helmholtz, Hölder, and Russell.[31] His striking conclusion is that only Russell fully adopted the representational position, driven as he was by his logicist conception of number. The point at issue here is the mirror image of ideas about number and magnitude discussed earlier. Now their use in science itself finally changed.

The break with science cannot alone account for the rise of mathematical modernism. I shall argue that it is impossible to consider this modernism merely as an intellectual shift, or as a deepening series of insights into (the nature of) mathematics. It is too broad and deep for that, although the striking continuities and rich successes of what might be called traditional or classical mathematics throughout the period must not be forgotten. In that respect the situation is similar to Schoenberg's remark

---

[29] See also Rowe 1997.

[30] See among others Archibald 1995, Hunt 1991, Kuhn 1987, Jungnickel and McCormmach 1986, Pyenson 1985, Warwick 2003.

[31] See Michell 1993, Michell 1999.

that much good music remains to be written in C major, or the continued production of good novels with a classical structure.

1.3.2 Professionalization

Rather, what we see in mathematics is a change in attitude of the whole mathematical profession. Most visibly in Berlin, but also in Göttingen and elsewhere, they had their own ethos. This distinct professional position was strongly conducive to the development of an autonomous body of ideas within and about their subject, which, as we have seen, increasingly declined to make outward reference for its basic terms. Mathematics was placed in a complicated rather than a naive relationship with the day-to-day world. It was not derived from the world in any simple way, and it was not necessarily applied to it in any simple way. Nor, however, was it at ease with itself but, as the whole story of the rise and fall of the foundations shows, it was an anxious, self-conscious creation. Most crucially here, it was the de facto view of more and more of the profession.

The motor for these changes was the growth of the professoriat.[32] In general, and nowhere more so than in Germany, university professors in the nineteenth century formed a distinct group with a certain social standing among the middle classes and, accordingly, like many a similar semiautonomous group, they acquired certain responsibilities. They were also more numerous in Germany than anywhere else. In his extraordinary book *Les Allemands* (*The Germans*) which was intended to alert France to the reasons for the rise of Germany, Fr. Henri Didon noted that there were twenty-two German universities, employing more than 2,000 teachers and with more that 25,000 students who were, moreover, free to move at will between the universities (pp. 72–74).[33]

To varying degrees, the professors had to set and maintain standards, teach students, recruit new members of staff, and conduct research. Sometimes life was only too cozy, as was often the case in Cambridge colleges in the period. Many universities were small, their mathematics departments very small, and the life accordingly restricted. Klein found the library of the University of Erlangen when he arrived there in 1872 "antediluvian"[34] and there were no funds for the purchase of books. In other places, such as Paris, the teaching loads were so high it is remarkable that bright young mathematicians got any research done at all. Senior professors had a considerable degree of autonomy, they could usually choose what to teach and when, and in many places midcentury no structure cajoled or helped them to continue research.

Nonetheless, for those who heeded the implicit call to take a proper, professional position, there was much to be done and much to be gained by doing it. Mathematics

---

[32] For the case of mathematical physics in nineteenth-century Germany, see Jungnickel and McCormmach 1992 and, for the case of Königsberg in rich detail, Olesko 1991.

[33] Didon is an interesting figure: a founder, with Pierre de Coubertin, of the modern Olympic movement; the author of a very popular life of Christ (*Jesus Christ*, 1891) that he wrote while ordered on a seven-year exile in Corsica by the Dominicans for his modernist theological tendencies, and a belligerent opponent of Dreyfus in 1898.

[34] See Klein 1923.

could be promoted as a separate discipline, not merely the servant of science; mathematicians could form a separate department, and establish their own constituencies in the schools, their own links (not always comfortable) with engineering. The late nineteenth century also saw the creation of many specialist learned societies. The first of these was the London Mathematical Society, founded in 1865. The Société Mathématique de France (the French Mathematical Society) was founded in 1872 as a response to the disaster of the Franco-Prussian War (with the geometer Michel Chasles as its first president); the Deutsche Mathematiker-Vereinigung, or DMV (the German Mathematical Society) was founded, after earlier attempts had failed, in 1890 with Georg Cantor as its first president. These brought with them new, specialist journals, and, in the French case, a renewed commitment to top-level research, largely seen as catching up with the Germans. The New York Mathematical Society was founded in 1888 on the initiative of T. S. Fiske, then a graduate student at Columbia University, who had been impressed by the London Mathematical Society and the Royal Astronomical Society on a visit to London. It was refounded as the American Mathematical Society in 1894, following the enthusiasm generated by the International Mathematical Congress held in Chicago in August 1893 in conjunction with the World's Columbian Exposition. The Italian Mathematical Society was not founded until 1922, but this may be because the Circolo matematico di Palermo (the Mathematical Circle of Palermo) carried out most of the relevant functions.

These societies mostly cared for research; teaching, especially school teaching, was often discussed in other venues and published in other journals. Nor were major issues of policy formulated here—the major universities were not about to surrender what influence they had with government to a national body. The societies' journals carried original papers, reports of meetings, and some of them (and other new journals of the period) carried abstracts of recently published papers and books. Some carried news of appointments, titles of lecture courses in various universities, and details of prize competitions. They capitalized on, and helped to strengthen, a feeling that had been growing with the nineteenth century that research was a duty of a professor, a feeling that animated both the sciences and the humanities. By providing outlets for the publication of research, the societies helped advance the individual researchers, the universities for whom they worked, and, it was implied, the intellectual needs of the nation.

The learned societies also traded with one another, and by the early twentieth century quite a network was in place, which further encouraged authors to publish abroad. The first International Congress of Mathematicians was held in 1897 in Zurich (neutral territory between the French and Germans), the second in 1900 in Paris, the third in 1904 in Heidelberg, and since then the mathematicians have met at four-year intervals whenever they could.[35] The intellectual world this created was curiously national and international at the same time: international in that mathematicians in one country were likely to identify with mathematicians in another, to want to read their work and be read by them, to travel and meet them (or, at the least, exchange photographs); national, in that mathematicians were no better than anyone else at decrying or defying the nationalist temper of the times.

[35] As described in Barrow-Green 1994 and Lehto 1998.

Teaching, of course, is a private matter, endured by staff and students alike in the absence of any structure to determine and drive up standards. Then as today, some mathematicians seemed to speak directly to the minds of their audience, others, as Newton was said to have done, to the walls. But research is another matter. It is published and open to scrutiny, and indeed the ethos was that it was only acceptable if it was published and scrutinized.[36] Publication promoted competition, a sense of what was important and who was doing it. Colleagues, who were also rivals, were as likely to appear abroad as at home. It became much harder to wallow in ignorance. Once something was done, it became harder to ignore it. A more rigorous piece of calculus, a novel geometry, a new part of algebra—all had to be learned, adopted, assimilated, and then extended and made to yield further new results. Or one was no longer, really, a mathematician.

But I do not claim that the modernist shift was caused by the mere existence of an autonomous profession of mathematicians. The first section of this book is devoted to establishing a panoply of internal reasons for modernism in mathematics. Nor was the modernist shift simply the view of a few mathematicians, however gifted. It is very different from Poincaré's conventionalism, which was largely rejected outside France,[37] and Brouwerian intuitionism, which has always been a minority pursuit. It became the orthodoxy, and to this day the difficulty of the transition from school to university mathematics reflects the magnitude of the change. Mathematics, for mathematicians, is simply not what it was or what many people today still take it to be. Or perhaps more accurately, even the popular sense of the abstract, artificial nature of mathematics does not prepare students for the shock of the real thing.

What, therefore, confronts the historian is a change in the philosophy of mathematics that embraced the whole mathematics profession. There is automatically a professional aspect to the modernist shift, and I argue that without the autonomy of the profession this shift would not have taken place. Indeed, insofar as such claims have any meaning in history, I argue that without this professional autonomy the modernist shift could not have taken place. Modernism in mathematics is the appropriate ideology, the appropriate rationalization or overview of the enterprise. That different mathematicians bought into this with different degrees of enthusiasm is exactly what one would expect, but it became the mainstream view because it articulated very well the new situation that mathematicians found themselves in. By 1900 leading figures could see that they were bereft of easy appeals to nature, that their own practice was no longer to be described in anything like traditional Kantian terms, and their place in cultural life was therefore in question. They were not despondent, but argued that they had instead a new and valuable role: modernism gave them a new legitimacy just as the old one seemed either preempted, bankrupt, or dishonest.

Indeed, I argue that the unparalleled extent of public debate about the nature of mathematics between 1900 and 1914, which involved many leading mathematicians, represents the shock wave as the professional acceptance of modernism reached the general educated public. The most visible sign of this debate was excitement over the possibility of non-Euclidean geometry, including the possibility of

---

[36] Industries might well conduct research, but they did not selectively publish their results and claim status for them; they turned them into products.

[37] As is well described in Walter 1997.

space having four dimensions, which did much to promote the later wave of enthusiasm for Einstein's theories of relativity. If the leading, and most eagerly read, figure in all this was Poincaré, the extent to which popular scientific journals took it up was extraordinary. The same is true of discussions of the nature of mathematical knowledge and mathematical intuition, in which Poincaré was joined by Mach in Austria and Enriques in Italy. Again, especially in France and the United States, popular interest was intense, as the success of the *Monist* and Paul Carus's Open Court Press demonstrates.

A relevant point I have not been able to explore but wish to note here is that in both Germany and France the teaching of geometry in schools was reformed along similar lines at the start of the twentieth century. In France after 1902 the study of Euclid's *Elements* was dropped in favor of treating geometry as a physical science based on the study of rigid body motion, and concrete experience was stressed; Latin was no longer essential for higher education and the importance of science was promoted. Mathematicians supported these moves. Borel, for example, argued that mathematics must be made theoretical and practical or one day it will not be taught. In Germany, Klein's energetic participation in the so-called Meran reform movement, which had begun in 1905, promoted functional thinking and geometric intuition in German schools,[38] Klein's books entitled *Elementary Mathematics from an Advanced Standpoint* date from this period. The International Commission on Mathematics Instruction (ICMI) was established in 1908 at the International Congress of Mathematicians in Rome, when it was resolved that "the Congress, recognizing the importance of a comparative study on the methods and plans of teaching mathematics at secondary schools, charges Professors F. Klein, G. Greenhill, and Henri Fehr to constitute an International Commission to study these questions and to present a report to the next Congress." Klein was its first president and served until 1920, when the ICMI dissolved in the acrimony following the First World War. In 1924 these reforms were reversed in France: the former emphasis on the humanities was restored, mathematics and science were cut back, and the earlier reform was denounced as too German and utilitarian.[39]

---

[38] Tobies 1981, chap. 8.
[39] The information on France is taken from H. Gispert, 1900–1930, Mathematical France, identifying breaks and continuities, http://www.cirm.univ-mrs.fr/videos/2007/exposes/02/Gispert.pdf.

2
BEFORE MODERNISM

2.1 Geometry

2.1.1 Projective Geometry

Among the many lasting consequences of the French Revolution was its effect on higher education. Although the École Polytechnique was and is very different from a modern university, its creation marks the decline of the learned academy as a central focus for research, and the start of the system of high-level teaching coupled with the production of new knowledge. In its early days this, the first of the Grandes Écoles, was animated by Gaspard Monge's vision of the central place of mathematics in the advanced education of the citizen.

When the revolution began, Monge was drawn into radical politics.[1] After the fall of the monarchy he became minister of the navy, a post he held for only eight months until he resigned in April 1793 under criticism for being too moderate. He retreated to the safety of numerous public commissions, and in March 1794, when he was forty-eight, he was appointed to the commission responsible for creating a new École Centrale des Travaux Publics. He became the instructor there in November of that year, teaching so-called revolutionary courses; in his case, descriptive geometry. The subject, his own invention, was a method of depicting three-dimensional figures on two planes (typically plan and elevation). The school, soon to be called the École Polytechnique, got off to a confused start initially, but by June 1795 it was running smoothly, and in 1804 it became a military school and the students military cadets.[2]

In due course, Monge was appointed director of the École Polytechnique, and eventually Napoleon, whose support he enjoyed, appointed him senator for life. He retired from the École Polytechnique in 1809, and went into something of a decline on Napoleon's defeat in 1812, but in 1815 he rallied to Napoleon's side one last time in the regime of the Hundred Days. He died in 1818, and many of his current and

---

[1] See the contemporary memoir (Dupin 1819) and the biography (Taton 1951)
[2] See Belhoste 2003.

former students defied the wishes of the Bourbon authorities to attend his funeral and pay tribute, for Monge had been a gifted teacher of mathematics who inspired his pupils.

After his death, Monge became a name to invoke. His legacy connected service to the state with a passion for teaching, especially elementary but original geometry. Descriptive geometry was hardly a research subject, but it was essential to the training of an engineer, and it suffered the strange fate of being progressively marginalised at the École Polytechnique, where the syllabus was increasingly controlled by professional mathematicians. Their rationale was that they provided a rigorous general education, leaving the specialist *Écoles* to finish the students' education. But if descriptive geometry was not a research subject, it spawned a subject that not only had much more life to it, but one was to challenge every idea about geometry before the century was over. This was projective geometry, and its inventor was a pupil of Monge: Jean Victor Poncelet.

Poncelet had been taken prisoner in Russia on the retreat from Moscow, and had survived only by luck. The winter of 1812 had been so cold the mercury in the thermometers froze (a temperature of $-39°C$). To keep his spirits up, he tried to remember the mathematics he had been taught at the École Polytechnique, which he had graduated from in November 1810. There he had acquired a taste for the work of Monge (and the other geometers, Lazare Carnot and Brianchon), but now, completely cut off in Saratov, he occupied himself summarizing all he knew of the mathematical sciences in the notebooks he was allowed. He found that he could only remember the elementary ideas, but not the complicated and laborious methods and the abstract and spiny proofs that had been introduced into mathematics.[3]

On his return to France in 1814 he set about becoming a mathematician. He realized that Monge's simple projections of figures onto horizontal and vertical planes could be hugely generalized, using the way a point source of light casts shadows on a plane. He found that this idea was not entirely new; it had certainly been known to Girard Desargues and Philippe de la Hire in the seventeenth century, and also to Newton, but any tradition that derived from their work was by now moribund. He found, as these writers had before him, that by means of such projections an ellipse and even a parabola or a hyperbola can be made to look like a circle. Properties of such figures may therefore pass over to properties of circles, and whenever that is the case their proofs should be much easier, because circles are easier to understand than ellipses, parabolas, and hyperbolas (the so-called conic sections). In this way much of the theory of conic sections, at that time and even today the most widely taught part of coordinate geometry, is unified, an outcome a mathematician views with pleasure.

But what properties of figures are amenable to this kind of treatment? Not properties to do with length: an object and its shadow may clearly have different lengths. Nor are angles a property of the required kind: a right-angle looked at obliquely may take any size at all. On the other hand, the shadow of a straight line is again a straight line, so being a straight line is a projective property. If a straight line crosses a curve, the shadow of the line crosses the shadow of the curve, and if the originals cross in, as it might be, six points, so too do their shadows. More generally,

---

[3] See Poncelet 1862–64, ix. For more on the history and mathematics described here, see Gray 2006c.

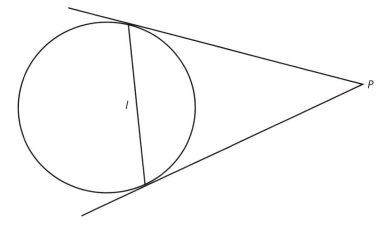

FIGURE 2.1. The line *l* is the polar line of the point *P*, its pole.

the number of points in which two curves cross is a projective property. Again, if a straight line touches a curve, their shadows also touch: being a tangent to a curve is also a projective property. There are other projective properties, but these formed the basis of Poncelet's realization that he had discovered a new branch of geometry, one concerned with properties of figures that do not require the concept of distance. Accordingly, he called it nonmetrical geometry. For a time the English called it descriptive geometry; today it is called projective geometry. It is worth savoring this name, in order to get a sense of the novelty Poncelet was proposing. The term "geometry," after all, is commonly taken to mean measurement of the Earth. Its business is with length and angles. From this point of view nonmetrical geometry is almost an oxymoron, a contradiction in terms.

Poncelet naturally wanted his new geometry to obviate all "spiny" considerations, such as those which split proofs up into separate cases. It is a measure of the extent to which he succeeded that Michel Chasles, his successor as the leading projective geometer in France, could say of Poncelet's *Traité* that it proceeds without a word of calculation.[4]

2.1.1.1 POLE AND POLAR

This involved him in some awkward arguments of his own. Figure 2.1 shows a circle, a point *P* outside the circle, the two tangents from the point to the circle, and the line *ℓ* joining the two points of tangency. The entire figure is made up of projective properties, if one allows that the circle may, upon projection, become some other kind of conic section. The French term for this, introduced by Gergonne, is that the point *P* is the pole of the line *ℓ* and the line *ℓ* is the polar of the point *P*. So to a point one associates a line, the polar line of the point. It is a remarkable fact, known to the seventeenth-century geometer la Hire, that if the pole *P* moves on a line *m*, the corresponding polar lines *ℓ* all meet in a common point, *Q*, as figure 2.2 suggests.

[4] Chasles 1837, 215.

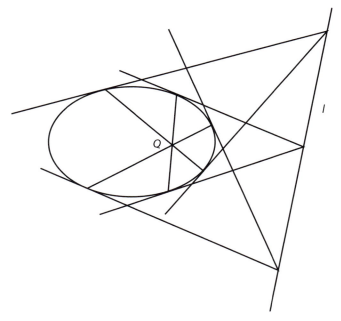

FIGURE 2.2. The polars of points of *l* all pass through Q, the pole of *l*.

This means there is a way of associating a polar line to a point inside the circle. The names "pole" and "polar" were taken from the geometry of the sphere where, to every point one can associate a great circle, which is the equator obtained by thinking of the given point as the North or South Pole.[5]

This gives two methods for associating a polar line to a point when one is given a circle, one for points outside the circle, and another for points inside it, but that was too "spiny" for Poncelet. He preferred to argue that the construction of the polar line of a point inside the circle was obtained by his "method of continuity," which allowed him to start with a point outside the circle and move it inside. Faced with the evident fact that in the one case the polar line meets the circle and in the other it did not, Poncelet argued that one could extend the meaning of terms like "meet," so that in each case the line that it produced did indeed "meet" the circle. This approach was never to find favor, but it derives only too poignantly from his desire to give proofs devoid of subtlety and sophistication. His formulation of geometry was marred by his insistence that his counterintuitive notion of meeting was truly elementary, and could not be written out of the theory in favor of algebra. What commended it to other former students of Monge at the École Polytechnique, such as Dupin, Brianchon, Olivier, and Lamé, and to other geometers, such as Gergonne, was that it grounded its arguments on objects which you could imagine as physical objects: points, lines, etc., as Monge had taught.[6] Algebra, as Poncelet's critic Cauchy observed forcefully, allowed one to say that a line which does not seem to meet a circle

[5] Duality has its origin in the work of Euler on spherical triangles; see Chemla 2004.
[6] See Daston 1986.

may yet meet it in points whose coordinates are complex numbers, but talk of imaginary meeting points was unsatisfactory, because it was not physicalist in that sense. So geometrically inclined mathematicians were left with a challenge: recapture as many as possible of Poncelet's results and thereby vindicate the belief that he had indeed discovered a new field of geometry, without lapsing into his dubious talk of lines meeting curves when they are visibly disjoint. It was a challenge they had to face without Poncelet, for by the early 1830s he abandoned geometry for the theory of machines (where he did notable work on the design of water wheels). The challenge was met in a variety of ways. Michel Chasles in France and August Ferdinand Möbius in Saxony did so by independently introducing the concept of cross-ratio. Chasles' treatment was more geometric, Möbius's more algebraic.

2.1.1.2 DUALITY

Poncelet was not the first to notice that la Hire's result allows the construction of the polar line of a point to be reversed, so that given a line ℓ (and a circle) one obtains a point. In this case the construction is as follows. Take every point of the given line ℓ and construct the corresponding polar: the common point of these polars is the pole of the given line l. Moreover, if one starts with a point P, constructs its polar line, and then constructs the pole of that line, one obtains the original point P. So, following Poncelet, one can speak of a pole and its polar as twinned, for which the technical word is dual.

Poncelet was, however, the first to appreciate the force of this observation (a distinction he shares with Gergonne).[7] Poncelet realized that the dual of three points on a line is three lines meeting in a point, and conversely, the dual of three lines meeting in a point is three points meeting on a line. One may therefore take any true statement just involving the words "point," "line," "collinear," and "concurrent," replace the word "point" by "line" and "line" by "point," "collinear" by "concurrent," and "concurrent" by "collinear," and obtain another true statement. One may therefore take a succession of statements proving a particular theorem, and make this switch to obtain a succession of statements proving some new theorem. It might happen that the new theorem is in fact the old one reworded, but that is unlikely. In general, one expects by this method to have almost doubled the theorems in projective geometry without having done any extra work.

This simple switch of point with line, collinear with concurrent, is called *duality*. As much as any feature of nonmetrical or projective geometry, it accounted for the appeal of the subject. It was built into textbooks often by writing pages of text in two parallel columns, so that students could see the switch being made. Its implications for the philosophy of geometry were not to be drawn for sixty years or more, but its implications for the new geometry were appreciated from the first. It exemplifies the way in which a subject with strong research possibilities attracts the allegiance of mathematicians in a way that a merely pedagogic subject does not. Only the research-oriented topic is propitious for a group seeking to establish a degree of independence from those (engineers, architects, scientists) who would confine mathematics to a purely service role.

---

[7] The discoveries led to a painful argument over priorities, which need not concern us here.

First at the École Polytechnique, then in many German universities, then in universities in Italy and in Britain, projective geometry was to become the heart of geometry. The main reason for this is that projective geometry was seen as more basic than Euclidean geometry. Any projective theorem is also true in Euclidean geometry, but many Euclidean theorems are meaningless in projective geometry. It was also true that projective geometry had its own charm, and this rested on the application of duality. Duality was seen as an unexpected but fundamental property of points and lines. Gergonne, when he introduced it, spoke of it in these terms:

> In each science there are certain elevated points of view where it is enough to stand there to embrace a great number of truths at a glance—which, from a less favourable position, one could believe were independent of one another—and which one realizes accordingly are derived from a common principle, often even incomparably more easy to establish than the particular truths of which it is the abridged expression.
>
> It is with a view to confirming these considerations with some quite remarkable examples, on common points and common tangents to plane curves (etc.), that we are proposing to establish here a small number of general theorems offering an infinity of corollaries, among which we restrict ourselves to pointing out the simplest or those most worthy of notice.[8]

Trouble, when it came in projective geometry, and it came with the introduction of duality, was therefore trouble the mathematicians had to manage and turn to their advantage. The paradox arose when duality was applied not to configurations of points and straight lines, but to curves. It was first noticed by Poncelet and resolved by the German mathematician Julius Plücker in the 1830s, in a way that opened up the study of curves defined by polynomial equations.[9] This confirmed mathematicians in the importance of bringing projective ideas into algebraic geometry, and this in turn strengthened the desire of mathematicians to give a truly projective account of the foundations of geometry, independent of all reference to Euclidean, metrical ideas.

The mathematician who did most to bring this about was Carl Christian Georg von Staudt in his *Geometrie der Lage* of 1847. There he endeavored to produce what could be regarded as a projective reworking of Euclid's *Elements* based not on a primitive undefined concept of length but on the idea that configurations of four points on a line have a property that projective transformations cannot alter, just as pairs of points do in Euclidean geometry. He was strikingly successful in this endeavor, and a signal influence on the young Felix Klein and his Erlangen Program (see §3.1.1).

2.1.2 Non-Euclidean Geometry

I do not know how to meet . . . the calmness with which people, who are incapable of grasping the simplest geometrical statement, pronounce upon the most complex problems of the Theory of Space in the sure conviction of superior wisdom.

—Helmholtz to Lipschitz, March 2, 1881, quoted in Koenigsberger 1906, 267.

---

[8] Gergonne 1827.
[9] See Gray 2006c, chapter 14.

Projective geometry developed in Paris in the 1820s in the very center of the mathematical world. Its growth, largely in Germany, is one way to measure the rise of Germany as an intellectual center that eventually surpassed France. The situation with the discovery of non-Euclidean geometry could not be more different, because it was to be kept almost literally at a distance for a whole generation, until all with a claim to have discovered it were dead. A proper understanding of how difficult non-Euclidean geometry was to discover and then for others to accept will shed light on the full force of the challenge it posed to all previous conventional ideas about geometry.

The first discoverers were Nicolai Ivanovich Lobachevskii in Kasan in Russia and János Bolyai, in Hungary/Transylvania.[10] Many would argue the case for Gauss, although I shall not, but in the present context the precise details do not matter.[11] The contribution of non-Euclidean geometry to the Modernist formulation of geometry came only in the second half of the nineteenth century, after the publication of Gauss's *Nachlass* and many of his letters to astronomers had shown that he was at the very least favorably disposed to the idea and that his eminent colleagues had been in agreement with him.

2.1.2.1 LOBACHEVSKII

The Russian mathematician N. I. Lobachevskii studied at the University of Kasan and went on to become a professor there and for many years the well respected rector of the university.[12] In 1829 he published the first of several articles in the *Kasan Messenger* describing an alternative geometry to Euclid's. Lobachevskii deeply believed that the Euclidean foundations were flawed. In his view terms like "line," "surface," and "position" were obscure, and certainly not fundamental. Rather, geometry should be based on ideas about bodies and the motion of bodies. This echoes the ideas of d'Alembert in the *Encyclopédie*, whose philosophy of geometry was well adapted to the application of geometry and the calculus to Newtonian mechanics. Lobachevskii argued that ideas about straight lines are derived from a consideration of bodies, and when this is done carefully, lines need not be as Euclid had said they were.

Lobachevskii proposed an alternative geometry in which, as he showed, the angle sum of a triangle is always less than two right angles, and the angle sum gets less as the triangle gets bigger. This suggested to him that one could use the parallax of stars to attempt to see if space was in fact non-Euclidean. In Euclidean geometry the parallax gets smaller as the star is farther away, becoming arbitrarily small if stars are to be found far enough away. But in the new geometry the parallax cannot fall below a certain level, determined by the diameter of the Earth's orbit, so if measurements of parallax always exceed a certain non-zero amount, this would be evidence in favor of the validity of the new geometry. Lobachevskii took advantage of the latest astronomical work to consider various stars, but found the observations suggested that to all intents and purposes space was Euclidean.

---

[10] Some short passages in this section repeat what I have said elsewhere, in Gray 2004a, and 2006c, I hope not to an egregious extent. For a general account of the history of non-Euclidean geometry, see Gray 1989.

[11] I have argued that Gauss did not discover a mathematical theory of non-Euclidean geometry in Gray 2003.

[12] Biographical information is taken from Lobachetschefskij 1899.

Sadly for Lobachevskii, what he wrote in French and German was unconvincing. He provided a string of deductions from a hypothesis that differs from the parallel postulate, but did not directly confront the question of whether this can consistently be done. His Russian accounts are more satisfying, but Russian authorities such as Ostrogradskii in St. Petersburg savagely denigrated his work, which they did not understand. His German booklet (1840) got only one review, which was a travesty. His only success came when he sent a copy of it to Gauss in Göttingen. Gauss immediately acclaimed Lobachevskii's work and in 1842 had him made a corresponding member of the Göttingen Academy of Sciences, but this was to be the only reward Lobachevskii was to receive for his discoveries in his lifetime.

What gave Lobachevskii his conviction that his geometry was not self-contradictory was not just his novel starting point, with its emphasis on motion as the central idea in geometry, but also his finishing point. Lobachevskii expressed his theorems in the language of trigonometry and the calculus. He deliberately sought out formulas because he deeply believed that geometry was about measurement, and that measurements, numbers, are related to one another by formulas. In turn, the validity of these formulas was a matter of algebra, whatever might be their geometrical significance. In turning from geometry to trigonometry, Lobachevskii uncannily echoed the approach of his co-discoverer, János Bolyai.

2.1.2.2 BOLYAI

János Bolyai was born in Klausenburg, Transylvania, Hungary (now Cluj, Romania) on December 15, 1802, and moved with his parents to Maros-Vásérhely (now Târgu-Mures, Romania) in April 1804, when his father Farkas became professor of mathematics at the Evangelical Reformed College there. He read the first six books of Euclid's *Elements* under his father's direction, and in due course moved on to Euler's *Algebra*, before persuading his father to let him attend lectures at the college when he was twelve years old. From 1818 to 1823 János studied at the Royal Engineering Academy in Vienna, which trained cadets for military service. He then served as an engineer in the Austrian army for ten years, and this seems to have given him some time for mathematics. In 1833, weary of the military life, he retired on a pension as a semi-invalid to Maros-Vásérhely.

While in the army, his interests had turned to the parallel postulate, doubtless because of his father. As early as 1820 he began to think that his failures to prove the parallel postulate might arise because in fact the parallel postulate was not true. He switched direction and henceforth attempted to show that there could be a geometry independent of the parallel postulate. He wrote to his father that "one must do no violence to nature, nor model it in conformity to any blindly formed chimera; but on the other hand, one must regard nature reasonably and naturally, as one would the truth, and be contented only with a representation of it which errs to the smallest possible extent."

His father was alarmed, and entreated him to leave the science of parallels alone. "Learn from my example: I wanted to know about parallels, I remain ignorant, this has taken all the flowers of my life and all my time from me."[13] The idea János

---

[13] This and the next three quotes are from Stäckel 1913, 81.

proposed struck him as worthless. "I admit that I expect nothing from the deviation of your lines." But happily, the son did not listen to his father, and on November 3, 1823, he could write to say that he was confident of success:

> I have not yet made the discovery but the path that I am following is almost certain to lead to my goal, provided this goal is possible. I do not yet have it but I have found things so magnificent that I was astounded. . . . All I can say now is that I have created a new and different world out of nothing. All that I have sent you thus far is like a house of cards compared with a tower. I am as convinced now that it will bring me no less honour, as if I had already discovered it.

His father advised him to publish his results as soon as possible, and suggested it appear as an appendix to a work on geometry that he had been writing for some time. But when János visited his father in February 1825, he was unable to convince him. His father worried about an arbitrary constant that entered the formulas his son had found. This is potentially troubling, for might not further work show that the constant is not arbitrary but can be calculated, and indeed calculated in two different ways that would yield two different answers? That would be a fatal contradiction and the new geometry would not be possible. They disagreed about this until finally, in 1829, they agreed to publish anyway. The two-volume work, entitled *Tentamen juventutem studiosam in Elementa Matheosis purae, etc.* was published by the College in Maros-Vásérhely in 1832. A copy was sent to Gauss, but it was lost in the chaos of a local cholera epidemic, and another was sent.

On March 6, 1832, Gauss replied:

> If I commenced by saying that I am unable to praise this work, you would certainly be surprised for a moment. But I cannot say otherwise. To praise it, would be to praise myself. Indeed the whole contents of the work, the path taken by your son, the results to which he is led, coincide almost entirely with my meditations, which have occupied my mind partly for the last thirty or thirty-five years. So I remained quite stupefied. . . . On the other hand it was my idea to write down all this later so that at least it should not perish with me. It is therefore a pleasant surprise for me that I am spared this trouble, and I am very glad that it is just the son of my old friend, who takes the precedence of me in such a remarkable manner.[14]

Farkas was pleased that the great geometer had endorsed his son's discoveries, but the son was appalled, and it was to be almost a decade before he could be convinced that Gauss had not dishonestly claimed priority. He made no further attempt to publish his work, and it began to sink into obscurity.

2.1.2.3 THE SIGNIFICANCE

It is not always clear from letters and accounts what is at stake here, and why the very idea of a geometry different from Euclid's should arouse such passions. It is of course true that anyone teaching the subject today would point out that non-Euclidean geometry is a metrical geometry which differs from Euclidean geometry in just one of its initial assumptions but in many of its consequences. It therefore offers a plausible

---

[14] Gauss 1900, 220–224, quote on 220.

but novel description of physical space. But this was not a nineteenth-century perception, for two reasons. One is that no one ever thought it was likely that non-Euclidean geometry would be a useful description of small regions of space. It was easy to calculate, as Riemann did in 1854, that regions of space in which any departure from Euclidean geometry would be detectable were unfeasibly large.[15] The significance of non-Euclidean geometry was, therefore, never that it posed a practical alternative geometry of space. The second reason is that today we are all familiar with many novel geometries of space. Throughout the nineteenth century, non-Euclidean geometry was the sole alternative, even though Riemann pointed out in 1854 (published in 1867) that there was no reason to suppose that space was not finite but unbounded, like a sphere. So it easy for us to accept non-Euclidean geometry, but a responsible nineteenth-century reaction would have been to suspect that it rested on a (possibly deep) error.

In the first half of the nineteenth century, Euclidean geometry was a well-entrenched orthodoxy. More weight was probably attached to Kant's opinions at the end of the nineteenth century than at the start of it, but he had emphatically endorsed the idea that we necessarily construct space as Euclidean. It was also the space of Newtonian physics. It was the geometry dinned into one at school. The idea that all this might be wrong was to prove exciting and disturbing. Euclidean geometry was a repository of truths about the world that were as certain as any knowledge could be. If it failed, what sort of useful knowledge was possible at all? But the excitement had to wait until more convincing accounts were given than Bolyai and Lobachevskii had been able to provide.

2.1.2.4 GEOMETRY

The new geometry was also, almost literally, unthinkable, and its ultimate reception has a lot to do with how anyone might come to think of such a geometry at all. We may focus on the problem by summarizing Gauss's thoughts on the matter. Gauss studied at the University of Göttingen from 1795, and while there he became friends with Farkas Bolyai, the father of János. When Farkas Bolyai returned to his native Hungary in 1798, Gauss wrote to him of his own researches into the parallel postulate, cautioning: "Only, the path which I have chosen does not lead to the goal that one seeks, and which you assure me you have achieved, but rather makes the truths of geometry doubtful."[16] Farkas Bolyai took his own road, however, and in 1804 sent Gauss his defense of the parallel postulate, only for Gauss to find a crucial error.

Gauss's misgivings continued to deepen as he worked though arguments purporting to prove the truth of Euclidean geometry, only to find mistakes, for example in several of Legendre's arguments, which he criticized in a letter to Gerling in 1816.[17] In 1817, he wrote to the astronomer Olbers (after whom the paradox is named) to say: "I am becoming more and more convinced that the necessity of our geometry cannot be proved. . . . Perhaps only in another life will we attain another insight into the nature of space, which is unattainable to us now. Until then we must

---

[15] See Scholz 2004.
[16] Gauss 1900, 159.
[17] Gauss 1900, 168–169.

not place geometry with arithmetic, which is purely a priori, but rather in the same rank as mechanics."[18] By 1829 he could write to Bessel that "my conviction that we cannot base geometry completely a priori has, if anything, become even stronger," but that he probably would not publish his ideas in his lifetime because he feared the howls of the Boetians (a tribe the ancient Greeks had regarded as particularly stupid).[19] Bessel replied that he regretted this modesty on Gauss's part, and in April 1830 Gauss wrote back to say that "it is my inner conviction that the study of space occupies a quite different place in our a priori knowledge than the study of quantity . . . . we must humbly admit that if Number is the pure product of our mind, Space has a reality outside of our minds and we cannot completely prescribe its laws a priori."[20]

Gauss's letters do not support his claim that he knew what János Bolyai knew. His ever-deepening conviction falls short of the possession of a solid proof, and it seems to have been confined to a study of the geometry of two dimensions. At some stage he despaired of showing that there was an absolute logical necessity to geometry, and came to believe that it could not be known a priori but had an arbitrary element in it, and could at best be a system of empirically valid truths akin to mechanics. But he was also not absolutely—mathematically—certain on the point. Most people who have written on the subject have tried to explain why he did not want to publish his ideas. He surely found the prospect of many weary months of controversy with people who would simply misunderstand his ideas unappealing. Perhaps he had simply not brought his ideas to the high level he regarded as necessary for publication, and Gauss kept himself to high standards. Sartorius von Waltershausen, who knew Gauss personally and published a memoir about him in 1856, said that "Gauss always sought to give his researches the form of a completed work of art . . . and would therefore never publish a work before it had attained this completely worked-out form that he desired. After a good building has been put up, he used to say, you can no longer see the scaffolding."[21] The puzzle arises if one assumes that Gauss was clear in his mind that non-Euclidean geometry could exist, and must therefore have had some reason for not publishing. In fact, there are reasons to suggest Gauss was not so clear.

Two issues are involved, and though they are related, they are distinct. Only by confusing them is it possible to argue that Gauss knew everything that Bolyai and Lobachevskii had to tell him, and it is unlikely that Gauss was confused in this way himself. One issue is the nature of physical space, and by the 1820s Gauss had what one might call a scientist's conviction that another geometrical description of space was possible. Bessel was similarly convinced. The other issue is the mathematical description of this new, three-dimensional space, and that was lacking. Gauss only possessed accounts of a two-dimensional geometry, and that is not enough. One may, after all, produce an ordinary two-dimensional sphere, but that does not show that physical space is to be modeled on the three-dimensional analog of the sphere. There is no mathematical difficulty in producing such an object, it is the presumed analogy that is wholly spurious.

[18] Gauss 1900, 177.
[19] Gauss 1900, 200.
[20] Gauss 1900, 201.
[21] Sartorius 1856, 82.

What F. K. Schweikart, a professor of jurisprudence at Marburg,[22] sent to Gauss via his colleague Gerling in 1818 is remarkable, but it is no more than a hunch that a novel two-dimensional geometry is possible. Gauss agreed. In 1824 Schweikart's nephew Taurinus rejected the same geometry because on it the measurement of lengths reduces to that of angles, and so there is an absolute measure of length. This was an old observation, made by the English mathematician John Wallis in the seventeenth century and the Swiss mathematician and polymath Johann Heinrich Lambert in the eighteenth, and did not detain Gauss; but Taurinus was looking for a way out and seized on this, among others equally inadequate. The inevitably flawed defenses of Euclidean geometry and Gauss's investigations of the foundations of elementary geometry share one feature: they are about a geometry of two dimensions.

How do we justify the study of two-dimensional Euclidean geometry? By claiming that it has all the essential features of Euclidean geometry in three dimensions, but it is easier. The passage from three dimensions to two and back is facilitated by the prior belief that three-dimensional space is to be described by Euclidean geometry. If the nature of three-dimensional space is not clear, if the status of the parallel postulate is contentious, then no discussion of a two-dimensional geometry can resolve the matter. There might be a two-dimensional geometry that differs from Euclid's but no three-dimensional equivalent. In order to show that space can be described by non-Euclidean geometry, it is necessary to produce a three-dimensional non-Euclidean geometry.

Gauss never did this. It was the independent achievement of Bolyai and Lobachevskii, resting though it did on a mere assumption. Without a three-dimensional mathematical description, Gauss was deprived of the one key ingredient that would have brought him adequate certainty. There is further evidence that Euclidean geometry exercised more influence over Gauss than his mathematical insights might suggest. How do we speak so easily of geometry on the surface of a sphere? Because we can see the sphere in Euclidean three-dimensional space. We can draw curves on it, imagine curved rulers on it with which we measure lengths, and so on. All the basic geometrical terms for discussing geometry on the sphere are derived (the mathematical term is induced) from the surrounding three-dimensional space. Now, in the 1820s Gauss had made a remarkable discovery, one that struck him so forcefully that he even called it the *Theorema Egregium*, the exceptional theorem. This says that on any surface in space there is at each point a number which measures how curved the surface is. This number is the same at every point of the sphere, of course, and works out to be the reciprocal of the square of the radius. For a plane this number is zero everywhere; the plane is not curved at all—in Gaussian terms it has zero curvature. The remarkable thing is that this number can be calculated from measurements taken only on the surface. No reference is required at all to quantities that take you off the surface. So although we can step off the Earth to see that it is round, we could discover it was round even if the planet was perpetually shrouded in the densest fog.

Gauss discovered an intrinsic property of a surface, today called its Gaussian curvature. But it was left to Riemann a generation later to draw the full lesson from this discovery: the fundamental geometrical properties are those which can be de-

---

[22] Gauss 1900, 180–181.

fined in intrinsic terms. And, because they are intrinsic, they make no reference at all to a surrounding space, and so there need be no surrounding space. Questions about the properties of one geometrical object sitting inside another are entirely legitimate, but entirely separate, questions, and Riemann was the first to argue for a distinction between them. Gauss never took this step fully, and what seems to have impressed him most when he was the senior examiner of Riemann's *Habilitation* in 1854 was precisely Riemann's grasp of the idea that all geometries should be formulated intrinsically.[23]

However, when a three-dimensional geometry of space is the object of investigation it is clear that it must be described in intrinsic terms. A three-dimensional non-Euclidean space cannot be treated as if it is a surface sitting inside three-dimensional Euclidean space. To create three-dimensional non-Euclidean geometry it is necessary to start where you will have to remain, outside Euclidean three-space in some other space entirely. Gauss's failure to do that is evidence that even he found the idea that Euclidean space is the source of all geometrical properties difficult to shake off, and what he found difficult others quite generally found impossible. Indeed, without a three-dimensional intrinsic approach to geometry, non-Euclidean geometry is literally unthinkable. There is no mental space for it.

2.1.3 Acceptance: Riemann and Beltrami

If we wished to refer the space of the painters to geometry, we should have to refer it to the non-Euclidean scientists; we should have to study, at some length, certain theorems of Riemann's.

—Albert Gleizes and Jean Metzinger, Cubism, 1912, in Chipp 1968, 212.

Acceptance of non-Euclidean geometry began in the late 1860s, with the publication of Riemann's *Habilitationsvortrag* (henceforth referred to as his *Lecture*) and Beltrami's *Saggio*. Riemann seems to have had no interest in the topic of non-Euclidean geometry as traditionally conceived, that is, as a study of the question, "Are the assumptions of Euclid plane geometry other than the parallel postulate consistent with any assumption about lines that produces a geometry other than Euclidean geometry?" Note that the question of the independence of axioms was hardly raised in the discussion above. There are indeed reasons, to which I shall return, for suggesting that to see the search for non-Euclidean geometry in this axiom-based way is an artifact of mathematical modernism that distorts the historical record. Be that as it may, Riemann never discussed the problem in this axiomatic fashion, and he never mentioned either Bolyai or Lobachevskii by name. He may have seen Lobachevskii's name in a journal he borrowed from the Göttingen library, and one does not know what Gauss said to him in conversation, but the topic may simply not have come up between them. What Riemann saw as fundamental in geometry was the idea of intrinsic properties.[24]

---

[23] The *Habilitation* in nineteenth-century German academic life was a necessary and sufficient condition to become a university teacher, but it did not entitle you also to be paid.

[24] English readers may like to note that there is an English translation of the second edition of Riemann's works, Riemann 2004.

FIGURE 2.3. Bernhard Riemann.

Riemann had a recipe for creating a geometry in his *Lecture*. Take a space, that is, a set of points.[25] Grant yourself the ability to measure distances between points, so that given any two points you can talk about the shortest path between them, and measure angles between curves that cross. Then you have a geometry. Most likely, and this was Riemann's preferred way, you have this ability because you have a way of applying the calculus to your space. The details were, and are, technical, and for present purposes they do not matter. It did his work no favors that he had to present it to the philosophy faculty, which made it long on verbal descriptions and short on the formulas that would speak to mathematicians, and Riemann was a poor expositor. But it is immediately clear that there is no restriction to two or three dimensions. Indeed, Riemann was prepared to contemplate spaces of infinite dimension. Nor does Euclidean space play a privileged role. In fact, it plays no role at all. Geometry no longer starts with Euclidean geometry.

In a passage in the *Lecture* that is often overlooked, Riemann set out his description of the simplest geometries, and he did so in a way which made non-Euclidean geometry visible for the first time, and therefore evidently consistent. Suppose for simplicity one takes a three-dimensional manifold, in Riemann's sense of the term. There are two ways one can measure distances in it. One can imagine

[25] Riemann called such a set of points a *Mannigfaltigkeit*, thus the word "manifold"; interestingly, Cantor's name for his sets of points was also *Mannigfaltigkeit*. Gradually, "manifold" became the word for the structured object and a mere set such as Cantor discussed became, as Cantor also called them, a *Menge*, meaning a collection or aggregate.

moving a very small one-dimensional ruler around and adding up the number of times it is used to measure a length (as it might be, the distance from London to New York in meters). Or, one can imagine moving a very small three-dimensional rigid body around, with a ruler engraved on its edges. Now, if we assume, as we always do in real life, that the rigid body does not deform as it rotates no matter where it is, then the space is isotropic (the same in all directions). If we can put the rigid body anywhere without it distorting, the space is said to be homogeneous. Conversely, spaces that are isotropic and homogeneous may be given a geometry in which the small rulers are actually small rigid bodies. It is evident that we believe that physical space is isotropic and homogeneous, and that we use small three-dimensional rigid bodies to make measurements; a purely one-dimensional ruler is a mathematical fiction.

Riemann now gave a description in coordinates of an n-dimensional isotropic and homogeneous space. It is easy to misread the passage as describing what the simplest n-dimensional geometry would look like in mathematical terms, but Riemann discusses them in terms of (a suitable generalization) of Gaussian curvature. The value of the generalized curvature at each point is to be the same if the space is isotropic and homogeneous, and depends on a single parameter that varies from space to space. If the value of this parameter is zero, Euclidean n-dimensional geometry results. If it is positive, n-dimensional spherical geometry results. But if the value of the parameter is negative, the result is a description in coordinates of n-dimensional non-Euclidean geometry. The space is the interior of the $(n-1)$-dimensional sphere (the n-dimensional ball), but the way distance is measured is not the standard Euclidean one.

What Riemann did was to produce a map of non-Euclidean space in exactly the way familiar geographical atlases depict the Earth on a flat page. All such maps distort: a distance of one centimeter on the page will mean different things depending on where it is on the map. Just so, Riemann's "map" of n-dimensional non-Euclidean space distorts intrinsic non-Euclidean distances as it depicts them on the n-dimensional ball. But by depicting them in a systematic, mathematically described way, it demonstrates the consistency of the new geometry.

Riemann knew enough of the historical context to know what he was doing. The *Lecture* opens with a reference to a darkness at the heart of geometry that has been there since the work of Legendre.[26] This is a clear reference to the eminent French mathematician's failure to resolve the problems of parallels despite many attempts. Then, toward the end of the *Lecture*, Riemann described the three two-dimensional geometries of constant curvature and observes that they are distinguished by the angle sums of triangles. As he put it, in one (corresponding to Euclidean geometry) the sum of the angles in a triangle is always two right angles. In the second, the angle sum always exceeds two right angles: this is spherical geometry. In the third, the angle sum always falls short of two right angles: this is non-Euclidean geometry. This is the trichotomy familiar to anyone interested in non-Euclidean geometry and the parallel postulate since the time of Saccheri in the first half of the eighteenth century. Consistent with his foundational approach to geometry, Riemann accepts spherical geometry, while to a strict investigator of the parallel postulate it is of no interest,

---

[26] Adrien-Marie Legendre (1752–1833) was one of the leading French mathematicians of his day. He worked on mathematical analysis, number theory, geometry, and statistics, and often clashed with Gauss.

because it contradicts not one but two axioms of Euclid (in spherical geometry lines cannot be extended indefinitely). This underlines that Riemann took a fresh, and altogether different approach to geometry, and did not proceed from the assumption the Euclid had got it all more or less right, so to speak.

2.1.3.1 BELTRAMI

By the time Riemann's *Lecture* was published in 1867, the account of non-Euclidean geometry that was to become much better known was waiting in the wings. This is the *Saggio* (meaning essay or attempt) by Eugenio Beltrami, which was held up because Beltrami was worried that his supervisor, Luigi Cremona, was himself worried that the *Saggio* might rest on a vicious circle.[27]

Beltrami's very simple idea was adapted from a familiar process: the depiction on the flat pages of an atlas of the curved surface of the Earth. This can be done in many ways, none of them perfect, all of them entirely intelligible once the rule for constructing the map is stated.[28] We can say that the pages of an atlas depict or describe the Earth. It is easy to learn how to read off from a map that a distance, on the map, of some ten inches corresponds to a trek of some 20 miles, or, on a very different map, a distance of some 3,000 miles. Beltrami took a standard map of a sphere onto a disk—think, if you wish, of the sphere as the Earth. This map gave him a formula for converting distances on the disk into distances on the sphere. He now changed this formula, but continued to interpret the disk as a map—a mathematical description—of something. It was no longer a description of the sphere, so what was it describing?

Beltrami had deliberately changed the formula in such a way that the shortest journey on the mystery surface between any two points appeared on the disk as a straight line. His formula gave him a way of calculating angles between these shortest journeys. He could do trigonometry on "triangles" made up of three such journeys. He found, when he examined the matter, that the angle sum of these "triangles" was invariably less than π, and the Gaussian curvature of the mystery surface was constant and negative. While other mathematicians had found similar things, only Beltrami saw the whole picture: the disk before him was a picture of non-Euclidean two-dimensional space.

He took his discovery to his former supervisor and mentor, Cremona, the leading geometer in Italy at the time. Cremona worried that the proof of Beltrami's claims, the basis of his trigonometry, might depend on the truth of Euclidean geometry, in which case the whole argument was nonsense. What Beltrami needed, and soon found, was the reassurance of the Riemannian view of geometry. This shored up his discovery in two crucial respects. First, a set of points, such as the points of a disk, and a formula for distance make a geometry; there was no need for Euclidean geometry. Second, there was no need to exhibit the mystery surface in our three-dimensional space. Riemann had been very clear: one question is to exhibit a geometry, another (and certainly interesting) question is to exhibit that geometry in our three-dimensional space, or any other space, for that matter. Beltrami did not have to know that no model of non-Euclidean two-dimensional space can be made in three-

---

[27] See the letter from Beltrami to Genocchi, quoted in Boi, Giacardi, and Tazzioli 1998, 10.
[28] See Feeman 2002.

dimensional Euclidean space that is anything as good as the sphere;[29] he only had to know that exhibiting a surface like that was not necessary. His disk was as good a proof of the existence of non-Euclidean geometry as you can get. It was quite a useful model, too, because shortest journeys (what mathematicians call geodesics) appear as straight lines in the disk. When, a year later, he constructed *n*-dimensional non-Euclidean geometries in a similar way, the post-Riemannian mathematicians could ease into accepting the new discoveries, and that is what they did.[30]

In this they were helped by the discovery and publication of the relevant portions of Gauss's *Nachlass*. After his death in 1855 it was discovered that the greatest mathematician of his time had long believed that non-Euclidean geometry was possible, had corresponded with his fellow astronomers such as Bessel and Olbers about it and found that they agreed, had made quite some detailed study of it, and had personally endorsed the remarkable, and almost completely forgotten, discoveries of Bolyai and Lobachevskii. Volumes of his correspondence were published on this and many other topics, and editions of his collected works carried extensive extracts from these and his notebooks. Suddenly accepting non-Euclidean geometry put one in respectable company: Beltrami, Riemann, and now Gauss. One can only wonder what might have happened if Gauss had energetically promoted these ideas from the 1830s.

2.1.4 Professional Aspects

Between let us say 1867 and 1880, when Poincaré began his mathematical career with a striking new use of non-Euclidean geometry,[31] the mathematical world learned to accept non-Euclidean geometry.[32] They did so quite quickly, after a prolonged period when it had seemed impossible to accept. There were exceptions, of course, most notably Cayley in England, and Bertrand in France fell for and published a "refutation" of non-Euclidean geometry long after he should have seen through it at once. These responses were all the more remarkable because Cayley[33] was the leading pure mathematician in Britain for most of his long life and instrumental in establishing a place for mathematics at Cambridge outside the realm of applied mathematics and natural science; and Bertrand[34] had risen to the top of the mathematical profession in France and had become the permanent secretary of the Académie des Sciences in 1874. But the climate had changed. There can be no doubt that the crucial feature bringing about this change was the description of non-Euclidean geometry afforded by Beltrami's *Saggio*, with its underpinnings in Riemann's philosophy of geometry. Helmholtz's rapid endorsement of non-Euclidean geometry in a public lecture in 1868 must also have helped spread the word.[35] There may also have been a generational effect, as younger mathematicians in several

---

[29] That was only proved much later, by Hilbert in 1901; see Hilbert 1901b.

[30] In this his second paper (1868b), Beltrami explained the relationship between his description of non-Euclidean geometry and Riemann's, thus preceding Poincaré's account by over a decade.

[31] See Poincaré 1997 and Gray 2000a.

[32] See Voelke 2005 for a detailed account.

[33] Cayley 1865; see also Richards 1979, Richards 1986, and for a full-length biography of Cayley, Crilly 2006.

[34] See the introductory essay by Gray and Walter in Poincaré 1997.

[35] Helmholtz 1870, in Helmholtz 1977, 1–38.

countries picked it up without much difficulty. The professional aspect must also be taken into account.

János Bolyai's chosen method of publication was an *Appendix* in Latin to his father's two-volume textbook on geometry published in Hungary. Dismayed and angered by Gauss's response, he never sought to publish again. Had he enlisted Gauss's support, had he rushed the *Appendix* to, say, Crelle's newly founded journal, things might have been different, but he did not. Most mathematicians probably never heard of him, and Gauss did nothing much to help. Lobachevskii did try harder—he wrote in French in Crelle's *Journal* and his booklet of 1840 is in German—but with little more success. No one came forward to rescue their work from its one signal failing before the late 1860s.

We can get a sense of who might have done so from the tables published in the fiftieth volume of Crelle's *Journal* in 1856. From these tables we see that 111 Prussian authors had been published, alongside 41 other Germans, and a smaller number of other nationalities (mostly French, Russian, German, Scandinavian, English, and Italian). There were, incidentally, nearly 12,000 pages in German, about 4,500 in French, 2,400 in Latin, 112 in English, and 88 in Italian. This gives a total for the number of research mathematicians of about two hundred, evidently biased to the German-speaking world. Doubling the numbers to count in the French-speaking world, which was more vigorous in the period, we arrive at a total of some four hundred or five hundred active mathematicians in the thirty years covered by the table. However many of these would have been capable of responding to the work of Lobachevskii, or even interested in doing so, the fact is no one did. Making due allowance for the mundane nature of many of the articles published, and the number of mathematicians who published very little, we can guess that only a small number of mathematicians could have responded, even if they had been ordered to do so.

In 1867/68 the first abstracting journal in mathematics appeared, the *Jahrbuch über die Fortschritte der Mathematik* (henceforth *Fortschritte*). It was edited and produced in Berlin, and it was apparent from the first volume that there was now a much greater number of mathematicians at work. This number continued to grow as the century wore on, forcing the editors into double volumes and many delays. Its very existence is one way to document the emergence of a large-scale, university-based mathematical profession. Crelle's *Journal* had been set up to promote research in mathematics, especially in Prussia. When he set it up, Crelle knew very well that outside the University of Berlin there was everything to play for. There was very little mathematics going on in Göttingen, for example, despite the presence of Gauss, who was in any case professionally and actively an astronomer, a surveyor, and a physicist. The rise of the University of Königsberg as a center for mathematics was led by Jacobi in the 1820s. The situation in the southern German states was generally even weaker. By 1867 the German university scene was not only larger, it was better focused. The University of Berlin had entered the era of Kummer, Kronecker, and Weierstrass, and would soon reach a stage where audiences of two hundred for Weierstrass's lectures were not unusual (it helped that he never wrote them up as a book, so students had to attend the lectures or miss out entirely).

In Göttingen, too, things were improving. After Gauss's death in 1855 a number of distinguished mathematicians began to drift in and out of the university. Riemann and Dedekind had already been there as students. Dirichlet came as Gauss's successor, and he was undoubtedly the leading mathematician of his day, and a great force for rigor in mathematical arguments. After his death at the age of fifty-four, the university attracted Clebsch, but he died in 1872. Above all, until his health collapsed in 1862 and he tried to spend as much time as possible in Italy on medical grounds (he died there in 1866 at the age of thirty-nine), Riemann was a presence at Göttingen. There was no dispute that he was a mathematician of extraordinary ability. This is not to say that he was easy to understand, or that he set trends that other, lesser, mathematicians automatically followed, but he was a mathematician that others listened to. Their expectation was that while what he said, and still more how he said it, was likely to be hard, even almost unintelligible, it was also likely to be profoundly right. Lobachevskii had been a lonely provincial, easy enough to dismiss. Riemann was an insider, the insider's insider, one might say, in a growing profession. However distantly, mathematicians could and did set their compass by him.

In Italy, the unification of the country proved a decisive moment for the growth of mathematics. The struggle for unification is generally said to have begun in 1815 after the Congress of Vienna restored Austrian power over the fragmented country, and to have ended with the establishment of Vittorio Emmanuele II, the king of Sardinia, as the king of a united Italy in 1861. Further changes came with the addition of Venice in 1866 following the Prussian defeat of Austria, and after the withdrawal of Napoleon III's troops during the Franco-Prussian War, which allowed Rome to become the capital of Italy in 1871.

Many notable mathematicians, for example Cremona, had been prominent in the struggle for unification. This set the stage for a dramatic improvement in the mathematical life of the country, and one subject they took up energetically was geometry. Cremona was the leading figure here, too, a professor of engineering chiefly and best remembered for his work in geometry, and later a senator. Under his influence Italy produced a steady stream of major geometers, some of whom in the early days sought out Riemann, which was a further reason for Riemann going to Italy (oddly enough, Beltrami does not seem to have been one of them). Inspired by Beltrami's presentation, Italian mathematicians had no trouble adopting non-Euclidean geometry.

In France too the situation was about to change. The catalyst here was the traumatic defeat of the Franco-Prussian War. In every sphere of life the French asked: How could the country that Napoleon led to the largest empire Europe had seen since the Romans wind up sixty years later with the enemy at the gates of Paris? The reforms that followed saw a growth in the universities, especially outside Paris, and a new generation asserted themselves in French mathematical life, which had indeed begun to stagnate. Among them were two geometers: Gaston Darboux and Jules Hoüel. Darboux was the more original, while Hoüel performed the inestimable service of tracking down and translating the relevant works of Bolyai and Lobachevskii. Lobachevskii's booklet of 1840 was translated into French by Hoüel in 1866, who followed it with a translation of the *Appendix* by Bolyai the next year, and in 1869 set a translation of Beltrami's *Saggio* before a French audience. Meanwhile, Bolyai's *Appendix* had been translated into Italian by Battaglini in 1868. In the major mathematical countries, therefore, there was a solid institutional lobby, if that is not too strong, for non-Euclidean geometry. Expert opinion could speak with one voice, the originals could be easily consulted, the one glaring flaw in the work of Bolyai and Lobachevskii was repaired. It was now dissent that was marginalized, not discovery.

In the light of later developments, the immediate response to this new situation was remarkably quiet. After Felix Klein showed how to relate non-Euclidean geometry to projective geometry in the early 1870s, as will be described below, the next significant advance in the subject came in the early 1880s, when Poincaré connected non-Euclidean geometry to topics in the theory of functions of a complex variable. It was a piece of wonderful mathematics, profound and unexpected in equal measure, and it drew new mathematicians into the non-Euclidean world. But it was a technical tool there, and in the face of the evidently Euclidean nature of space for all practical purposes, mathematicians did not wish to draw any lessons from it of a more philosophical kind. When they did, the plunge into modernism was inevitable, as we shall see.

2.2 Analysis

2.2.1 What to Look For in a History of Mathematical Analysis

Mathematical analysis, loosely defined as the rigorous calculus and its rigorous generalizations, is both a core discipline of modern mathematics and its most extensive one. That being so, mathematical analysis poses a particular problem for this book, because its concerns, its methods, its insights are famously difficult. Yet the subject cannot be passed over here, for various reasons.

One reason is precisely its extent. No case can be made for modern mathematics undergoing a large-scale change of any kind that does not test its arguments on analysis. Another reason is that the real difficulties in this domain of mathematics set in precisely with the rigorization of the calculus and the creation, therefore, of mathematical analysis. Many of these changes took place before the period where I argue mathematics took its modernist turn, so the rigorization of the calculus is a challenge to the case argued in this book. It suggests that much mathematics has always been severely abstract, and was certainly so by the 1830s. To show that modernism is something different, it is first necessary to examine analysis in this earlier period. A third reason is that there is a real sense in which the history of analysis does challenge the thesis argued here. Grant, if you will, that our ideas of geometry, number, algebra, and logic all changed decisively around 1900. It still must be conceded that in mathematical analysis, of all the domains of mathematics, the continuities are strong. Accordingly, I do not claim that after modernism all was changed utterly. A degree of continuity is only to be expected, and is apparent in all of the areas where the modernist thesis is usually applied: in painting, music, literature, and architecture. But it is still necessary to address these continuities in order to depict accurately the transformations that happened at the same time.

To analyze the history of mathematical analysis here, it is therefore necessary to address these challenges directly, where necessary to concede the force of tradition, and where appropriate to draw out the modernist case. It is enough to admit the sheer extent of analysis. Enough of it will be visible here to all but the most selective of readers. Before addressing the two remaining substantial problems alluded to above, it may be helpful to alert readers as to what they should look out for within mathematical analysis.

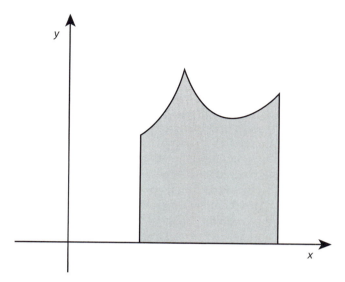

FIGURE 2.4. Area under the graph of a function.

What Newton and Leibniz did in the seventeenth century, when they independently invented the calculus, was to talk about variables, and how as one variable changed, another did as well. One of the many improvements to their work that Euler introduced in the eighteenth century was the concept of function. If the value of one variable depends on another, the dependent variable is said to be a function of the independent variable. At its simplest, the function idea is: put in a number, and get a number out. Euler had very little sense that there might be functions to which the processes of the calculus do not apply: that was the agenda slowly revealed by the nineteenth and twentieth centuries.

There are two fundamental processes in the calculus: differentiation and integration. Both are applied to functions by means of ingenious arguments that involve successfully better approximations tending in the limit to a definite value. The differential calculus deals with rates of change. It is based on a series of arguments that allow one to talk of change at an instant by approximating it with changes over shorter and shorter intervals. If this change is produced by motion (change of position), the differential calculus finds velocities; if the change is change of position along a curve, the differential calculus finds tangents. More theoretically, the differential calculus offers a solution to the paradoxes of Zeno. The integral calculus deals with areas and more generally with counting and measuring. The key case is of an area defined in the plane by the x-axis, two vertical segments standing at the endpoints of an interval on the x-axis, and the graph of a function (see figure 2.4) defined between those end points. The integral calculus specifies the area determined by the function and the position of the endpoints. If they are fixed, the area is a number; if the endpoints are allowed to vary, the answer is a function depending on the endpoints.

The processes of differentiation and integration are asymmetric in one respect: it turned out to be routine to differentiate a function, but not to integrate it. However, eighteenth-century mathematicians believed they are symmetric in another respect: integrate a function and then differentiate it, and the result is the original function. Equally, differentiate a function and then integrate it, and the result is the original function (plus an arbitrary constant). This symmetry, deservedly called the fundamental theorem of the calculus, said to mathematicians of the eighteenth century that integration could be redefined as the reverse of differentiation. Augustin Louis Cauchy, for good reasons, persuaded mathematicians to revert to defining integration as an area-finding process, so he had to prove the fundamental theorem of the calculus, and that turned out to be a complicated story. We need not go into it, but we do need to know that an underlying reason for the asymmetry between differentiation and integration emerged.

Now, a theory, any theory of anything, proceeds at a certain level of generality. *All* objects of such-and-such a kind have *this* property: all mammals have hair (even whales!). As a theory, the calculus makes statements about all functions, or all functions of a particular kind. Inevitably, mathematicians found themselves forced to distinguish classes of functions, just as biologists distinguish species. They found, in particular, that many functions can be integrated, but many fewer can be differentiated. The process of differentiation applies to a much smaller class of functions than integration, but certainly every differentiable function is integrable.

A fundamental intuition that almost everyone has is of continuous change. Certainly, a differentiable function is continuous. Even Cauchy, who was the first to give a rigorous account of continuity in strictly mathematical terms, plainly thought that if a function was continuous it was likely to be differentiable, and his contemporary Ampère (after whom the amp is named) even offered a (flawed) proof. More precisely, it would seem that a continuous function is differentiable except at a few points where it has a sharp corner. We shall see in due course that this naive intuition is simply wrong: in the second half of the nineteenth century it was discovered that there are continuous functions that are not differentiable at any point where they are defined.[36]

So, in the nineteenth century, calculus was a theory. It may help to orient the reader to note that, according to this theory, differentiable functions are continuous, but not all continuous functions are differentiable. Every continuous function is integrable, but not every integrable function is continuous.[37]

The other major figure in this part of the story is that of Joseph Fourier.[38] For thoroughly scientific reasons to do with heat diffusion, he had powerfully promoted the idea that any function might be represented in a novel way, as an infinite sum of sines and cosines. Series like this are called Fourier series in his honor.[39] A function f is represented as an infinite sum $\sum_{n=0}^{\infty} a_n \sin(nx) + b_n \cos(nx)$. Here the coefficients

---

[36] Actually, this was discovered by Bolzano much earlier, but only appreciated in the 1860s; see Russ 2004.

[37] To see why, note that the way integration is defined, a function that is continuous everywhere except at a point where it takes a "wrong" value is integrable and has the same integral as the corresponding function that is continuous everywhere.

[38] Joseph Fourier (1768–1830) combined a career in mathematics with one in politics, survived an arrest during the French Revolution to become one of the leaders of Napoleon's expedition to Egypt, and in 1822 became the permanent secretary of the Académie des Sciences in Paris.

[39] In 1758 Daniel Bernoulli had promoted this way of writing any function that could represent the shape of a vibrating string, but his ideas were not generally accepted.

a_n and b_n depend on, and are determined by, an integral involving the function f. It was a brilliant idea, and still lies at the core of numerous technologies. It corresponds to analyzing a sound in terms of its component frequencies. But unlike Cauchy, who was an enthusiast for mathematical rigor, Fourier was an applied mathematician, and he made sweeping claims for his way of representing functions without offering any proofs. It was left to the next generation of mathematicians to check it out, and they did so in ways that are entirely reasonable—and that anyone may find intelligible, unless they get too close.

For example, given a function f, is it true that $f(x) = \sum_n a_n \sin(nx) + b_n \cos(nx)$ for every value of x? In other words, is the representation correct for each value of x? At the other extreme, are there functions for which the representation is completely useless, for example, non-zero functions for which the coefficients in the representation are all zero? If so, will there be different functions with the same representation? Are there functions for which the very method of determining the constants a_n and b_n fails? It turned out that the answer to all these questions is "yes," so attention focused on the question of when the representation of a function is unique. For example, Cantor showed in 1870 that if two Fourier series are continuous and converge to the same sum, then their coefficients are equal.[40] He then investigated the conditions under which this equality continued to hold even though the continuity or the convergence of the series is not insisted on at every point.

Let me insist for a moment that these questions are entirely reasonable and arise anytime anyone has an imperfect process for making a picture of something. The uniqueness of the coefficients of a Fourier series is what we all demand of fingerprints. We only need to know that mathematicians asked and answered such questions, or on occasion how they answered these questions, if that knowledge helps us with our historical questions.

Now we can begin to see what we need to know about the history of mathematical analysis. If we grant that the mathematical analysis of the first half of the nineteenth century was abstract, was the level, even the nature of that abstraction ratcheted up as the century drew to a close? I shall argue that it was, and if the reader will grant me a small measure of credence the mathematical methods may be suppressed and the rapid evolution of the frightening beasts of mathematical analysis may be appreciated from a safe distance. It should be possible to demonstrate that standards of rigor rose, and why, and that as they did the objects of mathematical analysis, and some of the fundamental approaches to mathematical analysis, changed in a characteristically modernist way. If so, did the motivation for introducing the new type of abstraction derive from work that had gone before, or were there other factors? How strong are the continuities, how marked the changes? I shall argue that the primary motivation was derived from earlier work, so I need to establish the force of that tradition, and its novelty. To argue for the force of tradition I must look at the prevailing problems and methods sufficiently to prove my case. To argue for the changes, I must demonstrate them: changing foundations for the calculus, especially the idea of number; the clean separation of mathematical analysis from its applications in geometry and mechanics; the nonintuitive nature of its fundamental concepts; the use of characteristically modernist devices, such as novel forms of

---

[40] That is: if $\sum_{n=0}^{\infty} a_n \sin(nx) + b_n \cos(nx)$ and $\sum_{n=0}^{\infty} a'_n \sin(nx) + b'_n \cos(nx)$ are continuous and have the same sum, then $a_n = a'_n$ and $b_n = b'_n$.

axiomatic reasoning. We shall see that, while in the eighteenth century the simplicity of the rules of the calculus compensated for the fact that there was no satisfactory account of why the calculus worked, when finally rigor and intellectual clarity were attained the intuitive aspects of the calculus were lost. Mathematical analysis, as the rigorous calculus is called, remains a subject that divides mathematicians from scientists and engineers who use even sophisticated mathematics but proclaim a dislike, bordering on hostility, for what they see as the nitpicking rigor of the mathematicians. It is a paradigmatic piece of modernism. It meets a valued goal, it is willing to put rigor before immediate intelligibility, it separates an elite who appreciates it from a wider public that does not, it rescues what it sees as the best of the previous tradition. Above all, it boasts that it does not rely on intuition.

2.2.2 Cauchy

The creation of modern analysis is often taken to start with Cauchy.

The historian Judith Grabiner wrote:

> The most important figure in the initiation of rigorous analysis was Augustin-Louis Cauchy. It was, above all, Cauchy's lectures at the École Polytechnique in Paris in the 1820s that established the new attitude toward rigor and developed many characteristic nineteenth-century concepts and methods of proof. . . . And, after Cauchy had initiated the new rigorous calculus, many mathematicians—including Cauchy himself—made further contributions in the same spirit. . . . But all the accomplishments are based on the revolution begun earlier, and in important ways on the work of Cauchy, who created the mathematical climate which made them possible.[41]

Cauchy provided the first rigorous formulation of the calculus that mathematicians then and since have been able to agree was adequate. He defined what it is for a function to be integrable, to be continuous, and to be differentiable, using careful, if not altogether unambiguous, limiting arguments.

That a function can be differentiated is not, and was not in 1820, the most fundamental property a function can have. There was no clear statement of what the most basic property was, but there was a widespread agreement that the useful functions, those that could be used to represent aspects of the real world, were continuous. This was taken to mean that their graphs had no jumps. The rationale was that physical objects do not disappear here and immediately appear over there, but move in a continuous fashion from one location to another. One might even push this argument a little further, and say that what is true of position was also true of velocity, of acceleration, and so on. On this account, a moving point has a velocity at every moment of time (described by its tangent), and this velocity varies continuously, indeed in such a way that the moving point has an acceleration at each moment of time. The tangent is found by differentiation, the acceleration by differentiating again. The upshot was that everyone supposed a function can be differentiated arbitrarily often, for which the useful technical term is "analytic." In the common mathematical opinion around 1820, functions were naturally analytic, with

---

[41] Grabiner 1981, 2–3.

the caveat to be noted below. The most fundamental property of functions, continuity, was not what made the calculus work, but it was taken to imply that functions have properties that do make the calculus work. We shall see that the most dramatic implications of mathematical analysis precisely concerned the failure of continuity to supply foundations for the calculus.

The caveat was that trivial exceptions to the rules are to be acknowledged, but regarded as indeed trivial. A curve might have a corner in it, where it did not make sense to speak of it having a tangent, or it might cross itself, in which case it had two distinct tangents. A sudden impulse, such as a blow, might instantaneously give an object a velocity. But even then it was possible to imagine that the velocity rushed swiftly through intermediate values. So the common opinion more precisely was that functions were analytic unless they obviously and trivially were not.[42]

Cauchy's highly novel approach to rigorizing the calculus does not rest so much in his definitions as his use of them. As Grabiner (1981) argued, the definitions themselves are wordy and vague, but Cauchy used them in ways that often give a precise indication of what he meant and how he departed from the views of his predecessors. Moreover, as Grabiner went on to argue,[43] he used them extensively. He built up the whole of the calculus using these precise ideas, and he did so in a systematic way, great care being taken to avoid hidden assumptions, illicit appeals to intuition, and vicious circles. Nonetheless, there are mistakes and confusions in Cauchy's work, and we shall see that they are very significant.

The central concept in Cauchy's formulation of the calculus was that of a limit. He took the idea of a variable quantity tending to zero as basic, and he called such a quantity an infinitesimal. This is the vaguest part of his theory. Quantities, in Cauchy's theory, were numbers and magnitudes, so one can read this as meaning that an infinitesimal quantity is a finite quantity that can be taken to be arbitrarily small. There is a hint of a dynamic process about it that might help or hinder its application; one might imagine the quantity tending to zero much as a damped pendulum tends to the vertical position. But infinitesimals were commonly invoked in accounts of the calculus, when they were taken to be very small quantities, perhaps smaller than any finite quantity. Did Cauchy have that in mind? Did he think that he was capturing the essence of such talk but expressing it in a strictly finite way? He did indeed write that a variable having zero as a limit becomes infinitely small, but his use of the term suggests that given any quantity greater than zero (in symbols, given any $\varepsilon > 0$), the values of the variable can always be assumed to be finite and less than this given quantity ε. This does not rule out the idea that there might be finite magnitudes and old-style infinitesimal ones, but Cauchy never specified what he took magnitudes to be. This is partly because philosophy was not his strength, and partly because at the time saying what quantity is was not a matter of definition, but of apprehending what is given in nature. This task can call for great delicacy and skill, or it can be taken for granted. Cauchy took the latter view. On the other hand, his use of infinitesimals in the course of his proofs did much to promote an arithmetical treatment of the fundamental ideas of the calculus.

---

[42] An indication of how this worked in practice is given by Ampère's purported proof that a function is always differentiable, see Lützen 2003, 176.

[43] See also Bottazzini 1986, chapter 3.

Based on his definition of infinitesimal, Cauchy defined limiting processes, and so defined what it is for a function to be continuous. Again, his definition is wordy, but his repeated use of it was helpful in explaining what he meant. He wrote:

> The function $f(x)$ will be a continuous function of the variable x between two assigned bounds if, for each value of x between those bounds, the numerical value of the difference $f(x + a) - f(x)$ decreases indefinitely with a. In other words, the function $f(x)$ is continuous with respect to x between the given bounds if, between those bounds, an infinitely small increment in the variable always produces an infinitely small increment in the function itself.[44]

This is not the clearest statement one could hope to read. Let us see what it says about a simple continuous function, such as $f(x) = x^2$. We must say on what range the function is considered: let us say between $x = 1$ and $x = 2$. Our function is then continuous between the bounds 1 and 2 if the numerical value of the difference $f(x + a) - f(x) = (x + a)^2 - x^2$ decreases indefinitely with a. This difference is $2ax + a^2$, and it can be made as small as we please simply by choosing a small enough. So we see that the function $f(x) = x^2$ is continuous between $x = 1$ and $x = 2$, a result we have known all along, but have now given something like a rigorous proof.

As we shall see, however, the concept of continuity is very subtle and functions that are merely continuous may fail to have the properties one wants. As Cauchy was himself to notice, often a theorem can only be proved if an extra property is assumed, which is a strengthened form of continuity, and it is the strengthened form that provides the results. Cauchy, understandably, made mistakes, and the mathematical community took as much as fifty years before it obtained clarity on the issue. A hint of his confusion can be seen in his view that a function is continuous between two bounds, whereas a function may fail to be continuous at a single point. His ideas were refined over a generation or so until continuity was defined at a point, not on an interval, and a function was said to be continuous on an interval if it was continuous at every point in the interval. This can seem like a distinction without a difference, but when the words are replaced with symbols it became apparent that Cauchy's formulation is ambiguous and had on occasion led to error.

> *Suppose, as in our above example, we wish to make $2ax + a^2$ less than some positive ε (we shall assume x and a are positive). The values of a that make $2ax + a^2 < \varepsilon$ plainly depend on x. This was not clear in the verbal formulation. Does it matter? This is not clear until the mathematical arguments get delicate—and then, perhaps surprisingly, it matters.*

Cauchy then defined the derivative of a function as the limit of a difference quotient. He took the term from Lagrange but the idea from Lacroix, and so from tradition. Cauchy could then have defined what it is for a function to be differentiable, but he did not. He preferred to speak of continuous functions and the limits of their difference quotients "if they exist." In fact, Cauchy had the habit of stating theorems for continuous functions, and in the course of the proof adding assumptions to ensure that the function was differentiable. This is further evidence that the conse-

---

[44] Cauchy 1821, 34.

quences of continuity, what it implied and what it did not, were hard to discern, not just for Cauchy himself, but also, as we shall see, for his successors.

2.2.2.1 CAUCHY'S DEFINITION OF THE INTEGRAL

As Cauchy's successors appreciated, his new analysis rested on several distinct features:

> *Technically*, on the use of εs, δs, and Ns, as we shall see briefly below, which was consistent with and supportive of an arithmetical account of the calculus, but unclear about the nature of the magnitudes involved.
>
> *Philosophically*, on an unspoken realist assumption that mathematical objects are given in nature but must be described and handled carefully.
>
> *Mathematically*, on a systematic and seemingly very general theory of the processes underlying the formulas of the calculus that was considerably more rigorous than anything that had preceded it, but that nonetheless left the subtleties of continuity unexplored and sometimes equivocated between continuity and differentiability.[45]

Cauchy followed Fourier in defining the integral of a function and not treating it as an antiderivative. This meant that he had to prove that an identifiable class of functions can be integrated, and his proof that a continuous function has an integral is justly acclaimed as one of his great successes, because it returns to the original insight that an integral is an area, and does so in such a way that eventually the relation between tangents and areas, differentiation and integration becomes a theorem. But it also serves to demonstrate how Cauchy added ingredients to his proofs as he became aware that much still needed to be done. Cauchy started with a continuous function f defined on the interval $[a, b]$. The value of the integral is going to be the width of the interval, which is $b - a$, times the average height of the function f. Cauchy used the continuity of the function to approximate the integral by an expression of the form $(b - a) f(a + \theta(b - a))$ for some $0 \leq \theta \leq 1$, which is to say that he multiplied $b - a$ by the value of the function at some point intermediate between a and b, which is to say he invoked the intermediate value theorem.[46] This was not a result he proved in the main body of his *Cours*, where he took it for granted as a property of magnitudes, but it is a result he "proved" in a note he appended to it. Or rather, he took it for granted that if an increasing and a decreasing sequence of magnitudes ultimately differ by an arbitrarily small amount, then they converge to a common limit. From a realist perspective, this is unproblematic. If the nature of magnitude is problematic, however, then so is this argument.

Cauchy also gave a purported proof of a false result that is of great interest.[47] It has occupied many historians of mathematics, who have sought to explain how such a great mathematician could make such a mistake, and have sometimes done so in order to suggest that Cauchy had at least a glimpse of their own favorite theory. It seems more likely that even Cauchy had failed to realize just how subtle the concept

---

[45] Cauchy 1821.

[46] The intermediate value theorem says that a continuous function defined on an interval takes every value between the values it takes at the endpoints of the interval.

[47] On the mistake, see Lützen 2003, 168–169 and 181–184.

of continuity, as he had defined it, was, and that his way of writing about it was in a specific way too vague.[48]

In 1823 Cauchy claimed that a convergent sequence of continuous functions converges to a continuous function (as he put it, the sum of a series of continuous functions is a continuous function). This is wrong. He argued that because the infinite sum exists, as by hypothesis it does, the contributions of the tails (all the infinitely many terms after the first N for an arbitrary N) must become arbitrarily small. The sum of the first N terms is obviously continuous, and a standard continuity argument, using the fact that the tails contribute arbitrarily little, finished the proof. In 1853 Cauchy admitted that his "theorem" was false, and that counterexamples from the theory of Fourier series existed. By then Abel, Seidel, and perhaps even Stokes had formulated restricted versions of the theorem which they had been able to prove, but what stimulated Cauchy to return to print was a criticism of his work by the young French mathematicians Briot and Bouquet. He now refined it by stipulating that the tails behave in a fashion that guarantees convergence.

The condition he introduced became known as uniform convergence and can be briefly explained. The epsilon-delta formulation of the calculus says that a function is continuous at a point x_0 if for all ε there is a δ such that some condition holds. For a function to be continuous at every point of an interval, it is therefore not only likely that these ε's and δ's will vary from point to point, they may even do so in undesirable ways (they might get very small very quickly as x tends to the ends of the interval, for example). A function is said to be uniformly continuous on an interval if the choice of δ depends on ε, but not on the point within the interval.

The subtleties of continuity may have defeated Cauchy because his language in many places in the *Cours* and elsewhere displays a curious asymmetry. He speaks of a function failing to be continuous "at a point," say, where it has a jump. But he speaks of functions being continuous "on an interval." The best evidence that he genuinely thought of continuity as a property that functions have on intervals, rather than at points, and then, if at all, at every point of an interval, is the existence of mistakes like this one. But it is also the case that his proof—that every continuous function on an interval has an integral—makes tacit use of the strengthened form of continuity alluded to above.

Cauchy is often said to be among the first to undermine the idea that functions are generally analytic. It would be more accurate to say that he undermined confidence in the idea that being analytic was enough. The basic belief, true in all specific cases ever considered before him, was that an analytic function is represented by its Taylor series. The techniques of the calculus, going back to Newton and Leibniz, permitted mathematicians to take an analytic function $f(x)$ and write is as a power series (called, for reasons we need not discuss, its Taylor series): $f(x) = a_0 + a_1 x + a_2 x^2 + \cdots + a_n x^n + \cdots$, where the coefficient a_n depends on the nth derivative of f at $x = 0$: $a_n = \frac{1}{n!} \frac{d^n f}{dx^n}(0)$. However, in 1823 Cauchy gave this example of a function that is infinitely differentiable, but not represented by its Taylor series: the function is defined to be e^{-1/x^2} when $x \neq 0$ and 0 when $x = 0$. Cauchy showed that although it is clearly not 0 when x is not 0, its Taylor series is identically 0 (every coefficient $= 0$).

---

[48] The idea that a great mathematician only makes interesting mistakes is in any case hardly defensible.

Cauchy also drew a strong consequence from his fundamentally arithmetic approach to the calculus. Because all of its theorems rested on an arithmetic of inequalities, he required that their applications be restricted to contexts where they made arithmetic sense. So, for example, a power series such as $1 + x + x^2 + \cdots$ only makes sense when $|x| < 1$, and Cauchy insisted that studying this power series otherwise was nonsensical. Formal manipulation of such divergent series was placed off limits, and any use for them was hindered for a long time. Any talk of extending the domain of interpretation of symbolic expressions indefinitely was suspect to Cauchy (thus his criticisms of Poncelet, quoted above) and in place of confidence in the "generality of algebra" came a security grounded in numbers.

2.2.2.2 FIRST RESPONSES

Cauchy's foundations for the calculus were not an immediate success. He published his new ideas in the lecture courses he was giving at the École Polytechnique in the 1820s, and his student audience, military cadets at the École Polytechnique, stamped their feet in an egregious breach of discipline, for which they were reprimanded. But Cauchy, too, was told to play down the details in order to get students quickly onto the applications of mathematics. It was, after all, the *École Polytechnique*, a training ground for the later specialist engineering schools. But the lecture notes had to be distributed—it was a condition of lecturing that the courses be properly written up—and so his new approach to the calculus was there for all future instructors to measure themselves against. By and large, mathematicians rallied to the cause.

Prominent among these was the German mathematician Peter Lejeune Dirichlet, who studied for some time in Paris before returning to Germany, where he was to teach mathematics at Berlin and Göttingen. He did more than import French standards to his home country: he significantly improved the level of rigor that could be attained in mathematical arguments.[49] In particular he was the first to spell out what had to be done to turn Fourier's methods into rigorous mathematics, and to achieve partial, but only partial, success in that matter. He noted that he had had to impose restrictions on the kind of function that could be represented. These, he said, needed to be lifted, but "for this to be done with the all the clarity one could desire requires some details related to the fundamental principles of infinitesimal analysis."[50] He promised to supply them on another occasion, but he never did.

One of his major achievements is associated with what, in hindsight, can seem a triviality. This is the function $f(x)$ which takes the value c when x is rational, and the value $d \neq c$ when x is irrational. He introduced this function as an example of one which is not continuous on any interval, however small, and indeed does not have an integral, so any attempt to represent it by a Fourier series must fail utterly. His point was not that such functions are interesting in themselves, still less that they arise in

---

[49] In 1905, in an address at Göttingen to mark the centenary of his birth, Minkowski praised Dirichlet by saying that the modern era in the history of modern mathematics may be said to have started with the other Dirichlet principle: to solve problems with a minimum of blind calculation and a maximum of insightful thought. See Minkowski 1905, 149–163, quote on 163. This is insightful about Dirichlet, but it took a long time for the lesson to be drawn generally.

[50] Dirichlet 1829, 169.

any important way, but that they show that the condition that a function be integrable is not vacuous. At a time when all examples of functions were given in some analytic form or other, his example of a function must have seemed stark indeed.[51] It was the first indication that the concept of function could be much broader than the examples that normally sprang to the mathematician's mind, and shows that the "input-output" view of a function is much broader than the view that a function is essentially given by a formula. Such functions might be artificial, but they exist and stand as a challenge to the mathematician's ability.

If we put the pieces together, we see that by the 1830s the best mathematicians knew that there were arithmetically based definitions using limiting arguments for all the fundamental concepts of the calculus (continuity, differentiability, integrability, convergence). These definitions had been built up into a systematic theory capable of delivering precise proofs of many results hitherto taken for granted or only provided with what could now be seen to be mere sketches of an argument. Such mathematicians knew that a function need be neither continuous nor integrable, that while a continuous function is integrable there are elementary examples which show that an integrable function need not be continuous, that continuity of a function does not even guarantee certain results intuitively associated with continuity and certainly did not imply differentiability, and that a function need not agree with its Taylor series. The situation was that new methods had put the foundations of the calculus beyond criticism for the first time, and simultaneously the new concepts were pulling away from the intuitive understandings they had been formulated to defend. Moreover, as Dirichlet implicitly pointed out, one could not stand up as a professional mathematician and say one understood the calculus until the implications of the new ideas had been properly and fully worked through, and that was far from being the case. Just how far, even Dirichlet could not have guessed.

There were two arenas in which this tension would be played out. The first was research, especially into the field of the calculus of several variables, and for reasons to do with the technical nature of that material I defer discussion of it to the next section. The second arena was teaching, at least in those universities which prided themselves on the quality of their material, and after the 1850s that meant Berlin, and in particular the magisterial figure of Karl Weierstrass.

2.2.3 Weierstrass

The young Weierstrass had learned the theory of elliptic functions from Gudermann, and in so doing had come to appreciate the importance of uniform convergence.[52] He went on to solve one of the major outstanding problems in mathematics of the day, extending the theory of elliptic functions to a new theory of hyperelliptic functions. This made his name, and in October 1856 he became an associate professor at the

---

[51] Independently, Lobachevskii had had similar ideas.

[52] Elliptic functions may be thought of as the natural generalization of the trigonometric functions to functions of a complex variable. The hyperelliptic and Abelian functions Weierstrass went on to construct are still further generalizations. They all arise naturally when attempting to integrate a function defined via a function of two variables, just as the trigonometric functions arise when studying the circle with equation $x^2 + y^2 = 1$. The elliptic functions also had numerous uses in applied mathematics.

Industry Institute in Berlin, moving to the University of Berlin as a professor in 1864. Once established there, he settled into a two-year cycle of lecture courses, on the theory of analytic functions, elliptic functions and their applications, and the theory of Abelian functions (a complete generalization of hyperelliptic functions). He lectured at Berlin for thirty years, continually revising these lecture courses, which became his principle means of disseminating his ideas. This placed Berlin at the center of advanced mathematics, and audiences of two hundred were not uncommon—at least at the start of a course. Those who survived, if they took their higher degrees at Berlin, went on to become part of an extensive network of Berliners across Prussia and even beyond, and many more mathematicians, taking advantage of the German tradition of *Lern- und Lehrfreiheit* (freedom to learn and teach), studied in Berlin for a year or so. They might well receive lithographed copies of Weierstrass's courses, thus enabling them to move adroitly among the forest of formulas that form the bulk of his work. The best among them kept in touch with him by letter, hoping for the call back to Berlin as a professor (Fuchs and Schwarz were lucky; famously, Cantor was not).

Weierstrass has acquired the reputation of being the arch rigorist, the man who put the edifice in place, and upon whom great confidence could and did rightly rest. Through his example, inculcated into his students, the practice of rigorous analysis was apparently installed in mathematics. This idea, naturally popular among the Berliners and with historians since, needs qualifying in a number of respects. Weierstrass's interests were overwhelmingly in complex function theory. The large area of real function theory, the theory of Fourier series and integration, remained oddly underdeveloped at his hands. Weierstrass was part of a trend that ran strongly throughout the nineteenth century that regarded complex function theory as more important. In Berlin, he had the support of Kronecker, who dwelled on the interaction of elliptic function theory with number theory, a major branch of mathematics in German eyes. In Weierstrass's case, his emphasis on the complex side came with a curious set of preferences for some techniques over others.

Complex analysis for Weierstrass was a theory of power series within their circle of convergence. To put the same point another way, it was formal algebra and uniform convergence. As Weierstrass put it, in a much-quoted letter to Schwarz of October 3, 1875, emphasizing the importance of algebra:

> The more I meditate upon the principles of the theory of functions,—and I do this incessantly,—the firmer becomes my conviction that this theory must be built up on the foundation of algebraic truths, and therefore that it is not the right way to proceed conversely and make use of the *transcendental* (to express myself briefly) for the establishment of simple and fundamental algebraic theorems; however attractive may be, for example, the considerations by which Riemann discovered so many of the most important properties of algebraic functions, That to the discoverer, quâ discoverer, every route is permissible, is, of course, self-evident. I am only thinking of the systematic establishment of the theory.[53]

Differentiation was admitted, via the usual limiting arguments, and shown to apply term by term to a power series. But integration was not admitted, and he went to

---

[53] Weierstrass 1895, vol. 2, 235.

increasing lengths to exclude it from his theory. This forced him into awkward contortions, because in complex function theory the link between a function and its series expansion was already, and is still, provided most naturally by the Cauchy integral theorem. As the name implies, this theorem rests on a remarkable, one might say, characteristic, property of complex functions that many mathematicians were prepared to regard as both simple and profound. To eschew it was a bold move indeed.

Indeed, Weierstrass's analysis of singular points illustrates his conceptual strengths and weaknesses very well. Roughly speaking, the challenge was to make sense of functions like $f(z) = 1/z$. As z gets nearer and nearer to 0, the value of this function increases without limit, and it is natural to say that therefore the value of $1/z$ at $z = 0$ is infinite (written ∞). Similarly, as z gets larger and larger, the value of $1/z$ gets smaller and smaller, and so it is natural to say that the value of $1/z$ at $z = \infty$ is 0. On the other hand, this sort of approximating argument breaks down entirely for functions like $f(z) = e^z$ (the exponential function) even though it is natural to hope that it takes the value ∞ when z tends to ∞. Likewise, there is no value at zero for the function $f(z) = e^{1/z}$. Weierstrass was the first to clarify this, and make the important distinction between points z_0 where $\frac{1}{f(z_0)}$ can be defined to be zero, and points z_0 where $\frac{1}{f(z_0)}$ cannot be defined at all. The first kind of point is called a *finite pole* of the function, the second an *essential singular point*. This removed a fundamental muddle from the subject and enabled a lot of work to be done. In the case of a function with a finite pole, the key theorem is that such a function has a Laurent series,[54] and the only proof Weierstrass knew that the coefficients of the Laurent series are uniquely determined by the function involved the Cauchy integral theorem. He therefore gave no proof of this important result in his lecture courses, saying only that it was well known—the gap was only filled as late as 1884 by Mittag-Leffler and Scheefers. But when it came to functions of several complex variables, Weierstrass's analysis of singular points was simply wrong, and many of his pronouncements of any depth were offered without adequate proof. Ironically, it was to turn out that the next generation to study complex functions relied heavily on Poincaré's generalization of the Cauchy integral theorem to the functions of several variables.

Weierstrass also disagreed fundamentally with his immediate predecessor in German mathematics, Bernhard Riemann, over the nature of a complex function.[55] Riemann had advocated the principle that a complex function is a function of a complex variable that satisfies a certain pair of partial differential equations. Weierstrass found this too vague,[56] and defined a complex function by a power series convergent on some domain. These definitions turned out to coincide, but they reveal an important difference that will be discussed further below. This difference cropped up in other ways. Riemann's approach was heavily conceptual, and he expected the formulas to follow from his conceptual analysis. It was to turn out posthumously that he was not always correct, and one of those most active in puncturing this Riemannian bubble was Weierstrass. Riemann was often visionary, and his arguments sometimes fell short of proofs. Weierstrass, conscientiously lecturing to the

---

[54] A Laurent series is a power series involving some negative powers of the variable.

[55] Tappenden 2006 argues persuasively that this disagreement has significant implications for our understanding of Frege's work.

[56] Dugac 1973, 116.

next generation of mathematicians, would not and could not allow himself such laxity. That said, his conceptual framework was less satisfactory than Riemann's and often led him into error.

If Weierstrass was not some impossible paragon of rigor, he was nonetheless its most powerful advocate. He brought the ε, δ, N form of analysis into the definitions, and deployed them throughout, eliminating many of the obscurities that had attended Cauchy's usage. He made a clear distinction between pointwise and uniform continuity, and established pointwise continuity as the primitive concept. He showed precisely where uniform convergence was needed in the integration and differentiation of power series. What Cauchy had begun was brought to the level of a standard for the field, one that any competent mathematician could learn, and henceforth had to master. The diffusion of the new standards was inevitably slow. First the Berliners took it up, then Camille Jordan among the French, and Poincaré. Italian mathematicians, several of whom studied in Berlin, also appreciated it, as did the Americans who passed though Berlin and Göttingen. Textbooks in real analysis naturally remained in Cauchy's style for some time. Complex analysis was taught in Germany for many years in either the Riemannian or the Weierstrassian fashion, until there was a rapprochement toward the end of the century.

2.2.4 George Green and Potential Theory

The research arena came to be dominated by the topic of potential theory. This is a topic in which the intuitions of the physicists continually challenged the abilities of the mathematicians, and a gap opened up that acted as a goad for the production of ever more rigorous mathematics. Informally, a potential function is one whose partial derivatives give the components of a force in the coordinate directions. They had been used extensively by Laplace, because they simplified his analysis of problems having to do with gravitational attraction, and Laplace had shown that the potential function associated to Newtonian gravity satisfies the partial differential equation that today carries his name.

When it was shown that the electrostatic force likewise obeys an inverse square law, the techniques of potential theory were applied in this domain, too. But in this context physicists found it makes sense to imagine the charge lies on the surface of the body, and so the question of what the forces are inside the body arises. Laplace's equation does not apply, and the modified law is named for its discoverer, Poisson.

A crucial development came with the discovery that there was a rich theory of potential functions. This was the achievement of the English mathematician George Green, and later, independently, of Gauss. It came at a price: the interests and opinions of mathematicians and physicists were put permanently askew.

George Green was born on July 14, 1793, and was largely self-taught.[57] He read what he could of the relevant literature in has native town of Nottingham: Laplace's *Mécanique Celeste*, and some papers by Poisson, Biot, and Coulomb. In 1828 he wrote his first and single most important work, entitled "An Essay on the Application of Mathematical Analysis to the Theories of Electricity and Magnetism."[58] In the

---

[57] See Cannell 1993.
[58] Green 1828, reprinted in Green 1871.

"Essay," which ran to nearly eighty pages, Green had the happy idea of coining of the term "potential function." He then formulated what today is called Green's theorem. This relates an integral of one function taken over a volume to another taken over the surface enclosing that volume. A mathematician sees this theorem as generalizing the fundamental theorem of the calculus to several variables. A physicist sees Green's theorem as relating a flux across a surface to the quantity of material inside it.

Green's functions, the third and deepest insight in the "Essay," derived from Green's interest in physics, because it is only there that his insight makes intuitive sense. He introduced the functions to solve the differential equation satisfied by a potential function. His idea was to solve the equations in extreme cases, where the solution is easy to discover. In particular, Green considered the case where the potential was caused by a single charge at an isolated point. So the potential function satisfied Laplace's equation everywhere except at the point. Green assumed that the potential vanished on the boundary of the body, and that the potential increased like $1/r$ as one neared the point charge. He then showed how to solve the equation for such a curious function by an attractive use of the consequences of Green's theorem that he had developed in the earlier part of his "Essay." He was aware that his mathematics was not rigorous, but gave a plausible limiting argument to show how such potential functions can be defined. Such examples have tended to strike mathematicians as likely to make sense on physical grounds, and physicists (such as Maxwell) as likely to be amenable to rigorous mathematics.

Green died in 1841, and his "Essay" was not appreciated in his lifetime, although his later papers earned him a modest reputation. That the "Essay" became known at all was the work of William Thomson, later Lord Kelvin, who had picked up a stray reference to it and finally tracked it down the day before taking his degree at Cambridge and leaving for Paris in January 1845. Thomson was entranced. He took the "Essay" to Paris, and showed it to Joseph Liouville and his friend Charles Sturm. They too were excited by it, and so too was the German Leopold Crelle, who immediately accepted it for publication in his journal (it was published in three installments between 1850 and 1854).

The combination of ideas, their elegance, and their clear presentation even when rigor lay out of reach impressed mathematicians, and rapidly ensured that Green's name was henceforth securely attached to his discoveries, even when some had by then been discovered by others. From Crelle, news of the essay passed to Dirichlet, and from him to Riemann. Green's functions have played a prominent role in mathematical physics ever since. Likewise, among the physicists William Thomson remained a staunch advocate of them, as was Maxwell.

There are many other situations where a point source of influence is a natural object to study. To mathematicians among Green's eventual readers the example of a complex function was irresistible. Cauchy showed during the 1820s that much of their theory follows from the distribution of their poles, and for later writers, such as Riemann, Green's theorem was the natural way to formulate the Cauchy integral theorem. To a physicist a single force concentrated at a point might represent an impulse. A load on a beam might be concentrated at a point. These extreme situations have the advantage over more general ones of simplifying the attendant mathematics, which is why Green's functions have become such a powerful idea. More plausible situations, where the electricity is distributed through an entire body, for example, are solvable by a two-stage process. First, the problem is approximated as a problem with finitely many point charges and solved simply by adding the solutions. Second, the continuous distribution of charge is regarded as infinitely many point charges distributed in some way, and the use of addition is replaced with integration.

The nineteenth-century physicist had no difficulty in seeing that any distribution of charge on any surface would yield an unambiguous potential function. The relatively few mathematicians inclined to rigorous analysis could accept that this would be the case, but needed to give the result the cloak of a decent proof. That proved to be exceedingly difficult for good reasons. The precise point at which this became acute was identified by Gauss in his study of the Earth's magnetic field in 1838. Given a distribution of magnetism throughout the body of the Earth, Gauss looked for an equivalent, fictitious distribution on the surface of the Earth. He found that there always was one, and indeed that it was unique, but by an unsatisfactory argument. He deduced it from the claim that a certain integral, involving an unknown function, necessarily took a minimum value. The function that minimized this integral then defined the sought-for distribution.

The problem with the argument is that all that was known was that the integral was bounded below (meaning that all of its values are greater than or equal to a certain number). It does not follow that the integral takes a least value, any more than it would follow from the observation that the values of the function $f(x) = 1/x$ when x is positive are all positive and therefore bounded below by 0, that there is a value x_0 of x such that $f(x_0) = 1/x_0 = 0$. The high degree of plausibility of the result on physical grounds seems to have blinded mathematicians to the weakness of the argument, and among those who did not spot the error was not only Gauss, but also Dirichlet, in a set of lectures in 1856/57.[59] Dirichlet's presentation was influential not only because of his standing as a mathematician, but because Kirchhoff and Riemann were among his students. It is in recognition of this debt that Riemann, after reviewing the use of it by Gauss, Thomson, and Kirchhoff, called the principle that guarantees the existence and uniqueness of a function satisfying the Laplace equation and given boundary conditions after Dirichlet.[60] For Dirichlet had, as he put it, "informed me that he had been using the principle in his lectures since the beginning of the 1840s, if I'm not mistaken."

Riemann based his own treatment of the fundamental ideas of complex function theory on a variation of Dirichlet's approach, which he adapted to a more general situation. In that context he introduced the term "Dirichlet principle" to refer to the general principle that guaranteed the existence of potential functions under certain conditions. The problem he set out to solve (known as the Dirichlet problem) is this: find a function which

1. is defined on the whole of a disklike region,
2. takes prescribed values on the closed curve bounding the region, and
3. satisfies the Laplace equation in the disklike region. Such a function may be thought of as defining a surface in space that is bounded by a given curve in the space. An impression of such a surface is given in figure 2.5.

---

[59] Published as Dirichlet 1876.
[60] Riemann 1857, 97 in 1990, 129.

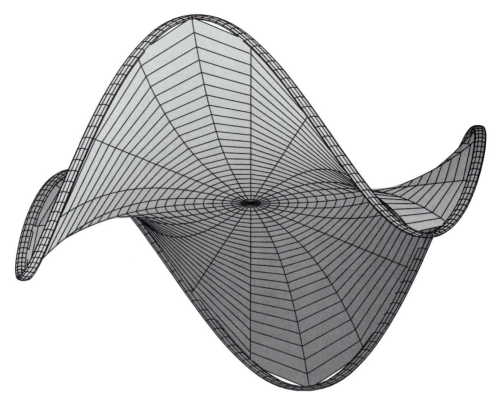

FIGURE 2.5. A surface of least energy spanned by a given curve.

Riemann's argument was that one may write down an integral that evaluates the energy of any function that satisfies (a) and (b). The terms in the integral are sums of squares, so they, and therefore also the value of the integral, are necessarily positive or zero. The value of the integral is therefore always greater than a certain minimum, and there is therefore (by the Dirichlet principle) a value of the integral that is its least value. He could show, by a much more elementary argument we need not discuss, that the corresponding function is unique. The details of his method now permitted him to write down the function he sought.

If that had been all Riemann did, his argument would have been open to the same objections as Gauss's, but he did not naively apply the Dirichlet principle. In his paper of 1857 Riemann explained that in his 1851 paper he had been seeking to generalize this principle to cases where prescribed discontinuities are allowed. He arranged that the functions he considered depended on a parameter. "The totality of these functions," he wrote in §16, "represents a connected domain closed in itself, in which each function can be transformed continuously into every other, and a function cannot approach indefinitely closely to one which is discontinuous along a curve without [the Dirichlet integral] becoming infinite." So the desired consequence was supposed to follow by analogy with the result that a continuous function on a closed bounded set is bounded and attains its bounds.

Riemann was clear that the Dirichlet principle needed a proof. The problem was that he offered a proof, and it shortly collapsed, refuted by his former student Prym. Prym's work was shortly to be superseded by Schwarz's much more extensive treatments, but Prym's example is interesting because it showed, as he said, that even when Dirichlet's problem can be solved, Riemann's approach to it via the Dirichlet principle might be in error.[61] Prym thereupon exhibited a function satisfying all of (a), (b), and (c) above but for which the Dirichlet integral was infinite. The reason, as his formulas make clear, is that Prym's function oscillates infinitely often near a specified point on the boundary of the disk. The conclusion was not only that the Dirichlet principle was inadequate, but that continuous functions have highly unexpected, counterintuitive properties.

In 1870, Weierstrass published a different example to show that an integral that depends on a function may be bounded below but never attain its minimum. His integral could be made to take arbitrarily small values, but never the value zero. He drew the conclusion that "Dirichlet's method of proof thus leads in this case to an obviously false result" (1870, 54). It is striking that the conclusion needed to be pointed out, but it seems that his was a decisive intervention—evidence, it would seem, that a mixture of physical intuition and mathematical naivety was capable of leading mathematicians astray.

2.3 Algebra

2.3.1 Algebraic Number Theory

Algebraic number theory was one of the prominent new disciplines of mathematics in the nineteenth century, chiefly, although not exclusively, in Germany, where the subject may be said to have been created by Gauss.[62] He was followed by a string of eminent German mathematicians: Dirichlet, Eisenstein, Kummer, Kronecker, Dedekind, Weber, and Hilbert. As a subject, algebraic number theory is the study of polynomial equations in several variables, and the existence or nonexistence of integer or rational solutions to them. Crucially for our purposes, the concept of "integer" was transformed by the perceived needs of research in this area and generated a profound ontological debate that ran throughout the century.

Fermat's last theorem is undoubtedly the most famous and for that reason the most important example of a problem in algebraic number theory, even though it survived the entire nineteenth century and was solved only recently.[63] Fermat had been investigating Pythagoras's theorem, which says that in a right-angled triangle with sides x, y, and z, $x^2 + y^2 = z^2$. He was able to show that the sides could not all simultaneously be integer squares, which rules out the existence of (non-zero) integer

---

[61] Prym 1871.

[62] More specifically, by his *Disquisitiones Arithmeticae* of 1801, recently given a full-length historical account that should set a new standard for accounts of the history of number theory: *The Shaping of Arithmetic* (Goldstein, Schappacher, and Schwermer 2007). I regret that I was not able to make proper use of this work in this book.

[63] See Wiles 1995 and Wiles and Taylor 1995.

solutions to $x^4 + y^4 = z^4$.[64] In a marginal note to his copy of Diophantus, he claimed that there were no integer solutions to any equation of the form $x^n + y^n = z^n$, for any integer n greater than 2. He gave no details of how he had achieved this, and no one since was able to do so until 1995; the claim became known as his last theorem.[65] More positively, toward the end of his life he claimed, in a letter to Huygens of 1659, that he could show that no cube was the sum of two cubes.[66]

At the start of the nineteenth century, the theorem had been established for $n = 3$ (by Euler), for $n = 5$ (by Dirichlet and Legendre), and for $n = 7$ (by Lamé). These proofs were increasingly difficult, and it became clear that a new approach was needed if the general problem was ever to be solved. On March 1, 1847, Lamé proposed to the Académie des Sciences in Paris to draw a general lesson from the cases of $n = 3$ and $n = 5$. These had been solved by introducing numbers of the form $a + b\sqrt{-3}$ and $a + b\sqrt{5}$, respectively, and treating them as if they were integers. Lamé's idea was that these new kinds of "integer" were still capable of being prime, so one can speak of their factors, and show that the prime factors of any such "integer" are unique. This is in general false, as this simple example of integers of the form $a + b\sqrt{-5}$ shows: $3.7 = (1 + 2\sqrt{-5})(1 - 2\sqrt{-5})$, but Lamé had a trick he hoped would save the whole approach.[67]

He claimed correctly that the known proofs for $n = 3$ and $n = 5$ worked by introducing the new "integers," which permit one to factorize the polynomial $x^n + y^n$, and to show, by considering their factors, that there could be no solutions of the equations with such "integers." It follows that there can be no integer solutions of the usual sort, either. He therefore proposed to tackle the general case by introducing factors of the form $x + yr$, where r is some complex number. As soon as he presented it to the Académie, Liouville pointed out that the theory of prime factorization for such integers was not established. Cauchy then stood up to indicate he thought the approach was promising, and indeed he then tried, unsuccessfully, to carry it out. Finally, however, on May 24, Liouville read to the Académie a letter from the German mathematician, Ernst Eduard Kummer, which showed that the approach was hopeless. When $n = 37$, the theory of unique factorization into primes is false.

Kummer had been led to his discovery by his interest in another problem of algebraic number theory: the search for a generalization of Gauss's theory of quadratic and biquadratic reciprocity to higher powers.[68] It was thus firmly anchored in that tradition, which was growing rapidly with the work of Dirichlet and Jacobi. It is this connection, which explains why his discovery and the new techniques he invented to deal with it became so central to the growth of the subject; Gauss in particular had not regarded Fermat's Last Theorem as an important problem. In their investigations, Gauss and Dirichlet had already been led to discuss new "integers" of the form $m + ni$, where m and n are ordinary integers and $i = \sqrt{-1}$ is the square root of minus one, and to extend the concept of prime to them. What Kummer

---

[64] Fermat to Huygens, Fermat 1891, 340–341.

[65] Fermat 1891, 291. The name is doubly wrong: Fermat never proved it, nor was it the last of his claims to be proved: one remains.

[66] Fermat to Huygens, in Fermat 1891, 340–341.

[67] The "integers" 3,7, $(1 + 2\sqrt{-5})$, and $(1 - 2\sqrt{-5})$ are all prime.

[68] The details need not detain us; see Edwards 1977, from which the above story about Lamé is taken.

did was to consider expressions (today called *cyclotomic integers*) of the form $a_0 + a_1r + a_2r^2 + \ldots + a_{n-1}r^{n-1}$, where r is a complex number such that $r^n = 1$, and the a_i are integers. This raises no ontological problems beyond those raised by complex numbers in general. For certain values of n, however, there are real problems in pursuing a theory of prime cyclotomic integers, precisely because, although they can be factorized completely, their factors are not unique.

It is here that the ontological debate started. Kummer's ingenious solution was to invent still further types of "integer," with which to factorize the factors so far obtained, and for which the new factors are prime and factorization into primes is unique. The interesting ontological problem that arises is the status of these new integers. One mathematician and historian of mathematics has commented lucidly, with the example of the number 47 in mind, that

> the test for divisibility by the hypothetical factor of 47, . . . , is perfectly meaningful even though there is no actual factor for which it tests. One can choose to regard it as a test for divisibility by an *ideal* prime factor of 47 and this, in a nutshell, is the idea of Kummer's theory. [In general] he found methods for testing for divisibility by prime factors of [a prime] p, tests which continued to be defined even when there was no actual factor for which they tested. He took these tests to be—by definition—tests for divisibility by "ideal prime factors" of p and built his theory of ideal complex numbers on the basis of these ideal prime factors.[69]

In Kummer's own words:

> Because ideal complex numbers, as factors of complex numbers, play the same role as actual factors, we will denote them from now on in the same way as these, by $f(\alpha)$, $\phi(\alpha)$, etc., in such a way that $f(a)$, for example, will be a complex number satisfying a certain determined number of characteristic conditions for ideal prime factors, except for [the condition of the] existence of the number $f(a)$.[70]

Kummer then proved that ideal prime factors, as he had defined them, had the required properties. For example, the ideal factors of a product are the products of the ideal factors of each term of the product, one number divides a second if the factors of the first divide the factors of the second, any cyclotomic integer has a finite number of ideal factors and these are essentially unique. Thus unique factorization was recovered on introducing the new ideal factors of a cyclotomic integer.

Other important problems in contemporary number theory also suggested to Kummer that his ideal numbers were important. The central section of Gauss's remarkable book, the *Disquisitiones Arithmeticae* of 1801, was devoted to the theory of quadratic forms. These are expressions of the form $ax^2 + 2bxy + cy^2$, where a, b, and c are integers. The principal question is: for given a, b, and c, what integers n can be written in the form $ax^2 + 2bxy + cy^2$? In other words, when are there integer solutions to the equation $ax^2 + 2bxy + cy^2 = n$? The study of this problem led Gauss, as it had led Lagrange and Legendre before him, to the question of simplifying quadratic forms. Gauss's most significant contribution to that question was to show how quadratic forms with the same value of $ac - b^2$ (called the discriminant) can be multiplied together to give a third quadratic form with the same discriminant. This

[69] Edwards 1977, 106, emphasis in original.
[70] Kummer 1851, §VI, quoted in Edwards 1980, 342.

is one of the earliest examples of objects that are not numbers being multiplied together, and later became of central importance in the growth of finite group theory.

This chapter in Gauss's work was found to be very hard, and in an attempt to simplify it, Kummer proposed to connect it to the study of complex integers of the form $x + y\sqrt{-D}$, where $D = ac - b^2$. He recognized that here again it would be necessary to bring in his ideal prime factors. However, he did not pursue that line of inquiry in detail; it was his successor Dedekind who did so.

Dedekind's intervention gave this branch of mathematics a truly modernist turn, so it is better to consider it later, when we can connect it, as he did, to his better-known reworking of the concept of a real number. Let us therefore conclude this section by noting the emergence of a new branch of mathematics that attracted almost all the leading German mathematicians of the day, and that stretched the familiar concept of integer and the multiplicative properties of integers (divisibility, prime factor) to breaking point.

2.4 Philosophy

2.4.1 Kant

Many nineteenth-century arguments about the nature of mathematics can be seen as responses to Kant's philosophy: extensions of it, reinterpretations of it, breaks with it, emancipation from it, rediscovery of it. Kant had argued, in the preface to the second edition of his *Critique of Pure Reason* (1787), that mathematics and physics were the two sciences in which reason yields theoretical knowledge. Accordingly, they have to determine their objects a priori, and by a priori knowledge Kant meant knowledge that was absolutely independent of all experience. A priori knowledge carries with it the implication that it cannot be otherwise, and it is therefore strictly universal. Kant further claimed that mathematics yields theoretical knowledge purely, whereas physics is at least partially concerned with sources of knowledge other than reason. But mathematics also differs from logic, in which reason needs to deal with itself alone. Mathematics required a step more momentous than the rounding of the Cape of Good Hope. Its true method was to bring out what was necessarily implied in the concepts that the mathematician had himself formed a priori, and had put into the figure in the construction by which he presented it to himself. So the mathematician forms concepts a priori, and these concepts and their implications are latent in the construction which presents the figure so that it can be reasoned about. This is in contrast to the mathematician inspecting what he discerned either in the figure or in the bare concept of it, and from this, as it were, reading off its properties. For, if the mathematician is to know anything with a priori certainty, he must not ascribe to the figure anything save what necessarily follows from what he has himself put into it in accordance with his concept.[71] The figure is a means to an end, and that end is a priori certainty.

Science followed, according to Kant, but only much later, when the twin approach of reason and experiment was discovered. But metaphysics lagged behind, and Kant

---

[71] Kant 1787, B, xii.

FIGURE 2.6. Immanuel Kant.

saw his mission to propel it finally on the path of science. Since each of the two previous breakthroughs had been the revolutionary achievement of one, or at most a few great thinkers, Kant was making no modest claim here. His idea—and he compared himself to Copernicus—was to make the objects of metaphysical knowledge conform to our intuition. Metaphysics reformulated in this way, said Kant, would necessarily be unable to make known the thing in itself, because that would go beyond the limits of experience, but within those limits it could acquire exhaustive knowledge of its entire field. Kant offered a compromise: all we can know is appearances, but about appearances we can know everything, including an understanding of how knowledge is possible at all.

Let us unpack this a little further. A particular challenge for Kant was to explain how increase in knowledge is possible. Logical reasoning is trustworthy, but if all completely reliable truths are analytic, or true in virtue of their form, as the term is sometimes glossed, then any deduction from a reliable truth merely brings out what was there all along, and cannot count as an addition to knowledge. Experience does increase knowledge, but it can seem to be a less than completely reliable source, tainted as it is with the possibility of error. Kant therefore looked for a way of describing certain types of knowledge as both reliable and cumulative. Absolutely reliable knowledge he called a priori. It is valid independently of experience, to which it is logically prior. A priori judgments, he said, are universally valid and necessary, and they are the only judgments of this kind. Analytic judgments are a priori. For example, a judgment such as "All bodies are extended" is a priori, because it could not be otherwise.[72] But it is analytic, because being extended is part of being a body. What was needed, he said, was a complete enumeration of all the types of a priori knowledge, together with their underlying principles.

For Kant, part of acquiring knowledge is having an intuition. Intuition is a technical term in Kant's epistemology. It means a direct intellectual apprehension of an object, and as such intuitions are distinct from concepts, which are general and deal with objects only indirectly. Intuitions belong to the perceiving mind and are akin to sensations; concepts belong to the thinking mind and facilitate understanding. Intuiting something is passive, understanding is active. In Kant's view, some concepts are, as he put it, grounded in intuition, and it is not difficult to see how they can apply to the intuitions that ground them. But there are concepts whose applicability to intuitions is obscure, and for Kant the means by which we count was one of these. His solution was to distinguish between two kinds of intuition, the empirical sort, and a pure kind derived from the structure of experience. The structure of experience is something the human mind supplies, which we nonetheless intuit directly. Experience of the world forces the intuition of this structure, and because the knowledge it brings is of the world, it is synthetic in nature. In fact, experience of the world forces us, said Kant, to resort to the concepts of time and space. Thus we come to the famous Kantian idea of the synthetic a priori. And, as paradigm examples of synthetic a priori truths, Kant offered those of mathematics. As he put it, "*All mathematical judgements, without exception, are synthetic.*"[73]

As already indicated, Kant believed that we express ourselves in judgments. A judgment, in subject-predicate form, might imply that the predicate is contained in the subject, or that it lies outside it and is merely somehow related to it. The former judgments are analytic, the latter synthetic. New knowledge must be drawn from synthetic judgments, but to have the requisite degree of certainty must be a priori.

Let us consider an example drawn from arithmetic. It is in the act of enumerating objects that we come to realize that $7 + 5 = 12$, because an act of (sensible) intuition is needed to realize the numbers 7, 5, and 12. This shows that arithmetic is synthetic. Moreover, it is only the objects capable of being the objects of a sensible intuition that can be counted. Quite how the structure of experience required for arithmetic is the pure intuition of time has baffled all of Kant's commentators, and may for the moment be sidestepped. But we should attempt a plausible summary of his views before picking up a consequence of them. Kant explained our ability to count and to reason with numbers by hypothesizing a structure to our ability to have sensible intuitions. Human experience is determined by this structure, without which counting would be impossible.

[72] Kant (1787—and passim—A7, B12) relied on the principle of contradiction here.
[73] Kant 1787, A10, B14; emphasis in original.

The structure does more than make counting possible. It makes it capable of being correct, and because all that is required is the apprehension or intuition of objects qua objects without regard to their nature, we can only count (correctly) in one way. This gives arithmetic its truth. Arithmetical truths for Kant were all of the form just illustrated $(7 + 5 = 12)$. Nothing but the appropriate intuitions were needed, he felt, to generate the proofs.[74] This was unlike the situation in geometry, where intuitions about space were often, if not always, required. Kant came to believe that these intuitions about space were encoded in Euclid's *Elements* in the form of theorems, which are general, rather than in the problems, which are existential in nature. The role of the latter is to make possible the constructions through which the mathematician typically proceeds. It is only constructions that are needed in arithmetic, in Kant's opinion, because they generate the numbers that intuition requires. It follows, on this account, that arithmetic has no axioms. All of which tends to make it seem impossible that there could be any "non-Euclidean arithmetic," a system of "numbers" sufficiently close to the familiar ones but somehow different.

One final Kantian position worth noting, because it was to be contested later, was his view that our knowledge is necessarily limited. In a charming image, Kant suggested that to imagine that knowledge can grow beyond its limits was a mistake, such as a dove might make if she imagined that air resistance was merely a hindrance and she might fly more easily through space, and such as Plato had made.[75] He insisted on this observation precisely because mathematics was such a shining example of how far we can progress in a priori knowledge; however, he pointed out, this was knowledge of objects that can be exhibited in intuition.

There was a sufficient degree of obscurity in Kant's theory of arithmetic for other theories to have some chance of success. When Kant argued that the role of intuition was still more palpable when large numbers were involved,[76] Frege, by contrast, was to object that very large numbers simply cannot be given in intuition. How indeed do we know that $123,456,789 + 987,654,321 = 1,111,111,110$? Surely not by counting with dots. If the answer is: "By appeal to certain rules," then, how do we know those rules are always true?

Nor are the truths of geometry analytic. That the straight line between two points is the shortest curve joining them is synthetic, said Kant, because it too requires an act of intuition. He did, however, concede that there are a few analytic truths at the foundation of geometry, such as "the whole is greater than the part" because they can be derived from the principle of contradiction.

Kant therefore set himself, among other tasks, that of demonstrating how pure mathematics is possible, and for this he unveiled his transcendental aesthetic. Intuition is what is happening when a mode of knowledge relates to objects. There must be an object. This object affects us through our sensibility, which is what yields intuitions. These intuitions are then thought about through the understanding, and from the understanding arise concepts. The effect of an object on the faculty of representation is sensation and the corresponding intuition is empirical. A representation

---

[74] It is an odd fact about Kant's views on mathematics that all his arithmetic examples are of this form, even though there are theorems in number theory. The examples are much less substantial than the geometrical results he refers to.

[75] Kant 1787, A5, B9.

[76] Kant 1787, B16.

that does not arise from a sensation is a pure representation, which give rise to pure intuitions. A pure form of intuition exists in the mind a priori as a mere form of sensibility. Finally, the transcendental aesthetic is the science of all principles of a priori sensibility.

Space, it emerges, is not an empirical concept, because for each of us our representation of space is presupposed in any account of what lies outside us. Space is a necessary a priori representation underlying all intuitions of what lies outside us. We cannot imagine the absence of space, only space empty of objects. It is a pure intuition because we can only imagine one space. To speak of diverse spaces is only to speak of diverse parts of space, and these parts are not the constituents of space, but are simply already in it. It follows that geometric propositions can only be derived from intuition (and not from concepts alone) and that the derivation is a priori.

The task of the transcendental aesthetic is to show how a concept from which knowledge flows is enabled to do so: to explain the concept is to explain how it is productive of knowledge. What then, asked Kant, is our representation of space that enables us to have geometric knowledge? He answered: it is the form of all appearances of our sense of what is outside us ("outer sense"). It exists in the mind a priori as a pure intuition, prior to all experience, and indeed must be in our mind for us to have a sensation of the outer world. Kant gave a similar account of time, except that time is an a priori condition of all appearance whatever, whereas space is limited to outer appearance: "Time and space are, therefore, two sources of knowledge from which bodies of a priori synthetic knowledge can be derived. (Pure mathematics is a brilliant example of such knowledge, especially as regards space and its relations.) Time and space, taken together, are the pure forms of all sensible intuition, and so are what make a priori synthetic knowledge possible."[77]

Knowledge therefore arises, according to Kant, from two fundamental sources: our capacity to receive representations, and our power of knowing an object through those representations by the spontaneous generation of concepts. This spontaneous power is called the understanding, and it is this faculty that enables us to think. "Thoughts without content are empty, intuitions without concepts are blind." and "The understanding can intuit nothing, the senses can think nothing."[78] Only through their union can knowledge arise, and that union is expressed in judgments.

By comparison, Kant's thoughts on logic were less original. He singled out a pure general logic having to do only with principles a priori, and nothing to do with content. There should, said Kant, even be a transcendental logic, which should determine the object, scope, and objective validity of the knowledge that belongs to pure understanding and concerns itself with the laws of understanding and of reason solely insofar as they relate a priori to objects. When different representations are put together in one act of knowledge, Kant said they had been synthesized, so he spoke of a pure synthesis if the original intuitions were pure, as for example those of space and time. Such synthesis is what gives rise to knowledge. From his analysis of the logical functions used in all possible judgments, Kant derived the twelve possible pure concepts of the understanding he called, following Aristotle, the categories. Those most obviously involved in mathematics are those of Quantity: Unity, Plurality, and Totality.

[77] Kant 1787, A39, B56.
[78] Kant 1787, A51, B75.

From this dense, and not always convincing, body of ideas people took parts to agree with, parts to disagree with. It is not, indeed, immediately clear what it would mean for someone to proclaim themselves a Kantian, or to call for a return to Kant, nor is it clear what it would be for someone to be called a Kantian. As we shall see, even the key terms of the Kantian lexicon were to be taken up with different meanings.

Perhaps the most central of Kant's claims was that convincing metaphysics can be had, but only at a price: that of accepting there are limitations on human knowledge. All things are forever unknowable in themselves; we can only deal with appearances. As for the nature of that knowledge, its elusive central feature is that some of it is synthetic a priori. We process our representations of appearances through our built-in and directly intuited structuring of experience. It is because we can have (indeed, must have) these pure intuitions that we can have synthetic a priori knowledge. The other candidate for impeccable knowledge, analytic truths, does not allow for the growth of knowledge. It is, as it were, fenced in while the realm of the synthetic a priori is limited only by our experience. For our purposes, it is interesting that almost the paradigm example of synthetic a priori knowledge was mathematics, cast in the categories of Quantity. In view of the very high status of such knowledge, the examples of it that Kant supplied acquire a high status; it would be difficult for them to be impugned without the whole of Kant's theory seeming to come into question. Kant simply took it for granted that space is correctly described by Euclidean geometry. He might have noticed that his friend and correspondent Johann Lambert was less confident, but he did not, and later generations aware of the plausibility of non-Euclidean geometry were to agonize over what else was mistaken in Kant's scheme of things.

2.4.2 Two Post-Kantians: Herbart and Fries

Once the nineteenth-century was underway, Kant's ideas ceased to be the mainstream, as fashions for romanticism and Hegelianism swept German philosophical circles. The thread that matters here runs fitfully through Jena, where Fries taught at one time or another, and where in due course Frege was to live and work, and through Königsberg and Göttingen, where Herbart lived and worked.

2.4.2.1 HERBART

> Mathematics is the ruling science of our time; its acquisitions grow daily, though noiselessly. He who does not befriend it, will have it his enemy in the future.
>
> —Herbart 1877

Herbart first studied philosophy under Fichte at Jena in 1794, before going to work as a tutor at Interlaken in Switzerland from 1797 to 1800. There he met the educational reformer Johann Heinrich Pestalozzi, who was then in his early fifties and just enjoying his first success as an educational reformer. The Helvetic Republic in Switzerland, which had been imposed by the French in the aftermath of the French Revolution, had invited Pestalozzi to reorganize higher education, but instead he had

FIGURE 2.7. Johann Friedrich Herbart.

thrown himself into caring for those orphaned by the war, apparently to considerable effect. In 1800 he established a new school at Burgdorf, and in 1801 he published the book that set out the main principles of his educational philosophy. These include that instruction should start with the familiar, and proceed to what is new in step with the gradual unfolding of the child's development.

Herbart published the first of his educational books in 1802 with the title *Pestalozzis Idee eines A B C der Anschauung* (Pestalozzi's Idea of an ABC of Intuition) and followed it with a second in 1806, which also echoed Pestalozzi in many ways. He returned to Göttingen in 1802 and was appointed an extraordinary professor there in 1805. In 1808 he became Kant's successor as professor at Königsberg, dividing his time between philosophy and pedagogy until 1833, when he returned as professor of philosophy to Göttingen, where he remained until his death in 1841. While at Königsberg he wrote his major work, the two-volume *Psychologie als Wissenschaft neu gegrundet auf Erfahrung, Metaphysik, und Mathematik* (1824–25) (or) *Psychology as Science Newly Founded on Experience, Metaphysics, and Mathematics)*.

Almost every word in the title could detain us profitably. That the apex of the work is a theory of psychology; that experience and metaphysics are placed on a par; that mathematics is so important; all these aspects are significant in assessing Herbart's response to Kant and his influence on his successors. As the title makes clear, he saw philosophy as fundamental to psychology, not the other way round. Herbart was critical of attempts to interpret Kant in psychological terms. Herbart's psychological theory was marked by a distrust of introspection, and a corresponding desire to supplement it with something in order to make an intelligible whole. The missing part was supplied by his own transcendental shift that led ultimately to his metaphysical theory of the ego and the soul. This structure in place he could then turn back to psychology, and explain, to his satisfaction, how thoughts and feelings were ultimately reducible to basic presentations (*Vorstellungen*, as he called them).

In Herbart's theory, these presentations led a dynamic life that generated the activity of the mind; nothing was innate, everything was learned from experience. The details of this dynamical system were to be understood through mathematics, and Herbart's *Psychologie* is full of mathematics, although we may wonder how productively. In Part I of the book there is an account of the statics of mental processes, modeled on contemporary applied mathematics, followed by an even more mathematical account of the mechanics of mental processes. In the statics, various mental elements or qualities hold each other in check. In the mechanics, stronger ones drive processes according to differential equations that determine the final states, or at least the next states, of the mind. Momentary sensations are captured by the statics, the workings of memory and perception by the mechanics. Generally, sensations weaken with time, but they may be reinforced by new stimuli. In Part II all this is put to work to describe how, on this theory, we see a line, parallel lines, and intersecting lines, and then surfaces. This is far from being a passive act, as sensations are compared so that the visual space can be constructed. Nor is it a trivial consequence of the mathematics that has already been introduced, because the mind constructs points and a special process of interpolation that Herbart explicitly compared with the idea of continuity in mathematics is introduced to explain the appearance of the continuous nature of space.[79]

He regarded every sensation as operating over time and being formed of a sequence of momentary stimuli. Left to itself, a sensation will inevitably diminish as it jostles with others already at work in the mind until we are unaware of it, but it can be repeated and combine with others. Residues of these stimuli will be retained in the mind, and if the first member of a sequence is later recalled, it tends to summon up the others with it, in a way that depends on the original intensity of the stimulus. Thus, unconscious memory is involved in perception. The perception of space, Herbart believed, derives from sequences that can be read in either direction. This eliminates the sense of direction (or time) from them and presents the sequence as coexistent or atemporal. We acquire such experiences through our senses in many different ways: by our eyes, our fingers, our ears. These are woven together into the presentation of two-dimensional space; depth, Herbart believed, was an inferred quality.[80] It is different with our perception of time, because time series are inevitably to be read in only one way, but series of presentations of both kinds share the feature that they are infinitely divisible. We can imagine residua in our minds that are closer together than any residua actually present.

[79] For a recent account of Herbart's arguments, see Boudewijnse, Murray, and Bandomir, 1999, and for its reception see idem, 2000. For a discussion of how Herbart's ideas relate to those of Kant, and then to Riemann, see Banks 2005.

[80] See Lenoir 2006 for a more detailed account.

When we focus our attention on a moving object and see it against varying backgrounds, these spatial series of backgrounds first come associated with the object. But as we track other objects against other backgrounds that generate yet more spatial series, the mind recognizes similarities in the spatial series of backgrounds, and so forms an initially crude psychological concept of a spatial configuration and so of space itself. In the same way, other types of presentations gradually cohere, and Herbart went on to give an account of how the residua of presentations in our mind become arranged in our mind. This grouping together he called *apperception*.[81] It is a process of assimilation that generates concepts, and these concepts can then be treated by the mind independently of the contexts in which they were originally laid down.[82] Thus logic operates on concepts—there is no suggestion that the mind is fundamentally psychological—and we can form judgments. In humans, but not in animals, a further process operates: language puts judgments into words and so gives them a lasting form, and indeed language suggests the possibility of making new judgments.

Herbart disagreed with Kant in a number of ways. He was concerned that the historical change in concepts that had occurred was a strong argument against the Kantian categories. He distrusted the transcendental faculties, which he thought smuggled psychology into philosophy. He saw no reason to claim that knowledge of appearances fell short of knowledge of the thing in itself. Most awkwardly for those who wanted to adopt his ideas as part of a reform of education, he elaborated a philosophy of the Real, which he saw as made up of monadic-like essences akin to Leibniz's but differing from them in being causally interconnected.[83] Herbartian monads are not, however, spatially or temporally interconnected. The Real consists of discrete points, the mind generates the concept of a continuum because it can postulate continuous relations between its residua. In making the space of the monads intelligible, the mind necessarily imported the concept of continuous variation. It is this intelligible space that serves as the source of our intuitions of space with its familiar geometric properties. But Herbart's space is not an a priori form of intuition; it is a derived and constructed one. Indeed, Herbart's hostility to all forms of a priorism was to be an important influence on Riemann, as we shall see.

2.4.2.2 FRIES

The other philosopher in the Kantian tradition who gave a central place to mathematics was Jakob Fries. This was partly as a result of his religious upbringing among the Herrnhuter,[84] who regarded mathematics as a safe topic, unlike philosophy. He

---

[81] The term derives from Descartes, *Les Passions de l'ame*, and came to mean a state of focused awareness. Leibniz regarded it, none too helpfully, as the conscious reflection on the inner state of the monad in *Principes de la nature*. Kant divided the concept into empirical and transcendental apperception, which makes it possible for all perceptions to have a meaning. Herbart regarded apperception as endowing us with the power of reflection, and is fundamental to his pedagogical theory.

[82] Apperception works differently in different people, so that one may see or hear what another fails to appreciate.

[83] It was, of course, possible for enthusiasts to drop or skip the philosophy. For a late example, see Stout 1930.

[84] See Gregory 2006.

admired the clarity of mathematical reasoning, but nonetheless found his way to Kant's *Prolegomena*, and by the time he left the religious group to go to Leipzig he was a confirmed Kantian, but with his characteristic turn to psychology (his term, which does not agree with the modern sense). He now regarded a critique of reason as a general theory of psychology, or philosophical anthropology, as he would later call it. He moved to Jena, where he read and admired Newton's *Principia*, and in 1805 he became a professor at Heidelberg before returning to Jena in 1816. Politics intervened, and for some years he was forbidden to teach philosophy.[85] Fries turned instead to mathematics and physics, publishing his *Mathematische Naturphilosophie* (1822).

Fries's writings on the philosophy of science carried a degree of credibility because of his extensive work on the theory of probability, physiological optics, and experimental physics. Gauss, for example, defended his own interest in Fries's book against a student by remarking, "Young man, if after three years of intense study you have progressed to where you understand this book, you can leave the university with the conviction that you have made use of your time better by far than the majority of your fellow students."[86] Fries's philosophy of science argued for the possibility of an a priori natural science and developed a philosophy of applied mathematics. He defended Kant's synthetic a priori by making a distinction, not always understood, between the language of a theory and the meta-language used to describe it. He was also successful for a time in acquiring followers, among them Apelt and Schlömilch, and although interest then lapsed for a time until, as we shall see, Leonard Nelson then led an active neo-Friesian school at Göttingen after 1900 with the support of David Hilbert.

Fries recognized that the synthetic a priori could not be defended on grounds that themselves were synthetic a priori, but if they were only defended on a posteriori empirical grounds, such an argument could, at best, only give empirical certainty, which is necessarily inadequate to the task. He argued, however, that accounts and explanations of metaphysics were, so to speak, couched in a meta-language. It was therefore appropriate for them to be empirical, even though it was the synthetic a priori nature of metaphysics that the arguments were establishing. He spoke systematically of transcendental deductions "exhibiting" but not proving philosophical principles. He sought to develop a theory of reason (which would deal with synthetic a priori knowledge) and to exhibit the universal validity of its metaphysical principles.

It is for this reason that Fries called his own theory psychological and anthropological. Just as the scientist, he argued, generalizes inductively from well-attested evidence to laws, the philosopher, said Fries, may allude to laws of thought that are certain causal processes for the generation of particular thoughts. These laws, he claimed, have the same universal status as any scientific law. Like Kant, he settled for restrictions on the kind of knowledge that could result: in his case, a subjective theory of truth according to which something is true if it is consistent with the rest of one's

---

[85] In 1819 one of Fries's students, Karl Sand, assassinated August Friedrich Ferdinand von Kotzebue, who was not only an antiliberal dramatist, historian, and journalist, but a Russian agent. Metternich used the assassination and the student unrest of the time, with which Fries was associated, to block reform in Germany. We should also remember that some of Fries's views were violently anti-Semitic.

[86] Quoted in Gregory 1983, 188.

consciousness. That said, the truths of which one is aware were broadly familiar, in particular and without question, they were those of Euclidean geometry. That did not mean, however, that the critical philosopher had nothing to say about Euclidean geometry. As Gregory has shown, Fries did believe that attempts on the parallel postulate had all proved fruitless, and had ended at best in the production of a postulate equivalent to the parallel postulate itself. Here he relied on the discussion of A. G. Kaestner, a Göttingen professor whose student Klügel had written a thorough account of twenty-eight attempts. Other of Fries's comments on specific axioms of Euclid's seem at best misguided.[87]

Fries's critique of Euclid's *Elements* was not that of a mathematician, however, but that of a philosopher. He advocated new axioms for the *Elements*, couched in what he termed discursive concepts. They constitute, he said, our immediate knowledge. In geometry, the discursive concepts are figure, position, direction, and construction. "Figure" includes such intuitive concepts as body, surface, line, and point. "Position" and "direction" between them include motion, rotation, straight, angle, curve, and parallel. Motion is a change of position, rotation a change of direction. A straight line has the property that all its parts lie in the same direction. Organizing the elementary geometrical ideas in this fashion put Fries in a position to say some positive as well as some inadequate remarks. His axioms for position and direction begin well:

Through two points one and only one line is possible;
Through a straight line and a point outside it one and only one plane is
   possible; and, for example,
Two distinct straight lines through a point always make an angle.

Gauss, intermittently throughout this period, was asking similar questions with a similar disquiet about Euclid's *Elements*.[88] He felt that some definitions presumed too much and were really to be seen as theorems, and he wrote in his private notebooks about topics such as how to define a plane. If it is defined as all the lines perpendicular to a given line at a given point, for example, does it follow that the line joining any two points of this plane lies entirely in the plane? In this and some other cases, Gauss's writings of this kind are more plausibly seen as critiques and reworkings of Euclid's *Elements*, and not as part of a systematic attempt on the parallel postulate.

But, as critical readers may have begun to suspect, Fries's discursive concepts were likely to lead him into error. He claimed that parallel lines (those having the same direction at different points) cannot meet because if they did they would form an angle, which would entail that they had different directions at that point. Similar loose talk about what it is to have the same direction at different points led him to believe that he could show that the angle sum of a triangle was two right angles without the use of the parallel postulate, which was therefore established as a theorem. Such arguments are nothing more than mathematical mistakes, which others were to make after him, concealed in philosophical language.

---

[87] See Gregory 1983, 193.
[88] See the numerous items in Gauss 1900. The ones discussed in this paragraph can be found on 193–195.

2.4.3 Mathematicians and Scientists as Philosophers of Mathematics

It is characteristic of the period that a number of mathematicians got drawn much more deeply into philosophical issues than is usually the case, most notably Hermann Grassmann and Bernhard Riemann. They form a contrasting pair, for Grassmann alienated his audience and was little read in his lifetime, whereas Riemann emerged as the most important, if obscure mathematician of his day. That said, his explicit philosophizing was left unpublished at his death, leaving unclear for many years just how profoundly it had affected his mathematics.

2.4.3.1 GRASSMANN

Grassmann's reputation today is as a mathematician, but in his time he was probably better known and was certainly more successful as a linguist and in particular as an expert in Sanskrit. He wrote two major works that concern us here, his *Die Lineale Ausdehnungslehre* of 1844, and *Die Ausdehnungslehre* of 1862.[89] The latter was both an extension of the former and a reworking of it in more mathematical, less philosophical dress, because, as Grassmann put it, of "the difficulty . . . the study of that work caused the reader on account of what they believed to be its more philosophical than mathematical form."[90] He regarded the first publication as "an utter failure" (1995, 19), and noted in the foreword to the second edition of his *Linear Extension Theory* in 1878 that its reception only began in 1867 with a favorable notice by Hankel and, more importantly, by Clebsch.

The philosophy that mathematicians apparently found so impenetrable pervades the first work, which is always seeking to justify the novel mathematics it describes by appeals to first principles. In the reworked version of 1862, Grassmann more often simply proposed rules for the new mathematical operations he wished to study, and then got on with it. Where formerly he had attempted to establish his new branch of mathematics independently from the ground up, he now presumed at least the elementary branches of mathematics. What had often been presented discursively was now presented in a theorem-proof style deliberately modelled on Euclid's *Elements*. If we pass over the discussion of the calculus with which he ended the second book, the subject matter in each case is abstract quantities, called *extensive magnitudes,* which may be added (like vectors) and also multiplied together in various ways. Most of his readers probably took the vectorial interpretation, because vectors were coming in to mathematics at that time (Hamilton had coined the word in 1843), although it is not forced. On this reading, a mathematician would find the addition elementary, but the various multiplications novel. It should be noted, however, that when Grassmann published his book in 1844 all work on the "geometric addition of displacements" was unknown to him.

The difference in the reception of the two works is interesting in itself. It suggests that Grassmann's foundational ideas were thought to be irrelevant, but it may also be that they were inimical. He had attempted to define mathematics not as a science of

---

[89] For Kannenberg's English translations, see the bibliography.
[90] See Grassmann 1862, in Grassmann 2000, xiii.

magnitude (because combinatorial mathematics and even arithmetic did not fit that bill) but as a science of "the particular existent that has come to be by thought."[91] This was in contrast with logic that dealt with the general, rather than the particular. He then asserted that geometry did not belong to mathematics, so defined, because the concept of space is given, and not produced by thought. He challenged anyone who believed otherwise to prove from the laws of pure thought alone that space must have three dimensions, which he said could evidently not be done. Indeed, most of his book concerned an arbitrary finite number of basic quantities and hence dimensions. Therefore the relation of geometry to mathematics depended on the relation of space perception to pure thought. However, he said, space perception does not emerge only from the consideration of solid objects; it is a fundamental perception of the sensible world.

Grassmann therefore attempted to separate out what he saw as two activities: the study of particulars generated by thought from the study of space. This was a radically new view of geometry, which mathematicians saw as a paradigm entrance point for the study of mathematics. Geometry might have its own problems, even foundational ones, but it had not been suggested that one could approach these by so markedly abandoning the study of space. He made matters worse by a still more obscure pair of distinctions. One lay between the equal and the different. By means of what is different, one existent can be compared with another; by the idea of equality, one may collect various existents under the heading of the same universal. Apparently the algebraic existent arises from the equal, the combinatorial form from the different. It is no better with the other distinction, between intensive and extensive magnitudes. Somehow, intensive magnitudes are algebraic, while extensive magnitudes, which involve the concept of separation, are combinatorial. It is unfortunate to say the least that a book on the topic of extensive magnitudes, which in its abstract form Grassmann took himself to be introducing, the basic idea was so obscurely defined.

After developing the theory of pure mathematics for almost forty pages, Grassmann broached the topic of geometry, which, he claimed, lacked a scientific starting point and which, as a theory, needed complete restructuring. As he observed: "I cannot make such a claim, which threatens a structure sanctified for a millennium, without illustrating it on the most decisive grounds."[92] He then gave the example of problems with the definition of a plane. All accounts known to him, he said, assumed tacitly that a line having two points in common with a plane lies completely in the plane. It was obviously indefensible to make such an assumption tacitly, but to prove it from any definition of a plane required a lot of work, thus showing how much has been smuggled into geometry inside the definition, which was also an unacceptable procedure. (It is interesting to see how closely Grassmann here agreed with Gauss, whose unpublished remarks, later reprinted in Gauss's *Werke*, vol. 8, he cannot possibly have known.)

To put matters right, Grassmann first separated out purely logical items, such as those to do with equality, from postulates drawn from the perception of space. The most important postulate of the latter kind he regarded as the assertion that space is homogeneous and isotropic. To this he added that space has three dimensions. Then, by a blunt assertion about constructions involving positions and directions that recalled Fries's views, he deduced that one may construct parallelograms and speedily

[91] Grassmann 1844, Grassmann 1995, 24.
[92] Grassmann 1844, 1995, §21.

deduced results equivalent to the parallel postulate. Or, perhaps one might say, he showed that a three-dimensional vector space may be regarded as a space in which Euclidean geometry applies, and if the perception of space permits one to describe it by his theory of extensive magnitudes then space is indeed Euclidean. Either way, Grassmann was unaware of the complexity of dealing with the concept of "having the same direction" and slid without realizing it into too glib a theory of geometry.

On the other hand, Grassmann's work contained some distinctly modern features that undoubtedly contributed to its poor reception. In 1844 M. W. Drobisch, a friend and former student of Herbart's and by then a professor in Leipzig, persuaded the Fürstlich Jablonowskischen Gesellschaft der Wissenschaften to set one of its regular prize competitions on Leibniz's idea of a logical characteristic.[93] Möbius, who was also at Leipzig University, had recently received a personal copy of Grassmann's *Ausdehnungslehre* and he persuaded Grassmann to enter the competition. This Grassmann did, with an essay entitled *Geometrische Analyse . . .* , and he won. The prize was presented to him by Drobisch at the opening ceremony of the new, or more precisely, refounded Königliche Gesellschaft der Wissenschaften zu Leipzig on 1 July 1846. But in correspondence, even Drobisch and Möbius admitted to finding Grassmann's work difficult. The stumbling block was that it did not ground geometry in intuition, but on the concepts of magnitude, and, worse, on objects which Grassmann treated as magnitudes but which, in Möbius's opinion, cannot be so regarded.[94] Grassmann's emphasis on magnitudes as fundamental concepts, and his Leibnizian, tacitly anti-Kantian position, mark him out as a modernist "avant la lettre"; small wonder that he was little read.

2.4.3.2 RIEMANN

> As has been said of the poet Coleridge, so it could be said of Riemann, he wrote little, but that little should be bound in gold.
>
> —Van Vleck 1914, 116

Although it is unlikely that he ever read Grassmann's *Extension Theory*, one who would have appreciated its attempt to build up the right kind of mathematics before applying it to the study of space would have surely been Bernhard Riemann. Arguably, Riemann would have found Grassmann's creation unduly impoverished for the task, but the mere distinction would have struck a chord. Riemann's philosophy of mathematics is in many ways enigmatic, not least because of its delicate relation to his profound mathematics. There is a strong connection to Herbart, although another name that appears in Riemann's surviving philosophical notes is that of Fechner. Not the sober-sided Fechner inclined to materialism, but the mystical side, the Fechner who sought to show that plants have souls and who wrote the *Zend-Avesta* (1851), subtitled *On the Things of Heaven and the Beyond*. "*Zend-Avesta*, a truly life-giving word," wrote Riemann, and this pantheistic piety is just the sort of thing to appeal to a shy young man from a poor church background.[95]

---

[93] On prize competitions in mathematics in the nineteenth century, see Gray 2006a.
[94] See Lenoir 2006.
[95] Riemann 1990, 546; see the opening pages of Laugwitz 1999.

The emphasis on the soul in Riemann's jottings on psychology and metaphysics is a commonplace of the time with Herbartian roots. "With every simple act of thinking," Riemann wrote, "something permanent, substantial, enters the soul," apparently as a unit but actually containing an inner manifoldness, which Riemann termed a "mind-mass." "All thinking is accordingly the formation of new mind-masses," he went on, adding that this takes place in the cerebrospinal system in consequence of physical, chemico-electric processes.[96]

Riemann himself set a high, and precise, value on Herbart's work. He wrote (1990, 539) that he could agree with almost all of Herbart's earliest research, but could not agree with his later speculations at certain essential points to do with his *Naturphilosophie* and psychology. Elsewhere in the Nachlass he wrote that "the author is a Herbartian in psychology and epistemology (Methodology and Eidolology) but with most of Herbart's *Naturphilosophie* and the related metaphysical disciplines (ontology and synechology) he cannot agree."[97] Synechology is a general theory of continuous serial forms, and covers space, time, and motion, in particular intelligible space, the mental construct that makes the explanation of matter possible. That there was nonetheless a deeper influence is something that will become apparent shortly.

Herbart's name appears in the notes for the first time during a long passage on the theory of knowledge. Here Riemann characterized natural science as the attempt to comprehend nature by precise concepts. Such concepts determine what is probable, and if predictions based on these concepts fail, the concepts must be modified. In this way our comprehension of nature becomes ever more complete, even as it recedes ever farther behind the surface of phenomena. Herbart, said Riemann, had shown that all concepts serving to comprehend the world, even those whose origins are lost, arose by the transformation of earlier concepts, and so need not be derived a priori, as with the Kantian categories. It is only because concepts originate in comprehending what is given in sense-perception that, said Riemann, *their significance can be established in a manner adequate for natural science*" (emphasis in original).[98]

Riemann agreed entirely with Herbart that the Kantian categories presumed too much. That space was an empty vessel into which the senses ought to pour their perceptions was an idea Herbart called "completely shallow, meaningless and inappropriate."[99] Nonetheless, space for Herbart remained intrinsically three dimensional; we were not, it seemed, capable of forming a higher-dimensional idea. Herbart saw spectra everywhere: qualities came in bundles, they could vary, as for example the sensation of color or sound. Riemann, Scholz has argued, saw spectra only seldom in the day-to-day life of the mind, but very often in mathematics and physics, and then by no means always restricted to three dimensions.[100]

Riemann came to make a crucial distinction between the space concept and a particular intelligible space that is used to describe real events. A conception of the world is correct, he wrote, "when the coherence of our ideas corresponds to the

---

[96] See Riemann 1990, 3rd ed., 541.
[97] Quoted in Scholz 1982, 414.
[98] Riemann 1990, 544.
[99] Quoted in Scholz 1982, 422.
[100] Scholz 1982, 423.

coherence of things," and this coherence of things will be obtained "from the coherence of phenomena."[101] Put another way, there must be a coherent set of ideas, an
internally self-consistent set, to be matched somehow to the coherent phenomena.
Riemann did not show much interest in Herbart's description of how the coherent
system of ideas about space is generated from experience. He went straight to the
generation of geometric concepts in mathematics. He allowed that the corresponding
phenomena need not be continuously variable (he even allowed that space could be
discrete, perhaps mindful of Herbart's opinion that it was only intelligible space that
was continuous, perhaps more simply thinking of the familiar dichotomy between
numbers and magnitudes). Aware of complex numbers as a species of quantity, and
of the possibility of thinking geometrically about any number of variables, he dispensed with the dimensional restriction.

Herbart had laid great store by establishing philosophy as a science. It would exist,
ideally, in a to-and-fro relation with the sciences in which speculation about concepts
led to an ever-deepening process of education. If there was a distinction, it was that
philosophy dealt with concepts while the sciences dealt with what is given. Mathematics, in Herbart's view, was closer to philosophy than to science. As we have seen,
this is very close indeed to Riemann's view of science, which was heavily conceptual.
As Scholz put it, recalling Riemann's interest in Herbart's 1807 article:

> The idea of clarifying conceptual structures stood at the centre of Riemann's investi-
> gations throughout his mathematical work, whether it was complex function theory,
> geometry or integration. This was true to such an extent that one might be tempted to
> read Herbart's note as providing a characterisation of mathematics as Riemann himself
> would have given: A science dealing with concepts generated to solve problems arising in
> the attempt to gain knowledge and to clarify connection between already established
> knowledge.
>
> [. . . Riemann's interest] seems to have been the result of his desire to *clarify his own
> perception of mathematics in the mirror of philosophy.*
>
> This is all the more likely since a very close relation between mathematics and
> philosophy was suggested by Herbart in the very article summarised by Riemann.[102]

Scholz concludes that Riemann's views on mathematics "seem to have been
deepened and clarified by his extensive studies of Herbart's philosophy. Moreover,
without this orientation, Riemann might never have formulated his profound and
innovative concept of a manifold."[103] To which one can add, as Scholz has pointed
out on various occasions, that Gauss's deepening commitment to the geometry of n
dimensions was surely another influence of comparable importance.

2.4.4 Kronecker's Foundations for Arithmetic

In 1887 the friends and colleagues of the influential theologian, neo-Kantian
philosopher, and historian of philosophy Eduard Zeller celebrated the fiftieth

---

[101] Riemann 1990, 555.
[102] Scholz 1982, 426, emphasis in original.
[103] Scholz 1982, 426.

anniversary of the award of his doctorate. Zeller, the originator of the term *Erkenntnistheorie* (the theory of knowledge, or epistemology), was by then professor of philosophy at the University of Berlin, where he had gone as the successor of Trendelenburg. Among those asked to contribute were the mathematician Leopold Kronecker, and Zeller's friend Hermann von Helmholtz, the doyen of German scientists. Interestingly, both spoke about the nature of number, and both from a point of view that was already showing its age and its weaknesses. Kronecker's paper (1887), "Ueber den Zahlbegriff" (On the Idea of Number), revised on its second publication in his *Mathematische Werke*, has been much quoted by historians of mathematics, because it is one of the few places where he was explicit about his philosophy of mathematics, which is often taken to be of an antimodern kind. It opens with Jacobi's humorous parody of Schiller's "Archimedes und die Schüler," which Kronecker took from a letter of Jacobi's to Alexander von Humboldt in 1846, shortly after the discovery of Neptune, and I venture to put into English as follows.

> An eager young man came to Archimedes,
> Ordain me, he said, in the holy art
> that served astronomy so well and
> found a planet beyond Uranus.
> Holy you call it, said the wise man, and so it is,
> but so it was before the Cosmos was explored
> before it served astronomy so well and
> found a planet beyond Uranus.
> What you see in the Cosmos is only the holy reflection,
> in the Olympic family reigns eternal number.

This is more or less the position Kronecker endorsed. He also endorsed Gauss's dictum that "mathematics is the Queen of the sciences and arithmetic the Queen of mathematics," along with the then-recently published observation of Gauss that arithmetic stands in the same relation to geometry and mechanics as mathematics does to science. Not only does each assist the other, but in each case the theoretical status and the validity of the former exceeds the latter. Arithmetic is the pure product of the human mind, whereas space and time require a knowledge of reality. Kronecker drew a line between arithmetic and algebra, on the one hand, and geometry and mechanics on the other; analysis was to go with algebra and arithmetic, but in a special sense. All of mathematics was to be arithmetised, as he put it, using purely and simply the idea of number in the strictest sense, without modification and extension of the idea to irrational numbers or continuous quantities.

Kronecker then sketched out how this could be done. Ordinal numbers arise on encountering a set of objects and enumerating them: "first," "second," and so on. Forget about the ordering, which is arbitrary in any case, and the cardinality of the set is obtained. Think about several sets at once in appropriate ways and the familiar rules for addition and multiplication of positive numbers are obtained. There is rather a lot tucked out of sight here: the set of numbers under consideration at any one time (Kronecker called it a *Schaar*, or family) is tacitly taken to be finite. Kronecker simply did not say if this was purely in order to get started, or in response to a position about the existence of infinite sets. Can one start with ordinals, or are cardinals really prior? A lengthy philosophical literature grew up around this one point, which we shall discuss below.

To define subtraction and fractions Kronecker borrowed a technique from Gauss about congruences he (Kronecker) had used to great effect in his advanced research. For subtraction he introduced an indeterminate x and studied expressions of the form $m + nx$ congruence modulo $x + 1$, which is equivalent to treating x as meaning -1. This neatly explains what negative integral quantities are. More complicated congruences dealt with rational numbers. Kronecker was well known by then for his dislike of the theories of the real numbers introduced by Cantor and Dedekind, and favored by his Berlin colleague Weierstrass. With the example of real roots of a polynomial equation in mind, Kronecker gave a simple numerical rule for finding sequences of nested intervals of rational numbers shrinking down arbitrarily close to any real root. Then, he said, "The so-called existence of real irrational roots of algebraic equations is based purely and simply on the existence of intervals of this kind." In other words, not only could any careful talk about real numbers be rewritten as talk about approximations by families of rational numbers, but there was no need to suppose that such meant anything other than statements about families of rational numbers. The concept of a real number, for Kronecker, was otherwise an empty one.

2.4.5 Helmholtz's Foundations for Arithmetic and Geometry

2.4.5.1 ARITHMETIC

Helmholtz (1887) took up the non-Kantian theme of axioms for the natural numbers from a suitably epistemological standpoint. He presented his ideas as a continuation of his work on the axioms of geometry, which had shown, he said, that the axioms of geometry are not a priori but are propositions to be confirmed or refuted through experience. This, he gallantly added, did not eliminate Kant's view of space as a transcendental form of intuition, but merely excluded one particular, if popular, interpretation of it. Nonetheless, if axioms of space had to be treated empirically, then so too must those of the form of intuition of time, which, in the Kantian scheme of things, meant number. Helmholtz proceeded to follow the path taken by the Grassmann brothers, who "have got further with this investigation than the other arithmeticians whose work is known to me, while at the same time pursuing philosophical viewpoints." He aimed to show that what he called Grassmann's axiom, that $(a + b) + 1 = a + (b + 1)$, was indeed the correct basis for the theory of addition, and that for the application of this theory it was necessary to have the concepts of magnitude, alikeness in magnitude, and unit. As will be discussed more fully below, he also cited the mathematician Paul du Bois-Reymond approvingly, and noted that Schröder in 1873 had correctly observed that it was a problem for psychology to explain how the mind assigns a unique cardinal number to a set of objects no matter how they are enumerated, while what empirical properties make a set enumerable at all must also be defined.

Arithmetic, said Helmholtz as he began his explanation, is a method founded upon purely psychological facts and teaches the logical application of a symbolic system. It explores the different ways of combining these symbols and is rescued from being nothing more than a game of chess by its applicability. Now, to be applicable,

FIGURE 2.8. Hermann von Helmholtz (photo from 1891).

there must be an objective sense in which two objects may be said to be alike in some respect, and when the attributes in respect of which they are alike may be additively combined. There is, he went on, a lawlike series that has a recognized sense of direction because it unfolds in time, and which is that of the positive whole numbers regarded as ordinals. This leads to the axiom that of two different numbers one is necessarily smaller than the other (it is reached sooner in the series).

Because one can start natural numbering at a number a and number up to b, one can give meaning to the expression $(a+b)$; this follows from the fact that each number has a unique successor.[104] Helmholtz therefore deduced Grassmann's axiom, noting that more work needed to be done to make a and b play symmetrical roles in the theory, by establishing that $(a+b)=(b+a)$. The proof that $(a+b)=(b+a)$ is interesting, because it appeals to mathematical induction. First, Helmholtz showed, following Grassmann, that $a+1=1+a$. He did so by showing that if this equation holds for a it holds for $a+1$, and that it holds for $a=1$. Therefore it holds for every a. The logical and philosophical status of mathematical induction was to become a major concern in later debates about the natural numbers.

Moreover, every number can be reached by numbering far enough from any starting point, so given any two numbers a and b there is a unique number c such that

---

[104] Helmholtz denoted the successor of a by $(a+1)$; cf. Peano on successors below, §3.3.4.

$(a + b) = c$. The number which must be added to a to get c Helmholtz said could be denoted $(c - a)$, thus introducing subtraction, but only in the easy case, it would seem, when a smaller number is taken from a larger one. Genuine subtraction is to be defined, apparently, by going backwards from 1 to 0, from there to -1, and so on. Helmholtz continued in this Grassmannian fashion defining multiplication, introducing the distributive laws, and finishing up with exponentiation.

What of alikeness? Two objects that are alike with a common third must themselves be alike. If A and B are alike, then B and A are alike. With this concept of alikeness, measurement becomes possible—of distances, time intervals, weights, pitches, and so on. By thinking through what is involved in measurement, Helmholtz argued, it becomes clear when it is additive (when, for example, the quantities being measured can be combined by juxtaposition). The appropriate rule for combining the measured numbers follows from the corresponding natural law governing the relevant behavior of the objects, and rules can be found for scalar and vectorial quantities. Although Helmholtz did not make this clear, this discussion was presumably intended to explain how and when addition is the appropriate way to combine measurements. (We shall return to this discussion below, §5.3.7.)

It is hard to see from these accounts how the principle of induction can be defended, and it is accordingly not clear what we actually know about numbers. Does this knowledge, for example, extend to the laws of arithmetic (distributive, commutative, etc.)? Nor is there any account of how we might come to know complicated truths of arithmetic, such as Goldbach's conjecture or Fermat's Last Theorem. As one might suspect, Cantor's opinion of Kronecker's and Helmholtz's ideas was very low. He wrote: "They adhere to an extreme empiricist-psychological standpoint with a harshness that one would not think possible if one had not met it twice here in flesh and blood."[105] For them, he said, numbers were signs first of all, but not signs for ideas that rest on sets, but signs for the subjective counting process, and this, in his opinion, inverted the entire process of thought that should rather go from number to counting. Cantor argued that theirs was an old, long-running error, and, to rub the point in, he doubted if treating numbers as signs could ever connect with the eternal number praised in the Schiller parody.[106]

2.4.5.2 GEOMETRY

Alongside arithmetic, geometry and spatial perception stood to be interrogated, and here Helmholtz was a leading protagonist. As we shall see (§3.1.4) he argued that non-Euclidean geometry was just as intuitive, in the Kantian sense, as Euclidean geometry. To some critics, this simply showed that he had failed to grasp the full import of the Kantian a priori, to others this showed how dangerously open to psychologism the Kantian a priori was, while to yet others (Lange, for example) this showed how fruitfully open to psychologism was the Kantian a priori. Helmholtz himself saw his work as renewing Kant's work, with an emphasis on epistemology and a rejection of metaphysics, although of course he did not endorse Kant's exclusive attitude to Euclidean geometry.

---

[105] Cantor 1887–88; see Cantor 1932, 382.
[106] Husserl 1970, 198, dismissed Kronecker and Helmholtz as nominalists.

Helmholtz saw his philosophical essays as an integral part of his scientific work. He sought naturalist or scientific explanations for the workings of the mind and framed psychological laws for the unconscious processes as well as the conscious ones. Spatial perception was, he argued, learned through experience. It gave knowledge, but not because the perceptions accorded with reality (that would have implied a non-Kantian direct acquaintance with things in themselves or a metaphysical argument to make the connection). Rather, it gave knowledge because "sensations are only signs for properties of the external world, whose interpretation must be learned through experience."[107]

In "The Facts in Perception" (1878, 138–140), Helmholtz argued that it is the lawlike nature of phenomena that make perception possible. This law-governed aspect of the world may reside, he conceded, in its appearance, but it is on the basis of appearances that we say, for example, that things which remain alike as time goes by are substances, and the relationship which remains alike between altering magnitudes we call a law. Laws that permit unambiguous correct predictions we call causes. Helmholtz believed that causal laws had to be accepted as a presupposition of knowledge of an external world, although toward the end of his life he came to believe that there could be no empirical justification for such laws, and they must be accepted on faith as a precondition of our ability to make sense of appearances.[108] In 1878 he had come to think that "the law of causality actually is an a priori given, a transcendental law" that cannot be proved from experience.[109] This seems to make it a necessary preliminary to thought, but Helmholtz held back from such a ringing Kantian endorsement, perhaps because he hoped for some more pragmatic way of grounding science. On the other hand, without some assumption of lawlike behavior, not even measurement is possible and the whole scientific enterprise grounds to a halt.

Helmholtz, unlike Kant, saw the possibility a priori of three geometries (Euclidean, non-Euclidean, and spherical). He therefore believed, unlike Kant, that there was an irreducibly empirical element in our coming to know that space is correctly described by any one of these (Euclidean being the result). We would come to realize this through the operation of our senses and our experience of bodies, our own and other, solid bodies. In this matter he placed an emphasis on epistemology that was not found in Kant's work.[110] But Helmholtz was not a hard-line empiricist, and indeed he was happy to urge, as he put it, that "space can be transcendental without the axioms being so." In this sense he did not contest Kant's transcendental argument that something must be in place for us to be able to talk about the geometry of space at all, merely that this something delivered one geometry rather than three.[111]

[107] Quoted in Hatfield 1990, 169.

[108] For an interesting account of how Helmholtz's color theory was affected by the criticisms of Grassmann, see Lenoir 2006.

[109] Helmholtz 1878, 142, quoted in Hatfield 1990, 215, in the course of a discussion of Helmholtz's evolving views upon which these paragraphs are based.

[110] Friedman 2000a has an interesting argument that, whereas Kant took as his fundamental act of imagination our being able to be anywhere in space, Helmholtz started from the idea that we can put a rigid body anywhere in space. These are certainly distinct ideas, valuably distinguished, but it seems to me that Kant's truly fundamental idea was that space is as Euclid and Newton had said it was: three-dimensional and infinite. He therefore attributed to intuition enough abilities to enable the transcendental argument to deliver (Euclidean) geometry.

[111] The Facts in Perception, Appendix 2, in Hertz and Schlick 1977, 149.

2.4.6 Erdmann and Tobias

2.4.6.1 ERDMANN

In 1877 the twenty-six-year-old philosopher Benno Erdmann published his *Die Axiome der Geometrie*, subtitled a philosophical enquiry in the Riemann-Helmholtz theory of space. Its introduction gives us a window onto the debates initiated by the ideas of Riemann and Helmholtz. There was, said Erdmann, a great tangle of ideas. Some, including most mathematicians, connected the empiricist, psychological assumptions with the theory of knowledge that speaks of the objective reality of space. Only Herbartian mathematicians, led by Drobisch, opposed this, because Herbart had distinguished intellectual space from sensual space. Some, such as Liebmann, found confirmation of their belief that our spatial intuition was purely subjective and entirely phenomenological. Wundt was more careful. He recognized that the new theory was the result of a psychological analysis of an empirical kind, but withheld complete consent hoping to put these conclusions to epistemological use. Lange, and the mathematicians Klein and Baltzer, suggested that there were no philosophical implications at all in the new geometry.[112] Dühring, Tobias, and Becker dismissed the mathematics, Tobias and Becker from an adherence to Kantian idealism that said to them that all this mathematics was invalid and self-contradictory.

In Erdmann's opinion the new theory had positive consequences for psychology, because it contributed to the new empirical conception of space; but it had only negative consequences for the theory of knowledge, in that it excluded the rationalist conception of space as a form of sensual awareness. To establish these claims to his satisfaction, he reviewed the journey from Bolyai and Lobachevskii to Riemann and Helmholtz's construction of an axiom system for Euclidean geometry. We have, he said, a unique intuition of space. To have alternatives, without which no discussion is possible, Erdmann allowed geometries that are constructed, like coordinate geometry, from arithmetic. He pulled out the importance of the idea of rigid body motion and offered three characteristics for space: it is a three-dimensional manifold; the same at every point (and so of constant curvature); and indeed infinite and of zero curvature. Finally, he concluded that the implications of the Riemann-Helmholtz theory were that it showed that our spatial intuition, as so characterized, can be subsumed in the idea of magnitude of a particular kind, the more precise investigation of which was a matter of logic. In being so subsumed the empiricist theory of space was strengthened, and the rationalist position much weakened, but not completely excluded.

Erdmann's book was reviewed by the mathematicians Axel Harnack, Eugen Netto, and Max Noether.[113] Harnack found the mathematics frequently sloppy, but the philosophical conclusions largely sound, although he did not agree with Erdmann or Helmholtz that we can have intuitions of other than Euclidean space. He regretted that Erdmann had not discussed the notion of the space element before deciding our space was three dimensional since, after all, it is four-dimensional as a space of lines (see §3.1.2). The argument that space has constant curvature was, he said, clearer in

---

[112] Richard Baltzer was the first person to write about Lobachevskii sympathetically in his *Die Elemente der Mathematik* (1868); the book has many historical notes.

[113] Harnack 1878, Netto 1877, and Noether, M. 1878.

Helmholtz's work. The derivation of the third characteristic of space from the idea that a line can only meet itself in one point was flawed by Klein's example of single elliptic or projective space, but a different theorem could be established to much the same effect. Netto, on the other hand, found both that the mathematical arguments were sketchy (for example, that space has zero curvature), and that philosophical issues were overlooked. Erdmann merely asserted space was a magnitude (a grave weakness in his position in any case) but, asked Netto (in a way reminiscent of the Kantian antinomy that space cannot be either finite or infinite, but for that reason perhaps unfairly), how can something be a magnitude that can neither increase nor decrease? The whole discussion would be different if only the author had known of Cantor's work. Among other problems, Netto raised the issue that if the axioms for geometry were entirely empirical, how could geometry differ from the other sciences as it did in being general, necessary, and invariable?

Like Harnack, Noether observed that even the ascription of three dimensions to space was problematic until one had decided on the fundamental element: when this is taken to be a point then of course space has three dimensions, but if the line is fundamental then space has four dimensions (because it takes four coordinates to specify a line in space, see below, §3.1.2).

As Torretti notes, in his summary of Erdmann's book, Erdmann's mixture of muddled mathematics and inconclusive philosophical argument is only too typical of the period.[114] But one may deplore confusion or try to understand it. In the 1870s the issues were extremely unclear. Not only did the referees cited above disagree with one another, their criticisms were not always fair or sound. The rift that had opened up between modern geometry, physics, and the philosophy of these enterprises made it very difficult for anyone to sort out three kinds of things: spaces, mathematically defined; spaces with some chance of being useful descriptions of the space we live in; the space we live in. Indeed, we should add a fourth: our conception of the space we live in. Erdmann found his way to a philosophical position about geometry that was closer to Riemann's than many people found possible, even if his account of how we come to know that space is three dimensional was unsatisfactory. That was a topic with much more life in it, and one to which Lotze, Helmholtz, and Poincarè would discuss with the so-called theory of local signs, as we shall see below.

2.4.6.2 TOBIAS

Another who felt compelled to assess Riemann's mathematics in the light of Kantian philosophy was Wilhelm Tobias, who is remembered these days, if he is remembered at all, for having disagreed with Helmholtz sufficiently strongly for Helmholtz to respond.[115] The best thing about Tobias's book of 1875 may be its title: *Grenzen der Philosophie constatirt gegen Riemann und Helmholtz, vertheidigt gegen von Hart-man und Lasker* (The Boundaries of Philosophy Established against Riemann and Helmholtz, Defended against von Hartman and Lasker). Tobias agreed that Riemann's mathematics was entirely logical and unobjectionable as such. His objection to Riemann's work was that it proposed a range of a priori geometries of which Euclidean three-dimensional geometry was but a special case, and indicated that

---

[114] Torretti 1984, 264–272.
[115] See Helmholtz 1977, 2.

some empirical work was needed to determine the correct geometry, but nowhere discussed how such a synthesis would be possible (p. 61). This was for Tobias the intellectualization of experience, and a failure to appreciate the role the pure intuition of space played in the acquisition of knowledge.

If Riemann as a mathematician concealed his philosophical positions, the same was of course not true of Otto Liebmann, and next Tobias turned to attack him as a fighter for Riemann and Helmholtz.[116] Liebmann's error, said Tobias, was to defend the idea that a mathematically well-defined n-dimensional manifold was an intuitively accessible space. This distinction between the logical and the intuitive (in Kant's sense of the term) was to prove a recurring theme; it was characteristic of the neo-Kantians at the turn of the century. In analyzing intuition and the limitations on human knowledge that it entails, Tobias concluded that the fundamental problem was down to the vexed Kantian *Ding an sich* earlier identified by Liebmann himself, and which Tobias presented as a boundary idea (*Grenzbegriff*) that can neither be overcome or ignored. The implication is that these Riemannian manifolds are beyond the boundary of what can be known.

Tobias's book did not start the debate its author must have hoped for, still less did it win it. But it has its place in the history of the philosophy of mathematics, not only for promoting the idea that a logically impeccable theory may need something else if it is to count as knowledge, but for documenting a range of writers that well-educated Germans read: Wundt, Haeckel, Helmholtz, Riemann, Müller, and many others as they returned to Kant.

2.5 British Algebra and Logic

For many centuries it had been widely agreed that syllogistic logic was the definitive form of reasoning, to which all other forms could be reduced, and that Aristotle's theory was complete. Kant, in his *Logic* (p. 23) said that logic "has not gained much in *content* since Aristotle's times, and indeed it cannot, due to its nature." However, logic, conceived of as the way of making valid deductions, enjoyed a revival in the early nineteenth century.[117] Gergonne in France, George Bentham and Augustus De Morgan in England, William Hamilton in Scotland, Bolzano in Bohemia, Twesten and Drobisch in Germany all wrote at length on the tangled and tricky nature of deductions involving statements like "All A are B" and "Some C are not D," and all advocated one or another form of symbolic method for handling such statements and a classification of the different types. The problem, which goes by the name of "quantification of the predicate" was taken to be about extending the syllogistic logic of Aristotle to deal with propositions of the form, "Some men are immortal," which arises by negating the familiar claim, "All men are mortal."

Authors also began to point out that not even Euclid's *Elements* was in remotely syllogistic form. De Morgan, who took up the subject of quantification of the

---

[116] Liebmann, n.d., 337.

[117] A brief account of the philosophical background in Britain and Germany will be found in Peckhaus 1999.

predicate in the 1840s, and was led to make remarks on mathematical relations, albeit none of great import, gave this example of a nonsyllogistic form of reasoning: "A man is an animal, so the head of a man is the head of an animal." Jevons pointed out that the statement that "*All* equilateral triangles are *all* equiangular triangles" could only appear in syllogistic form as "All equilateral triangles are equiangular," which does not even say the same thing.

2.5.1 Boole

Boole's *Mathematical Analysis of Logic* (1847) is a remarkable distillation of a number of themes. It opens with this remark about logic: "If it is lawful to regard it from *without*, as connecting itself through the medium of Number with the intuitions of Space and Time, it was lawful also to regard it from *within*, as based upon facts of another order which have their abode in the constitution of the Mind" (p. 1). In that spirit he offered a "Calculus of Deductive Reasoning." Boole in these words juxtaposed a Kantian account of the possibility of logic as something forced upon the Mind by the way of the world, with an account of logic as an innate activity of the Mind. He then set to work to extend British symbolical algebra to logic. He regarded algebra as an activity in which symbols were dealt with entirely according to stated laws of combination, quite independent of any meaning they might have. Since, as a matter of fact, mathematics concerned itself with magnitudes, he accepted that symbolic arguments had been treated as coextensive with arguments about magnitudes, but he wished to break with this limitation and extend mathematical analysis to include a calculus of logic. He set himself these standards: "We might justly assign it as the definitive character of a true Calculus, that it is a method resting upon the employment of Symbols, whose laws of combination are known and general, and whose results admit of a consistent interpretation" (p. 4). It was not enough, however, that a calculus of logic be symbolic, general, and interpretable. He regarded logic as possible because of "the existence in our minds of general notions, our ability to conceive of a class, and to designate its individual members by a common name." And he immediately went on: "The theory of Logic is thus intimately connected with that of Language" (pp. 4–5).

The first fruits of his endeavors in the direction of a new calculus were to find logical arguments of a nonsyllogistic form. He was also led to contemplate the relation between logic, as a branch of philosophy, and mathematics. The Scottish philosopher William Hamilton was at the time in a public argument with De Morgan in which he contended that the study of mathematics was simultaneously dangerous and useless. Philosophy, he contended, was the science of real existence and real causes; it answers the question "why?," whereas mathematics is credulous in its premises. To this, Boole suggested that he might agree with those who contended that such a philosophy was impossible. Better to study laws and phenomena, and regard "the nature of Being and the mode of operation of Cause, the why, [as being] beyond the reach of our intelligence" (pp. 12–13). But, he said, even if he did not regard the philosophical quest as hopeless, such a philosophy did not include logic, which should be classified instead with mathematics, because logic, like geometry, could be given axiomatic foundations.

Boole then developed a symbolic calculus which he presented in an extended setting six years later in his *Laws of Thought* (1853), and though he carefully pointed out that this work was not a republication of the old one, being much more general and wider in its applications, it did overlap with it. It is therefore sensible to describe what Boole did only in the later, larger, and much more influential work. That logic concerned itself with the laws of thought was not an original idea of Boole's. He was not even the first to use the phrase as the title of a book. But he had an original view on what it meant to speak of laws of thought. He defined them as the fundamental laws of operations of the mind by which reasoning is performed. Those who doubted the existence of such laws were asked to consider the evidence presented in the book. Those who accepted already that a science of intellectual powers is possible were asked to consider how knowledge of it could be obtained. Plainly, the answer must be by observation, but the mind's observation of the mind raises special problems. The science of external nature used many observations to approach the truth, but its claims could never be certain, only more or less probable. The science of the mind dealt with truths (valid deductions) apprehended as such in a single clear instance. That being the case, the task of the scientist of the mind was not to discover (probable) laws, but to arrange the general laws of logic in an appropriate deductive framework. In this regard he felt that the syllogisms of traditional logic could be analyzed into simpler forms, and therefore should not be regarded as basic.

Rather, he said, he would accomplish his task with his symbolic calculus, and he pointed out that the symbolic rules he found to apply were very nearly those that also applied to arithmetic. The only point at which the laws of logic and of arithmetic differed he felt was the germ or seminal principle of logic. As for the scope of logic, he argued that logic dealt with propositions, and a conclusion in logic expressed an implied relation between the premises. A useful logic should therefore permit elimination of those terms which are not wanted in the conclusion, and it should be able to deal with any kind of admissible proposition.

Boole then set to work to deliver what he had promised in his preface. He began with language, which was generally admitted, he said, to be an instrument of human reason and not merely a medium for the expression of thought. The elements of language are signs or symbols, such as words. A sign he defined to be "an arbitrary mark, having a fixed interpretation, and susceptible of combination with other signs in subjection to fixed laws dependent upon their mutual interpretation" (p. 25). He emphasized the arbitrary nature of the sign, and that it should also have a fixed interpretation. They came in three types: literal symbols, such as x, y, etc., representing things as subjects of our conceptions; signs of operation such as $+$, standing for operations of the mind which combine or resolve conceptions into new ones; and the sign of identity, $=$.

Appellative or descriptive signs either named things or some quality of things, so they could be either nouns or adjectives. Concatenation of symbols produced a sign naming the things with all the properties of the concatenated parts (so if x stands for sheep and y for white, then xy stands for white sheep). Their laws of combination were commutative, $xy = yx$ "like the symbols of algebra" (p. 31). But in the logical case it was also true that $x^2 = x$. The symbol "$+$" he introduced to stand for the exclusive "or," and the symbol "$-$" to stand for taking complements, so if (to give his example) x stands for all men, and y for Asiatics, then $x - y$ stands for all men except Asiatics.

The laws governing the symbols were advanced as inductive generalizations from the workings of language (chiefly, but not exclusively, English). Boole then attempted to derive them from the operations of the human mind, by considering how the mind deals with classes of objects with particular properties. This led him to the useful device of considering a universal class for a given discussion, embracing all the objects under discussion. He denoted this class by the symbol "1," and the class Nothing by the symbol "0." So "$1-x$" represents all the objects under consideration that are not x. This duly led Boole to the following proposition (p. 49):

> That axiom of metaphysicians which is termed the principle of contradiction, and which affirms that it is impossible for any being to possess a quality, and at the same time not to possess it, is a consequence of the fundamental law of thought, whose expression is $x^2 = x$.

Let us write this equation in the form

$$x - x^2 = 0$$

whence we have

$$x(1-x) = 0;$$

both these transformations being justified by the axiomatic laws of combination and transposition (II. 13).

Boole then showed how a great many statements of the form "things with these properties are the same as things with these (seemingly other) properties" can be expressed symbolically. That done, he set to work to establish the theory of elimination he had advocated in his preface as an analogy to the solution of algebraic equations. This inevitably led him, via an obscure concept of division, to devise a symbol for "all, some, or none," because it can easily happen that logical equations have indefinite solutions. Boole's choice for this symbol was $\frac{0}{0}$. For example (p. 105), he expressed the statement "No men are perfect" as

$$y = v(1-x),$$

where y represents men, v is an indefinite class, and x perfect beings. He eliminated v and arrived at $yx = 0$, which states, "Perfect men do not exist." From this he wrote

$$x = \frac{0}{y} = \frac{0}{0}(1-y),$$

"No perfect beings are men," and

$$1 - x = y + \frac{0}{0}(1-y),$$

"Imperfect beings are all men with an indefinite remainder of beings, which are not men."

Boole dealt in this way at some length on what he called primary or concrete propositions, assertions respecting facts, as he put it, before turning to secondary propositions, those that relate propositions to other propositions. He turned these secondary propositions into ones of the former kind by the device of interpreting the proposition "If the proposition X is true, then the proposition Y is true" as saying that the time during which X is true is time during which Y is true. So (p. 165) Boole introduced the convention that x represents the act of mind by which we fix the portion of time for which X is true (and that that is what it means to say x denotes the time for which X is true). He could then write the proposition "X is true" as $x = 1$, the contrary proposition "X is false" as $x = 0$, and the proposition "If X is true then Y is true" as $x = vy$.

To demonstrate the fertility and generality of his method, Boole then showed how many celebrated passages in philosophy can be expressed in his symbolic language, including theological arguments from Plato, Clarke, and Spinoza. This led him to the conclusion that Spinoza's *Ethics* was imperfect, and that it is "impossible, therefore, by the mere processes of Logic, to deduce the whole of the conclusions of the first book of the *Ethics* from the axioms and definitions which are prefixed to it." Boole felt that his examination of theology compelled the view that it was futile to establish entirely a priori the existence of an Infinite Being, and it was better to fall back on the argument by design.

In chapter 15 Boole took up the topic of the syllogism, to which Whately and John Stuart Mill attached such importance, claiming that all correct reasoning can be expressed in syllogistic form. Boole disagreed. Insofar as syllogistic reasoning is a method of elimination, he felt that all elimination could be effected by syllogisms but only after the original statements had been reexpressed using nonsyllogistic methods. And inasmuch as syllogistic reasoning always eliminates, he felt that reasoning need not always be eliminative. Instead, Boole advocated organizing Logic so that it was derived from a set of simple axioms, such as Leibniz's principle of contradiction.

Boole then moved into an extensive discussion of probability before returning, at the very end of his book, to discuss some of the philosophical issues his analysis of logic had raised. He felt that science proceeded by an appreciation of order in the universe. The fundamental concepts of science (here he naturally cited Whewell) seemed to him to be neither intellectual products independent of experience nor copies of external things, but, while they have necessary antecedents in experience they also required some power of abstraction for their formation. Reflecting upon his laws of thought led him to admit that they were of a different kind. External nature conformed to (correct) laws inexorably, but the mind frequently violated the laws of thought. These laws were "the laws of right reasoning only" (p. 408). But still, they showed that the human mind has the capacity to ascend from particular facts to the general propositions of science. Finally, Boole contended that the laws of thought are the same kind as the laws of the acknowledged processes of mathematics. But he did not claim that mathematics was a mere part of logic, still less that mathematics could be regarded as all of knowledge (that would exalt the faculty of reasoning over those of observation, reflection, and judgment).

Boole's achievement was not, therefore, what Russell said it was in his oft-cited aphorism, "Pure mathematics was discovered by George Boole in his work published in 1854."[118] But nor was it to languish in obscurity, as some historians have suggested.[119]

---

[118] Russell 1901, rep. 1963, quotation on 59.

[119] A trickle of problems even passed through the *Educational Times*, so the topic had some currency.

Both views may be different ways of exaggerating its originality. In fact, it was read, in Britain at least, but with difficulty. The method of elimination was a particular problem, and it was much refined by Stanley Jevons, whose interest in logic had been awakened by De Morgan when Jevons studied under him at University College London. He left there, however, without taking his degree in order to embark on a career in business, and shortly took a job working in the Australian Mint, returning in 1859 to complete his education.[120]

Jevons replaced Boole's exclusive "or" with the inclusive "or," thus abolishing the problem of understanding expressions of the form $x + y$ when x and y have elements in common (particularly acute in the case when $x + x$ is to be understood). An eloquent example of how much he simplified the process of elimination was his demonstration that it could be mechanized. After some ten years of work, he exhibited a logical machine at the Royal Society of London in 1870 (a description was published in the *Philosphical Transactions of the Royal Society*, 1870, and in his *The Principles of Science*, 1874). It became known as Jevons's logical piano. A more interesting disagreement between the two is that whereas Boole had seen his mathematical logic as a new form of mathematics that was not restricted to the study of quantity, Jevons saw "the Mathematics as rather derivatives of Logic."[121]

It is sometimes suggested that it is surprising that the British line of research into logic begun by De Morgan and Boole did not immediately lead to a powerful school of logicians. However, in the strictly British setting one need look no further than the fragmentary organization of mathematics in Britain, with Cambridge at its head. There is nothing to explain; in fact, the development of logic closely resembles that of invariant theory or even analysis. As noted, Boole's work was taken up by Jevons, and by Venn, who was not first rate but kept the subject alive. By 1900 it had reached Whitehead and Russell, where it joined with other influences to be sure, but the flowering of logic cannot be denied. Internationally, the situation was also propitious. Through Jevons it passed to Schröder in Germany. It was taken up directly by C. S. Peirce. It was eventually to reach Poland. There would be no difficulty in listing branches of mathematics which did no better for their first fifty years.

2.5.2 The Americans: Peirce and Ladd

The American story is much harder to tell than the British one, dominated as it is by the extraordinary figure of Charles Sanders Peirce (fig. 2.9). Peirce's father Benjamin Peirce was a professor of mathematics at Harvard, with a particular interest in algebra. Charles, who was born in 1839, read Whately's book when he was twelve and discovered a passion for logic that never deserted him, and was then educated by his father in philosophy while still at school and for many years afterwards. He acquired good Latin and Greek and a love of Shakespeare, all of which showed up in his choice of neologisms in later life. He also acquired the habit of thinking of himself as a genius, which coupled to a secure position in Boston society made him intolerant of superficial or muddled thinkers. He became a considerable dandy, but while at Harvard began to suffer from disabling facial neuralgia, which he kept at bay with

---

[120]Jevons's main work was on economics. He died by drowning just before his forty-seventh birthday.
[121]Jevons 1864, 5.

FIGURE 2.9. Charles Sanders Peirce.

heavy doses of morphine and, later, of cocaine.[122] He graduated with a poor degree from Harvard, whether from this illness or a disdain for college regulations, and drifted around Cambridge and Boston for some further years picking up various bits and pieces of training.

In 1862, when he was twenty-three, he married Zina Fay, an active feminist intellectual of Puritan stock, and they lived in Peirce's father's house for eight years in straitened circumstances. Peirce was by now immersed in the study of philosophy, and as a result he was invited to give the prestigious Lowell lectures in February 1865. William James wrote to his sister Alice that he could not understand a word of them, an experience many were to repeat down the years.[123] In January 1872 he founded the short-lived Metaphysical Club with Oliver Wendell Holmes, Chauncey Wright ("the Cambridge Socrates"), and William James among its early members. James had by then mastered the art of standing up to Peirce, skimming him, as it were, for ideas, and not attempting complete comprehension. Of this group only Wright could follow Peirce into mathematics and logic: together they discussed

---

[122] See Brent 1993, 40.
[123] See Menand 2001, 203.

philosophy, the philosophical implications of science, and ideas about causation. Wright was also an influence on Peirce's move into the United States Coastal Survey, where, helped by his father who was the superintendent, he began work in the summer of 1872, all the time maintaining his interest in logic and philosophy.

In 1870 he traveled to England, where he met and impressed de Morgan and other British logicians. Upon his return to America, his father appointed him to an influential position in the Coast Survey, and after a few years of failure he went on to do important work on the survey of terrestrial gravity, showing a sensitivity to problems with instruments that was acclaimed in 1877 at the International Geodetic Survey in Berlin and on in his other trips to Europe in 1875, 1880, and 1883.[124] However, toward the end of the decade his marriage fell apart, his expenditures began greatly to exceed his income, and his father and protector died in 1880. Bereft of his influence and that of Zina, his behavior began to furnish ammunition for those who disliked him, among them Charles William Eliot, later the president of Harvard, and the astronomer and mathematician Simon Newcomb, who was head of the Nautical Almanac Office and the Naval Observatory.

Finding that his career was blocked at the Coast Survey, Peirce accepted a lectureship at the newly founded Johns Hopkins University in Baltimore in 1879. The five years he spent as a lecturer at Johns Hopkins saw the publication of *Studies in Logic by Members of the Johns Hopkins University* (1883). During these years, Peirce's reputation in Britain stood high. Clifford called him the greatest living logician,[125] Jevons hailed him in *Nature* in 1881 as the leading logician of the day,[126] and Venn reviewed the *Studies* in *Mind* in 1883 in these terms: "The volume seems to me to contain [a] greater quantity of novel and suggestive matter than any other recent work."[127] Finally, despite the great success of the collaborative work on logic "within a span of seven years, he was suddenly and unexpectedly dismissed from Hopkins [and] forced to resign from the Survey for failing to complete the important geodesic researches for which he was solely responsible."[128]

Peirce's first major work on logic was his "Description of a Notation for the Logic of Relatives" (1870).[129] Where Boole had stressed the close analogy of the laws of thought with those of algebra, Peirce wished to extend algebra so that it could deal with relations (or relatives as Peirce unfortunately called them, a fact he later regretted.[130] Most of Peirce's work is afflicted with the outpourings of his chaotic, original energy, and this paper is no exception. Odd analogies jostle with confusing, even irrelevant, pieces of other topics to the detriment of his message. In this he reads oddly like an autodidact, who is afraid to separate his ideas out and present them in a more conventional way so that they can be understood. The confusion is deliberate, and derives from deep psychological factors in Peirce's makeup.

Peirce took the study of relations from De Morgan, but he discovered many laws that the corresponding algebra must have. One may say that if De Morgan discov-

[124] Brent 1993, 138.
[125] Brent 1993, 119.
[126] Quoted in Brent 1993, 138.
[127] Quoted in Brent 1993, 138.
[128] Brent 1993, 139.
[129] See several essays in Houser et al. 1997 for detailed accounts of Peirce's work on logic.
[130] Merrill 1997, 160.

ered that there was something to say about the logic of relations, it was Peirce who showed that there was something to study. His work is altogether deeper. The problem is that many of the depths Peirce encountered are philosophical rather than mathematical, and his philosophy is muddled.

In 1880 Peirce took up the study of syllogisms which he had set aside some years before, with a view to seeing how it could be quantified.[131] It is typical of Peirce, but not out of keeping with the ideas of the day, that the paper begins with a brief discussion of how thinking arises in the brain, and offers definitions of belief, judgment, and thought. There follows a lengthy investigation of the syllogism, built round Peirce's sign for existence: $-<$. He wrote

$$\text{Griffin} -< \text{breathing fire}$$

"to mean that every griffin (if there be such a creature) breathes fire; that is, no griffin not breathing fire exists," and

$$\text{Animal} \overline{-<} \text{Aquatic}$$

"to mean that some animals are not aquatic, or that a non-aquatic animal does exist." Peirce saw clearly that Boole's work lacked a good analysis of existence. Boole had dealt with this by means of his symbol v, but most later writers found his approach unsatisfactory. Peirce began by extending Boole's operations for classes to binary relations. He introduced the relative product of two relations, which contains existential statements implicitly. If l is a relation (Peirce suggested "lover of") and w is a class ("woman"), then lw stands for "lover of woman." This must be unpacked. With Peirce, let $l(i, j)$ stand for "i is a lover of j" and $w(j)$ for "j is a woman," then $lw(i)$ means "i is a lover of a woman," or, more formally, "There is a woman j such that there is an i such that i is a lover of j." Among the logical operations next introduced is exponentiation, which works like this: $l^w(i)$ means "i is a lover of every woman," or more formally, "For every j such that j is a woman, i is a lover of j." The examples of "servant of" and "lover of servant" do make one wonder what latent content these examples conveyed, if only to their author.

But his study of the syllogism was inconclusive, and in due course he was surpassed by one of his students, Mitchell, whose work was published in the Studies (1883).[132] This book shows that Peirce had successfully adopted the style of working favored by J. J. Sylvester, the head of the Department of Mathematics at Johns Hopkins, and encouraged his students as a group to carry out original research.[133] The preface to that work indicates some of the distance they covered. Christine Ladd (by then Mrs. Ladd-Franklin) provided an article which "may serve, for those who are unacquainted with Boole's 'Laws of Thought' as an introduction to the most wonderful and fecund discovery of modern logic." Her essay is on logical notation and syllogistic reasoning, with application to problem solving, and it cites some

---

[131] Peirce 1880.
[132] See the discussion in Brady's essay in Houser et al. 1997, §4, and, for Peirce's response, §5.
[133] On Sylvester, see Parshall 2006.

twenty works written on logic since Boole's book of 1854, roughly one every year and a half, so it is indeed a good introduction to a flourishing field.[134]

In 1885, Peirce made the breakthrough to the productive introduction of quantifiers into logic. He used the symbols Σ and Π, sometimes with subscripts, as he put it "in order to make the notation as iconical as possible, we may use Σ for some, suggesting a sum, and Π for all, suggesting a product. . . . If x is a simple relation, $\Pi_i \Pi_j x_{ij}$ means that every i is in this relation to every j, $\Sigma_i \Pi_j x_{ij}$ that to every j some i or other is in this relation, $\Sigma_i \Sigma j \ x_{ij}$ that some i is in this relation to some j." He referred to his symbols as quantifiers, and the indices as pronouns, thus bringing out a linguistic analogy. Note that the quantifiers, like functions in mathematics, must here be read from right to left. Peirce then showed how to extend this analysis to deal with several relations at once.[135]

Peirce's later papers in the *American Journal of Mathematics* (volume 4 in 1881 and volume 7 in 1885) are on different if related topics. The first is, as its title says, on the logic of number, and gives Peirce's account of the natural numbers, along with his delightful example of the deduction from "Every Texan kills a Texan" the conclusion "Every Texan is killed by a Texan."[136] Hollywood would have loved it. The second of these was subtitled "A contribution to the Philosophy of Notation," and sets out his account of signs—icons, indices, and tokens: icons resemble, indices point, and tokens are general. At one point he thanked his "friend, Professor Schröder" for detecting an error in his account of the syllogism. He hazarded a remark that pulls together so many of the themes in this book that it deserves to be quoted in all its characteristic obscurity: "The distinction between *some* and *all*, a distinction which is precisely on a par with that between truth and falsehood, that is, it is descriptive, not metrical."[137] Peirce returned to the themes of mathematics, logic, philosophy, and language many times in later years, but whatever treasures these writings contain, nothing can disguise the fact that by then he was a marginal figure.

Christine Ladd is surely the most interesting of Peirce's students. She is perhaps best remembered today for her later work on color vision and perhaps for her feminist advocacy and example, but her work on logic is also valuable. She was born in 1847, graduated from the Wesleyan Academy in Wilbraham, Massachusetts, in 1865 and entered the second class of Vassar College, graduating from there in 1869. Then she taught in schools for nine years, an activity she came to hate, and finding a scientific career barred to her because women were not allowed in laboratories, she turned to mathematics. She applied to Johns Hopkins, which had not previously

---

[134] Russinoff 1999 makes the point that without a theory of semantics no analysis of syllogisms could be complete, but Ladd's was the best possible. She also quotes Royce at second hand saying "there is no reason why this should not be accepted as the definitive solution of the problem of reduction of syllogisms."

[135] Ferreirós has reminded me that it was mathematical analysts such as Weierstrass in the 1860s who first appreciated the fact that the order of the quantifiers can be decisive. Consider the expression $y - x^2$ and ask when is it positive. To say that for all ($\forall$) x there exists ($\exists$) a y such that (:) $y - x^2 > 0$ is true, because the choice of y may vary with the given x and one can always choose a y greater than a given value of x^2, thus making $y - x^2 > 0$. But to say that there exists ($\exists$) a y for all ($\forall$) x such that (:)$y - x^2 > 0$ is false, because now the choice of x may vary with the given y and one can always find an x^2 greater than a given value of y, thus making $y - x^2 < 0$. So the statements $\forall x \exists y: y - x^2 > 0$ and $\exists y \forall x : y - x^2 > 0$ are quite different.

[136] Peirce 1881, 90.

[137] Peirce 1885, 195.

admitted women, but J. J. Sylvester was the professor there, and he had noticed her many contributions to the *Educational Times*, so he persuaded the university to admit her (although at first they only agreed to let her attend his lectures). She was finally granted a stipend for 1879–1882, and finished her PhD in 1883 with a thesis on mathematical logic. The degree was granted in 1926, and she attended the ceremony aged seventy-nine. She was never able to attain a university teaching position commensurate with her talents. Among her many jobs, she worked from 1901 to 1905 as an associate editor for logic and philosophy on Baldwin's *Dictionary of Philosophy and Psychology*.[138]

Ladd-Franklin's most important work in logic was her paper in the *Studies* volume, pages 17–71, in which she gave a test for the validity of any argument couched in syllogistic form. Her method was to express the claims of such an argument in algebraic terms and then to reduce it to a certain algebraic form, and she asserted that only valid syllogisms reduced to a form of a particular kind. Her test was this. A valid syllogism is an argument of the form *a* and *b* imply *c*. Given an argument of this form that purports to be valid, write any universal proposition it contains in the form "No *a* is *b*" and particular propositions in the form "Some *a* is *b*." Take the contradictory of the conclusion. The argument is a valid syllogism if and only if exactly two of these three propositions are universal and if the only term common to the two universal propositions has unlike signs (one negative and one affirmative). Applied to the familiar syllogism "All men are mortal, all Greeks are men, therefore all Greeks are mortal," the test generates these propositions: "No non-mortals are men," "No non-men are Greeks," and "Some Greeks are not mortal." The only common term in the two universal propositions is "men," which occurs once as "men" and once as "non-men," so the test correctly says that the syllogism is valid (as are the other two syllogisms that this triad generates). Ladd-Franklin's test for a valid syllogism is correct, and both C. I. Lewis and Quine said so, but she did not prove it. Her argument depended on the claim that every inconsistent triad has a certain form, a point noted and corrected in Russinoff (1999).

2.5.2.1 ALGEBRAIC LOGIC BY 1880

Boole's work in logic, as refined by Jevons, produced a simple way of writing and analyzing complicated statements of a particular kind. Peirce's contribution was to bring in the use of the quantifiers "for all" and "there exists." This enabled the process of deduction to be formalized. Much could now be written succinctly and precisely that had hitherto been a matter of long and often ambiguous verbal expression. In this sense the achievements of these logicians was linguistic rather than philosophical. It could therefore easily seem, as Boole had thought, that logic was a branch of mathematics, or, as Jevons preferred, that mathematics was an outgrowth of logic. This might raise a question that interested Peirce of the deeper relations between mathematics and logic, but deeper investigations were necessary if a satisfactory answer were to be given.

---

[138] Even in death she is not properly served. The Columbia University Web site notes that the university has seven thousand items in its Christine Ladd-Franklin and Fabian Franklin Papers. These could surely enrich the history of mathematical logic.

2.6 The Consensus in 1880

The preceding account has highlighted just those aspects of mathematics that will detain us for the next few chapters, but it is worth stressing how the world of mathematics looked to most mathematicians in 1880, in order to show how strong and unexpected the modernist move was.

Berlin in 1880 was the undisputed center of mathematics, and the leaders there agreed that the core discipline was analysis, indeed, complex analysis, around which were closely situated the fields of number theory and algebra. A hardliner such as Kronecker explicitly placed geometry outside the core of mathematics,[139] and the evidence of Weierstrass's two-year lecture cycle, as well as the hirings made in the department, suggest that he never disputed that judgment. Felix Klein was not to attack this orthodoxy until 1895, by which time Weierstrass had retired, Kronecker was dead, and Klein's position at Göttingen was secure; and even then it took the form of a defense of geometry wrapped up in a plea for intuition as an aid to discovery.[140] Klein was also promoting what he saw as Riemannian ideas that had been largely neutralized by the heavyweights in Berlin, and thereby his own credentials as Riemann's heir.

Another generational shift was taking place in Paris. The young Poincarè and Picard took to Riemann's ideas in a way their mentor, Charles Hermite, had always resisted.[141] Poincaré was soon to swing toward a lifelong, but not exclusive, interest in mathematical physics, which can be seen as his attempt to keep alive a French mathematical tradition that had largely been driven out in Germany. In Italy, geometry was particularly well studied. In Great Britain, the dominant tradition put physics first and mathematics second.

Nowhere in Europe was it doubted that analysis stood center stage. Enthusiasm for number theory was more marked in Germany, geometry, and mathematical physics stood higher elsewhere. No major voices were raised for the idea that philosophy, logic, and mathematics should come together in a way that would renovate each field and bring them into fruitful inter-action. Set theory and its novel methods, abstract axiomatic thinking with its emphasis on structural and formal symbolic reasoning, a disdain for applications outside mathematics and a corresponding growth of new types of pure mathematics—none of this was expected or being planned for. Those changes were unexpected, but they were to be massive and permanent.

---

[139] See Boniface and Schappacher 2001.
[140] Klein 1895.
[141] See Gray 2000.

3

MATHEMATICAL MODERNISM ARRIVES

3.1 Modern Geometry: Piecemeal Abstraction

3.1.1 Projective Geometry: The Kleinian View

Beltrami's disk model of the geometry of Lobachevskii gave it mathematical rigor and respectability. Where there had been neglect, distrust, hostility, now there was a quiet chorus of agreement among mathematicians: there was, after all, a geometry that differed from Euclid's only over the definition of parallels, and it was the intrinsic geometry of a disk with constant negative curvature. Beltrami (1868b) showed that the same approach could be made to work in any number of dimensions. Non-Euclidean geometry had arrived among the mathematicians.

There was still to be a quarter-century of debate about what it was that had been discovered, and what its implications were. One of the first to advance this debate was the young Felix Klein. Klein was born on April 25, 1849, in Düsseldorf. The way mathematics was taught at school left him cold, but he was a precocious child and able to enter the University of Bonn in 1865, when he was only sixteen and a half. He went intending to study mathematics and natural sciences, but what mathematics and physics there was at Bonn was at too low a level to catch his interest. He was saved by becoming Plücker's assistant around Easter 1866. His job was to set up and carry out demonstrations accompanying Plücker's lectures on experimental physics. He also assisted Plücker with his mathematical research, to which Plücker had recently returned.

Two years later, however, in May 1868, Plücker died unexpectedly. Klein's research into Plücker's new geometry (to be described below; see §3.1.2) was by then sufficiently advanced for him to complete his doctorate in December 1868, and he was immediately invited by Clebsch in Göttingen to take on the task of publishing Plücker's posthumous geometrical works. Klein moved into the circle of algebraic geometers inspired by Clebsch, and physics took a back seat, because the work struck him as too pedantic; all his life Klein was no lover of detail. As was the custom among

FIGURE 3.1. Felix Klein.

students working for their *Habilitation*, he traveled, in his case to Berlin in autumn 1869 to attend the seminars of Kummer and Weierstrass, and it was while he was there that he heard for the first time from Otto Stolz of Innsbruck of the existence of non-Euclidean geometry. He immediately conjectured that it was closely connected to the so-called Cayley metric, although Weierstrass opposed this idea.[1]

He also went to Paris, but the Franco-Prussian War disrupted his plans, and while on military service he caught typhus and nearly died. Once he had recovered he went back to Göttingen, where he habilitated in January 1871. Then, on Clebsch's recommendation, he was appointed as a full professor of mathematics in the small University of Erlangen in autumn 1872, at the astonishingly young age of twenty-three. When Clebsch died suddenly in November of diphtheria, a number of Clebsch's special followers, who were often older than Klein, followed him to Erlangen, and so Klein found himself among the leading German geometers.

The famous "Erlangen Programm" (or program) was not the inaugural address Klein gave on becoming a professor at Erlangen.[2] That was on the importance of cultivating pure and applied mathematics together, so as to preserve the connection between mathematics and related fields of knowledge, such as physics and technology. Klein also spoke about developing visual perception as an equal factor with logical abilities, and on mathematical imagination and the spontaneity that flows it. This was his way of saying what needed to be done at Erlangen. The Erlangen Program is a written text that circulated at the time, in which he gave a uniform account

---

[1] Klein 1921–23, vol. 1, 51, and Klein 1926–27, 152.
[2] See Rowe 1983.

of the existing directions of geometrical research and sorted them into a system. In the 1890s it became famous as a retrospective guideline for his research.

What Klein called the "Cayley metric" was an adaptation of Cayley's way of starting with projective geometry and smuggling in the idea of Euclidean distance. Cayley had not realized the generality of the idea. It was Klein who saw that it could be used to relate projective and non-Euclidean geometry. He spelled this out in detail in two papers entitled "On the So-called Non-Euclidean Geometry" in 1871 and 1873, and more schematically in the Erlangen Program. The "so-called" was to head off criticisms from philosophers. As Klein wrote about the effect of the article of 1871 toward the end of his life:[3]

> With this report I stirred up many contradictory opinions, initially from the philosophical side. No less than Lotze had already provided the cue, that any non-Euclidean geometry was a nonsense. To this came an ineradicable misunderstanding, which still plays a role today among philosophers and popular writers, and that I therefore cannot leave unmentioned. It is unfortunately connected to the entirely intuitive expression "measure of curvature". [Klein defined it, denoted it K, and observed that K is constant in non-Euclidean space.] This purely immanent mathematical theorem was given a transient meaning by philosophers and mystics in an entirely inadmissible way, as if space was endowed with in some intuitively concrete property. In this connection much was written and squabbled over, since space necessarily needs a new dimension in order to be "curved." The Mathematical Society in Göttingen participated in such discussions for a year. Compare Blumenthal's verse:
>
>> Die Menschen fassen kaum es
>> Der Krümmungsmass des Raumes.
>
> [Literally, People can scarcely grasp the measure of curvature of space. Perhaps one might allow:
>
>> People cannot keep ideas in place
>> About the curvature of space.]
>
> All these distorted expressions, now for us, now against, were of consequence: they caused us many great difficulties.[4]

To realize a description of non-Euclidean geometry within projective geometry, Klein considered the region inside a conic in the plane, and observed that to define the distance between two points A and B inside the conic, one could use the cross-ratio of the four points A, B, C, and D, where C and D are the two points where the line AB meets the conic. The cross-ratio of the four points is not altered by a projective transformation, and therefore not by a projective transformation mapping the conic to itself and its interior to the interior. It remained to find a way of making the cross-ratio define a distance.[5] Once this was done, Klein had shown how to define projective geometry using only projective ideas (and some deeper assumptions of a topological kind that no one could articulate at the time).

---

[3] Klein 1926–27, 152–153.

[4] Harman 1982, 96, reproduces a postcard from Maxwell to Tait written in 1874, after Clifford had published his translation of Riemann's paper, in which Maxwell disagrees with the "space crumplers."

[5] Klein defined the distance between two points A and B inside the conic in terms of the logarithm of the cross-ratio of the four points A, B, C, and D.

Klein took great pleasure in making non-Euclidean geometry a consequence of projective geometry, for he felt that geometry had become too fragmented. There was Euclidean geometry, non-Euclidean geometry, projective geometry, and a geometry to do with inversion in circles. Klein did not mention, and perhaps at that stage did not know of Möbius's affine geometry, but he took the opportunity to include references to his friend Sophus Lie's ideas. Klein belonged to a mathematical tradition that dislikes too much diversity and seeks to find a suitable, possibly abstract, unifying standpoint. As his creative powers declined, his unifying desire grew, and starting in 1894 he organized what became a twenty-three-volume *Encyklopädie der Mathematischen Wissenschaften (Encyclopedia of the Mathematical Sciences)*; the first volume appeared in 1904. This sought to provide a series of overviews of all of mathematics and its applications. Klein's rationale was that what Gauss had been able to do on his own at the start of the century now required a team to accomplish, but was no less necessary.[6]

Klein saw his Erlangen Program as providing an underlying unity for the fragmented discipline of geometry. It depended heavily on an innovative use of the group concept, which was not then widely known and which he had gone to Paris to learn.[7] Mathematicians had long used transformations of figures to replace a figure with an equivalent but simpler one, or to choose more convenient coordinate axes. Klein shifted attention from figures to transformations. These transformations were maps of the whole space to itself (for example, rotations of the plane, which turn it bodily in one direction by a given amount). So a motion of a figure in space is thought of as moving the whole space with it, much as mathematicians regard a rotation of the coordinate axes. Then he shifted attention again, from individual transformations to the class, or group, of all transformations. He presented a philosophy of geometry which said that it was about groups as well as shapes. On this view, a geometric property was one that was invariant under all the operations of the group associated to that geometry. So the cross-ratio of four points on a line is a projective property because cross-ratio is invariant under all projective transformations, that is, all the transformations of the projective group. Accordingly, it does not need to be built up out of lengths, which are not, indeed, projective invariants. This shift of focus resolved what had begun to seem like a paradox: projective geometry is more fundamental than Euclidean geometry, yet its basic idea, cross-ratio, seems to depend on the definition of length.[8]

Klein exploited the idea of one group being a subgroup of another. He recognized that one could start with a space and a group, say projective space and the group of

[6] The principal editor, Wilhelm Franz Meyer, marked the appearance of the first complete volume of the work with an essay (Meyer 1904) in which he endorsed the Kantian position that mathematics was an example of pure a priori knowledge and sought to argue that all the new mathematics since Kant's time only confirmed this idea. He made his way to the conclusion that mathematics was both a logical science and an aesthetic art that spun new conclusions out of the same fundamental perceptions that Kant had identified, by sticking to the superficialities of every topic he took up and avoiding every novelty and objection to the threadbare philosophy he espoused. Walther von Dyck, a former student of Klein, wrote a much more informative essay on how the *Encyklopädie* had come about for the opening volume.

[7] Klein's presentation, as he himself noted when the work was reprinted in the 1890s, was inadequate, inasmuch as the only property of a group that he insisted upon was closure.

[8] This was a nontrivial shift: Klein correctly wrote in 1921 that Cayley and some other mathematicians never got over their suspicion that this argument somehow contained a vicious circle. See Klein 1921, vol. 1, 242, and Klein 1926–27, vol. 1, 153.

all projective transformations. Then one could select a figure in a space and consider the subgroup that maps that figure to itself. In this way, one obtains a new geometry, which one may reasonably call a subgeometry of projective geometry. For example, by selecting a conic in projective space, Klein obtained non-Euclidean geometry as a subgeometry of projective geometry. This proved to be a fundamental way to interrelate geometries and so to encompass all of geometry in a hierarchy, with projective geometry as the mother geometry, and non-Euclidean, Euclidean, and inversive geometries as special cases. By December 1871 he could claim "that there are as many different ways to treat the study of a manifold as there are ways to construct continuous groups of transformations, and that the Euclidean and non-Euclidean metrics are just as certainly subsumed by the projective treatment as their 'groups' are contained in the entire group of projective transformations by a suitable choice of coordinates"[9]

The Erlangen Program was distributed as a pamphlet when Klein gave his inaugural address. Perhaps copies were also sent to individual mathematicians and libraries. But it was not well distributed, and it does not seem to have had much impact at the time. Nor did it significantly guide Klein's own work in the 1870s. Klein mostly worked on topics other than the projective hierarchy of geometry, until in the 1890s he found himself again at Göttingen. Now he was a senior professor with an established career, a reputation, and the ambition to build an empire. Foreign mathematicians visiting Göttingen were willing to translate his papers, and so the Erlangen Program was put into English, Italian, French, Polish, and Russian and presented as a prescient research program uniting geometry and group theory.

These two subjects had progressed considerably in the intervening twenty years, and had indeed grown closer together. Accordingly, the question of the influence of Klein's Erlangen Program has been much debated. The older view, often advanced by mathematicians such as Garrett Birkhoff, is that it was of great influence. Group theory is a staple of the modern curriculum, and many places that teach geometry teach the Kleinian view of geometry—the view that various different geometries, all the ones where we represent the transformations with matrices, form into a hierarchy. The more recent view, advanced by a number of historians, for example, Tom Hawkins and Erhard Scholz, with much better evidence is that the solid work establishing group theory between 1870 and 1890 was done by Camille Jordan, Sophus Lie, and even Poincarè without the Erlangen Program having any impact, while Klein, while he certainly liked to bring groups into his study of geometric problems, did not implement or advance the view of the program.[10]

3.1.2 Projective Geometry: Rigor, Duality, Novel Spaces, Novel Ingredients

Von Staudt's book of 1847 had given projective geometry its first truly autonomous foundations independent of Euclidean geometry. The only weak part of his work was

---

[9] Klein 1892a.

[10] For an interesting pair of contrasting views of the situation (from which papers giving the original, more strongly contrasting views can be found) one may consult the essay by Birkhoff and Bennett in Aspray and Kitcher 1988, 145–176 and Hawkins, 2000, 34–42.

his proof of what is called the fundamental theorem of projective geometry, which says that if a projective transformation maps three points on one line to three points on another, then its effect on every other point of the first line is determined. Klein drew attention to this problem, and a number of mathematicians addressed it.[11] The issue is what can be said about lines. In synthetic projective geometry from Poncelet to von Staudt, a line is a primitive concept immediately accessible to intuition. However, in the course of trying to prove the fundamental theorem, it turned out that there could be many kinds of line. Apart from lines with only finitely many points, there were lines with, one might say, too many points—many more points than a Euclidean geometer would suspect. These extra points could create a line in which the so-called Archimedean property does not hold. This asserts of two quantities A and B that if $A < B$ then there is a natural number n such that $B < nA$. The possibility of non-Archimedean projective geometries was explored by Veronese and then by Hilbert (see below, §4.1.2 and §5.3.5.). It was a costly success for the synthetic geometers because it destroyed belief in the truly fundamental nature of the line and its value as a primitive concept.

Elementary geometry was also finally starting to look insecure. The compliment paid to Euclid's *Elements* by mathematicians of all cultures, which consisted of putting right the imperfections that had turned up while regarding the whole structure as fundamentally sound, was ceasing to be paid. In the preface to his book, Moritz Pasch (1882, 2), who had studied under Weierstrass and Kronecker at Berlin in the 1860s and was by then a professor of mathematics at the University of Giessen, wrote: "Elementary geometry cannot only be reproached for its difficulties, but also for incompletenesses and obscurities, which the ideas and proofs still retain in extended measure. The repair of this defect is an incessant struggle, in manifold ways, and if one examines the results one can come to the opinion that the struggle is hopeless." His criticism of elementary, intuitive geometry from the standpoint of late nineteenth century criteria of rigor was typical. People were finding faults, so much so that this whole branch of introductory mathematics seemed riddled with flaws and perhaps beyond repair. It was to this anxiety that Pasch responded with his book.

Pasch chose to reformulate geometry by concentrating on the simpler case of projective geometry. He rejected the idea of reducing geometry to analytic geometry on methodological grounds, judging that proofs in analytic geometry are heavily arithmetical; but proofs in geometry are only occasionally so. He then took up plane projective geometry with the aim of formulating rigorously every fact that he felt was necessary, starting with the undefined or primitive concept of the straight line segment between two points. He listed the properties of line segments he felt were necessary to assume without proof under the heading of *Grundsätze* (Fundamentals). All these, he said, were immediately grounded in observation, and he based himself on Helmholtz's 1870 paper, "On the Origin and Significance of the Geometrical Axioms" at this point.[12] Eight Grundsätze were needed to establish the theory of line segments, starting with "there is always a unique segment joining any two points."

Pasch's program was to lay down Grundsätze until logical reasoning without further appeal to sense perceptions could produce enough other results as theorems, which Pasch called *Lehrsätze* (Theorems). Four more Grundsätze enabled Pasch to

[11] See Gray 2006c, appendix A.
[12] Pasch 1882, 17.

conduct plane geometry (the eponymous Pasch axiom—that if a line enters a triangle through a vertex it exits through the opposite side—is the fourth of these; p. 21); three-dimensional geometry followed in the same manner. Pasch presented rigorously the so-called points at infinity in projective geometry, or ideal points, to use Klein's term, by asserting a pencil of lines through a point and a pencil of parallel lines are equivalent.

Projective transformations and congruence, which relate to the fundamental theorem of projective geometry, were handled by the same division into Grundsätze and Lehrsätze. Pasch pointed out that his axioms for congruence permitted him to coordinate projective space using a projective net, much as Möbius and then von Staudt had proposed.[13] Pasch concluded his account by showing that all elementary projective geometry was now established: duality, geometry with respect to a fixed conic (including Klein's version of non-Euclidean geometry), and cross-ratio.

We have already seen that duality was an attractive feature of plane projective geometry, even in the otherwise unconvincing hands of Poncelet. Duality permitted mathematicians to exchange the words "point" and "line," "concurrent" and "collinear," and in this way obtain from one statement of a theorem and its proof the statement of another theorem *and* its proof. So striking was this that mathematicians often wrote their books in two columns when appropriate, one for one version of an argument and the other for its dual. They appreciated that this facility was a fundamental feature of plane projective geometry; Cremona even called it a logical fact. It was prior to any particular theorem in the subject.

There was also a principle of duality in three dimensions, but it was more subtle. As explained earlier (on §2.1.1) the geometric version of duality in the plane says: fix a conic, take a point, draw the tangents from this point to the conic, draw the line joining the points of tangency. This line is the dual (with respect to the conic) of the original point. The corresponding algebraic version says to the point with homogeneous coordinates $[a, b, c]$ there corresponds the line with equation $ax + by + cz = 0$. The corresponding versions of these arguments in three dimensions start similarly (and were known in the geometric version to Brianchon as early as 1806): fix a quadric, take a point, draw the tangents from this point to the conic, draw the plane through the points of tangency. This plane is the dual (with respect to the conic) of the original point. The corresponding algebraic version says to the point with homogeneous coordinates $[a, b, c, d]$ there corresponds the plane with equation $ax + by + cz + dz = 0$.

The catch question, avoided by mathematicians before Möbius, is: What about lines in space? The dual of a line turns out to be another line. Plücker's main research interest when Klein joined him was to study the space made of all lines in space. This is a slippery beast on first acquaintance. One way to grasp a line in space is to fix an origin O and a usual set of coordinate axes in space, to note that the line, let us call it ℓ, has a direction, given by a vector of unit length **d**, and a point, C on it which is its closest point to the origin. So to single out a line, one can start at the origin, go out to this closest point, C, and then face in the direction of the line.

How much information is needed for this? To get from the origin to the nearest point involves specifying a vector OC in space—three coordinates are needed. To

---

[13] Later mathematicians have preferred weaker axioms for which this is not always the case.

specify the direction of the line involves two more coordinates. But the vectors OC and $\mathbf{d}$ are at right angles to one another, because C is the closest point on ℓ to O, so there is an equation between the coordinates of OC and $\mathbf{d}$. The result is that it takes four coordinates to specify a line in space.[14]

Several significant points must be noted here. One is that a collection of familiar objects forms a four-dimensional space, a fact of great delight to the German educational reformer Rudolf Steiner.[15] Another is that the lines of one space have been taken as the fundamental entities (the "points," one might say) of another space. A third is that the study of lines in space took off as a subject. It had deep contemporary connections to the study of light as well as proving to be a challenging and rewarding subject of study.[16] We shall not follow that story here, but it is well to note that line geometry, as this subject became called, was a significant topic of research.

Another important use for spaces of higher dimensions was to mimic the key idea of plane projective geometry, which is that a shadow of a figure may be easier to understand than the figure itself, or, conversely, a complicated figure may be the shadow of a simpler one. Thus a curve in the plane that crosses itself may be the image of a curve in space that does not cross itself (as is the case with knots, for example).

In 1882 the Italian mathematician Guiseppe Veronese, who was visiting Klein at the University of Leipzig at the time, published a paper in which he generalized the method of projection to arbitrary dimensions. In this paper he showed that some surfaces with singularities could be studied by regarding them as projections of nonsingular surfaces in a higher-dimensional space. Since surfaces and their singularities are much harder to understand than the singularities of curves, this approach promised to be a real breakthrough, and so eventually it was, but not until the late 1930s. Veronese's best example of his new method was well chosen, however, because it concerned a surface known to be of interest to geometers: Steiner's Roman surface, so called because Jakob Steiner discovered it while in Rome in 1844. Veronese showed that it was a nonsingular surface in five-dimensional projective space, so its singular appearance in projective three-space was, as it were, a corollary of it being forced into three dimensions when it properly belonged in more.

Another use for higher dimensions that caught on is Plücker's idea of taking fully fledged geometric objects in one setting as the elements of another set. One final example must suffice, which is easier to explain if less potent in its effects. A conic in the plane is given algebraically by an equation of the form $ax^2 + hxy + gxz + by^2 + fyz + cz^2 = 0$. Evidently this defines nothing if all the coefficients a, b, c, f, g, h vanish. Equally evidently, the same locus is defined by the given equation and any equation of the form $kax^2 + khxy + kgxz + kby^2 + kfyz + kcz^2 = 0$, where k is any non-zero number. Indeed, one may say that a conic is defined uniquely by the ratio of the six numbers $a:b:c:f:g:h$. But this is merely to say (to a nineteenth-century projective geometer) that a conic is defined uniquely by the point $[a, b, c, f, g, h]$ in five-dimensional projective space. In this setting one can discuss the set of curves that are really lines (for example, where $a = b = c = 0$) or parabolas, and give answers in terms of shapes in five-dimensional projective space. In the jargon of the time, the

---

[14] The space of all lines in projective three-dimensional space turns out to be a subset of five-dimensional projective space but is not a projective space itself.

[15] Rudolf Steiner (1861–1925) was no relation to the mathematician Jakob Steiner.

[16] As described in Atzema 1993.

idea of making up projective spaces whose points were themselves lines or curves in some other space was known as changing the space element: the point or element of one space might also stand for a whole object in a simpler space.

The point is not to follow the mathematics beyond this point, but to note what is going on. Even a nineteenth-Century mathematician found five-dimensional space hard to think about, and did not feel comfortable with thinking of the same object as, at the same time, a conic in one space and a "point" in another. They dealt with this difficulty by regarding the higher-dimensional space abstractly. Arguments about it should be governed by the rules that define a four- or five- or higher-dimensional projective space. This put pressure on mathematicians to define such a thing suitably, and then to allow that these abstract things were open to many different realizations.

The mathematicians who took the lead in this exploration of higher-dimensional projective spaces were mostly Italian, and Corrado Segre was prominent among them. Segre graduated from Turin in 1883 at the age of twenty with a thesis on line geometry and higher-dimensional spaces that was much influenced by Klein's ideas and had a significant influence on the Italian mathematical community. It opens with these words:

> The geometry of n-dimensional spaces has now found its place amongst the branches of mathematics. And even if we consider it aside from the important applications to ordinary geometry, that is even when the element or point of that space is not considered as a geometrical element (and not even, which amounts to the same thing, as an analytical element made of values of n variable quantities), but as an element in itself, whose intimate nature is left undetermined, it is impossible not to acknowledge the fact that it is a science, in which all propositions are rigorous, since they are obtained with essentially mathematical reasoning. The lack of a representation for our senses does not matter greatly to a pure mathematician. Born, as it were, out of Riemann's famous work of 1854 . . . n-dimensional geometry develops along two separate lines: the first one deals with the curvature of spaces and is therefore connected with non-Euclidean geometry, the second one studies the projective geometry of linear spaces . . . and in my work I am to focus on the latter. This path opens for keen mathematicians an unbound richness of extremely interesting research.[17]

In due course, Segre's views were to lead him into conflict with his Turin colleague Guiseppe Peano, so it is worth articulating their differences. Segre appreciated that the key to a rigorous theory of higher-dimensional projective spaces was a rigorous theory of higher-dimensional vector spaces, and such a thing was proposed in full detail by Peano (1888) only a few years later. It was more rigorous than Segre's version, which was marred by trivial mistakes, but as Avellone et al. point out, it was also presented as a synthesis of preceding work, and it came at the end of Peano's paper. It was not presented as a basis for further work, whereas Segre was making an analysis of these spaces fundamental to a whole program of future work. Indeed, by 1891 Peano turned his back on geometry in more than three dimensions because it connected to no practical hypotheses.[18] Rigor, or its lack, was not a problem; the two divided over the purpose or value of the enterprise. Segre set out the basic tools for work on the geometry of higher dimensions and then took up the study of algebraic

---

[17] Quoted from Avellone et al. 2002, 374–375; see also Segre 1884, 3.
[18] See Avellone et al. 2002, 384.

curves and surfaces from that point of view, hoping to elucidate the difficult study of their singularities. He did not succeed completely, but he gave mathematicians a reason, practical enough on their own terms, for taking his lead. The abstract geometry that Peano and his school went on to develop was very different, as we shall see.

Segre's finest student, Federigo Enriques, was only twenty-two when, in January 1894, he began to lecture on projective and descriptive geometry at the University of Bologna. Like his mentor, he combined a modern attitude to the fundamental terms in geometry with a sensitivity to their historical origins and to how they can be understood: "As to those intuitive concepts, we do not intend to introduce anything other than their logical relations, so that a geometry thus founded can still be given an infinite number of interpretations, where an arbitrary meaning is ascribed to the elements called "points." [However,] we think that the experimental origin of geometry should not be forgotten while researching those very hypotheses on which it is founded."[19] Or, to quote from the appendix to his *Lezioni di geometria proiettiva* published four years later:

> We have tried to show how projective geometry refers to intuitive concepts, psychologically well defined. . . . On the other hand, however, we have warned that all deductions are based only on those propositions immediately inferred from intuition, which are stated as postulates. From this point of view, the geometry that we have presented, looks like a logical organism, where the elementary concepts of point, line and plane (and those defined through these) are simply elements of some primitive logical relations (postulates) and of other logical relations that are inferred. The intuitive content of these concepts is totally irrelevant. This observation originates from a very important principle that affects all modern geometry: the principle of replaceability of geometrical elements. [. . . .] projective geometry can be considered as an abstract science, and it can therefore be given interpretations different from the intuitive one, by stating that its elements (points, lines, planes) are concepts determined in whichever way, which verify logic relations expressed by the postulates. A first corollary of this general principle is the law of duality of space.[20]

The theme of intelligibility was one Enriques was to develop in later life. It is one he shared with Poincaré and the other great popularizers of mathematics of the period, and we shall return to it (see below, §5.4.4). However awkwardly it sits with the tendency to present projective geometry as a ruthlessly abstract subject, it should not be dismissed as an attempt, well meant but doomed, to soften the blow. Enriques felt deeply that both aspects of geometry were essential to it: the logical structure, and the intuitive feel. Proclaiming them as he did is a mark of shock of the transition from the old to the new.

Nor was Enriques alone in this, for his position was close to Klein's. Klein had always argued for the intuitive in mathematics; his touchstone for understanding something was to make it geometrical (even his few forays into number theory emphasized the geometrical features of the subject) and he advocated geometry as a way to be creative in mathematics. By the 1890s he was the leading professor at Göttingen, eager and able to advance mathematics on his terms, which were not that

---

[19] Bottazzini, Conte, and Gario 1996, 142, n. 5.
[20] Enriques 1898, 347–348.

different from those he had proclaimed twenty-five years earlier at Erlangen. This helps explain why a German translation of Enriques's book came out in 1903, with a foreword by Klein himself. Klein praised the book in these terms:

> Over the last two decades Italy has been the true centre of advanced research in the field of projective geometry. Among the specialists, this is well known. . . . But the Italian researchers have gone far beyond also on a practical level: they have not disdained to draw some didactic conclusions from their own studies. . . . And this is all the more desirable in Germany since our didactic literature has lost all touch with recent research achievements. . . . The presentation is always intuitive, but completely rigorous, as it could only be after the clever researches on the foundations of projective geometry presented in earlier essays by the same author.[21]

Klein did not have to promote Enriques's book; there were, as he noted, other Italians who were much more fiercely abstract. But Enriques, a young man of the next generation, shared many of Klein's views. As we shall see, Enriques was to be drawn further into Klein's great schemes for things, and the abstract tendency was not. Battle lines were being drawn.

3.1.3 Non-Euclidean Geometry

In 1875 Poincaré graduated from the École Polytechnique—he only came second, brought down because his inability to draw cost him marks on descriptive geometry.[22] Even so, he could enter the finest of the specialist schools, the École des Mines in 1875, embarking on a path to a teaching career taken earlier by some of the professors Poincaré most admired, including Camille Jordan and Alfred Cornu.[23] It was at the École des Mines in 1878 that he presented his doctoral thesis to the faculty of Paris. It was on the subject of partial differential equations, and although Darboux said of it that it contained enough ideas for several good theses, he also judged that it should be much more rigorous, and he urged Poincaré to tighten it up. Poincaré decided not to bother. The thesis permitted him to give a course in analysis at the Faculté des Sciences at Caen, and he was officially released from his duties as a mine inspector on December 1, 1879.

The next year was a busy one for him. The current prize competition of the Académie des Sciences in Paris asked for any essay improving some point in the theory of differential equations.[24] As was the custom, the closing date was two years after the problem was announced, and in this case Charles Hermite, who had set the problem, was chairman of the panel of judges. Hermite and Bertrand, the permanent secretary of the Académie des Sciences since 1874, between them more or less controlled appointments in French mathematics; they could decide who was to

---

[21] Enriques 1902, iii.

[22] I am indebted to Scott Walter for the information that this story probably derives from Appell 1925, 28, who also suggested that visionary reasoning may have been a cause. Bellivier noted in his biography (1956, 150), that Poincaré lost crucial points in topography, drawing, and architecture.

[23] Jordan was an analyst and the leading figure in the development of group theory in the 1870s. Cornu was a French physicist of the generation before Poincaré, who did important work on optics and spectroscopy, and the determination of the speed of light.

[24] On prizes in mathematics, see Gray 2006a.

get the call to Paris and who was to languish in the provinces. Poincaré had already submitted one entry to the competition, but on June 14, 1880, he withdrew the first essay in favor of another, presumably to concentrate on the theme of the second one.

As the year progressed, Poincaré added three lengthy supplements to his essay, each marking significant progress. He did not win the competition, and was only awarded second prize.[25] He did get the call to Paris, however, and went to the University of Paris in a junior position in 1881; here he became professor of mathematical physics in 1886. He was elected to the geometry section of the Académie des Sciences in 1887, the first of his generation of mathematicians, and he was elected to the Académie Française in 1908.

What had captured Poincaré's interest was non-Euclidean geometry and the role he saw it could play in the study of functions of a complex variable, and so in the theory of differential equations. The moment of discovery has even come down to us; it is the celebrated occasion when Poincaré boarded a bus at Coutances while on a field trip for the École des Mines. Poincaré was to describe the circumstances in detail in 1908 when he talked to the Société de Psychologie in Paris on the psychology of discovery in mathematics, and those aspects will accordingly be discussed below (see §6.2.1).

The mathematical aspects are equally interesting (see §6.2.1). Poincaré had begun to work for the prize competition with the example of a well-known, important differential equation. His approach led him to study a network of triangles, which, at least in particular cases, might grow in a simple fashion with no awkward overlaps. This was difficult to prove because the triangles were of different shapes, but the growth of the triangles determined the nature of the functions that solved the differential equation (and which he called Fuchsian functions for reasons that need not concern us). Almost certainly what happened as he boarded the bus was that he saw a mental picture of these triangles but for the first time connected it to the Beltrami disk in non-Euclidean geometry. The result was that what had been a difficult process to follow, as one triangle gave rise to another, became much easier: each new triangle, he could now see, was congruent to its predecessor. Congruent, that is, in the sense of non-Euclidean geometry: corresponding sides have the same non-Euclidean length, corresponding angles the same size. As he put it in 1908: "At the moment when I put my foot on the step the idea came to me, without anything in my previous thoughts having prepared me for it; that the transformations I had made use of to define the Fuchsian functions were identical with those of non-Euclidean geometry."

This realization dominated the work on the supplements he went on to write. As he said in the first one:

> There are close connections with the above considerations and the non-Euclidean geometry of Lobachevskii. In fact, what is a geometry? It is the study of a group of operations formed by the displacements one can apply to a figure without deforming it. In Euclidean geometry the group reduces to rotations and translations. In the pseudogeometry of Lobachevskii it is more complicated.

---

[25] Halphen won the prize; see Gray 2000a. The supplements themselves were lost for a long time and only recently republished, in Poincaré 1997. In fact, neither Poincaré's thesis, nor the essay withdrawn in June, nor the second-prize essay was published until the first volume of Poincaré's *Oeuvres* (1928, 578–613).

> To study the group is therefore to do the geometry of Lobachevskii. Pseudogeometry will consequently provide us with a convenient language for expressing what we will have to say about this group.[26]

It is remarkable how clear Poincaré was in his own mind that to do geometry is to concern yourself with a group. This does not seem to be an idea he got from Klein, because it seems that he did not know Klein's work at the time—Poincaré was not, in any case, a very well read mathematician. Crucially, Poincaré connected geometry with groups of isometries, not with projective transformations, whereas Klein saw all geometry at this time through a projective lens. Group theory was well developed in Paris, where Camille Jordan taught, and it is likely that Poincaré picked it up there and then, of his own accord, gave it pride of place in geometry.

In due course, Poincaré published his ideas. He began to rework the essay, the supplements, and his subsequent discoveries as short papers for the *Comptes rendus*, and then the ambitious Mittag-Leffler signed him up to write the full-length versions for his new journal, *Acta Mathematica*. The first of five long papers occupied sixty-two pages of the first volume, and Poincaré signed off in 1884, leaving the field for others. What this shows is the huge appreciation the mathematical community had for his yoking together complex function theory (a central domain of contemporary mathematics), group theory, and the new non-Euclidean geometry. This was the moment at which non-Euclidean geometry paid its way; its mathematical value was thereafter never in doubt among mathematicians.

3.1.4 The Helmholtz-Lie Space Problem

What is called the Helmholtz-Lie space problem, or sometimes the Riemann-Helmholtz-Lie-Poincaré space problem, is the question: What possible geometries can there be (for a space of a given dimension) if the geometric properties are those that can be determined by the motion of rigid bodies? Helmholtz, taking his cue from Riemann's recently published *Lecture*, raised this question in 1868 and answered it in his lecture "On the Facts Underlying Geometry."[27] Helmholtz argued as follows: it must be possible to move any point of a given rigid body anywhere in space. Once one point of the body is fixed, the body must be able to rotate freely around the fixed point. These motions must preserve distances (because the body is rigid), and this led Helmholtz after some more detailed work to the most general possible expression for a Riemannian metric satisfying these constraints. In particular, it had to correspond to a space of constant curvature. He therefore claimed that in two dimensions, space must be either Euclidean or spherical.[28]

Beltrami immediately wrote to Helmholtz to point out that the possibility of non-Euclidean geometry had been overlooked, and Helmholtz agreed. This was the start

---

[26] See Poincaré 1997, 11.

[27] See Helmholtz 1868.

[28] In view of the striking similarities between these views and those of Poincaré, it is natural to suppose that Helmholtz's paper was a source for Poincaré's later ideas, but it seems there is no evidence that clinches the matter. It could even be that Poincaré's education brought him to Beltrami's *Saggio* or even Riemann's *Hypotheses* first.

of his interest in the significance of non-Euclidean geometry for any kind of Kantian philosophy of geometry. Some years later, Klein, who was an enthusiast for Helmholtz's work, urged his friend the Norwegian mathematician Sophus Lie to take up the question. A successful career had brought Lie to Leipzig, one of the major German universities. He had spent a number of years elucidating the ways in which, to lapse into technical language, groups can act on spaces. Put another way, Lie had begun to classify all the ways in which groups of transformations can be constructed and so to show how the idea of rigid body motion could be made into a mathematical one. Klein's own thinking about the question in the 1880s had isolated Helmholtz's assumption, in addition to the points mentioned above, that when two points are fixed, a rigid body in a three-dimensional space will return to its original position under motion about the axis they define (the so-called axiom of monodromy). Klein wondered if this axiom was necessary,[29] but he seems not to have known that he had kindled Lie's interest, and his lectures, which included a discussion of the Helmholtz space problem,[30] made only a brief mention of Lie's contribution. Lie felt he had been slighted, and the result was a painful controversy.[31]

Lie had indeed been able to work through Helmholtz's analysis with greater precision, and he found it erroneous. He published his own analysis of the problem in a lengthy section of the third volume of his *Theorie der Transformationsgruppen* in 1893, in which he showed that the monodromy axiom was superfluous, except in two-dimensional geometry, where it is essential. But it had also been expressed ambiguously by Helmholtz, as it turned out, and Lie severely criticized the lack of rigor in Helmholtz's presentation. Unhappily, Lie was by now entering a period of some instability, perhaps brought on or exacerbated by the pernicious anaemia that would eventually kill him, and he found Klein's recent remarks on geometry to be shockingly ignorant of his own work.[32]

Accordingly, in the preface to his own book Lie directed several polemical remarks at Klein (and to make matters even worse at a number of other mathematicians), the most wounding of which was: "I am not a student of Klein's nor is the reverse the case, even if it comes closer to the truth . . . I value Klein's talent highly and will never forget that part he had in accompanying my scientific efforts from the beginning. I believe, however, that he does not always distinguish sufficiently between induction and proof, between the introduction of a concept and its utilization."[33] However justified this critique of Klein's mathematics may be, it is not at all clear that Lie's work was any better in this respect. Most contemporaries found Klein's work quite lucid, though, to be sure, they were often uncomfortable with the holes he left in his arguments. Lie, on the other hand, had a reputation for being almost impossible to read, and his proofs were hardly the latest word in rigor, either. But be that as it may, the damage done by Lie's remarks was not to be undone, and his break with Klein was now out in the open for all to see.

Klein rallied remarkably well. Most significantly, in 1897 he wrote a report for the Kasan Scientific Society on the occasion of the first granting of the Lobachevskii

---

[29] See Klein 1890, 564.
[30] Klein 1892–93, 275, 353.
[31] This and the next two paragraphs draw on helpful conversations with D. E. Rowe.
[32] For a thorough discussion of this issue, see Stubhaug 2002, part 6.
[33] Lie 1893, 17.

Prize, later published in *Mathematische Annalen*.[34] Writing about Lie's work in the third volume of his *Theorie der Transformationsgruppen* on the Riemann-Helmholtz problem—the original source of their personal conflict and the very work in which Lie had sought to publicly humiliate him—he said that it not only met the formal conditions set by the society, it also "stands out so exceptionally among all other works that might possibly be compared with it, that there can hardly be a doubt as to the granting of the prize." He praised the extraordinary thoroughness and precision with which Lie treated the Helmholtz space problem, noting that it was a logical consequence of Lie's long and continuous work in the area of geometry. There can be little doubt that this glowing praise from an authority of Klein's stature virtually clinched the Lobachevskii Prize for Lie.

Lie's earlier contributions to what now became known as the Helmholtz-Lie space problem had also proved to be crucial to Poincaré's later thinking about geometry.[35] Poincaré (1887) argued that there are not very many geometries that can be candidates for the geometry of space. In fact, he said, Lie had shown that there are at most eight possible transformation groups, and after excluding five of these because they describe significantly counterintuitive properties, Poincaré narrowed the number down to the familiar three (Euclidean, spherical, and non-Euclidean). Poincaré was particularly pleased to see that an analysis of groups had implications for an analysis of space for, as noted elsewhere,[36] he always felt that groups came before spaces, precisely because our knowledge of space was derived from our knowledge of the behavior of rigid bodies. Indeed, for Poincaré all talk of space reduced to talk about rigid bodies. As he put it: "If, then, there were no solid bodies in nature there would be no geometry."[37]

3.1.4.1 SPACE FORMS

Klein's interest in the Helmholtz-Lie space problem had been quickened by his knowledge of the work of Clifford in 1873. There he described a surface with a novel geometry that in small regions is identical with small regions of two-dimensional Euclidean space, but globally has the shape of a torus. This had caught Felix Klein's attention because it was an example of a geometry in Riemann's sense of the term, which was at once very simple (being "locally" Euclidean) and quite complicated (being a torus, not a plane). He came back to it some years later in correspondence with Wilhelm Killing, a former student of Weierstrass's in Berlin, who had become interested in the foundations of geometry and the work of Sophus Lie.

Killing objected to Clifford's space on the grounds that in the usual examples (say, of a plane or a sphere) a rotation of a rigid body extends to a rotation of the whole space, but this is false in Clifford's geometry. Accordingly he rejected it because it did not "satisfy our experience," as he put it. Only after Lie's resolution of the Helmholtz-Lie space problem did they come to agree that the proper resolution was to

---

[34] Reprinted in Klein 1921, vol. 1, 1921, 384–401, see 384–385.

[35] Heinzmann (2001) makes the interesting point that at this stage Poincaré seems not to have known the work of Helmholtz, to which, philosophically, he was very close.

[36] See Poincaré 1898a, §22.

[37] Poincaré 1902b, 61.

regard constant curvature as a purely local condition and the free mobility of figures (on one interpretation, at least) as both a local and a global consideration.

Killing then took up the question in earnest and gave a lengthy analysis of how to produce globally different spaces with the same metric of constant curvature. The new spaces he called *space forms*, and he presented them as examples of possibilities for physical space, thus raising the idea that physical space may have a nontrivial topological structure for the first time.

In his paper of 1892 and his subsequent book, also called *Ueber die Grundlagen der Geometrie* (On the Foundations of Geometry) Killing engaged with both the mathematical and the philosophical aspects of the subject. Geometry, he began, like any system of ideas, proceeds from some assumptions. The fundamental ideas are those that cannot be reduced to any others, and they must be necessary for geometry, minimal in number, and capable of generating all geometry. The basic judgments or theorems are the basic facts (Grundsätze) of geometry, and they are rigid body, part of a body, space, part of a space, to occupy or cover a space, rest, and motion.

This differs little from what d'Alembert had written in the great French *Encyclopédie* a century earlier. Then, in a way that recalls what Pasch had done a decade earlier, he set out his Grundsätze in §2 of his paper, and deduced consequences from them. The Grundsätze are unremarkable: the first says that two bodies cannot occupy the same space; the seventh, for example, that if a body moves and part of it returns exactly to where it was before, then all of it does. The first seven Grundsätze alone implied that geometry essentially reduced to the theory of finite-dimensional transitive transformation groups, a fact that struck Killing as remarkable. He proposed that the subject so delineated be called "generalized geometry," and its branches "space forms in a general sense," as opposed to the proper space form that one obtained on adding Grundsatz number eight. The singular eighth Grundsatz says that if a point of a solid body in *n*-dimensional space remained at rest while the body moved, then no point of the body could sweep out an *n*-dimensional region of space.

All of this allowed Killing, after many pages of work, to show that on his definition of geometry the only candidates for the proper space forms were those with constant sectional curvature, zero, positive, or negative. Killing concluded his paper by exhibiting the proper two-dimensional space forms, which he showed were precisely the usual cases. He noted that Lie had classified the groups many years earlier and all that remained was to check that all the corresponding spaces could be real (rather then only definable over the complex numbers).

Another to move in these waters was the American mathematician Frederick Woods at MIT.[38] Woods took an approach similar to Killing's. A three-dimensional geometry should "accord with the facts of experience within the limits of observation."[39] This meant, he said, that it should be a Riemannian manifold admitting mobility of bodies, which he interpreted unusually as being a statement about transitivity of motions preserving geodesics. He concluded that the appropriate spaces were those of constant curvature, but, following Killing, he observed that the full story required an analysis of the possible discrete subgroups of the full group of motions. He only commented, however, on the parallelepiped with opposite faces

[38] See Woods 1902 and his Colloquium Address to the American Mathematical Society in Boston in 1903, Woods 1903.
[39] Woods 1903, 31.

glued together as a locally Euclidean space that is not, globally or topologically, Euclidean space but a flat three-dimensional torus.

A number of obscurities attended Killing's work that were not cleared up until the later introduction of topological ideas by Heinz Hopf in 1925, and the resulting picture is that the familiar geometries give rise to the new ones in a systematic way that also goes back to some other remarks of Klein's. An example will make clear what is meant. If a plane floor is covered by rectangular tiles, one can produce a space form by imagining the opposite sides of the rectangle are glued together. This produces a torus or inner tube. It is important that the edges are glued together in the imagination; this ensures that locally the geometry on the new space is Euclidean. If the tile is physically bent, the resulting geometry has variable curvature and is not wanted in this context. Klein had noticed that one can do this with figures in non-Euclidean geometry, and Hopf showed that the same family of ideas can be made to work in any number of dimensions. One starts with the sphere, Euclidean plane, or the non-Euclidean plane (or their analogs in higher dimensions), one then produces the analog of a tile (which for the mathematician is a group of motions moving the "tile" about) and glues the corresponding parts of the tile together in the imagination.

On the one hand, this work, and especially the abstract topology that Hopf brought in, seems impeccably modernist. But it is worth noting, as Epple[40] has pointed out, that Killing and Klein were both aware that the new spaces—space-forms, as they came to be called—might have implications for the nature of actual space. Klein in particular drew Schwarzschild's attention to the issue when the younger man was working on astronomical determinations of the curvature of space. Schwarzschild was persuaded and added an appendix to his paper suggesting how a nonstandard space-form might be detected through multiple repetitions of unusually characteristic objects. Mathematics might well have become abstract, but this did not mean it had become irrelevant.[41]

Mention should also be made of Bianchi's (1898) impressive classification of three-dimensional geometries, in which he used Lie's classification of the groups that can act on three-dimensional spaces to describe all the metrically different spaces there can be. These turned out to include several examples of spaces other than the familiar three (Euclidean, non-Euclidean, and spherical). The paper was a significant influence on Woods. Much later it was rediscovered by Abraham Taub (1951) and gave rise to the Bianchi cosmologies of modern general relativity theory. Even later, in the 1970s, the Bianchi geometries were again independently rediscovered by Thurston among his eight model geometries in three dimensions.[42]

3.2 Modern Analysis

3.2.1 What Are the Real Numbers?

It is once again Poincaré who set out most clearly where debates about the real numbers were to divide mathematicians and scientists. He began his essay "La

---

[40] See Epple 2002, upon which this section is largely based.
[41] See Schwarzschild 1900, 337.
[42] Thurston 1997.

grandeur mathématique et l'experience" ("Mathematical magnitude and experiment," in 1902a) with the blunt statement that "it is useless to appeal to geometry." Rather, it is the pure analyst (he cited Jules Tannery as a leading example) who has the answer to the question about the nature of the continuum, for he has "disengaged mathematics from all extraneous elements." The analyst, Poincaré explained, starts with the integers and intercalates more and more elements: first the rational numbers, then various kinds of irrational quantities. The Kantian antinomies, although Poincaré did not call them by that name, are avoided because "the real mathematical continuum is quite different from that of the physicists and from that of the metaphysicians." In particular, the physical continuum, as Fechner's work had shown, obeys the formulas $A = B$, $B = C$, $A < C$, which "is an intolerable disagreement with the law of contradiction, and the necessity of banishing this disagreement has compelled us to invent the mathematical continuum."

Poincaré went on that mathematicians succeed because they study not objects but only relations between objects. The objects may be changed provided the relationships do not: "Matter does not engage their attention, they are interested by form alone."[43] So the irrational numbers have a most unexpected definition, and Poincaré gave the one due to Dedekind, which will be described below.[44] This strange object, the mathematical continuum, is not obtained simply by putting finitely many new objects into each gap between two rational numbers, and then finitely many new objects between each gap that remains, and so on—as it were, going from centimeters to millimeters and so on. That would suffice for physicists, constrained as they are by the power of their senses and their instruments, but gaps would remain. The mathematical continuum has the property that any small piece of it can be magnified to resemble the whole, and to ensure this it is necessary to intercalate infinite sets of new elements. Although Poincaré did not use the term, he had in mind here that the mathematical continuum shall have no gaps; the irrational numbers as defined by Dedekind fill the gaps between the rational numbers.

What makes Poincaré's account particularly interesting is that he did not stop, as Dedekind and Weierstrass did, but enlarged the discussion by observing that the mathematician is not confined to this simple continuum but may define higher orders of infinitesimals, as du Bois-Reymond had shown (see the discussion below in §5.3.3). Then he added that if a somewhat blurred straight line and a curve are taken as examples of physical continua, their blurred regions may always overlap without the line and the curve intersecting. This is true to any level of precision, but it does not mean that the curve has the line as a tangent. On the contrary, mathematicians had shown how to construct curves that never have a tangent.[45] Faced with the clash of analysis and intuition, Poincaré wrote in one of his striking phrases, that "instead of endeavouring to reconcile intuition and analysis, we are content to sacrifice one of them, and as analysis must be flawless, intuition must go to the wall."[46] Intuition was rejected, but the claim is only that analysis *should* be perfect.

---

[43] Poincaré 1902a, 47, English translation, 20.

[44] Bizarrely, Poincaré attributed it to Kronecker throughout the essay.

[45] Indeed, Poincaré had been among the first to find and study such curves.

[46] English edition, 30. The original French may not have been so striking: "c'est á l'intuition que l'on donné tort' (58).

These observations are interesting because one concept that disappeared from analysis under Weierstrass's influence was that of the infinitesimal. It was first shut out by his creation in the 1860s of a concept of real number, and the identification of ordinary magnitudes with real numbers. This meant that the concept of a magnitude could be replaced by that of a real number, and the concept of a variable magnitude could be replaced by that of a real variable. In the event, Weierstrass's way of doing this was not adopted, and other ways of achieving this goal were preferred, but Weierstrass did identify the main features of the problem and its solution. He was clear that a system of "numbers" must be created that exactly captured the intuitive properties of finite magnitudes, and that they must be created out of rational numbers, whose existence and properties were taken to be more or less unproblematic. That is to say, positive whole numbers were collections of identical thought-objects, positive rational numbers were finite collections of units and aliquot parts ($1/n$ where n is a integer). Irrational numbers were made up as infinite sums of these rational numbers, or rather, as appropriate equivalence classes of them.

The details of this construction need not concern us. Of more interest is this major achievement in Weierstrass's creation of a theory of real numbers: his proof that every bounded infinite set of real numbers has a limit point. This result (called today the Bolzano-Weierstrass Theorem) encapsulates a key difference between Weierstrass's ideas and those of Cauchy. Cauchy was vague on this point, out of a belief that nature provided such a limit point. Its existence was somehow built in to the concept of magnitude. Weierstrass, who had constructed the real numbers out of more primitive concepts, knew that he had to prove that his creation had this fundamental property, and that is what he did. As Cantor (1883, 566) and Jourdain (1915, 14) pointed out, Weierstrass was the first to avoid assuming that the limits exist that are taken to define real numbers.

One may ask if Weierstrass was creating a class of objects where previously there had been none (but only a comforting illusion), or if he thought of himself as providing an improved definition of what was present in the world. On the first account he would say there had simply not been a foundation for the calculus. On the second account there had been foundations he was now making more precise and more amenable to the new level of mathematical precision. It seems that Weierstrass's answer would have been the second, with the proviso that it was only the new, mathematically precise system of numbers that could enter into proofs. Mathematics was no longer connected to intuition at the level of continuous magnitudes, but at the more basic level of the rational numbers, or even at the level below that, of the positive integers.

Weierstrass's ideas provoked quite some debate in Berlin. As early as 1870 Kronecker was moving toward a position critical of all talk of real numbers. Schwarz wrote to Cantor that Kronecker regarded the Bolzano-Weierstrass principle as an "obvious sophism" and expected to be able to produce functions "that were so unreasonable that, despite satisfying all of Weierstrass's assumptions, they would have no upper bounds." Schwarz and Cantor, then at the start of their careers, defended their mentor Weierstrass's views as indispensable for analysis, and indeed if Kronecker ever did seek such functions he failed to find any.[47]

---

[47] See Ferreirós 1999, 37

Weierstrass's clumsy construction was replaced by two others, one due to Dedekind, the other to Cantor (and Heine, independently). Dedekind came to his in 1858. This is his famous idea of what are today called Dedekind cuts. A cut, as he called it, is a division of the set of all rational numbers into two disjoint sets, such that every rational number in one set is less than every rational number in the other set. It is convenient to call two such sets L for "left" and R for "right," with every l in L less than any r in R. Dedekind showed how to consider cuts as numbers, calling the necessary work time consuming but not difficult.

He distinguished between cuts for which there is a rational number q, say, in L such that $l \leq q < r$ for all l in L and all r in R (or, similarly, for a q in R) and those for which there is no such rational number. He identified the former kind of cut with the rational number q, while he defined irrational numbers as cuts of the second kind. For example, $\sqrt{2}$ is not a rational number, but it is defined by the cut that assigns the rational number q to L if and only if either q is negative or $q^2 < 2$, and assigns the rational number q to R if and only if q is non-negative and $q^2 > 2$. Each cut is taken to define a real number, and Dedekind then showed that if one attempts to define cuts in the set of real numbers, the cuts are always of the first kind, so no new numbers can be obtained by iterating the construction.

It is worth noticing that Dedekind's construction of the real numbers produces them as infinite sets. Shortly (see §3.2.2) we shall see that Cantor showed that the real numbers formed an uncountable infinite set. This meant that the insight into the nature of the real numbers as mathematical objects was no better than the theory of infinite sets of different sizes, then in its infancy and shortly to become problematic.

Dedekind insisted that the real numbers he had defined were creations of the human intellect, and he went on to insist that the same was true of the rational numbers and even the integers. On this basis he spelled out, more clearly than Weierstrass had done, what the relationship was between the real numbers he had defined and magnitudes along a line. Dedekind saw that if real numbers and magnitudes were in a one-to-one correspondence, then it was necessary that whenever a line was divided into exactly two sets, such that every member of one set lay to the left of every member of the second set, then it was necessary that there was exactly one point that marked the separation of the sets.

Dedekind said that his account of the real numbers was produced with a view to giving rigorous foundations of the calculus without resort to the intuitive but obscure concept of magnitude, but it was rushed into print when he heard that Heine and Cantor were planning on publishing their own accounts, which Dedekind thought were inferior to his own. Heine observed that many of Weierstrass's as yet unpublished results would remain dubious until the nature of the real numbers was clarified. Heine then gave his account, which began with a purely formal account of the integers and the rational numbers, and then defined real numbers as (equivalence classes of) certain sequences of rational numbers.[48] He then showed that repeating this construction did not lead to yet more types of number. Then he used his newly defined real numbers to establish that a continuous function on a closed bounded set is uniformly continuous.

---

[48] Heine 1872.

Dedekind (1872, 11) then compared his system of cuts with the real line, which he characterized by the fact that "every point *p* of the straight line produces a separation of the same into two portions such that every point of one portion lies to the left of every point of the other." The continuity of the line he expressed by the converse, that any division of the line into two parts with every point of one to the left of every point in the other is made by severing the line at exactly one point. But then, in a passage not often cited, he added that he had no proof of this claim, indeed, "I am utterly unable to adduce any proof of its correctness, nor has any one the power." For, "If space has at all a real existence it is not necessary for it to be continuous; many of its properties would remain the same even were it discontinuous. And if we knew for certain that space was discontinuous there would be nothing to prevent us, in case we so desired, from filling up its gaps, in thought, and thus making it continuous; this filling up would consist in a creation of new point-individuals and would have to be effected in accordance with the above principle" (p. 12). This profound open-mindedness about the nature of space in general and straight lines in particular goes right back to the influence of Riemann (and behind him, of Herbart).[49]

Cantor was a colleague of Heine's in Halle, and his theory was almost identical with Heine's, as indeed was that of the French mathematician Charles Méray, which, unknown to the German mathematicians, he had published in 1869. The Franco-Prussian War may well have played some part in this lack of communication. The approach may be contrasted with that offered by two other mathematicians, Hermann Hankel and Johannes Thomae. Hankel gave a purely formal account of the integers and the rational numbers (which, indeed, Heine took over) but he despaired of giving a reductive account of irrational numbers, because he felt there were simply too many ways in which they could be created. Instead, he was content to leave the concept of an irrational number rooted in the intuition of magnitude. Thomae, another student of Weierstrass's and also at Halle in the 1870s, published a book on the elementary theory of complex functions in 1880, the year after he arrived in Jena. Here, and again at greater length in the book's second edition, he married Weierstrass's theory of rational numbers with the Cantor-Heine theory of the real numbers. However, and perhaps without realizing just what it could mean to have Frege as a colleague, Thomae then gave a philosophical interpretation of the new, extended, number concept.[50]

All numbers other than the natural numbers he labelled signs, and regarded as entirely free of content. Their conditions of existence were simply that they had been obtained by abstraction from the rules of arithmetic. He spelled out this position at greater length in the second edition of his book, published in 1898. Arithmetic, he said, was a game played with empty signs according to certain rules. The same, he observed, was true of chess, but whereas the rules of chess are arbitrary, the rules of arithmetic are such that numbers can be described by simple axioms in an intuitive way, which permits their use in the study of nature. The formal point of view, he went on to claim, dissolves all metaphysical difficulties.

Thomae may have thought he had gone far enough to meet Frege's criticisms by agreeing with him on the definition of the natural numbers, but if so he could not

---

[49] See Ferreirós 1999, 135–137.

[50] As noted earlier, this account follows Epple 2003, who notes (302, n. 4) that Wittgenstein had some sympathies with Thomae's position, despite all of Frege's criticisms.

have been more mistaken. Frege set out, as he later put it, to destroy it once and for all. Much of the attack came out in Frege's *Grundgesetze* and more in an unpleasant exchange in the *Jahrsbericht der Deutschen Mathematiker-Vereinigung* between 1904 and 1906; by which time, of course, Frege's own theory lay in the ruins to which Russell had reduced it. The lesser of Frege's criticisms was that Thomae had not bothered to spell out the rules completely and carefully. The burden of his objections was that Thomae irredeemably confused nonformal and formal considerations by a foolish use of abstraction. Abstraction, Frege pointed out, can confuse sodium bicarbonate with arsenic—both are white powders. The formal becomes the nonformal, figures can become signs, signs are confused with what they signify, science becomes a game. Aware also of the merits of abstraction, Frege broke into verse:

> Abstraction's might a boon is found
> While man does keep it tamed and bound;
> Awful its heav'nly powers become
> When that its stops and stays are gone.[51]

Frege reminded his readers that as long ago as his *Foundations of Arithmetic* he had inveighed against the attempt to found number on abstraction, but that was at a time when "even someone like Weierstrass could utter a farrago of balderdash when talking about the present subject."[52]

An irony of this polemic is that Frege was far from possessing a satisfactory theory of the real numbers himself. Frege never espoused a logicist theory of geometry, or of the real numbers. He did not believe the real numbers were reducible to the natural numbers, nor did he seek to reduce them to geometric quantities. Instead, he held to the Newtonian view that the real numbers were ratios of quantities. Accordingly, he outlined a theory of quantity, which will be discussed below (see §5.3.7), as part of a discussion of measurement. But his theory was developed in the shadow of Russell's paradox, and eventually Frege retreated to a geometric theory of the concept of real numbers.

Frege aside, there was an emerging orthodoxy by the end of the nineteenth century, which regarded the whole of the calculus as an outgrowth of the number concept. This was the express view of Jules Tannery in his *Introduction à la théorie des fonctions d'une variable*. Tannery was a solid mathematician, and very influential as assistant director of studies at the École Normale. This book, which no less a mathematician than Émile Borel said taught him how to think, was written to substantiate the claim that "one can build up all of analysis with the notion of integer number and the notions relative to the addition of integer numbers; there is no need to appeal to any other postulate, to any other given of experience. The notion of infinity, of which there is no need to make a mystery in mathematics, reduces to this: after each integer number, there is another." In the second edition, Tannery increased the amount of set theory, and went over entirely to Dedekind's method for defining the irrational numbers "because it relies directly on the theory of sets."

It was also the view of Hobson, the Cambridge mathematician who first tried to bring rigorous mathematical analysis to Britain, that the right place to start was with

---

[51] Taken from Frege 1906a, in *Collected Papers*, 343.
[52] Frege 1906a, in *Collected Papers*, 345.

a discussion of number.[53] After dismissing Kronecker's approach as "a species of Mathematical Nihilism" (p. 23) he too adopted Dedekind's approach. Like Tannery, he saw no reason to give an account of integer numbers; nor did the young Couturat in his own book *De L'infini mathématique* (On Mathematical Infinity), which he modestly regarded as explaining the more elementary parts of Tannery's book for the nonexpert reader. His view, as a recent highly successful graduate of the École Normale, was that Tannery had shown that "analysis thus conceived is merely an extension (prodigiously complicated it must be admitted) of elementary arithmetic."[54]

It is therefore amusing, and typical of the modernist shift, that just as two theories of real numbers were proposed that removed doubt from the foundations of the calculus (Dedekind's and that of Méray, Cantor, and Heine), differences of opinion opened up about the integers. Weierstrass had taken them as almost unproblematic. Dedekind located them in the free activity of the human mind. Hankel and Thomae took them to be empty signs governed by particular rules. Before we enter the aggressive world around Frege, we can pause to ask why anyone would want to say what the natural numbers are. Why was it not going to be enough to say, with Gauss, that arithmetic was a priori, or with Kronecker, that the good Lord has made the integers, even if all else is the work of man? Why were so many mathematicians straying into philosophers' territory?

Reasons internal to the development of the calculus are indeed few to find. The most important are two that involve Cantor. First, his research into the difference between a function and its Fourier series representation had forced him into a highly novel and delicate examination of the real number continuum. Second, out of this analysis, which already challenged him to think very carefully about the nature of the real numbers, came his celebrated discovery of a theory of infinite sets. The discovery of sets of differing ordinal types and differing cardinalities showed him that the distinction between succession and size was important, which it is not for finite sets and finite numbers. Even more importantly, the very existence of sets of differing cardinality challenged him to say what precisely a cardinal number was. The highly contentious claim to have discovered transfinite numbers required at the very least a clear account of finite numbers, so that the new ones could legitimately be seen as sufficiently numberlike to merit their name.

The question "What are the natural numbers?" did not become interesting to mathematicians until Cantor started to suggest that his new infinite sets had numberlike properties. And even then, the link between finite and transfinite numbers only raises questions for those mathematicians who accepted it. On the other hand, it is rather hard to imagine mathematicians taking seriously the question, "Why does $7 + 5 = 12$?" To analyze why mathematicians found nature of the natural numbers an interesting question, we shall find it helpful to resume the earlier discussion of algebra and of algebraic number theory in particular. But before we do that, we should consider how Cantor came to his own ideas, and what they were.

---

[53] See Hobson 1907.

[54] Couturat saw his own book as being largely in agreement with Husserl's *Philosophie der Arithmetik*, which he read just as he was finishing his own; see the note on p. 331. Some indication of its lack of depth may be indicated by Couturat's refusal to separate the concepts of cardinal and ordinal number as Cantor had done, which he felt (see p. 334) would needlessly complicate his book.

FIGURE 3.2. Georg Cantor.

3.2.2 Cantor's Introduction of the Transfinite

Georg Cantor is famous for introducing the study of infinite sets into mathematics. They came in on the back of two discoveries that are strikingly free of technicalities, but it also necessary to consider his related work on other matters if the controversies surrounding his work are to be understood, and the degree to which they became emblematic of the Modernist project.

As early as 1874 Cantor showed that the rational numbers are countable, but the real numbers are not. He was to reprove this result with the now-famous diagonal argument again in 1891, and because the argument is so compelling, and so adaptable to other uses, it can be given briefly here. To say that a set is countable is to say that its members can be put in a one-to-one correspondence with the natural numbers. Let us defer the proof that the rational numbers are countable to the next paragraph. As for the real numbers, it is enough to consider just the real numbers between 0 and 1. Each can be written as a decimal, in the form $0.a_1a_2 \ldots a_n. \ldots$. Suppose, for a contradiction, that the real numbers are countable. Then they can be arranged in a list. Now read diagonally down the list, and construct a real number by changing the nth decimal place of the number in the nth position in the list, say, according to this rule: if the nth decimal place of the nth real number is not a 7, write 7, and if it is a 7, write 3. In this way, a decimal number is obtained that differs from the nth real number on the list in the nth decimal place, so the new number is not on the list. But if the real numbers have been listed, then the number must be there. We have a contradiction, and so the real numbers cannot be countable. This result was the first clue that there could be a theory of infinite sets of different sizes.

*To show that the rational numbers are countable, we use the fact that the union of a countable collection of countable sets is itself countable. The collection is countable, so it can be enumerated, and let the ith set be denoted U_i. This set is countable, so its *elements may be enumerated thus: $u_{i1}, u_{i2}, \ldots, u_{ij}, \ldots$. Now enumerate the elements of the union of the sets U_i thus: $u_{11}, u_{12}, u_{21}, u_{13}, u_{22}, u_{31}, \ldots$. Clearly every element in the union is counted, and counted exactly once, this way. So to show that the rational numbers are countable, it is enough to show that all the rational numbers greater than or equal to zero and less that one are countable, because the set of intervals $\ldots$, $-1 \leq x < 0$, $0 \leq x < 1$, $1 \leq x < 2$, $\ldots$ is clearly countable. To enumerate the rational numbers between 0 and 1, we proceed much as we did with the countable union of sets. We arrange the rational numbers according to their denominators in this fashion: $1/1$; $1/2$, $2/2$; $1/3$, $2/3$, $3/3$; and so on. Strike out repetitions, such as $2/4 = 3/6$, etc. (which $= 1/2$) and enumerate what remains: $1/1$, $1/2$, $1/3$, $2/3$, $1/4$, $3/4$, $\ldots$.*

Cantor had also shown, in a paper Kronecker tried to keep out of Crelle's *Journal* in 1878, that there is a one-to-one map from the unit interval onto the unit square. This challenges naive ideas about the dimension of geometric objects. So paradoxical was it that even Cantor on discovering it remarked, in the letter telling Dedekind of his discovery, that "I see it, but I don't believe it," but the proof is quite easy to sketch.[55] Every real number in the unit interval is, as before, a decimal of the form $0.a_1a_2 \ldots a_n \ldots$. Every point in the unit square has coordinates given by two decimal numbers of that form. Pair the decimal $0.a_1a_2 \ldots a_n \ldots$ in the interval with this point in the square $(0.a_1a_3a_5 \ldots a_{2n+1} \ldots, 0.a_2a_4a_6 \ldots a_{2n} \ldots)$. Pair the point in the square with coordinates $(0.a_1a_2 \ldots a_n \ldots, 0.b_1b_2 \ldots b_n \ldots)$ with this point in the interval: $0.a_1b_1a_2b_2 \ldots a_nb_n \ldots$ by, so to speak, riffle shuffling the numbers. In this case (and, up to a point, in the previous case) it is necessary to work a little harder to get around problems posed by the ambiguities in decimal notation caused by repeating nines, but Cantor showed how to do that.

Ingenious paradoxes though these results are, they would not have changed mathematics if this were all that Cantor had done. To understand his enormous, controversial, and for him painful impact, we must look deeper.

Cantor's work may be conveniently divided into two phases. As noted, it began in the early 1870s with a profound study of how a function may differ from its representation as a Fourier series, which led him to his transfinite symbols. Some ten years later, as his interests in philosophy deepened, he sought successfully to regard these infinite symbols as numbers, and to open up the theory of transfinite sets. This in turn led to the famous paradoxes of the infinite, which were to prove the end of naive set theory.

3.2.2.1 ORDINAL NUMBERS

In the first phase, Cantor was interested in the question of when the Fourier coefficients in the Fourier series expansion of a function are unique. So he supposed he had a function given by a Fourier series that converged at every point of an interval, say the interval $[-\pi, \pi]$. He let the function be $f(x) = \Sigma a_n \sin nx + b_n \cos nx$, considered another series, $\Sigma a'_n \sin nx + b'_n \cos nx$, and showed that if this series converged everywhere in the same interval and agreed at every point with the first function, then indeed the Fourier coefficients were equal: $a_n = a'_n$, $b_n = b'_n$. This might seem obvious, but unless the convergence of the Fourier series is uniform and absolute, the claim that the Fourier coefficients of f are given by the familiar Fourier integrals is

---

[55] He even lapsed into French to make the point; see Cantor 1991, 44.

false. Instead, a more subtle argument is needed, one that picked up from Riemann's ideas in his work of 1857, which Cantor was pleased to discover and publish in 1871.

He then began to investigate what could happen if either the convergence or the equality of the series for each value of x was partially denied. He was soon able to show that the equality of the Fourier coefficients continues to hold even when the two series fail to be equal on an arbitrary finite set of points. This is not too surprising, for if a function f is replaced by a new function g that agrees with f everywhere except at one point, then the integrals of f and g will agree. It follows that if a function f is replaced by a new function g that agrees with f everywhere except at a finite set of points, then there integrals will still agree.

In 1872 Cantor published similar results when the trigonometric series either fail to converge, or do converge but fail to agree, on an infinite set of points. He had found a way to produce infinite sets that were negligible for the purposes of integration. This was uncharted territory. He wanted to know if this could happen and yet the function still have only one Fourier series (see above, §2.2.1). He found an ingenious construction for constructing ever more complicated examples of such "bad" sets. Prior to his work it had only been known that an arbitrary finite set can be "bad"; Cantor's iterative construction produced examples of infinite "bad" sets.

It is very hard to imagine that there are such point sets; that was one of the things Cantor had to establish, and his use of an iterative construction to obtain stranger and stranger mathematical objects was to prove useful in other parts of mathematics. Cantor also saw that it was possible to define a set that was the result of carrying out his iterative construction infinitely often, and indeed that one could then start again, and construct still more examples. To keep track of the sets he was constructing, Cantor introduced his infinite symbols. Given a set P, Cantor's construction produced what he called its derived set, which Cantor denoted $P^{(1)}$. The derived set of this set he called $P^{(2)}$, and so on. The derived set of the derived set . . . of the given set P he denoted $P^{(\infty)}$. He denoted its derived set $P^{(\infty+1)}$, and so on. Thus were created the infinite symbols, $P^{(\infty)}$, $P^{(\infty+1)}$, $\mathrm{P}^{(\infty+\infty)}$ $\mathrm{P}(\infty^{\infty})$, and many more.

At this stage that Cantor's work, although exceptional in its insight, was driven entirely by the problem at hand concerning trigonometric series. Its immediate origins are impeccably internal. What justifies its presence in this book is that other developments start here. Most were relatively unproblematic. Questions to do with the nature and properties of functions are readily turned into questions about point sets, so questions about integrability became questions about point sets. And when functions of more than one variable are at stake, the point sets are subsets of higher-dimensional sets, and can become much more complicated. Here beginneth measure theory (which deals with the definition and properties of the integral) and topology (which deals with the concepts of continuity and among its other achievements rigorizes the informal property of being in one piece). Ferreirós was perhaps the first historian of mathematics to stress with sufficient clarity that these aspects of Cantor's were accepted almost at once.[56] Poincaré, for example, was an early enthusiast, because he found strange point sets coming up in his own work, and he successfully proposed Cantor as a corresponding member of the French Mathematical Society on the strength of his work on point sets. A number of other mathematicians were studying

---

[56] Ferreirós 1996.

point sets and the theory of functions: du Bois-Reymond, Hankel, and Harnack, for example. While I shall certainly claim that measure theory and topology count as modernist disciplines (see §4.4.1.1 and 4.5.2), the case needs arguing, and their origins lie in classical, traditional mathematics of the mid-nineteenth century.

The much more problematic development came later, when Cantor began to press the case for an abstract theory of infinite sets and for his infinite symbols as some kind of new numbers that nonetheless formed a natural extension of the familiar integers. In 1883 Cantor published his *Grundlagen einer allgemeinen Mannigfaltigkeitslehre* (Foundations of a General Theory of Manifolds). Dauben, in his biography of Cantor (1979), notes that mathematicians had little difficulty in accepting the new mathematical subject Cantor offered them, but that the philosophical implications were to prove another story.

In his *Grundlagen* Cantor offered a way of regarding the infinite symbols as numbers. To mark their new status he dropped the use of ∞ as a symbol in favor of ω. He now proclaimed that there was a number, which he denoted ω, that expressed the natural order of the set 1, 2, 3, . . . and that successive addition of units produced new sets: $\omega + 1$, $\omega + 2$, and so on. The set of all these was another number, denoted 2ω, and still the process continued. Cantor was abstracting from the idea of counting objects in a given order. Each ordinal number, as he called these numbers, corresponds to a set, so by putting these sets together in various ways Cantor could mimic ordinary arithmetic and define addition and multiplication for his ordinal numbers. This was good evidence that they should be regarded as numbers, even though the familiar rules largely failed for them. For infinite ordinals α and β it is generally not true that $\alpha + \beta = \beta + \alpha$, or that $\alpha.\beta = \beta.\alpha$.[57]

Cantor's philosophical interpretation of his work was that it offered final conclusive demonstration of the actual infinite, and not merely the potential infinite. There were, after all, many different kinds of infinite sets, distinguished by their ordinal numbers, whereas previously one might have thought that there was just one potentially infinite set, one that was "larger than any finite set." This position incurred a remarkable degree of opposition. Much of it, Cantor felt, was the work of one man, Leopold Kronecker in Berlin. Kronecker was indeed influential, and after he had attempted to prevent a paper of Cantor's appearing in the *Journal für die reine und angewandte Mathematik* in 1878, Cantor had decided never again to send a paper to that journal. This was a painful decision for him, because the journal was the leading journal in the field, and very much a Berlin journal. It was his link to the university where he always hoped one day to be called back as a professor. Instead, Cantor took his papers to a new journal, Mittag-Leffler's *Acta Mathematica*. Gösta Mittag-Leffler was an ambitious Swedish mathematician, much attracted to the Berlin school of mathematics but shrewd enough to run some risks. He published all of Poincaré's long papers on complex function theory and non-Euclidean geometry, although Kronecker privately sent word that they would kill the journal; Mittag-Leffler repeated the story in an anniversary edition of the journal after Kronecker's death (vol. 39, p. iii), when it was clear how wrong he had been. Mittag-Leffler also took some of Cantor's new work, as did Klein's rival journal, the *Mathematische Annalen*, which Klein had been editing since 1876.

[57] For example, $1 + \omega$ is not equal to $\omega + 1$.

In 1884 Cantor had the first of the nervous breakdowns that were thereafter to assume an increasing role in his life. He recovered from this one in just over a month, and by the summer was well enough to attempt to mend relationships with Kronecker and to realize that, in the end, his work would have to speak for itself. But then he received a further setback. He had for some years been working on and off but without success on the continuum hypothesis (see §3.2.2) and that was taking its toll. However, he had some results on another topic, and in early 1885 he sent them to Mittag-Leffler for publication. To his distress, Mittag-Leffler turned them down. Mittag-Leffler gave as his reason that they would only damage Cantor's reputation and discredit his work, and he should wait until he had other results. If need be, Mittag-Leffler suggested, he should proceed as Gauss had done with non-Euclidean geometry, and be prepared to wait a hundred years. Cantor felt that Mittag-Leffler was only interested in protecting his journal from controversy and decided that *Acta Mathematica* too would have to manage without him.

Salvation, as Cantor saw it, came from religion, specifically Pope Leo XIII's attempts to combat atheism and materialism, so often attributed by the Church to science, by promoting a proper understanding of science.[58] The pope had endorsed the revival of Thomism in 1879 in his encyclical Aeterni Patris.[59] He argued (§2) that "false conclusions concerning divine and human things, . . . , have now crept into all the orders of the State, and have been accepted by the common consent of the masses." These errors stemmed from philosophy, and since we are rightly guided by reason, philosophy, guided by religion, must be employed to correct these errors, for (§9), "Faith frees and saves reason from error, and endows it with manifold knowledge." The scholastics, and Thomas of Aquinas above all, had organized Christian knowledge, and Aquinas had, moreover, not only worked through all philosophy but (§18) had "pushed his philosophic inquiry into the reasons and principles of things, which because they are most comprehensive and contain in their bosom, so to say, the seeds of almost infinite truths, were to be unfolded in good time by later masters and with a goodly yield." However, starting in the sixteenth century, philosophizers without faith had confused things, and, said the pope (§27), there is no better remedy than "the solid doctrine of the Fathers and the Scholastics, who so clearly and forcibly demonstrate the firm foundations of the faith, its divine origin, its certain truth." Moreover (§30), "It is well to note that our philosophy can only by the grossest injustice be accused of being opposed to the advance and development of natural science." In short, and contrary to the opinion of many in the late nineteenth century, there was no conflict between science and religion for both were true and philosophically defensible.

The neo-Thomist movement, as it is known, was active in France, Belgium, Italy, and Germany.[60] In Germany in particular, many Jesuit intellectuals led by Constantin Gutberlet, a professor of philosophy at Fulda, endorsed the actual Cantorian

---

[58] Cantor's deep religious beliefs were briefly discussed in Fraenkel's obituary of Cantor (Fraenkel 1932, 481–483) and then explored more fully in Dauben 1979.

[59] Available at http://www.vatican.va/holy_father/leo_xiii/encyclicals/documents/hf_l-xiii_enc_04081879_aeterni-patris_en.html.

[60] Paul 1979 shows that it was particularly successful in Belgium, where it was taken up at the University of Louvain, and in Germany, where in the Catholic south of the country there were twenty-three student associations having altogether 985 members by 1883.

infinite. For them, the consciousness of God guaranteed the existence of all consistent mathematical objects. Cantor appreciated the support he found in the neo-Thomist movement, which was much more receptive than the mathematical community and, as a deeply religious man, came to situate his theory of the infinite within what he regarded as the transcendentally larger realm of the theological infinite. He shifted his personal perspective too. No longer was he the creator of the transfinite numbers, he was merely a reporter.[61]

The published Cantor asserted somewhat different positions. In the *Grundlagen* he had discussed numbers as well defined in the mind, capable of changing thought, and therefore having an intrasubjective or, as he called it, an immanent reality. He had also called attention to their concrete reality when they are manifested in objects in the world. This was their transsubjective or transient reality. It was, he said, a major problem in metaphysics to understand the relationship between these two realities. It was brought about, he went on, through the unity of the universe, and this meant that it was possible to study just immanent reality, and in this lay the independence of mathematics. The essence of mathematics, Cantor concluded, lay in this freedom: "Because of this extraordinary position which distinguishes mathematics from all other sciences, and which produces an explanation for the relatively free and easy way of pursuing it, it especially deserves the name of *free mathematics*, a designation which I, if I had the choice, would prefer to the now customary 'pure' mathematics."[62]

3.2.2.2 CATHOLIC MODERNISM

The papacy enters the story of modernism in another way. In 1893, the encyclical *Providentissimus Deus*[63] cautiously acknowledged the work of Catholic theologians who were taking a historical and critical approach to the Bible, rather in the wake of such Protestant theologians as Adolf von Harnack (the brother of the mathematician Alfred). Included in their number were Maurice Blondel in France, Alfred Loisy who was influenced by Ernest Renan, and George Tyrell, SJ. This was delicate work; for example, the work of D. F. Strauss in Germany, whose book *Das Leben-Jesu* (1835) (or, *The Life of Jesus, Critically Examined*, translated into English by Marian Evans [George Eliot] in 1860), showed how quickly this activity clashed with Catholic dogma.[64] Pope Leo XIII wished to keep Catholic theology abreast of these Protestant debates without succumbing to them. The Catholic modernists whom he wished to rein in were part of a movement aimed at making explicit how the views of the Church could be reconciled with, or defended against, those of a post-Enlightenment era dominated increasingly by science. They placed more emphasis on being in tune with the times and the need of religion to evolve, as they saw it, and less on the interpretation of the Bible, which their historicism saw as by no means the infallible word of God. For these reasons they deserve to be called modernists. Some were drawn to the formal aspects of biblical scholarship, most maintained a complicated relationship with the day-to-day world of the Catholic Church, and they

---

[61] See Dauben 1979, 146. Out of sight in this book are interesting arguments between Gutberlet and Wundt that reflect their contrasting religious convictions.

[62] Cantor 1883, 182, quoted in Dauben 1979, 132.

[63] Available at http://www.papalencyclicals.net/Leo13/l13provi.htm.

[64] The twelve-volume edition of his collected works was edited by Eduard Zeller, on whom see §2.4.4

saw themselves simultaneously as radicals and as the true heirs of the prevailing tradition.

This proved too much of a challenge for the next pope, Pius X, who took steps to close the Catholic modernist movement down. Alfred Loisy had offered his *The Gospel and the Church* as a reply to Adolf von Harnack's work, but in 1903 five of his works were placed on the Index of Prohibited Books and in 1908 he was excommunicated. Tyrell was expelled from the Jesuits in 1906 for his historicizing, and was denied burial in a Catholic cemetery. In 1907 the pope denounced the modernist movement in his encyclical *Lamentabili Sane* of July 3, 1907.[65] The modernists were condemned because, "In the name of higher knowledge and historical research [they say], they are looking for that progress of dogmas which is, in reality, nothing but the corruption of dogmas." The last two of the sixty-five proscribed propositions are indicative of the whole, the first because it captures the essence of the critical position, the second because it hints at the perceived threat:

64. Scientific progress demands that the concepts of Christian doctrine concerning God, creation, revelation, the Person of the Incarnate Word, and Redemption be re-adjusted.

65. Modern Catholicism can be reconciled with true science only if it is transformed into a non-dogmatic Christianity; that is to say, into a broad and liberal Protestantism.

This was followed up with a longer denunciation, the encyclical *Pascendi Dominici Gregis* (*On the Doctrine of the Modernists*) on September 8, 1907,[66] in which (see §39) the movement was denounced as a "synthesis of all heresies." Modernism was strenuously driven out and scholastic theology reaffirmed.

Catholic modernism is an example of a modernism contemporary with mathematical modernism. I do not mean to imply that the mathematical modernists were up against similar opposition, or that they were drawn into these brutal confrontations. But it was one of the major intellectual, arguably anti-intellectual, debates of the time, and it points up the drama of the movement. Autonomous groups of professors rewriting the book, be it in history, mathematics, or theology, confronted at some level the forces of tradition with their expectations and demands. Doubtless, mathematicians had the least to worry about, but they knew that they were not the only ones engaged in making a new world according to new rules.

3.2.2.3 CARDINAL NUMBERS

Cantor was of course aware that ordinal numbers do not correspond to the most fundamental way of counting, precisely because they take note of the order in which objects are presented. The basic notion is simply the size of the set of objects. For finite sets, the ordinal way of counting and the "mere number" way agree. We pass easily from "first, second, third, . . . , ninth" to "one, two, three, . . . , nine." For transfinite ordinals, it is clear that many ordinal numbers correspond to sets of the same size, so a proper definition of the size of a set is needed.

---

[65] Available at http://www.catholic-forum.com/saints/stp06004.htm.
[66] Available at http://www.ewtn.com/library/ENCYC/P10PASCE.HTM.

To explain this, Cantor opened his *Beiträge* of 1885 with the words: "By an 'aggregate' [*Menge*] we are to understand any collection into a whole [*Zusammenfassung zu einem Ganzen*] M of definite and separate objects in of our intuition or our thought. These objects are called the 'elements' of M." This makes it very clear that Cantor was now dealing with abstract sets, not necessarily sets of points on the real line. He soon went on: "Every aggregate M has a definite "power," which we will also call its "cardinal number." We will call by the name "power" or "cardinal number" of M the general concept which, by means of our active faculty of thought, arises from the aggregate M when we make abstraction of the nature of its various elements *m* and of the order in which they are given." This is too vague, and merely gestures at a concept that requires proper definition if it is to win acceptance for infinite sets. Cantor had been more precise in a work he unfortunately withdrew from publication.[67] There he wrote:

> The power of a set M is determined as the concept [*Vorstellung*] of that which is common to all sets equivalent to the set M and only these, and thus also common to the set M itself. It is the *representatio generalis* . . . for all sets of the same class as M. Thus I take it to be the most basic (both psychologically and methodologically) and the simplest basic concept arising by abstraction from all particulars which can represent a set of a definite class, both with respect to the character of its elements and with respect to the connections and orderings between the elements, be it with respect to one another or to objects lying outside the set. Insofar as one reflects only upon that which is common to all sets belonging to one and the same class, the concept of power or valence arises.

From such uncertain beginnings, the rigorous theory of the cardinalities of infinite sets was to grow, not without controversy, as we shall see.

3.2.2.4 THE CONTINUUM HYPOTHESIS

Cantor attached the greatest importance to the one problem that was to elude him all his life. Indeed, it was one of the greatest sources of pressure on his mental strength. This was the continuum hypothesis, although he never called it by that name. It first appears in his *Beitrag* of 1878, and it comes out of a desire to link the useful theory of subsets of the real numbers to the resources of abstract set theory. It proposes that every infinite subset of the real numbers is either denumerable or has the cardinality of the continuum—there are no subsets of the real numbers of an intermediate size.

When Cantor showed that for any set S the set, denoted $P(S)$, of all the subsets of S is strictly larger, this opened the way to another way of thinking about the continuum hypothesis.[68] One can start with the set Q of all rational numbers, which is a denumerable infinite set, and form the set $P(Q)$. This set can be thought of as the set of all real numbers expressed in binary notation, so it has cardinality c, where c is the cardinality of the continuum. The question is: Is this the cardinality of the first non-denumerable infinite set, $\aleph_1$ or is it strictly bigger? The continuum hypothesis is the natural assumption that c is as small as it can be, so the cardinalities of $P(Q)$ and $\aleph_1$

---

[67] It is not even in his *Gesammelte Abhandlungen*, and was not found and published until 1970; see Grattan-Guinness 1970 and Dauben 1979, 151.
[68] Cantor 1890–91.

FIGURE 3.3. Paul du Bois-Reymond.

are the same. As with almost all of Cantor's ideas, the precise formulations, and even the overall aims of his theory, were to remain matters of contention for many decades.

It is well known that set theory became the foundations of mathematics accepted throughout the twentieth century. The impression is sometimes given that either the acceptance of these ideas was merely a matter of time, and objections to them were either technical or almost perverse. That this was not the case is nicely demonstrated by the philosophizing of his contemporary, Paul du Bois-Reymond.

3.2.3 THE PHILOSOPHY OF PAUL DU BOIS-REYMOND

The first volume of Paul du Bois-Reymond's *Allgemeine Functionentheorie*, and the only one to appear, was published in 1882. It has not had a good press, and indeed many of its arguments were inadequate even by the standards of the day, but it is an extraordinary book of particular interest to the emergence of modernism in mathematics. Moreover, some of the criticisms it received in its day were misplaced, and some of the reasons historians of mathematics have had for dismissing it derive from inaccurate and tendentious beliefs about the development of mathematics in the late nineteenth century.[69] That said, it provoked at least one sharp rebuttal, by the philosopher Benno Kerry, which shows how divergent views had become in the philosophy of mathematics.[70]

Paul du Bois-Reymond worked his way through physiology, the subject his older brother Emil became famous in, to mathematical physics to pure mathematics, especially real analysis, where he made a number of interesting discoveries, especially in the study of Fourier series. By 1882 he had also published some interesting ideas

---

[69] See the interesting discussion in Fisher 1981.
[70] On Kerry, see Peckhaus 1994.

on the rate of growth of functions, which led him to ideas about different kinds of magnitudes for measuring different approaches to infinity, and these magnitudes, he showed, were not only not Archimedean magnitudes but much more complicated than the familiar real numbers. As discussed elsewhere in this book (see §5.3.3), these ideas, although disputed by Cantor, were favorably received by Borel and Hardy, who reworked them into a rigorous theory.

Much of his work, his book included, was motivated by the following question. One often asks in mathematics if a certain expression has a meaning and, in particular, a numerical value. Typically, the expression is obtained from a sequence of meaningful expressions, so the question is whether this sequence has a limiting value. For example, the expression might be an infinite sum, $u_1 + u_2 + \ldots + u_n + \ldots$. Mathematicians, he said, try to understand an infinite sum as the limit of the sequence of finite, and therefore intelligible sums, $s_n = u_1 + u_2 + \ldots + u_n$, as n increases indefinitely. So the key question is this: Is there a sequence, or a family of sequences, against which any sequence can be measured and its convergence be decided? The ideal would be to have a way of comparing the given sequence $s_n = u_1 + u_2 + \ldots + u_n$, as n increases indefinitely, with a suitable "master" sequence $m_1, m_2, \ldots$ and say either "for all n larger than a certain number, $s_n < m_n$, so the sequence of s's converges," or "for all n larger than a certain number, $s_n > m_n$, so the sequence of s's diverges." The reader may like to know that such attempts will fail, but that is not the point at issue. The relevant point is that the sense of less than or greater than, the $<$ or $>$ in the imagined remarks, the "size," so to speak, of a sequence, has to be defined (and that can be done). But there is no reason to suppose that these sizes will be numbers, or that any two sequences will be comparable. Sequences might be so many incomparable kinds of fruit.[71]

Paul Du Bois-Reymond's *Allgemeine Functionentheorie*, published in 1882 when he was fifty-one, deals with two themes: philosophical questions in the foundations of the calculus, and certain types of infinite sets of numbers. They are therefore themes on which he could speak with some authority. The first of these was hung on a peg to which students and teachers of university-level mathematics can respond to this day: What sense are we to make of infinite, nonrepeating decimals? Given such a decimal, as it might be $3.14159265\ldots$, we can form a sequence of rational numbers: 3, 3.1, 3.14, 3.141, etc. These steadily increase, and all are less than any number that is obtained from this sequence by rounding the last digit up (we don't do this if the digit is a 9, of course). So we have $3 < 3.1 < 3.14 < 3.141 < 3.1415 < 3.14159 < 3.141592 < \ldots < 3.14159266$. Now, this sequence gives successively better approximations to π, but any infinite, nonrepeating decimal (say, $a_0.a_1 a_2 a_3 \ldots a_n \ldots$) is usually taken to define a number. The number it defines is the limiting value of the sequence of increasing approximations formed by the sequence $a_0, a_0.a_1, a_0.a_1 a_2, \ldots$. The question that du Bois-Reymond asked is: By what right may we conclude that such a decimal makes sense, and what, then, does it define? It is clear, as du Bois-Reymond well knew, that a satisfactory answer to this question is as good as an understanding of the Bolzano-Weierstrass theorem.

Du Bois-Reymond distinguished two positions, which he called the idealist and the empiricist. Both agree that certain things exist, such as arbitrary multiples and

---

[71] These non-Archimedean magnitudes are discussed further below; see §5.3.3.

submultiples of a given unit. The idealist is able to go to the limit, the empiricist is not. The idealist will have recourse to geometry, and talk of the length such a number measures on the line, even though they will never see such a length. The empiricist cannot agree that such lengths exist. Pushing such arguments to what may not be their logical conclusions, du Bois-Reymond found that the idealist is committed to such things as exactness in measurement, and to lengths that are infinitesimally small (and to infinitely large quantities). His argument here seems particularly weak—it was to the effect that the idealist finds infinitely many segments in a line. Were these to be entirely disjoint, the conclusion might follow. But, as noted above, du Bois-Reymond had other, and much better arguments for wanting to admit infinitesimals into mathematics. Nonetheless, the existence of a length of π is, for the idealist, a matter of how we form legitimate concepts, not a matter of exhibiting such a length *in concreto*. The empiricist rejects the idea of exactness in measurement, and finds that lengths can only be measured to within arbitrary fine levels of accuracy. Points are arbitrarily small regions. Numbers may be arbitrarily small or large, but are finite, never infinitesimal or infinite. Existence is existence in the mundane sense.

The idealist and empiricist positions are irreconcilable. To do analysis, du Bois-Reymond therefore proposed a neutral approach, set out under the motto of "Empiricist language, idealist proofs" (p. 156). In this spirit, one asks about the very existence of a limit, and the nature of the conditions that guarantee its existence. The decimal a that is the limiting value of a_0, $a_0.a_1$, $a_0.a_1a_2$, . . . is to be understood as a linear magnitude, du Bois-Reymond's term for quantities that (in the present context) can be compared with the rational numbers and with each other, so that if a and b are any two such, then exactly one of the following three conditions holds: $a < b$, $a = b$, or $a > b$. The decimal a is the linear magnitude with the property that members of the sequence get, and thereafter remain, arbitrarily close to it (in the epsilon-delta sense of close customary among mathematicians). The idealist may say this limit exists, the empiricist may say, with Kronecker, that the existence is nothing more than the existence of the rational numbers in the converging sequence—it matters not. What matters is that the talk of limits and limiting values has been made precise in a way each can, on different grounds, accept. This neutral approach resolved not only the question of the meaning of infinite, nonrepeating decimals, but, as du Bois-Reymond put it, broke the spell and allowed analysis to be the mistress of the house (p. 167).

The second half of the book moved on to consider the sorts of infinite subsets of the real line that can arise, to which du Bois-Reymond had been alerted by his study of Fourier series. This is rather recondite for our purposes and was neglected for over twenty years, until Hausdorff took it up and moved it slightly away from du Bois-Reymond's interests and into the study of point sets in the manner of Cantor. Hausdorff found du Bois-Reymond's work inspiring, but not sufficiently precise. But du Bois-Reymond had an interest Hausdorff did not share: giving some kind of scientific account of what his infinite sets (which he called pantachies) are. Here, the idealists and the empiricists again conflicted, for reasons rooted in their philosophies of mathematics.

The point sets are amusing (see §51.) The function $y = \sin\left(\frac{1}{x}\right)$ cannot be defined in a way that makes it continuous when $x = 0$. The function $y = \sin\left(\frac{1}{\sin\frac{1}{x}}\right)$ cannot be defined in a way that makes it continuous when $\sin\left(\frac{1}{x}\right) = 0$, which happens when $1/x$ is a multiple of π and therefore whenever x is of the form $1/n\pi$ for any n. There is no reason to stop there. Plainly there will be a still more complicated set of points at which the function $y = \sin\left(\frac{1}{\sin\frac{1}{\sin\frac{1}{x}}}\right)$ cannot be defined in a way that makes it continuous, and so on. Call the point set associated with the function with n occurrences of sin in its definition B_n. The point set B_n for each value of n is a subset of the point set B_{n+1}, and the nature (some might say, even the existence) of the union of the sets B_n is most at issue. Naive idealists and naive empiricists alike may wish to find some courage before proceeding. It is not the case that every number between 0 and 1 lies in one of these sets, so their limit is particularly challenging, and the author concluded that the infinite point set of all numbers between 0 and 1 existed only for the idealist. The empiricist, du Bois-Reymond believed, could accept only countably infinite sets, whereas what remains when the union of the sets B_n is removed is uncountable.[72]

Du Bois-Reymond's book is an intriguing mixture of advanced mathematics and muddled reasoning, and all the more interesting for that. He did not want mathematics to be the mere play of symbols, but if it were not, how could it be grounded, and within what system of ideas? If this attempt fails, can another be mounted, or has mathematics moved off into a realm of its own?

In 1890 the philosopher Benno Kerry's posthumous book *System einer Theorie der Grenzbegriffe* (A System for a Theory of Limiting Ideas) appeared. It represents his definitive criticism of du Bois-Reymond's ideas. It did not seem to Kerry that the existence of a limit in either the idealist or the empiricist senses that du Bois-Reymond had invoked mattered in mathematics, unless one wanted it to be true and applicable. Rather, all that was required was that a concept, such a limit, was clear and fruitful for making hypothetical judgments. For an infinite decimal to exist, in Kerry's view, it was enough that any two could be compared—and this was evidently the case. Du Bois-Reymond, Kerry noted, had an exclusively geometrical sense of existence in mind, but the way in which geometric points exist should not be confused with the way in which limits exist. Things exist in different ways, and failure to exist in one sense does not preclude it in another.

As for limits, he said, often their existence is undoubted: the sequence $1/n$ tends to 0 with increasing n, $1/3$ is the limit of $0.3333\ldots$, and so on. Some limits are taken to exist even when they are never seen—"a pure black color," for example. In the case of the physicist's perfect fluid, what matters is the applicability of the deductions, not their truth. Such limits are different from the "limit" of the sequence $1, 1 - 1,$ $1 - 1 + 1, \ldots$ because that "limit" is a self-contradictory idea.

Du Bois-Reymond's book bears witness to a deeply held belief that mathematics must be about something, its objects should exist, and should do so in a way closely akin to the way physical objects do. Existence should mean something like existence in space and time. Kerry's alternative was much more radical. Existence is freedom from contradiction, mathematical objects may exist in many ways and have merely to imply coherent conclusions. After the modernist transformation of mathematics, du Bois-Reymond's approach seems remarkably limited, but it was not, ultimately, that different from Frege's. Perhaps the whole idea of existence is most easily approached if one thinks of nonexistence. Talk—ordinary, mathematical, or

---

[72] The union of the sets B_n is the union of a countable union of countable sets, so it is countable, but the set of all real numbers is not countable, a fact Cantor had proved in 1874.

philosophical—of something that does not exist is taken to be thereby unconstrained and incoherent. So coherent talk must somehow be about things that exist. For some, the core sense of existence is the day-to-day one, from which one should not stray too far. The modernist sense is almost the most minimal possible: nothing more is required than that deduction continues. The meaning of "existence" is stretched thinner and thinner, until, you may feel, it becomes a shadow of its former self.

3.3 Algebra

> A theorem was considered according to Weierstrass to have been proved if it could be reduced to relations among natural numbers, whose laws were assumed to be given. Any further dealings with the latter were laid aside and entrusted to the philosophers. . . . That was the case until the logical foundations of this science [arithmetic] began to stagger. The natural numbers then turned into one of the most fruitful research domains of mathematics, and especially of set theory (Dedekind). The mathematician was thus compelled to become a philosopher, for otherwise he ceased to be a mathematician.
> —Hilbert's lecture notes for a course on general relativity theory, 1916–17.[73]

Algebraic number theory in the 1880s provided a rather evident disagreement between Kronecker and Dedekind about the nature of the new types of number being introduced into the subject.[74]

3.3.1 Dedekind

Dedekind agreed with Kronecker's mentor, Kummer, that ideal numbers were necessary, but he found Kummer's approach unacceptable. Their multiplication was especially difficult, and he remarked: "Because of these difficulties, it seems desirable to replace Kummer's ideal numbers, which are never defined in themselves, but only as the divisors of existing numbers ω in a domain σ, by a really existing substantive."[75]

These really existing numbers were defined as sets of complex numbers. For example, in Dedekind's theory, all the complex numbers of the form $p + q\sqrt{-5}$, which can be written in the form $m.3 + n(1 + \sqrt{-5})$ for some ordinary integers m and n, form an ideal number.

It was not just computational ease that Dedekind sought.[76] He also wanted an ontological foundation that was acceptable to him. He found it, as he had earlier for the theory of real numbers, in the naive concept of a set. There, as we saw, Dedekind defined a real number as a certain set of rational numbers. Here, an ideal number was defined as a certain set of complex integers. In each case, he had also to show

---

[73] Quoted in Corry 2004, 379.

[74] For instructive and detailed comparisons of the approaches of Dedekind and Kronecker that emphasise Dedekind's deepening commitment to a truly structural point of view, see Reed 1994, Corry 1996, and Avigad 2006.

[75] Dedekind 1877, §10.

[76] Indeed, Edwards has repeatedly argued that Kummer's theory, as revised by Kronecker, is easier to work with.

how his newly defined objects did what was expected of them. He was able to show how his ideal numbers enabled one to transfer the theory of quadratic forms to the study of complex integers, and in this way the concept of an ideal was launched on its way to becoming one of the central ideas of modern algebra. His first presentation of it came out in 1871, in Supplement X to the second edition of Dirichlet's *Zahlentheorie* (Number Theory), a year before he published his theory of real numbers, and he went on it make it clear that the two theories were very closely linked in his mind.

By presenting his theory of ideals as an appendix to the standard introduction to Gaussian number theory, Dedekind hoped to reach the widest possible audience. But he became disappointed with the result, and was convinced that almost no one had read it. He rewrote it for a French audience in 1876, and again for later editions of the *Zahlentheorie*. In 1882 Kummer's former student Leopold Kronecker presented a rival version, a divisor theory in the spirit of Kummer.[77] This theory is faithful to Kummer's emphasis on tests for divisibility and so is ontologically parsimonious, and it has always attracted a minority of influential adherents; but even Edwards writes that "it seems that no-one could read [it]."[78]

One way to point up the difference between Dedekind's work and Kronecker's is to start with a similarity. Both men were concerned with algebraic integers, the solutions of polynomial equations with integer coefficients and leading coefficient 1 (this includes familiar numbers such as $\sqrt{2}$ also the roots, for example, of $x^5 - 3x^3 - 4x + 8 = 0$). Kronecker preferred to work directly with the polynomial. Dedekind preferred to work with the ideals he had defined and their abstract properties: that of being a prime ideal, or an ideal that divides another ideal. This distinction between Kronecker's algorithmic and Dedekind's conceptual style of mathematics draws strong feelings from some number theorists, as long-running arguments attest. Quite apart from the renewed interest in algorithmic aspects of mathematics these days, Kronecker's position has long been argued for by Edwards.[79] Kronecker's preference for procedures that exhibited specific answers rather than abstract guarantees of existence was probably a deeper objection for him to Dedekind's approach than the latter's use of infinite sets, despite Kronecker's steady turn toward constructivism.

Dedekind, as he revised his theory, set more and more value by the generality and uniformity of his theory, which should apply to all sorts of rings of algebraic integers, and the fact that the introduction of prime ideals allowed him to establish a prime factorization theorem for ideals.[80] This circumvented a major problem that had opened up: algebraic integers cannot always meaningfully said to be prime, because they can be factored in essentially distinct ways. For example, among integers of the form $a + b\sqrt{-5}$, 21 can be factored as $21 = 3.7 = (4 + \sqrt{-5})(4 - \sqrt{-5})$, but the factors, although irreducible, have no common factors, so the factorization is finished.[81]

---

[77] Kronecker 1882. See also Edwards 1990.

[78] Edwards 1987, 19.

[79] See, for example, Edwards 2005 and the references therein.

[80] See Avigad 2006.

[81] This is one of Dedekind's (1876–77, 279 ff) examples, quoted in Corry 1996, 86. The substantial mathematical point here is that ordinary prime integers are prime in two senses of the word: they are irreducible—they cannot be factorized—and if they divide a product of two integers, they divide one of those integers. Unique factorization of ordinary integers is closely related to the fact that irreducible integers are prime, but in general an algebraic integer can be irreducible without being prime and unique factorization of algebraic integers fails.

This means that any theory of the multiplication of algebraic integers is threatened with terrible complexity, and generalization from the properties of ordinary integers is likely to fail. It is interesting to see that when the concepts of an "integer" being "prime" and "irreducible" diverged, the stronger and more traditional word went with the deeper structural property.

As Avigad and Corry have emphasized, Dedekind's successive revisions of his theory make it clear how firmly he identified with a structuralist approach and how unsatisfactory he felt any reliance of choices of elements to be. This not only denied algorithmic considerations a fundamental role, it drove his critique of the concepts he himself employed. As noted above, the immediate question in any version of algebraic number theory had to do with divisors, and Dedekind's first version of his theory dealt with that. He realized, however, that for the usual theory of integers the concept of division is derived from that of multiplication, and his second theory makes multiplication of ideals primary and derives division accordingly. That re-formulation nonetheless moved it further from the original problems.

Dedekind's approach is recognizably modernist in its emphasis on the need for internal definitions of objects, tailored to meet internally set goals of the theory. The concept of integer is stretched to include new integers, and the concept of prime is likewise stretched. Here, as with his other notable innovations in the study of numbers, Dedekind's key tool was his confidence in the use of infinite sets of more familiar objects. His famous cuts in the theory of real numbers, his chains in the definition of the natural numbers, and his ideals in algebraic number theory are all infinite sets of more basic elements. That said, Corry makes a very useful distinction between Dedekind's modernist tendencies and those of later, more structuralist, mathematicians. He writes (1996, 131): "Dedekind's modules[82] and ideals are not 'algebraic structures' similar to yet another 'structure': fields. They are not 'almost-fields' failing to satisfy one of the postulates that define the latter." Numbers, he continues, remained the focus of Dedekind's inquiries; modules and ideals were only tools. The structuralist formulation of modern algebra, due almost entirely to Emmy Noether and her school, lies beyond the scope of this book, but only because it places particular demands on the reader. It is almost a paradigm example of the modernist transformation of mathematics.[83]

The difference in the professional situation of Dedekind and Kronecker cannot explain the subsequent reception of their ideas. Kronecker was well established in Berlin, and while he attracted only a few students they were generally very good. Dedekind was in self-imposed isolation in Braunschweig, tied to tradition by his editions of Dirichlet's editions of Gauss's *Disquisitiones Arithmeticae*. Dedekind was a lucid writer, and Kronecker was a difficult one. Something of Dedekind's approach was saved for posterity, however, by Hilbert's adoption of it in his hugely influential *Zahlbericht* (*The Theory of Algebraic Number Fields*) of 1894. Hilbert had first learned number theory from Heinrich Weber, a close colleague of Dedekind's, and surely therefore learned it in the style of Dedekind. His account of the subject in 1894

---

[82] Another typically infinite set of objects.

[83] It is worth noting, too, that Dedekind first defined the set of natural numbers, and then defined a natural number as a member of this set. As noted in the Glossary, the later definition of vectors in mathematics proceeded the same way.

set the terms of research for over fifty years, and since Hilbert preferred Dedekind's ideals to divisors, their survival was assured.

The success of Dedekind's ideal theory and its philosophical underpinnings occurred in a high point of contemporary German mathematics, the algebraic theory of numbers. This was a tradition begun by Gauss, taken up by Dirichlet, Kummer, Kronecker, Eisenstein, and Dedekind, and it was to extend well beyond them. It was prominent in Berlin, the center of German mathematics for most of the nineteenth century. This gave both Kronecker's solution and Dedekind's great weight as examples of how existence questions could be treated. Because they arose from a genuine question in research mathematics, the outcome of which necessarily involved a point of mathematical ontology, mathematicians could not dismiss them as mere philosophy. When Dedekind's approach, as endorsed by Hilbert, became the mainstream one, it represented a victory for naive set theory, because the existence of an ideal number was established by presenting it as a set of more ordinary numbers (no doubts were expressed at the time about the existence of infinite sets of this kind). But the concept of integer was certainly changed. The natural numbers, 1, 2, 3, . . . and their negatives were now just one kind of integer among many. The term henceforth applied to any numberlike object that could be said to be prime (or not) to divide another exactly (or not), and so forth.

Research questions therefore drove mathematicians into making a choice between different ontological positions, and, following Hilbert, the one they mostly made identified some key properties of objects as integerlike. They did so in an arena where stakes were high: the theory of numbers. In German eyes, and to a lesser extent in those of mathematicians in other countries, number theory was where reputations were to be made. It was difficult, it was a subject with an honorable tradition, and it was rigorous. As Gauss had urged all along, it had the most surprising, deep, and obscure connections to other branches of mathematics. Moreover, the natural numbers were the archetypal mathematical objects, and if mathematicians did not care for them, who else would give them a home? In short, number theory was the paradigm pure mathematical subject. Naive set theory, and a new perspective on what constitutes an integer, had made it to center stage.

3.3.2 The Unity of Nineteenth-Century Mathematics

The similarities between the discovery of non-Euclidean geometry and the advance of algebraic number theory are quite marked, and recently Ferreirós (1999, 241) has suggested that Dedekind's shy friend Riemann was a powerful influence pushing Dedekind forward. With the discovery of non-Euclidean geometry, the question of the mathematical nature of physical space had become empirical. With the proclamation of Riemannian geometry, it was possible to argue that it had always been an empirical question. One might say that Riemannian geometry provided the ideology for a revolutionary change in geometrical ideas. In each case the needs of research provoked a fundamental shift in meaning, a radical enlargement of a basic concept with an accompanying change in ontology.

With changing attitudes to the nature and existence of mathematical objects came changes in the way proofs were regarded. The best-known example of this is the trend toward increasing rigor, but other criteria were often employed. For example, discussion of the appropriateness of a proof often hinged on a belief in the unity of mathematics. The nature of this unity was precarious as the nineteenth century progressed and the subject grew. Gauss on number theory is an emblematic case of a writer for whom the unity of mathematics was a palpable if elusive and mysterious fact that a good mathematician should learn to respect. He felt strongly that the hidden substantial and unexpected connections that existed between different parts of the subject should guide one's research, and especially one's choice of proofs. For that reason he often gave several proofs of the same theorem (four of the fundamental theorem of algebra, six of the theorem of quadratic reciprocity). On the occasion of one of the latter, he wrote:

> Proofs of the simplest truths lie hidden very deeply and can at first only be brought to light in a way very different from how one originally sought them. It is then quite often the case that several other ways open up, some shorter and more direct, others proceeding on quite different principles, and one scarcely conjectured any connection between these and the previous researches. Such a wonderful connection between widely separated truths gives these researches not only a certain particular charm, but also deserves to be diligently studied and clarified, because it is not seldom that new techniques and advances of the science can be made on this account.[84]

More programmatically, Hilbert remarked at the start of his *Zahlbericht* with not only number theory in mind, but also Riemannian function theory and the analogy between function fields and algebraic number fields, the theory of elliptic functions, and the arithmetization of analysis and geometry (notably, non-Euclidean geometry): "It finally comes down to this, if I do not err, that on the whole the development of pure mathematics principally came about under the badge of number."[85]

What, however, is being unified? Some authors claimed there to be an underlying similarity of methods, others of objects. Hilbert's motivating analogy was that in a very real sense there was a way in which questions in algebraic number theory could be thought of geometrically.[86] It was this analogy that not only led him to proclaim an underlying closeness between algebraic and geometric objects, but to develop powerful new methods in algebraic number theory suggested by the corresponding situation in geometry. Hilbert's success was striking, yet by 1900 the subject had grown so much that its unity could no longer be taken for granted.

Canons of appropriateness and conceptual clarity were linked to a new, more conceptual, less computational mathematics. With the changing perception of the objects of mathematics came new criteria for evaluating, governing, and directing their use. These did not replace the old ones. There continue to be mathematicians who are formidable masters of the formalisms, which themselves have proliferated. But the new conceptual and aesthetic criteria have often achieved paramount position at the level of explanation, overthrowing mere calculation as the best criteria for truth. For that reason, the explanations sought and proposed for the deepest aspects of a mathematical theory by Gauss at one end of the nineteenth century and Hilbert

---

[84] Gauss 1818, reprint 1981, 496.

[85] Hilbert 1897, 66.

[86] The analogy is too complicated technically to explain here. It was first suggested in an important paper (Dedekind and Weber 1882), and is described from a historical point of view in Klein's *Entwicklung* (1926, 324–334). It continues to inspire mathematical work to the present day.

at the other mark a momentous change in the way mathematicians think about their subject.

This change was built into the professional structure of mathematics, most notably in Germany. Gauss, although preeminent, had no students in the modern sense. His followers took their cues from his published work. Chief among these was Dirichlet, whose book on Gaussian number theory helped make it steadily more accessible. From him one is led to Dedekind. Dedekind's own work had to wait almost for Hilbert to find its audience, but he is also of considerable importance as a commentator and editor of the works of Dirichlet and Riemann. Dedekind also worked with Heinrich Weber, with whom he wrote the celebrated paper that first promoted the analogy between number fields and function fields. It was from Weber that Hilbert learned the confidence with abstract, nonconstructive existence proofs that marked his first important papers, on the theory of invariants.

The leading German university was that of Berlin, where Dirichlet and Jacobi taught for a while. Under the leadership of Kummer, Kronecker, and Weierstrass it grew remarkably in the second half of the nineteenth century, until audiences of two hundred were not unknown. Almost all German mathematicians studied here for a while. Complex analysis, including the theory of differential equations, was emphasized in Weierstrass's lecture courses, while Kummer and Kronecker were number theorists in the broad sense of the term. So although courses in applied mathematics were given, the emphasis was very much on pure mathematics, and indeed on algebra and number theory rather than geometry, which came to be bracketed by Kronecker (following Gauss) with mechanics.

Germany's experience at the start of the nineteenth century, culminating as it did in defeat by Napoleon, had convinced them that applied-led research was too narrow. The intellectual response was the philosophy of neohumanism, which argued that doing pure mathematics for its own sake was not only best for that subject but also best for those who would want to apply it. It was shared by mathematicians like Jacobi and builders of the scientific community like Crelle, himself an engineer, who filled the Academy of Sciences at Berlin and the pages of his *Journal für Mathematik* overwhelmingly with pure mathematicians and their works.

There was therefore a widespread social network, including many of the leading mathematicians of the day, which accepted and indeed advocated the revolutionary changes in algebra and number theory just described. In contemporary neohumanism they even had an appropriate ideology, or rather a family of overlapping ideologies, with which to rationalize their activities. But above all, this way of doing mathematics made sense mathematically. It derived from the problems presented by research, and it proved to be fertile, which provided a strong reason for continuing in that direction.

3.3.3 Kronecker

As we have seen, Dedekind spoke openly of creating mathematical objects, including numbers. With that in mind, Kronecker's much-quoted remark that "God made the integers, all else is the work of man" acquires fresh significance.[87] Kronecker was

---

[87] Weber 1891/92, 19, quoting from a lecture of Kronecker's of 1886.

willing to start from the integers but reluctant to create new objects, especially by using infinite collections; his remark is therefore a polemical antithesis to Dedekind's programme.

Dedekind espoused a more semantical, less purely syntactical approach to mathematical concepts. Kronecker's alternative, which was more Kantian, was an arithmetic paradigm. It was not well presented, and Hermann Weyl, who in this matter is a friend of Kronecker's, had to admit that "Kronecker's approach . . . has recently been completely neglected" but it is worth seeing beyond the difficulties.[88] We are helped to do so by the PhD thesis of the French mathematician Jules Molk, who studied under Kronecker in Berlin and became an enthusiastic advocate of Kronecker's point of view. Molk was born in 1857, and on his return to France wrote, with Jules Tannery, the standard French work on elliptic functions.[89] He then became the editor in chief of the French edition of the *Encyklopädie der Mathematischen Wissenschaften,* the *Encyclopédie des sciences mathématiques*, which made him France's answer to Felix Klein. He used his editorship of the *Encyclopédie* to publish revised, updated, and often considerably extended versions of the German originals, especially in the field of algebra. He died in 1914.

As is well known, Kronecker laid great store by explicit algorithmic procedures, but we should not lose sight of the sheer ambition of his project and why it held arithmetic in such high regard. His project was of enormous range. So far as possible, Kronecker wanted a common method for dealing with all the problems of mathematics that come down to properties of polynomials in any finite number of variables over some field, usually the rational numbers; but for his successors, at least, it could be the complex numbers or some pure transcendental extension of one of these. (So algebra and arithmetic were included, but analysis and differential geometry were not.) For Kronecker, the subject matter of mathematics included all of algebraic number theory, the theory of algebraic curves, and, insofar as it existed, the theory of algebraic varieties of any dimension. One problem these theories share is that of finding common factors. Kronecker sought to refine this to the benefit of all the various aspects.

His basic building blocks were two things: the usual integers and the rational numbers, on the one hand, and variables on the other. These were combined according to the usual four laws of arithmetic; root extraction was to be avoided in favor of equations (for example, the variable x and the equation $x^2 - 2 = 0$ is to be preferred to $\sqrt{2}$). He set out the thinking that led him to his general program in the preface to a paper of 1881.[90] The preface is a lengthy historical account indicating how much he had already proposed in lectures at the University of Berlin (and who his audience had included) and at a session of the academy in 1862. The guiding aim, which he traced back to 1857 (the date, one notices, of Riemann's paper on Abelian functions) was to treat algebraic integers,[91] but he encountered certain difficulties. These he resolved, he said, with the insight that it was a useless, even harmful restriction to pick on one root of an algebraic equation, but problems can be avoided by treating all the

---

[88] See Weyl 1940, iii.
[89] Tannery and Molk 1893–1902.
[90] Kronecker 1881.
[91] Roots of polynomial equations with leading term 1 and integer coefficients.

roots simultaneously. Kronecker did not say that algebraic integers do not exist, but rather that thinking of them in isolation from how they come about does not help. The productive way forward was to recognize that algebraic integers arise in families (as roots of an irreducible polynomial equation over the rationals, note that irreducible equations do not share roots). Their construction, rather than the concept, brings out the property that what is presented is the family of algebraic integers.

He discussed these results with his then-friend Weierstrass, who urged him to apply the same principles to algebraic functions of a single variable and if possible to the study of integrals of algebraic functions, taking account of all possible singularities. This set him on the road to a purely algebraic treatment, shunning geometric or analytic methods. He sent the first fruits to Weierstrass in October 1858, but then decided that Weierstrass's own results rendered his superfluous, and so he refrained from further publication. He was brought back to the topic by discovering how much his thoughts coincided with those of Dedekind and Weber (an agreement which did not, he noted, extend to the basic definition and explanation of the concept of a divisor). Therefore he presented his old ideas, abandoned in 1862, for publication in 1881.

Kronecker was trying to draw an analogy with considerable resonance. He worked with the integers and the rational numbers and as many variables as he wished. He could form polynomial equations whose coefficients were roots of other equations. He could therefore compare algebraic integers over the rational numbers and algebraic integers over the field of rational functions in one variable (ordinary integers and polynomials in one variable have strikingly similar theories here, thanks to the Euclidean algorithm). To make further progress, he introduced the concept of the divisor, which can be illustrated by listing some of the basic questions one asks about divisors: When does one object divide another, what is the greatest common divisor of two given objects, are objects that cannot be factorized further also prime?

The analogy is, however, troubled. Kronecker, and those who followed him, such as Molk and König, all gave the same example, due originally to Dedekind, because it is the simplest: algebraic integers of the form $m + n\sqrt{-5}$, where m and n are ordinary integers. The number $2 - \sqrt{-5}$ cannot be factored (so it is said to be irreducible), but it is not prime. Indeed, $(2 - \sqrt{-5})(2 + \sqrt{-5}) = 9 = 3.3$, but $2 - \sqrt{-5}$ does not divide 3.[92] So the algebraic integers 9 and $3(2 - \sqrt{-5})$ have no greatest common divisor: their common divisors are 1, 3 and $2 - \sqrt{-5}$, and neither of 3 and $2 - \sqrt{-5}$ divides the other.[93]

Kronecker and his followers, in pursuit of a profound analogy and therefore engaged in important mathematics, and confronted with real difficulties, felt the need to take a philosophical, even Kantian, stance in epistemology. This is clearest in a paragraph in Molk's thesis, which we may assume Kronecker did not disagree with.

---

[92] When the equation $(2 - \sqrt{-5})(x + y\sqrt{-5}) = 3$ is solved for x and y, the resulting $(x + y\sqrt{-5})$ is not an integer.

[93] It was exactly this problem that caused Dedekind to formulate his theory of ideals, specifically ideals that are not principal (generated by a single element), precisely to get around the problem.

Definitions must be algebraic and not only logical. It is not enough say "Either a thing is, or it is not". One must show what one wants to be and what not to be in the particular domain with which we are concerned. Only then do we take a step forward. If we define, for example, an irreducible function as one which is not reducible, that is to say which is not decomposable into other functions of a definite kind, we do not give an algebraic definition, we only state a simple logical truth. For us to give a valid definition in algebra it is necessary that it be preceded by an account of the method which allows us to obtain the factors of a reducible function by means of a finite number of rational operations.[94]

The point at issue is definitional; what is it to say that certain objects exist? The distinction is between the merely logical guarantee and the mathematician's warranty, which is harder to come by and signifies much more. If we change just one word in this statement, from "logical" to "philosophical," we have an utterance that is in many ways Kantian. The criterion for a mathematical object is that it comes with a construction. We are asked to deal with properties belonging to, but not contained in, a concept.

The reception of Kronecker's ideas in algebraic number theory is a complicated and indeed an ongoing story. His program drew a strong positive response in some circles. Best known among his followers are Hensel and Landsberg (Hensel knew Kronecker personally and worked actively on the edition of Kronecker's *Werke*; his joint book with Landsberg dates from 1902). As noted, it was passionately adopted by the French mathematician Jules Molk, although Molk had few new results to add. In 1904 the Hungarian mathematician J. König wrote the first textbook on Kronecker's theory. In 1911 Molk published in his *Encyclopédie* an extended version of Landsberg's original article of 1899 in the *Encyklopädie der Mathematischen Wissenschaften* on fields and algebraic varieties. Landsberg's essay, which was naturally from a Kroneckerian point of view, was amplified without departing in spirit from the original; it was brought up to date by Kurschàk and a rising star among French mathematicians, Jacques Hadamard.

It is clear, then, that Kronecker's approach grew out of attempts to solve mathematical problems. This is conveniently forgotten by those who present it as an arbitrary, even self-limiting philosophical position, one that Hilbert and others in the modern tradition rightly brushed aside. Such writers take their cue from Hilbert and Klein. Hilbert inclined more and more to a position of opposition to Kronecker, an opposition that dates from Hilbert's activities in the foundations of mathematics, if not indeed his successful use of nonconstructive existence proofs in his first work, on invariant theory. Felix Klein, in his *Entwicklung* (1926, 281), said plausibly of Kronecker: "He worked principally with arithmetic and algebra, which he raised in later years to a definite intellectual norm for all intellectual work." But he was unfair to continue (p. 284): "With Kronecker, who for philosophical reasons recognized the existence of only the integers or at most the rational numbers, and wished to banish the irrational numbers entirely, a new direction in mathematics arose that found the foundations of Weierstrassian function theory unsatisfactory."

---

[94] Molk 1885, 8.

3.4 Modern Logic and Set Theory

3.4.1 Some German Philosophers

German philosophers of the late nineteenth century, with the sole exceptions of Frege and Husserl, are usually only discussed by authors who are poised to refute them, or to record that even in their lifetimes they were refuted. English-language Russell scholars know the names of Erdmann, Lotze, and Trendelenburg because Russell passed them by. German-language scholars know them because Frege and Husserl took a considerable number of pages to distinguish their own work from that of these ponderous adherents of the old ways. Nor should it be much different, but an overview of what they were attempting will sharpen our perception of the relations of philosophy, language, logic, and psychology in the period.[95]

The decline of Hegelianism was not followed by another grand system that anchored German intellectual life, and this was a matter of concern to some philosophers. Friedrich Trendelenburg, who was a professor at Berlin from 1837 until his death in 1872, was among those who observed this decline and urged a return to the classics, Plato, and Aristotle. His main book, *Logische Untersuchungen* (Logical Studies) (1840, 3rd ed. 1870), is an attempt to reconstruct Aristotle's logic of purpose building on the insights of Kant, Hegel, and Herbart. He also wrote an essay, "On Leibniz's Project of a Universal Characteristic," which seems to have been how Frege found out about it, and this may also have been Frege's source for the word "Begriffsschrift" (Idea Script).[96]

Trendelenburg took from Leibniz the idea that a purely rational language would be valuable. Natural language is imperfect, he said, and does not carefully relate the sign and the concept of the signified idea. For this, a properly constructed language in which the shape of the sign captures the content of the concept would be preferred. What Frege took from this will be discussed shortly, but for later use note also that Trendelenburg was of a very different opinion from his predecessor Otto Gruppe. Gruppe did not lament the decline of system-building philosophy. Rather, he felt that the task was to clean out philosophy by means of a radical reworking of logic and epistemology, and the reform of logic would arise from a study of language. "Thinking," he wrote, "is not possible without language and language is not possible without thinking"—the second part of this aphorism seems strangely less secure than the first. Aristotelian logic failed, in Gruppe's opinion, because it did not reflect the way reasoning is conducted in a natural language.

Lotze's *Logik* of 1874 (first edition 1843) was another significant work of this interim period. Frege ignored his metaphysics, which lets us off the hook, but he shared Lotze's antipsychologistic standpoint, which Lotze presented, reasonably enough, as a return to Kant. He also argued in favor of the creation of a pure science of logic, and further that "mathematics really has its proper home in general logic."[97] Perhaps more importantly, Lotze is the originator of what is called the "theory of

---

[95] This brief account leans heavily on Sluga 1980.
[96] Sluga 1980, 49.
[97] Quoted in Sluga 1980, 57.

FIGURE 3.4. Gottlob Frege (c. 1920).

local signs" to explain our perception of space. This was to be taken up and developed by Helmholtz and Wundt, as we shall see below, §6.2.5.

Benno Erdmann, who came to prominence with his book on the philosophical meaning of the work of Riemann and Helmholtz on the concept of space, as discussed above (§2.1.3), then went on to write his *Logische Elementarlehre* (1892, 2nd ed. 1907). As Kusch (1995, 35–39, 49–53) makes clear, Erdmann is perhaps the archetypal psychological logician that Frege and Husserl sought to drive out.[98]

Two general points arise from this oversimplified survey. One is the range of issues that were entangled together in German intellectual life in the late nineteenth century. Logic and languages, both artificial and natural; logic and its relation, both to philosophy and to mathematics; the proper relation of logic to psychology. The other is the important places some of these philosophers occupied. It is these men, not Frege, who spoke for philosophy in the period, these men who views had to be worked around. Erdmann's insights into mathematics were not always accurate, Lotze's were downright obstinate, but Lotze had caused trouble for Klein when Klein had integrated non-Euclidean and projective geometry in the early 1870s. The philosophers interpreted, as they saw fit, the technical work of the mere mathematicians, it is they who spoke to an audience that wanted "understanding."

[98] Erdmann became a professor of philosophy in Leipzig, where he got to know Wundt and turned toward experimental psychology. In 1898 he was called to Bonn. There he founded the Psychological Seminar, which from 1901 was integrated as an independent division in the Philosophical Seminar. http://www.psychologie.uni-bonn.de/entpaed/berat/allgm.htm.

3.4.2 Frege

Frege's *Begriffsschrift*, said van Heijenoort on p. 1 of his splendid anthology *From Frege to Gödel*, "is perhaps the most important single work ever written in logic." In it Frege set out a number of themes that came to characterize the logicist movement, some technical, some philosophical. A scientific truth, he argued, is best apprehended through its proof, however it might have been discovered: psychological genesis has no role to play. Among the class of all truths, some may be proved by logic alone and some require the facts of experience. Here, and in his *Grundlagen*,[99] he sought to show that arithmetic is a branch of logic. In later years the belief that all of mathematics could be derived from logic, and any program to show how it could be done, came to be called *logicism*. The first to give it this name may have been Carnap, at a conference in Königsberg in 1930. Neither Frege nor Russell after him used that name, but they energetically pursued the program. As Frege was to note, it requires a considerable extension and revision of logic—just how much was to prove controversial—and we can also note that it requires a novel, and also controversial, idea of what constitutes mathematics.

If arithmetic is to be derived from logic alone, this would certainly require that the concept of ordering in a sequence be reduced to that of logical consequence, so that the ordering of the integers can be derived from logic. To ensure that nothing intuitive crept in unannounced, thereby bringing in some facts of experience, Frege derived his forbidding and unpopular *Begriffsschrift*. And, because Frege distrusted language—he saw it as "one of the tasks of philosophy to break the domination of the word over the human spirit by laying bare the misconceptions that through the use of language often almost unavoidably arise"—the *Begriffsschrift* was partly devised as a hard taskmaster.[100] He called it a "formula language for pure thought," and compared it to the microscope, an instrument generally inferior to the human eye but for select purposes a vast (and indispensable) improvement. Frege was optimistic that his ideography could be used to study the foundations of the calculus and, with less work, to geometry. It was, in any case, he reckoned, an advance in logic.[101]

His general theory of sequences was powerful enough to show that "pure thought, . . . , can, solely from the content that results from its own constitution, bring forth judgments that at first sight appear to be possible only on the basis of some intuition."[102] For example, one can introduce definitions that stipulate that this expression is to have the same content as that one. Such abbreviations are propositions, but they are not judgments, and so in particular are not synthetic judgments, although Kant had claimed that all judgments in mathematics are synthetic.[103]

Frege was generally hostile to Kantianism and sympathetic to Leibniz. He hoped, in vain, that his *Begriffsschrift* would contribute to realizing Leibniz's dream of a best

---

[99] See Frege 1893–1903, English trans, 1964, 29.

[100] Van Heijenoort 1967a, 7.

[101] It met with a mixed reception, as noted in Villko 1998: three largely sympathetic reviews, three less so. Frege was criticized for his notation, and for his lack of knowledge of previous work in logic—criticisms he spent the next three to four years answering.

[102] Van Heijenoort 1967a, 55.

[103] Discussed in Boolos 1998, 155, who shows that a vigorous Kantian reply to Frege's preferred refutation of Kant can be refuted, insofar as the matter can be resolved at all. The historical point to note is merely that Frege himself presented a good, if obscure, case.

possible language for thought. When Schröder criticized Frege for separating the logical and arithmetical symbols when Boole and others had been showing how profitably they can be made to blur together,[104] Frege replied that it had not been his intention to represent an abstract logic in formulas, "but to express a content through written signs in a more precise and clear way than it is possible to do through words. In fact, what I wanted to create was not a mere *calculus ratiocinator* but a *lingua characterica* [sic] in Leibniz's sense."[105] Leibniz's name was to become a common way of signaling a radically anti-Kantian position, reclaiming or re-creating a tradition precisely at the moment when breaking with the past.

Frege was, of course, a mathematician. This is evident the way he replaced traditional talk of subject and predicate with the idea of function and argument, which is fundamental to the production of the *Begriffsschrift*. His being a mathematician gave power and depth to his attempt to derive all of the basic concepts of arithmetic from logic. His contacts with other mathematicians locate him within a group of mathematicians and philosophers concerned to rethink the nature of mathematics, and with whom he had more agreements than his notoriously irascible letters and articles have led many historians of philosophy to suspect.[106]

Frege became an extraordinary professor at Jena the year Cantor became a full professor in Halle, some 80 kilometers away. He watched with interest as Cantor attempted to define numbers in such a way that his transfinite numbers could legitimately be accepted as numbers, because he (Frege) was himself attempting to define number in such a way that the concept could be seen to be purely logical. He shared with Cantor a hostility to the idea that mathematics is based on psychological considerations, such as feelings or sensations. Nor was it to be based on physical perceptions, and so it was not a branch of empirical science. It was, and here too he agreed with Cantor, about objects of thought that were to be presented to the mind clearly and distinctly. But Frege felt that the clarity of definitions in Cantor's work left much to be desired, and in an unpublished article of 1891 he satirized Cantor's approach with his typical ruthlessness, but also quite accurately. Cantor had not only argued that two sets should be considered to have the same (finite or infinite) number of elements if and only if they can be put in a one-to-one correspondence with each other, but that the cardinality of a set is the set of pure units with which it is in a one-to-one correspondence. This suggestion that every property of the objects of a set can reliably be abstracted away offended Frege: "If, for example, one finds a property of a thing upsetting, one abstracts it away. If one wants to order a stop, however, to this destruction, so that properties which one wants to see retained are not obliterated, then one reflects upon these properties. Finally, if one painfully misses properties of the thing, one adds them back by definition. Possessing such magical powers, one is not very far from omnipotence."[107]

As this quotation shows, Frege was clear that logic itself had to be rigorously improved before it could give birth to mathematics. In particular, he gave very careful consideration to what it is to define something and how this can be done precisely. The fruit of this consideration was his own definition of number.

[104] See van Heijenoort 1967a, 2.
[105] See van Heijenoort 1967a, 2.
[106] See Tappenden 2006 for a more positive assessment of Frege's mathematical position.
[107] Van Heijenoort 1967a, 98–103 and cited in Dauben 1979, 221–222.

3.4.2.1 FREGE ON NUMBER

It is worth dwelling on how Frege defined numbers for a variety of reasons. The very fact that he, and others at around the same time, even attempted to do so, rather than merely gesture at such a familiar concept, is characteristic of the modernist project. His own definition encapsulates exactly what is clever, indeed deep, about his project and also its fatal flaw, and without knowing that we cannot properly understand later developments. It will be instructive to compare his definition with others, such as those offered by Dedekind, Cantor, and Husserl, as well as preceding ones (due to Grassmann, Helmholtz, and Weierstrass). That said, no one has ever found a way to make Frege's ideas easy reading.

Frege took it for granted that there actually are numbers, specifically the numbers 0, 1 and the rest of the natural numbers. His aim was to capture exactly the properties of numbers that made them numbers, to show that those properties, including mathematical induction, were entirely logical in nature, and to express those properties in a form amenable to a precise logical formalism. The overriding aim was to establish, for mathematics and ultimately the sciences, the nature of correct reasoning, and to show that it is analytic. Now, because the idea of number is both fundamental and yet hard to grasp (much as time is), it is reasonable to examine sundry definitions or descriptions of the idea that have been offered. This Frege did, if only to reject them. For example, numbers cannot be empirical (as Mill had suggested) if we are to reason about them analytically as Frege wished. Nor are numbers properties of external things, because they depend on the concept under which the things are classified (Frege's example was Homer's *Iliad*: one book, twenty-four chapters, and a great many verses).

In view of the many inadequate approaches taken to the idea of number, it is also prudent to think about how one could grasp such slippery things, and here Frege developed quite an elaborate apparatus. Logic, for Frege, was about concepts, not the objects to which this or that concept refers. (Frege also spoke of objects belonging to a concept, or falling under a concept, to mean that they are examples of it.) Concepts are objective, unlike ideas. Frege insisted that one distinguish between concept and object, even in the case when there is a unique example of the concept. We may, Frege allowed, even have a concept when there are no examples of it (talking fish, for example). Number, Frege argued, is a property of a concept.

For Frege, as for any mathematician or philosopher interested in numbers, a key fact is that we can deal with arbitrarily large numbers, although we might be hard put to be sure that there are precisely 13,579,864,201 grains of sand in a given pile. We license our ability to talk about such numbers by subscribing to the principle of mathematical induction, so Frege set to work to formalize it. He had to define the number 0, the number 1, and to show how to get from one number, n, to the next, $n + 1$. He got off to a deliberately poor start, but a plausible one. Pick a suitable concept F (we can come back to that) and define the number 0 this way: the number 0 belongs to the concept F, defined as "For all x, x is in F." A suitable concept F might be "$x \neq x$." The number 1 is defined as being not the number 0 and as being such that for all x and y if x belongs to F and y belongs to F then $x = y$. The inductive step goes like this: $n + 1$ belongs to the concept F if there is a unique x such that x falls under F, and n is the number that belongs to the combined concept "F and not x."

Frege set this starting position up in order to criticize and refine it. One wants to be able to distinguish numbers from other objects, but Frege found fault with this approach because it does not permit one to say that Julius Caesar is not a number.[108] One also wants to say of two numbers whether or not they are they same. More precisely, suppose one person defines numbers as above with properties F, and another person defines numbers as above but with properties G. How can they be sure, with these definitions, that their numbers agree? Frege illuminated this problem with a geometric example, that of parallel lines in the plane. One wants to say that two lines are parallel if they point in the same direction, and then, in the plane projective geometry that Frege knew very well, that they define a point at infinity. So a definition of direction is needed (defining direction here is analogous to defining number). But what happens if one line is given by specifying a point on it and a direction, and another is given in some totally different way? It is very likely that the result will be a circular definition, defining direction in terms of parallel. Better, said Frege (not entirely obviously), to say that there are two concepts, "Line parallel to the line a" and "Line parallel to the line b," and that these concepts have the same extension. We know what the concept "parallel" means, so we know which lines are parallel to a or b and if these agree the direction of line a is the extension of the concept "Parallel to the line a."

Now, just as we know, given a line in the plane, all the other lines in the plane that are parallel to the given line, so, said Frege, given a concept we know all the objects falling under that concept. So, given a number defined with respect to concept F and a number defined with respect to concept G, we can decide whether these concepts have equinumerous extensions, meaning that there is a one-to-one correspondence between the two extensions. So, by analogy with his discussion of direction, Frege now said that the number that belongs to the concept F is the extension of the concept "Equinumerous to the concept F." According to Frege then, numbers are defined as soon as one has suitable candidates for the concepts F, and via the one-to-one correspondence idea there is no problem posed by a multiplicity of candidates. There has, of course, to be at least one suitable F, but the exposition above jumped the gun and offered such a concept, except for the inductive step. Frege defined the number $n + 1$ to be the number that belongs to the concept, "Member of the series of natural numbers ending with n." Frege has to define membership of a series, but he had done that in his *Begriffsschrift*.[109] This was Frege's vindication of the claim, made on p. iv of the *Grundlagen*, that "even an inference like that from n to $n + 1$, which on the face of it is peculiar to mathematics, is based on the general laws of logic," and so marks a significant step in Frege's logicist program.

To summarize: the claim is therefore that to each concept F there is a number. The number associated to the concept F is the extension of the concept "equinumerous with the concept of F." After some elementary deductions, Frege soon asserted in *Grundlagen* §73, that it was necessary to show that the number associated with the concept F is identical to the number associated with the concept G if the concept F is

---

[108] Frege 1884, §56.
[109] These technicalities are unproblematic and followed from his work in the *Begriffsschrift* on relations, where Frege showed how to define the ancestral of a relation.

equinumerous with the concept G. Following Boolos, this claim is nowadays called Hume's principle and stated in the form:

> The concept F is equinumerous to the concept G if and only if the extension of the concept "Equinumerous to the concept F" is identical with the extension of the concept "Equinumerous to the concept G."

The transition from a concept to its extension is deceptively easy; we shall see that it was to prove fatal to Frege's enterprise.[110] The argument Frege gave in favor of Hume's principle claim was skimpy to say the least, but these few pages at the end of the *Grundlagen* show that Frege recognized that he had to prove that his definition of a number allowed the familiar properties of numbers to be deduced, and at this he made a start. The full conclusive argument was to follow in the *Grundgesetze*, but he had grounds for cautious optimism.

It is perhaps more interesting to note that he explicitly recognized that his definition of number allowed for more than just the natural numbers. It allowed Cantor's transfinite numbers to be numbers, too, although Frege preferred his own definition to Cantor's "inner intuition." It is worth noting that even at the time of the *Grundlagen* Frege had the apparatus at hand to deduce the existence of infinite sets.

Cantor gave the *Grundlagen* a welcoming review, although he did not like the definition of number. He praised it for recognizing that the foundations of arithmetic needed a much more profound analysis than they had received hitherto, and that Kantian intuitions of space and time must be banished from the discussion, along with all matters of psychology. But he was less convinced by the suggestion that talk about the extension of a concept was the right way to go. Cantor felt, presciently, that the extension of a concept was generally completely undetermined quantitatively and only in special cases can be said to have a size, be it finite or infinite. The terms "number" and "power" (*Machtigkeit*) have to be defined independently and then brought to bear on the idea of extension. Cantor detected some slipperiness in Frege's use of a one-to-one comparison between sets, which at times inclined to define number as ordinal and at others as cardinal, a distinction he had been at pains to establish in his own *Grundlagen* (1883).

3.4.2.2 FREGE'S *GRUNDGESETZE*

Mention of the *Grundgesetze* leads us to the famous and devastating criticism of it by Bertrand Russell. The *Grundgesetze* admits a terrible paradox, and as a result is simply not coherent. The question therefore arises of how far back the contradiction in Frege's enterprise can be traced. In an important paper, Crispin Wright showed that second-order logic (as described in Frege's *Begriffsschrift*) combined with Hume's principle is consistent and does permit one to derive the Dedekind-Peano axioms. Now, the sketch Frege offered in his *Grundlagen* does not contain the crucial formal step (called Basic Law V) that vitiates the *Grundgesetze*. In fact, as Boolos then showed, Frege had actually established Wright's result, and as a result he suggested that it be known as Frege's theorem. Heck, Boolos wrote, then argued that

---

[110] For a discussion of earlier ideas about the relation between a concept and its extension, see Ferreirós 1996, especially §§2.1 and 2.2.

in fact Frege's *Grundgesetze* does offer a valid derivation of the laws of arithmetic, because it makes no use of the vitiating step, and gave reasons for believing that Frege was confident of this when writing the *Grundlagen*. The advance the *Grundgesetze* was intended to offer was therefore a purely logical derivation of Hume's principle, which had been more or less undefended in the *Grundlagen* and stood as an extra-logical assumption.[111]

It is worth bearing in mind that it was not until the 1980s that anyone noticed that Frege had derived the laws of arithmetic from logic as formulated in the *Begriffsschrift* together with the assumption of Hume's principle; this suggests that his *Grundlagen* was not widely or carefully read. It is indeed striking that Frege did not emphasize the point.

Nonetheless, there are significant obscurities in Frege's account in the *Grundlagen*. Even if we grant that it is clear what a concept is, we have the slippage from a concept to the extension of that concept. Apparently, to know a concept is to know its extension. But what is the extension of a concept? Frege nodded at some difficulties with this idea in a footnote, but concluded, "I assume that it is known what the extension of a concept is." He came to regret this, and the troubles with his Basic Law V shows how tricky the idea is. To oversimplify Russell's paradox (but not by much) consider the property of sets x that x is not a member of x. Let X be the set of sets x which are not members of themselves. Then if X is in X it is not in X and if X is not in X then it is in X—a paradox, indeed a veritable antinomy![112]

Russell's paradox showed Frege that his argument was in ruins, and that the fatal flaw was his assumption that every concept has an extension (or, to put this point in more modern language, that to every concept there is a set of objects that fall under that concept). He spent some time trying to get around the problem, but he was astute enough to see that although some of his suggestions might block one or another interpretation of the paradox. The paradox was Hydra-headed and would always defeat him.[113]

Difficulties with understanding such a presentation notwithstanding, it is clear that Frege's definitions do not create something that did not exist before. Rather, they aim to make precise something which exists independently of us but which we have not perhaps seen clearly before. Frege took the existence of numbers as given. His ontology is quite conservative; his aim is to present what is known in a new way, so that we can see how it is related to other things we know. Mathematical existence was not a topic on which Frege had original views—quite the contrary, as we shall see.

---

[111] The above information is taken from the introduction to the essays in part 2, Frege Studies of Boolos's *Logic, Logic and Logic*.

[112] To see how this connects to naive thinking about extensions (and here I follow in Boolos 1998, 173, reprinted in Demopolous 1995, 440), let R be the concept x: there is an F such that $x =$ the extension of F and F is false. Denote by $e(R)$ the extension of R. If it is not the case that $e(R)$ is a member of R, then since for all F, $e(R) = e(F)$ implies that $e(R)$ is in F, it follows that $e(R)$ is in R. But this in turn implies that for some F, $e(R) = e(F)$ and so $e(R)$ is not in F. The naive view of extensions now sanctions the conclusion that, for all x, $R(x)$ if and only if $F(x)$. It follows that $e(R)$ is not in R, which is a contradiction.

[113] There are many accounts of this oddly affecting moment in Frege's life and the history of modern logic. Two good ways in to the initial ramifications are the accounts of Giaquinto 2002 and Potter 2000.

3.4.3 Dedekind

In this Frege is a contrasting figure to Richard Dedekind, whose book *Was sind und was sollen die Zahlen?* (*What Are Numbers and What Are They For?*) was published in 1888, before he had read anything by Frege. It is worth looking at this book in a little detail, because in it Dedekind accomplished one of the seminal acts of mathematical modernism, that of redefining the natural numbers, hitherto a prime object of intuition.

The book itself is quite remarkable and can best be understood by working backwards. The aim was to give an account of the natural numbers. The preeminent feature of the natural numbers is that there is an initial one (either 0 or 1), and every number other than the initial number is obtained by adding 1 to the one before. So the sequence of natural numbers goes 0, 1, 2, . . . , n, ... The process of adding 1 to a natural number is often called the successor operation. Dedekind saw that any sequence of objects with the property that all but one are successors, and every one is obtained from that one by repeating the successor operation enough times, is a good candidate for the set of natural numbers: One, Two, Three, . . . ; Un, Deux, Trois, . . . ; Eins, Zwei, Drei, . . . ; !, !!, !!!, . . . ; all of them. There is no need to look for a defining property of 1, another of 2, and so on. All that matters is succession. So what is wanted is a set of undefined objects and a map from this set to itself that has the properties that the successor operator has.

So Dedekind carefully defined the basic operations on sets (which he called systems and regarded as composed of objects of our thought): membership, inclusion, equality, subset, union (forming a new set out of the elements belonging to at least

FIGURE 3.5. Richard Dedekind (1868).

one of several sets), and intersection (forming a new set out of the elements common to several sets). He defined a transformation as a map from one set to another, said it was "similar" if distinct objects of a set are mapped to distinct objects.[114] He said a set was infinite if it was similar to a proper part of itself (a subset B of a set A is *proper* if B is not equal to A). He called a set A with a transformation ϕ to itself a *chain*, because one can form A, $\phi(A)$, $\phi(\phi(A))$, and so on. Given a set A he defined A_0 to be the set of all chains that contain A. He showed it is a chain and called it the *chain* of A. To model the natural numbers he took an infinite set, and for the successor opera-tion he introduced a particular similar transformation from an infinite set to itself (certainly the successor operation is a similar transformation of the set of natural numbers). But now he had to do some work.

He called the set A *simply infinite* when the transformation ϕ is similar and the set A appears as a chain of an element not contained in $\phi(A)$. This element he denoted 1 and called the *base element*. This last requirement says that he will be looking at a set A whose members are 1, $\phi(1)$, $\phi(\phi(1))$, and so on. He then proved that every infinite set contains a simply infinite subset.

Now that he had a featureless, simply infinite set and a candidate for the successor operation defined without any even covert reference to the natural numbers, he was in a position to prove that he had indeed isolated all the essential properties of the natural numbers. This involved him in a lot of work, but enough has been done to show the way in which the natural numbers were entirely redefined. Accordingly, I shall not discuss how Dedekind's simply infinite set and its similar transformation deliver such results as the usual rules for inequalities between natural numbers and the result that for any natural number n the set of natural numbers less than n is finite. But it is worth noting that Dedekind established the validity of mathematical induction and hence the validity of the laws of addition and multiplication.

* Dedekind did this by showing (§ 126) that given an arbitrary transformation θ of a set Ω to itself, and an element ω in Ω, there is a unique transformation ψ of N to Ω such that $\psi(1) = \omega$ and—the inductive step—$\psi(\phi(n)) = \theta(\psi(n))$ for every n in N. Or, since ϕ is the successor operation, that $\psi(n+1) = \theta(\psi(n))$ for every n in N.[115] The key step here is the proof that the function ψ exists.[116] Dedekind established this in two stages, by an ingenious argument that is one of the first uses of the idea of recursion. First he showed that there is a transformation ψ_m defined for each m such that $\psi(n+1) = \theta(\psi_m(n))$ for all $n < m$, and then he showed that the transformation ψ can be defined this way: $\psi(n) = \psi_n(n)$. Now that the principle of induction is a theorem in this setting, the usual proofs of the validity of the commutative, distributive, and associative laws of addition and multiplication go over unchanged.

One further feature is also worth noting. Dedekind had to show that his setup not only described the natural numbers, but it essentially described *only* the natural numbers. This he also did. So we may add to the list of remarkable features in his

---

[114] In symbols, if $\phi : A \to A'$ and a and b are distinct elements of A then $\phi(a) \neq \phi(b)$. The modern term is *one-to-one*.

[115] This is worth unpacking. The claim is that it makes sense to say that $\psi(2) = \theta(\psi(1)) = \theta(a)$, $\psi(3) = \theta(\psi(2)) = \theta(\theta(a))$, and so on for all n. This makes the inductive or recursive character of the claim clear.

[116] See Potter 2000, 82–83.

little book that Dedekind not only introduced elementary set theory and the idea of maps between sets, and recursion, but the idea that a set may be described up to some notion of essential equivalence, for which the modern term is *categoricity*.

The book was reviewed by Hans Keferstein, an *Oberlehrer* (high school teacher) in Hamburg, who also reviewed Frege's *Grundlagen*, but he showed his failure to grasp the subtleties of the question and was drawn into a correspondence with Dedekind, the result of which is Dedekind's famous letter to Keferstein.[117] Although the letter was not published, it provides a clear and succinct expression of Dedekind's ideas, shorter and clearer than his book.

Dedekind gave priority to the idea of numbers as ordinals. He agreed that the number sequence is a sequence of distinct objects that come in a definite order, one after the other, such that every number has a unique successor, different numbers have different successors, and there is a unique number (the number 1) that is not itself a successor. But, he said, this does not characterize the numbers. There might also be some other objects. He did not give an example, but nothing so far excludes the presence of two objects *a* and *b*, placed after all the numbers, with the property that *a* is the successor of *b*, and *b* is the successor of *a*.

The truly difficult part of the work was to "cleanse the system . . . of such alien intruders." It was not enough to say that every number is reached eventually, and that numbers are precisely what are reached eventually. That is true, but it has to be proved, and to assume it is to close a vicious circle. The technical part of the solution to this problem, Dedekind wrote to Keferstein, was accomplished by his use of chains (one of the concepts Keferstein had wished to remove from Dedekind's presentation). He was happy to see, he added, that his concept of chain agreed in essence with Frege's way of defining the nonimmediate succession of one element upon another in a sequence. We need not follow Dedekind's argument here, but note that his use of chains secures the principle of mathematical induction (Theorem 126).

Dedekind and Frege did not agree on everything. At the head of his *Was sind . . .* (1888), Dedekind posted the motto: "Man always arithmetizes." In his opinion, numbers are not God-given but are the free creation of the human mind; Frege would not have agreed. Of what were numbers made? Here Dedekind moved closer to Frege and answered: out of sets, by means of logic alone. Therefore, and in particular, he wrote: "I declare already that I take the number-concept to be completely independent of the ideas or intuitions of space and time, that I see it as an immediate product of the pure laws of thought."[118] Dedekind regarded the idea of a set and of a mapping (such as the map from numbers to their successors) as purely logical.[119] For him, a set was any collection of things, and a thing is any object of thought. Moreover, a set is completely determined whenever it is known if any given thing belongs to it or not. A mapping between two sets *A* and *B* is a function (as mathematicians would say) or a rule, more informally, that associates to each element of the set *A* an element of the set *B*. It is allowed that the sets *A* and *B* may be the same.

As *Was sind . . .* makes clear, Dedekind was comfortable with the idea of infinite sets. In fact, it is infinite and not finite sets that are fundamental to his presentation. As Ferreirós (1999) noted (p. 233), this was a dramatic step to take at the time: "He

---

[117] Van Heijenoort 1967a, 98–103.
[118] Dedekind 1888, 335.
[119] See §3.4.3 and Ferreirós 1999, 226.

was defining the infinite through a property that Galileo, and even Cauchy, regarded as paradoxical, for it contradicted the Euclidean axiom "the whole is greater than the part.'" It cannot have helped that his attempt to show that infinite sets exist (Proposition 66) was particularly weak. As Dedekind put it:

§66. Theorem. There exist infinite systems.

Proof. My own realm of thoughts, i.e., the totality S of all things that can be objects of my thought, is infinite. For, if s signifies an element of S, then the thought, s', that s can be object of my thought is itself an element of S. If we regard this as the transform $\phi(s)$ of the element s, then the transformation ϕ of S, thus determined, has the property that $S' = \phi(S)$ the transform of S, is part of S; and S' is certainly a proper part of S, because there are elements in S (e. g., my own ego) which are different from any such thought as s' and are therefore not contained in S'. Finally, it is clear that if a, b are different elements of S, then their transforms a', b' are also different, and therefore that the transformation is a distinct (similar) transformation (§26). Hence S is infinite, which was to be proved. [Dedekind added a footnote: "A similar consideration is found in §13 of the *Paradoxien des Unendlichen* by Bolzano (Leipzig, 1851).]^{120}

Infinite sets are those that can be put into a one-to-one correspondence with a proper subset of themselves; finite sets are those that cannot. This gives a positive definition to what had hitherto only been defined negatively.

Frege responded to these ideas in 1893, and it should surprise no one that he was hostile. He found Dedekind's naive idea of a set unsatisfactory—not, however, out of a shrewd sense of its weaknesses, but from his own commitment to the entirely intentional point of view: everything should be stated in terms of concepts. So, for Frege a set could only be the extension of a concept, which was not to deny all merit to Dedekind's approach, merely to deny sets their foundational, and truly logical, role. That said, Frege's and Dedekind's views were rather similar, and were taken to be so until the whole question of the relation between a set and the extension of a concept went from being elementary to being very problematic indeed. But, as noted above, that was to be some years ahead, and in the meantime parallel versions of the logicist program could be pursued.^{121}

3.4.4 Peano

A third contribution in this direction came from the Italian mathematician Giuseppe Peano. Peano began his mathematical work in analysis and developed an acute sense of the need for mathematical rigor. He distrusted geometry, and came increasingly to distrust language.^{122} This was very clear in his account of the principles of arithmetic, his *Arithmetices principia* of 1889. He introduced several symbols into mathe-

---

^{120} I have slightly revised the translation in Dedekind 1930, 64, to make it more readable.
^{121} Parallel is Ferreirós's word (1999, 253).
^{122} One can note that the work of Peano and his followers made clear the arbitrary nature of the sign before it was seized upon by Ferdinand de Saussure, although not before Peirce. In that connection, I take this opportunity to say that Ferdinand had a brother René, who was interested in language and even wrote a paper in Esperanto; but I have found no influence of the mathematicians' debates on language described in chapter 7 on the celebrated linguist.

matics inspired by his reading of the work of Boole and Schröder, and some of them became standard, perhaps in an improved form. Since the presentation of mathematics was to play an important part in the reception of Peano's ideas, and indeed of modern mathematics generally, it is worth pausing to observe how Peano proposed that mathematics be written.

When dealing with propositions a, b, etc., Peano wrote $a.\supset.b$ for b is a consequence of a. He used $\cap$ for "and," followed Jevons and Schröder in using the inclusive "or," which he denoted $\cup$ used a solid bar, —, for negation, and Λ for the false or the absurd. When dealing with classes he introduced the sign ε to mean "is," and εK to mean "is a class," but he also used symbols such as a for classes, so $a \varepsilon K$ meant "a is a class." He wrote $b\varepsilon c$ to mean "b is a c," $b \cap c$ or bc for the individuals that are both b and c and $b \cup c$ for the individuals that are either b or c, $-b$ for the class composed of individuals that are not b. He had no use for a symbol that would denote all the individuals under consideration, but wrote Λ for the class that contains no individuals. He wrote $b \supset c$ to mean the class b is contained in the class c. There was also a special notation for definitions and theorems.

The notation has its defects—it is not completely clear when something is an element of something and when it is a subclass, for example—but it is clear that it is intended to allow a close parallel between the logical and the mathematical parts of an argument. As van Heijenoort pointed out, the most remarkable omission is that Peano gave no rules of inference for his calculus. For this reason, his presentation is best seen as an attempt to introduce a new mathematical language, not to amplify or refine mathematical logic. To emphasize the point, the book was also written in Latin. Using his notation, he proceeded to define the natural numbers. His five-part definition has become known as the "Peano axioms." Translated out of his symbolism, the first four parts assert:

- 1 is a number.
- If a is a number then the successor of a, written $a + 1$, is a number.
- For any two numbers a and b, $a = b$ if and only if $a + 1 = b + 1$.
- No number has 1 as a successor.

The fifth part recalls Dedekind's chain idea, and is what guarantees mathematical induction. It says that if k is a class containing the number 1, and the class k has the property that if x is a number belonging to the class k then $x + 1$ is in the class k then it follows that every number is in the class k.

Unlike Dedekind, who derived mathematical induction from simpler axioms, Peano merely asserted it. He took the laws of addition and multiplication to be definitions without realizing that it has to be proved that there are functions that describe addition and multiplication.[123] Unlike Frege, Peano did not attempt to analyze the logic of his arguments, but took it over as unproblematic. For this he was to be roundly criticized by Frege in 1896. But if the aim was to bring a Fregean microscope to mathematical writing by scrutinizing the role of language and minimizing its chances of smuggling in intuitive concepts, there is little doubt that Peano's symbolism far surpasses that of the *Begriffsschrift*.[124] No one, to my knowledge, ever

---

[123] See Potter 2000, 82.

[124] The postscript to Russell's famous letter to Frege, in which he identified a fatal flaw in Frege's system, was written in Peano's ideography, as Russell termed it.

wrote "*Begriffsschrift*." Peano was to attract a group of followers who were to go with his way of writing mathematics all the way to his simplified Latin and his avoidance, whenever possible, of words altogether. This was ultimately a self-destructive tendency, but it speaks volumes for the feeling that mathematics was to be purified by becoming a special mode of thought.

3.5 The View from Paris and St. Louis

The start of a century is often an excuse for a party, seldom for breast beating. The years 1900 and 1904 saw two major surveys of intellectual fields, in Paris and St. Louis.

The year 1900 was a time for a great number of celebrations of the new century. It was an opportunity to appraise the past and to sing the praises of the immediate future, and many people seized it. Paris played host to no less than six months of international congresses, of which those of the philosophers and the mathematicians were back to back, and several stayed for both. Famously, it was at the Congress of Philosophers that Russell met Peano and realized that there was a deep, subtle subject called mathematical logic that had a lot to offer him, and in which the Italians grouped around Peano were particularly strong.

Russell's own paper was on the virtues of the absolute theory of space over the relative theory espoused by Leibniz and Lotze. Shortly thereafter Peano read the

FIGURE 3.6. The World's Fair, Paris, 1900. Photo reproduced courtesy of Cambridge University Press.

paper on mathematical definitions that excited Russell's admiration. It was discussed by quite a number of people: Jules Tannery, Schröder, and Padoa. Burali-Forti's paper was on a similar theme: different logical methods for the definition of the real numbers. These were nominal, by postulates (the methods of Peano and Dedekind) and by abstraction (Cantor's method). Padoa's paper continued the impressive Italian presence. It was entitled in part "A Logical Introduction to Any Deductive Theory."[125] He described a theory as being a set of undefined symbols and unproved propositions, from which other propositions are derived logically and other symbols introduced by definition. He then proposed a test to see if a symbol cannot be deduced from the propositions of the theory: it is necessary and sufficient to find an interpretation of the system that "verifies the system of unproved symbols and continues to do so if we vary the meaning only of the symbol considered." So it is, to describe it in language that was coming in then but which Padoa did not use, a semantic test of a syntactic theory.[126] The idea is plausible, but Padoa offered it no defense because, indeed, he had no means to do so. Couturat then read Pieri's (1900) paper on 'Geometry Considered as a Purely Logical System." Geometry in this light is a hypothetico-deductive system based on two primitive ideas: point and movement. Primitive ideas, Pieri suggested, should be chosen to be those that are invariant under the largest group of transformations relevant to the science being studied.

And so it continued. Schröder had a generalization of the concept of number to discuss; Calinon proposed that the number continuum generates the geometric continuum (an idea long since surpassed); Lechalas reported on what it means when the same figure occurs in two different geometrical spaces; Hadamard counseled caution in the use of the method of generalization to solve problems.

What is striking at a congress of philosophers, and surely impressed Russell, who took his researches in a new direction on his return from Paris, was that the philosophy of mathematics was very little concerned with traditional philosophy, but very involved with the new methods of logic. Many of the Kantian themes are missing, shoved aside by the precision of Peano's notation and the idea that mathematical methods might be available to solve what had always looked like philosophical questions. Russell was going to take this idea a long way.

Russell did not stay for the International Congress of Mathematicians, but the Italians did. Some had papers to present, some stayed out of interest. The most long-lasting paper presented there did not please them at all. This was Hilbert's plenary address "Mathematical Problems," with its dramatic opening:

> Who among us would not be glad to lift the veil behind which the future lies hidden; to cast a glance at the next advances of our science and at the secrets of its development during future centuries? What particular goals will there be toward which the leading mathematical spirits of coming generations will strive? What new methods and new facts will the new centuries disclose in the wide and rich field of mathematical thought?

He made the challenge: "[The] conviction of the solvability of every mathematical problem is a powerful incentive to the worker. We hear within us the perpetual call:

---

[125] This part is anthologized in van Heijenoort 1967a, 118–123.
[126] Van Heijenoort notes that this claim seems to have slept for some thirty years before proofs were given; by then logic was much more precisely understood.

There is a problem. Seek its solution. You can find it by pure reason, for in mathematics there is no ignorabimus (we shall not know)."

Hilbert offered an analysis of the role of problems in shaping mathematics, both pure and applied, and a number of challenging problems grouped under twenty-three headings. As the years went by and Hilbert's fame grew, and that of Göttingen with it, these problems became famous, and the names of the solvers entered what Hermann Weyl called the Honors Class of mathematicians.[127] But on that day the impact was muted, even though Hilbert sensibly discussed only ten of the problems, leaving the full text of his paper to the printed version.

Significantly, the first of Hilbert's problems was Cantor's continuum hypothesis, which Hilbert observed was connected to the idea that every set can be well ordered. Then, for his second problem, he observed that he himself had been able to use arithmetic to show that various systems of geometry can be defined. This raised the question of finding a direct method to prove the compatibility of the arithmetical axioms, which in Hilbert's view included the Archimedean axiom. Because, in Hilbert's view, all it means for mathematical objects to exist is that they are described by a noncontradictory set of axioms, this means that "the proof of the compatibility of the axioms is at the same time the proof of the mathematical existence of the complete system of real numbers or of the continuum." After the talk, Peano pointed out that more was known about the problem of axiomatizing arithmetic than Hilbert seemed to recognize, and he went on to advertise the fact that Padoa would be reporting on the topic at the congress.[128] Indeed, said Peano, Padoa and others had in fact solved the problem, insofar as that was possible. Two years later Padoa published a paper in which he claimed that the only way to show that a set of axioms for the natural numbers is consistent is to exhibit a set of objects that exist and satisfy the axioms. But what could they be other than the natural numbers? There is a patent risk of a vicious circle, but to call for a proof of consistency without exhibiting objects struck Padoa as absurd. There is another way, but it took logicians many years to develop; Hilbert, it must be said, seems not to have engaged with Padoa's criticisms at all.

Four years later, the call came from St. Louis for another international congress.[129] The occasion was the centennial of the Louisiana Purchase, President Thomas Jefferson's acquisition in 1803 of land so vast as to double the size of the young American nation. It was also the start of the Lewis and Clark expedition to the West, which set off from north of St. Louis in May 1804 on its long journey to the Pacific Ocean.[130] Preparations, however, had overrun, and the exhibition opened a year late, on April 30, 1904. It went on to welcome 20 million visitors. Several of its buildings are today part of the campus of Washington University in St. Louis, and the

---

[127] See Gray 2000 for an account of the problems and their history, and Yandell 2002 for an account of the problems and the people involved.

[128] In fact, Padoa spoke on another topic altogether: Euclidean geometry.

[129] There was also an International Congress of Mathematicians in Heidelberg that year, but it was a much more technical affair.

[130] The sale marks the end of French ambitions in America, signaled by the successful slave revolt led by Toussaint Louverture, which liberated the most productive slave colony of all, Haiti, in the 1790s. It defeated not only French but also British and Spanish attempts to restore slavery. The sale of the former French colonies was partly conducted to spite the British.

FIGURE 3.7. The St. Louis World's Fair. Photo courtesy of Prof. Jeffery Howe, Fine Arts Department, Boston College.

cost of the exhibition has been estimated at around $15 million, about the same as the Louisiana Purchase itself.[131]

The intellectual side of the congress was organized by the Harvard philosopher and psychologist Hugo Münsterberg, a former student of Wundt's, who deliberately tried to break down barriers between academic disciplines and asked speakers to address the unity of the sciences. Inevitably, the attempt failed, although the journal *Science* (September 1904, 445-446) regarded it as a valiant attempt. It noted that some one hundred foreign delegates attended, among whom were the mathematicians and scientists Poincaré, Darboux, Picard, Boltzmann, Ostwald, and Bäcklund, and among the philosophers Höffding and Erdmann. Among the American philosophers and psychologists were G. Stanley Hall, Edward B. Titchener, James McKeen Cattell, James Mark Baldwin, and Josiah Royce. Max Weber spoke on sociology; Woodrow Wilson, then president of Princeton University, on political science.

Of these, Poincaré's contribution is the best remembered, for he came as close as he ever would to producing a theory of electrodynamics that would have rivaled Einstein's a year later.[132] Here, Poincaré proposed a principle of relativity in the form that "the laws of physical phenomena must be the same for a 'fixed" observer as for

---

[131] Information from *Meet Me at the Fair: A Centennial Retrospective of Psychology at the 1904 St. Louis World's Fair* by Ludy T. Benjamin, Jr., http://www.psychologicalscience.org/observer/getArticle .cfm?id=1603.

[132] Poincaré's St. Louis address, "L'État actuel et l'avenir de la physique mathématique," was first published in *Bulletin des sciences Mathématiques* (2) 28 (1904), 302–324, and reprinted in *La valeur de la science* (1905b, 170–211), where it was broken into three essays or chapters and given occasional paragraph headers; otherwise it is identical.

an observer who has a uniform motion of translation relative to him." From this he deduced that "from all these results would arise an entirely new mechanics which would above all be characterised by the rule that no velocity could exceed the velocity of light."[133] On the other hand, Poincaré was also willing to entertain the idea that gravity propagated a million times faster than light, and to contemplate, with more reluctance, the end of Newton's law on the equality of action and reaction. His address is an eloquent testimony to the perplexity felt by the best physicists as they contemplated the evidence of fast-moving electrons.

There were two essays on logic and two on the philosophy of science, one by Erdmann, but nothing that had the vibrancy of the Italian contributions in Paris four years before. Of the papers by mathematicians, most fall into the category of praise for famous men: superficial celebrations of major achievements. Pierpont surveyed the whole of the nineteenth century, and Darboux the development of geometric methods, and both must have gratified their audience. Pierpont ended up proclaiming: "We who stand on the threshold of a new century can look back on an era of unparalleled progress. Looking into the future an equally bright prospect greets our eyes; on all sides fruitful fields of research invite our labor and promise easy and rich returns. Surely this is the golden age of mathematics!"[134]

More bite was to be found in Edward Kasner's address.[135] In his opinion, the most striking development of geometry in the past decade was the critical analysis of the foundations, highlighted by Hilbert but most thoroughly worked upon by Peano and his followers. Mathematicians now knew of many geometries, obtained by varying the axioms suitably, and were faced with such questions as deciding on their simplicity. In the new subject of topology, the concept of a curve had quite escaped intuition, Riemann's call to separate intrinsic and extrinsic geometry had gone largely unanswered, and there were other, more technical difficulties across the subject. Kasner concluded that mathematicians would always have to balance the fact that their subject was ideal, self-created, and, as Cantor has proclaimed, "The essence of mathematics lies in its freedom" with their duty to deal with the unsolved question of the past and to keep their subject in touch with other fields.

Maxime Bôcher, a former student of Klein's who had been born and raised in Boston and became a professor of mathematics at Harvard in 1904, also spoke to more purpose.[136] He asked his audience to consider the question: What is mathematics? It was no longer the science of quantity. It had become too diverse for that. Some, such as Kempe, sought a hidden unity in the objects of mathematics; some (Benjamin Peirce among them) emphasized the methods of mathematics; yet others (Bôcher cited Russell) favored some mix of the two approaches. Recent advances in logic, due to Boole, Schröder, Peano, and, independently, Frege, had cleared the ground so that the concept of necessary conclusions was plain enough, but now Russell's paradox had to be confronted. A further point was that, as Bôcher put it (p. 121), "Since we are to make no use of intuition, but only of a certain number of explicitly stated premises, it is not necessary that we should have any idea what the nature of the objects and relations involved in these premises is." Consequently,

---

[133] Poincaré 1905b, 197.

[134] Reprinted in *The Way It Was*, ed. D. G. Saari, (American Mathematical Society, 2003), 25–48.

[135] Kasner 1904–5.

[136] Bôcher 1904/5.

different interpretations of the same formal setup may be given, and Bôcher quoted H. Wiener's example of a geometric configuration that may also be interpreted sociologically, when it yields such statements as, "In this community, any two men have one, and only one, woman friend in common."

Bôcher was sympathetic but brief with the view that geometry has been considered an experimental science. He noted that Poincaré objected to this view, but Bôcher himself was not persuaded by Poincaré's conventionalism. As for Kempe's view that mathematics was about sets of individual elements subject to certain relations, it coincided with Peirce's if restricted to exact, deductive mathematics. Russell's contribution was the argument, not yet conclusively established, that mathematics, being deduced from logic, was not about any hypothetical system but only those that may be said to exist or to be true.

If all this sounded dry and lifeless, Bôcher could only apologize. His heart was in the applications that gave the science its real vitality. Real mathematics, he was to end his speech by saying, was almost more an art than a science, intuition, experiment, even optimism should guide it past some logical weaknesses. But, he conceded, "It may perhaps be said that instead of inviting you to a feast I have merely shown you the empty dishes and explained how the feast would be served if only the dishes were filled" (p. 132), adding the footnote: "Notice that just as the empty dishes could be filled by a great variety of viands, so the empty symbols of mathematics can be given meanings of the most varied sorts." His plea in mitigation was that his subject was the *fundamental* conceptions and methods of mathematics.

Such, in his view, was the free creation of the human mind: a tightly rule-governed formal manipulation of symbols enlivened by its many, diverse, and valuable applications and capable of many interpretations—although the possibility of unintended ones was not mentioned.

4

MODERNISM AVOWED

4.1 Geometry

4.1.1 Abstract Italian Geometry

We have already seen that Italian mathematicians were energetic students of projective geometry, which some, such as Corrado Segre and Federigo Enriques, extended to n dimensions and treated abstractly, the better to allow it to be interpreted in a variety of ways. Another of Segre's students, who followed him in this work but was more sympathetic to the axiomatic approach, was Gino Fano. When Fano wrote about n-dimensional geometry, he tried systematically to show that each new postulate is independent of the previous ones, and in this way he made some interesting discoveries almost without noticing.

For example, Fano's fourth postulate asserts that each line contains more than two points. Its necessity is demonstrated by proposing a model where each line contains exactly two points (he took the three vertices of a triangle as points, and the three edges of the triangle as lines). This model satisfies the first three postulates but not the fourth, so the fourth postulate is independent. With these four postulates, Fano could construct the fourth harmonic point D, of three collinear points A, B, C. But is the fourth harmonic point D distinct from A, B, and C or not? Fano showed that it need not be by means of a model consisting of seven points and seven lines for which each line has exactly three points. (Take the three vertices of an equilateral triangle, the three midpoints of the edges, and the center of the triangle as points; take the edges of the triangle, the lines through the center to the opposite midpoint, and the three midpoints themselves as lines.)[1] Accordingly, another postulate is needed to ensure that there are harmonic series that do not fold back on themselves in this fashion.

That said, Fano missed the lasting significance of his new geometries, and saw them merely as counterexamples, whose sole purpose was to demonstrate the independence of the axioms. They were not presented as starting points for the de-

---

[1] This counterexample, presented as a projective plane over the field of two elements, is nowadays called a *Fano plane*.

velopment of new geometrical research, as was to be the view of the American geometers in the early 1900s.

The most forceful and innovative axiomatizer of geometry was the still somewhat neglected figure of Mario Pieri (1860–1913), who graduated from Pisa in 1884, and also went to Turin to study with Segre and Peano.[2] From Segre he learned that an axiomatic structure can be valuable even when completely independent of experience, which Peano had denied. From Peano, he learned rigor, and a formulation of geometry that blended the algebraic methods of Grassmann with the formalities of Pasch and further refined them with Peano's way of writing mathematics as a hybrid of logic and set theory.[3]

Pieri's work is marked by his complete abandonment of any intention to formalize what is given in experience. This was the first time this was attempted in geometry. Instead, he wrote that he treated projective geometry "in a purely deductive and abstract manner, . . . , independent of any physical interpretation of the premises."[4] Primitive terms, such as line segments, "can be given any significance whatever, provided they are in harmony with the postulates which will be successively introduced." Pieri presented nineteen axioms as the foundations for projective geometry, based on taking point, line, and motion (by which he meant congruence) as the primitive notions.[5] The premises and the methods were entirely independent of intuition. He made extensive use of Peano's notions of class and membership as a logical tool—he also wrote in the arid fashion of Peano—and in this way proved theorems without any recourse to external intuition or implicit perceptual, linguistic, or cultural experience. As he put it:

> A good ideographic algorithm is generally acknowledged as a useful tool to discipline thought, to eliminate ambiguities, mental limitations, unexpressed assumptions and other faults which are integral part of both spoken and written language, and that are so harmful to speculative investigation. Therefore it is vitally important to make use of the method of algebraic logic. Neglecting it, especially in this type of studies, seems to me a deliberate rejection of the most valid tool for the analysis of ideas that we have at our disposal today.[6]

Pieri went so far beyond Peano in allowing the mathematician to create a geometry that one can wonder what is geometric in Pieri's work at all (although usual projective geometry is a possible interpretation). This was surely not just because, as he knew very well, lines in one geometry may be taken as points in another, and so the basic terms must be capable of several interpretations. It is also because he felt very acutely the need for mathematicians to reason with scrupulous care, therefore he axiomatized a well-understood branch of mathematics rather than helping to ad-

---

[2] On Pieri, the best and most thorough account is Marchisotto and Smith 2007, which also has English translations of two of Pieri's memoirs: *Elementary Geometry Based on the Notions of Point and Sphere* (Pieri 1908) and *On the Axioms of Arithmetic* (Pieri 1907). Marchisotto and Smith discuss in careful logical detail how Pieri deduced all of the basic results of Euclidean geometry from his axioms, and give a detailed comparison of his treatment in 1908 with that of Hilbert; see also Marchisotto 1989, 1993, 1995, and 2006.

[3] As shown in Gandon, 2004.

[4] Pieri 1896a, 381.

[5] Pieri 1895; this account is based on Bottazzini 2001, 315.

[6] Pieri 1898, 4.

vance a new one. Whence the adoption of Peano's generally helpful notation, and also his rebarbative linguistic practices. Geometry, for Pieri, was not logic in the sense a logicist would endorse (there were too many primitive terms for that), but it was a branch of clear thinking, and logicists could easily see it as a great step in their direction (there were not so many undefined terms).

Initially, Pieri's work was very influential. Russell and Couturat regarded him as the founder of mathematics as a hypothetico-deductive science, and while there were earlier figures, few had his unrivaled attention to rigor. If he disappeared from the history of mathematics for decades, it is surely because he was eclipsed by the dominant and ever-growing influence of Hilbert and Göttingen. In addition, he attracted no students.

The language question merits proper examination (see §6.1) and is certainly far from trivial. For example, Klein wrote to him (and his remarks about readership were surely limited to Germans out of politeness):

> Is it not possible to express your thoughts in a simple language without Peano's symbols? My general experience shows that in Germany works written in this symbolic language have got a very limited readership; indeed they are rejected a priori. I do not intend to question the principle of this symbolic language, on the contrary I do believe that it can be very useful in purely deductive research such as yours, to avoid errors that could be all too easy to make. Although this can be of assistance to the researcher when he summarizes his results, he should however be able to express in ordinary language not only the results, but also the reasoning behind such results.[7]

It is indeed hard, if not impossible and most likely deliberately excluded, for someone to write "Peanian" and explain what they are trying to do, why it matters, and why they have done it in this way rather than some other way.

Pieri's work was also curiously self-limiting in much the way that Fano's was. The whole thrust of their work was to axiomatize real and complex projective geometry, and to derive it from as small a number of primitive notions as possible, thereby completing a program of work begun by von Staudt in the 1850s of giving projective geometry its own foundations independent of Euclidean geometry. In this they succeeded. But they did not do more. They did not take up the theme of finite geometry, even though Fano had exhibited examples of finite geometries (and there were famous configurations in the literature almost begging to be studied that way, such as the twenty-seven lines on a cubic surface and the twenty-eight bitangents to a quartic curve).[8] They did not investigate projective geometries with nonstandard coordinates, even though Veronese had attempted, with less than total success, to promote non-Archimedean geometry.

To this must be added a further, institutional factor, noted thus by Avellone et al.

> Klein, amongst others, has indicated another characteristic of the Italian situation as a fascinating strength as well as an obstacle to research. This is the too-direct connection between issues arising from teaching first-year undergraduates and the training of school

---

[7] Avellone 2002, 418.

[8] The first to study finite geometry as an autonomous subject were Beppo Levi (1904) and, at greater length, (1907), and Veblen and Bussey (independently of Levi) in two papers: Veblen and Bussey 1906 and Veblen 1907.

FIGURE 4.1. David Hilbert.

teachers. It was in this context that the question of the foundations took on its great methodological and cultural importance, but this development obstructed the recognition of the subject as part of advanced mathematical research."[9]

In all these ways, the Italians failed to accommodate themselves to a wave they could not have seen coming: Hilbert's *Grundlagen der Geometrie* (*Foundations of Geometry*) (1899).

4.1.2. Hilbert

In the early 1890s, Hilbert established himself as a leading number theorist with his extensive *Zahlbericht* (*Report on the Theory of Numbers*), in which he reformulated number theory along structural lines, giving a high priority to Galois theory as an organizing device. He then startled his Göttingen colleagues by dropping the subject in favor of geometry—not advanced geometry, but its seemingly elementary foundations. They were unaware that Hilbert had taught projective geometry at Königsberg, initially as a chore, but had then found new things to say.

Hilbert's curiosity had been piqued by discovering that the experts disagreed about which theorems followed from which others. In September 1891 he had heard a lecture by Hermann Wiener in which he raised the possibility of developing projective geometry axiomatically from Pascal's and Desargues' theorems.[10] On his return he waited in a railway station with his student Blumenthal, who much later recalled that Hilbert took the occasion to say, "One should always be able to say,

---

[9] Avellone et al. 2002, 365.
[10] These theorems are recalled in Appendix 1.

instead of "points, lines, and planes," "tables, chairs, and beer mugs."[11] Hilbert meant by this that words should not be used in an argument with any other meaning than those given by their definition, and the logical validity of an argument should depend on the structure of the argument and not on any unspecified meanings of the terms. Hilbert was far from the first to make this point, but he was to grasp its significance more profoundly.

In early January 1898 Friedrich Schur sent a letter to Klein which Klein passed on to Hilbert. Schur raised the possibility of deriving Pappus's theorem without using the Archimedean axiom (which says that given any two lengths $a < b$ there is always an integer n such that $na > b$). This suggested that there might be a new geometry, independent of the Archimedean axiom, and, thus inspired, Hilbert lectured on non-Archimedean geometry at Göttingen in the winter semester. As Blumenthal wrote, "This caused astonishment among the students, for even we older people, participants in the stroll through the number fields, had never known Hilbert to concern himself with geometrical questions. Astonishment and wonder were caused when the lecture began and a completely novel content emerged."[12]

At that time, the Göttingen professors were occupied with the idea of unveiling a statue to Gauss and Wilhelm Weber. Hilbert was invited to write up his ideas on geometry as part of the celebration, and this he did.[13] His *Grundlagen der Geometrie* was a great success. It has run to ten editions, seven in Hilbert's lifetime, and been translated into several languages. It came to exemplify what Hilbert's ideas about mathematics were, perhaps because they stood out with exceptional clarity in such an elementary arena (and notwithstanding a number of interesting flaws).

At the head of the work, Hilbert placed this motto from Kant's *Critique of Pure Reason*: "Thus all human knowledge begins with intuitions, proceeds from thence to concepts, and ends with ideas."[14] Most writers about Hilbert seem to have assumed that this is a polite, or perhaps even slightly self-aggrandizing nod at Hilbert's distinguished fellow Königsberger, but in fact it was a very precise hint to the educated reader of 1899 about Hilbert's approach to geometry. It comes from the appendix to the Transcendental Dialectic, in the section called entitled "The Final Purpose of the Natural Dialectic of Human Reason," and the key is to unpack the seemingly obscure distinction between concept and idea. Earlier in the *Critique* Kant had introduced a distinction between reason and understanding. Reason (A299, B 355) goes beyond understanding because out of it grows the capacity to discover why certain judgments are true. This transcendental aspect of reason has its own concepts, which Kant called "transcendental ideas." Reason, Kant argued, has a regulative role in science, set out at length in the third chapter of the Transcendental Dialectic ("The Ideal of Pure Reason'): it is what makes hypothetico-deductive, continually growing, theories possible.[15] The ideas of reason determine the procedures by which the un-

---

[11] Blumenthal 1922, 403.

[12] Blumenthal 1935, 402.

[13] It was published with Emil Wiechert's essay *Grundlagen der Elektrodynamik*. On Wiechert, see Corry 2004.

[14] *Critique* A702, B730.

[15] Contrary to Pace (Gardner 1999, 222), this does not make the theories scientific for Kant, although it does for us. For Kant, a science is apodictic, and so the operation of reason would be constitutive, but it can only be regulatory. I thank José Ferreirós for persuading me to disagree with Gardner here.

derstanding may gain knowledge of the objects of experience. Kant explicitly, and right before the quote in question, stresses that pure reason contains only regulative, not constitutive principles. If therefore, we take the motto to indicate that Hilbert is starting where Kant left off, in the realm of ideas—that is, regulative principles—then we can see that educated readers in 1899 would pick up the suggestion that Hilbert was going to consider the regulative rules governing geometrical concepts. And this is exactly what he does.

Hilbert explained in the introduction that his aim was to set out a new, axiomatic account of geometry that was complete and as simple as possible, to derive the most important theorems in elementary geometry and to explain the significance of the different groups of axioms into which his presentation was divided. He began: "We think of three different systems of things: the things in the first system we call points and denote A, B, C, . . . ; the things of the second system we call lines and we denote a, b, c, . . . ; the things of the third system we call planes and denote α, β, γ, . . ." Just like the tables, chairs, and beer mugs. There then followed five kinds of axiom which determine what one can say about these things. The axioms of incidence enable one to say such things as "this point lies on this line," those of order such things as "this point lies between these two." There follow axioms about congruence of line segments and congruence of angles, axioms about parallels, and finally axioms about continuity (which is where Hilbert located the Archimedean axiom).

At each stage Hilbert deduced theorems from the available set of axioms, which made it clear what axioms entail what theorems. He also endeavored, with considerable but not complete success, to show that the axioms are mutually independent and consistent.

The axioms of incidence, order, and congruence alone permitted Hilbert to establish a number of results about triangles, including Desargues' theorem in the plane. This in turn allowed Hilbert to move segments in the plane around (moving them to congruent segments) and so to establish what he called "segment arithmetic." This gave him ways analogous to addition and multiplication of combining segments to get others. This is, after all, how a surveyor lays out a square grid, and in Hilbert's case it followed that if the geometry was coordinated, the coordinates would have to obey certain rules. In fact, he showed that when Pappus's theorem is true the coordinates must lie in a system where multiplication is commutative, but Desargues' theorem can be true in a system where multiplication is not commutative. Indeed, Hilbert nearly showed what Hessenberg was to establish six years later, that Pappus's theorem implies Desargues' theorem.[16]

To study non-Archimedean geometry, Hilbert produced a non-Archimedean system of "numbers."[17] These new, non-Archimedean "numbers" contain the familiar real numbers as a subset. Hilbert then showed that the theory of proportion,

---

[16] Hessenberg 1905.

[17] Hilbert used series of the form $\alpha = a_0 t^m + a_1 t^{m-1} + \ldots$, where M is a positive or negative integer and the coefficients a_i are real numbers. The order relation is defined as follows: α is greater than, equal to, or less than 0 according as the leading coefficient a_0 is positive, zero, or negative; α is greater than, equal to, or less than β according as $\alpha - \beta$ is greater than, equal to, or less than 0. Any two such "numbers" can be compared, but they are not Archimedean because no matter how many copies of 1 are added to itself (to form $n.1 = n$), n is always less than t.

FIGURE 4.2. The unveiling of the Gauss-Weber memorial, Göttingen. Photo courtesy of Stefan Krämer, Gauss Society Göttingen, Städtisches Museum Göttingen.

similarity, and plane area are the same in Archimedean and non-Archimedean geometry.

Hilbert showed that the admission or exclusion of the Archimedean axiom has a decisive effect on the validity of Pappus's theorem. On the one hand, the axiom of Archimedes forces Pappus's theorem to be true. On the other hand, Hilbert could set up an analog of coordinate geometry in which the coordinates are drawn from an ordered but noncommutative system of "numbers."[18] The corresponding geometry is non-Pappian and a non-Pappian geometry is therefore necessarily a non-Archimedean one.

Hilbert also gave a convoluted example of a plane geometry in which Desargues' theorem was false, a result that had been suspected for some time but not conclusively established. His example was much improved by the American astronomer Forest Ray Moulton in 1902, when attending a seminar organized by E. H. Moore in

[18] Hilbert introduced this system of "numbers." These are series of the form $\sigma = \alpha_0 s^n + \alpha_1 s^{n-1} + \ldots$, where the α_i are as defined above, n is a positive, zero, or negative integer. So the α_i are power series in a variable t, and when one writes the expression out in full, or when one multiplies two "numbers" of this form together, the all familiar rules of algebra apply, except that $ts = -st$. The new "numbers" are ordered, like the α_i, by looking at the sign of the leading coefficient α_0, but they are noncommutative. They belong therefore with a non-Pappian geometry.

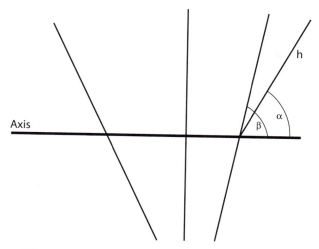

FIGURE 4.3. Straight lines on Moulton's definition.

Chicago on Hilbert's *Grundlagen*; Hilbert included Moulton's example in later editions of his book.[19] Moulton divided the familiar plane into two by a line he called the axis, as in figure 4.3, and then defined lines to be either a conventional line parallel to the axis, or a conventional line crossing the axis but sloping to the left, or a vertical line, or a new kind of line obtained from a line crossing the axis and sloping to the right. If such a line makes an acute angle, α, with the axis at a point P, it is replaced by a broken line composed of the half of the original line below the axis and a line segment above the axis, making an angle β at P with the axis, where $\tan \alpha = \frac{1}{2} \tan \beta$.

Even with this remarkable definition of a straight line, all the usual axioms of projective geometry are satisfied—for example, two lines meet in at most one point—but in this geometry, Desargues' theorem is false, as Moulton's figure, figure 4.4, shows. What gives this example its bite is that the incidence axioms for projective geometry in three dimensions are strong enough to imply Desargues' theorem. More precisely, if Desargues' theorem is stated for a figure in a plane, and that plane can be thought of as a plane in a three-dimensional space, then Desargues' theorem can be proved for that figure. It follows that Moulton's plane cannot be embedded in any three-dimensional space.

It is worth savoring this triumph of the axiomatic method, and noting its modernist characteristics. First, a certain set of axioms adequate to prove some theorems in plane geometry are written down. Then a precise set of objects is produced—the "points" and the "lines"—which satisfy those axioms. They need not do anything

---

[19] Moore had attended the unveiling of the Gauss-Weber memorial in Göttingen. Another student in the seminar was Oswald Veblen, who wrote his PhD thesis on twelve axioms, based on the notions of point and order, for Euclidean geometry. Veblen eventually became one of the leading American geometers and a major influence at the Institute for Advanced Study in Princeton, New Jersey.

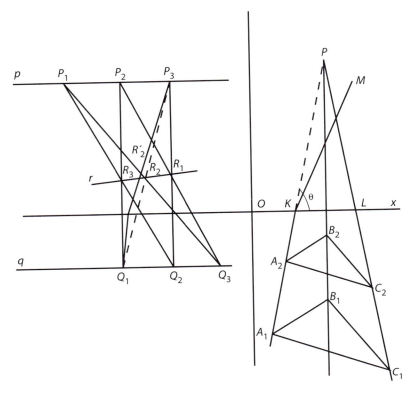

FIGURE 4.4. The failure of Pappus's and Desargues' theorems.

else. They may be counterintuitive, but they satisfy the axioms and so it is legitimate to talk about them in the context of plane geometry. Then it is shown that these objects fail to satisfy a certain theorem. This establishes that those axioms are not sufficient to prove the theorem. On the other hand, add just one axiom (which says no more than that space is three dimensional), and the resulting set does imply the theorem. It follows that there can be planes that cannot be regarded as subsets of a three-dimensional space. Such strange, counterintuitive planes could not have been discovered without the axiomatic method; without it, indeed, it is doubtful if Moulton's lines would have counted as lines. So mathematicians acquired a highly nonintuitive way of thinking about lines, and as a result obtained a novel view of planes, and also a striking sense of what axioms imply what conclusions.

The axiom system for plane geometry admits several exemplifications, or models, as they came to be called. This is a property an axiom system may or may not possess, and some interest came to attach to axiom systems that have an essentially unique model, and are satisfied by an essentially unique set of objects. This reflects a further modernist turn, as the study of axiom systems became a topic of mathematics in its own right. And Hilbert made repeated use of this device of choosing an axiom system with different models in order to show that such-and-such a theorem does not follow from the axioms. So much so that his former professor at Königsberg, Adolf Hurwitz, wrote to him, "You have opened up an immeasurable field of mathematical investigation that can be called the "mathematics of axioms" and goes far beyond the domain of geometry."[20]

The account of lines and planes given by Hilbert and Moulton is modernist in another way, in its attenuated relationship with the real world.[21] We may adapt an argument by Charles Rosen, who wrote: "In a Cubist painting . . . it is often hard to identify what object is represented until we have admired the painted surface. . . . Before modernism, . . . , the emotion or sentiment represented by music was evident at once. . . . With Schoenberg and Webern, . . . , and with Stravinsky (starting with the *Rite of Spring*) we must generally begin with a dispassionate understanding of the art and an appreciation of the technique in order to comprehend the emotional content. . . . In those works of the modernist movement considered difficult or hermetic, in short, the content is partially withheld from us until we have understood the technique. We could say "Moulton's work, and still more Hilbert's, cannot be understood as geometry before we have a dispassionate understanding of the art and an appreciation of the technique in order to comprehend the intellectual content, which is partially withheld from us until we have understood the technique.'"

Hilbert, like Mallarmé, could have said "I become obscure, of course, if one makes a mistake and thinks one is opening a newspaper."[22] The words of David Mumford, one of the leading algebraic geometers in the later half of the twentieth century, echo this observation.

> The 20th century has been, until recently, an era of "modern mathematics" in a sense quite parallel to "modern art" or "modern architecture" or "modern music." That is to say, it turned to an analysis of abstraction, it glorified purity and tried to simplify its results until the roots of each idea were manifest. These trends started in the work of Hilbert in Germany, were greatly extended in France by a secret mathematical club known as "Bourbaki," and found fertile soil in Texas, in the topological school of R. L. Moore.[23]

A further novelty in Hilbert's approach to axiom systems is also worthy of note. Most axiom systems hitherto had tried to codify what was already known. Hilbert saw, as Fano and Pieri had done before him, that the axiomatic method permitted one to *create*. Using it, one could recognize and study novel structures—non-Desarguean planes, for example—that were all but inaccessible with the conventional methods of the day. But only Hilbert embraced this possibility. Hilbert's message was increasingly taken by his followers to be that if mathematicians have a difficult problem they should try to axiomatize what they are studying, and work from the axioms. This exercise in codifying what one knows may well lead to the recognition that there is more going on, and so axiomatics becomes a process of discovery too.

Some later editions of Hilbert's *Grundlagen der Geometrie*, including the standard English translation (*Foundations of Geometry*, Open Court, 1971), carry an

---

[20] Quoted in Toepell 1986, 257.

[21] The sense of this abstraction being truly modernist is heightened by the amusing coincidence that another book on points, lines, and planes written (in fact later) in the period is *Point and Line to Plane*—by Wassily Kandinsky. See Kandinsky 1926.

[22] For both quotes, see Rosen 1999, 44.

[23] In Parikh 1991, xxvii.

axiomatization of non-Euclidean geometry that Hilbert first published in volume 57 of *Mathematische Annalen* in 1902. It had not been Hilbert's original intention to describe Euclidean geometry and non-Euclidean geometry as a contrasting pair of axiom systems, but, as the qualities of an axiomatic approach to mathematics were more and more vividly perceived, it became usual to see non-Euclidean geometry as exemplifying an axiom system different from Euclid's. It certainly is such a thing, but the original discovery was done in a different way and for different reasons. After all, a consistent axiom system different from Euclid's had been available all along: spherical geometry. Even to see the history of non-Euclidean geometry as the search for an axiom system differing from Euclidean geometry only in respect of the parallel axiom is to overstate the importance of axioms and to understate other questions about the nature of geometry: the role of transformations, the use of formulas, and its applicability to space.

4.1.2.1 STRAIGHTNESS AND SHORTEST DISTANCE

Hilbert's own emphasis on the creative powers of axiomatic geometry are evident in the famous speech he gave at the International Congress of Mathematicians in Paris in 1900. His fourth problem raised what might seem to be a question only a pure mathematician could ask, for it queried whether a straight line measures the shortest distance between its points. He noted that this theorem "has, indeed, been employed by many authors as a definition of a straight line," but he was looking, he said, for geometries that might stand next to Euclid's, and the triangle inequality (any two sides of a triangle are together longer than the third) had caught his attention. This was because it followed from his axiomatization of geometry that the inequality requires a theorem about the congruence of triangles and cannot be proved from results about segments and angles alone. He reduced the matter to nothing more than the isosceles triangle theorem that if two sides of a triangle are equal in length, the corresponding angles are equal.

How could this be false? Because of a fact generally overlooked in elementary geometry—that the corresponding angles in an isosceles triangle are equal in size only when we forget about the direction in which they are measured. Strictly, if one angle is measured clockwise, the other must be measured anticlockwise. We usually deal with this detail inconsistently, claiming both as students of mathematics and as people leading ordinary lives to ignore it or remember it as we see fit. Hilbert was asking if, in a plane geometry in which figures can be moved around and are congruent if they can be made to fit exactly, it necessarily follows that the definition of congruent can be extended to mirror symmetric figures. In fact, he already knew that the answer was "no," and remarkably a geometry that demonstrates this had been constructed as a powerful aid to researches in number theory by his friend Minkowski. As Hilbert put it: "One finds that such a geometry really exists and is no other than that which Minkowski constructed in his book, *Geometrie der Zahlen*, and made the basis of his arithmetical investigations. Minkowski's is therefore also a geometry standing next to the ordinary Euclidean geometry."[24] Hilbert concluded that because the idea of distance was fundamental to the use of geometry everywhere

---

[24] See Hilbert 1899, 12, the reference is to Minkowski 1896.

in mathematics, a further investigation along these lines was desirable. Thus, even the primitive concept of distance was dissolved by Hilbert into the methodology of axiom systems and implicit definitions—definitions that provide meaning from use, and not use from meaning.

4.1.2.2 POINCARÉ

Hilbert's *Grundlagen der Geometrie* was reviewed by Poincaré for Darboux's *Bulletin* in 1902, and the review was immediately translated, with Poincaré's permission, by Huntington for the *Bulletin of the American Mathematical Society*, where it appeared in 1903. The extent of the agreement, and the nature of the disagreement, between the two leading mathematicians of their day is instructive.

Poincaré began by observing that non-Euclidean geometry had once scandalized us (and even today, he said, the Académie des Sciences in Paris receives several "proofs" a year of the truth of the parallel postulate, which of course it refuses to publish). Some now had swung the other way, and supposed that experiment could decide if space was slightly curved or not. "Needless to add that this was a total misconception of the nature of geometry, which is not an experimental science"—a reference to his conventionalist philosophy of geometry, as will be discussed below. But from an axiomatic standpoint there is no reason for the parallel postulate to be the only one singled out for critical analysis, and indeed it is no longer alone. Mathematicians, he reminded his audience, have clarified the number concept, enriched it with complex numbers, enlarged the operations of arithmetic with quaternions and their noncommutative multiplication—"a revolution in arithmetic quite comparable to that which Lobachevskii effected in geometry," said Poincaré (1903, 3) with a degree of exaggeration. Cantor had enlarged our conception of the infinite, Italian mathematicians have "endeavored to construct a universal logical symbolism and to reduce mathematical reasoning to purely mechanical rules." And now we have Hilbert's new approach to geometry.

Poincaré reviewed Hilbert's axioms in the order in which Hilbert had introduced them, commenting astutely that "we must find out whether, if we put these axioms into the reasoning machine, we can make the whole sequence of propositions come out. If we can, we shall be sure that nothing has been overlooked. For our machine cannot work except according to the rules of logic for which it has been constructed; it ignores the vague instinct which we call *intuition*."[25] How much of a criticism this was of Hilbert's approach we shall see later.

Poincaré noted that the rules Hilbert wrote down do not give us the usual geometry: "His space is not our space, or at least is only a part of it" because his axioms

---

[25] Poincaré 1903, 12. Poincaré noted that when the novel geometries arrived, Hilbert's work takes an arithmetical turn. Poincaré listed four groups of concepts:
   (i) the usual laws of arithmetic except for the commutative law of multiplication;
   (ii) the axioms of order and the usual rules for inequalities;
   (iii) the commutative law of multiplication;
   (iv) the axiom of Archimedes.

Numbers satisfying (i) and (ii) shall be called Desarguean, Pappian if they satisfy the (iii) as well and non-Pappian if they do not, and independently of these, Archimedean or non-Archimedean if they do or do not satisfy (iv). Here I have written Pappus for Pascal in line with modern practice and accurate historical attribution.

are strictly those of ruler and compass geometry, and so they do not permit an angle to be trisected. Hilbert's logical concerns, as we have seen, drove him to introduce a non-Archimedean geometry, in which lengths are non-Archimedean magnitudes. Poincaré commented: "At first blush the mind revolts against conceptions like this. This is because, through an old habit, it is looking for a visual image. It must free itself from this prejudice if it would arrive at comprehension."[26]

Poincaré had only two criticisms of this work. One was that Hilbert seemed to neglect the way in which the allowable transformations in each geometry form a group. This was quite unlike Lie's way of proceeding with the geometry of manifolds, and of course ran counter to Poincaré's insistence that groups come before geometries, which are constructed out of them. The other was the absence of interest in psychology. "The logical point of view alone appears to interest him. Being given a sequence of propositions, he finds that all follow logically from the first. With the foundation of this first proposition, with its psychological origin, he does not concern himself. . . . His work is then incomplete; but this is not a criticism which I make against him. Incomplete one must indeed resign one's self to be."[27] The lament, one might say, of the man rooted in the world against the world of fantasy, or the reproach of the semanticist to the siren singing of syntax.

4.1.2.3 ENRIQUES

One the many interesting features of Klein's monumental and inevitably incomplete *Encyklopädie der Mathematischen Wissenschaften* in twenty-three volumes is that the essay on the "Principles of Geometry" is not written by Hilbert, his Göttingen colleague and world authority, but by Enriques.[28] Enriques (1907) took the opportunity to develop his thought in an organic way. He supplied lengthy, and not always accurate, historical digressions, which were very important to his ideas about how research in mathematics can also be done. In this spirit, he drew together the work of Veronese and Levi-Civita on infinitesimal magnitudes with the axiomatic geometries of Hilbert. He contrasted Hilbert and Peano in these terms: Hilbert studied any system of postulates, without regard to physics or psychology, provided they had mathematical interest; Peano emphasized formal, logical considerations. It was Hilbert and his followers who were, therefore, more and more abstract.

Enriques discussed the work of Hilbert and his followers carefully and at length. Among these was Max Dehn, a gifted mathematician who had already made his name by solving one of Hilbert's twenty-three Paris prize problems.[29] Dehn had gone on to do extensive work on the consequences for geometry of denying the Archimedean postulate. He found two new geometries that can only exist in a non-Archimedean setting, a result that even Hilbert found surprising. Enriques discussed this work at the end of the essay and noted its implications for the related study of the parallel postulate, thus further strengthening the idea that Euclidean and non-Euclidean geometries are best studied axiomatically.

---

[26] Poincaré 1903, 5.
[27] Poincaré 1903, 23.
[28] Hilbert contributed just one essay, on algebraic number fields: Hilbert 1900.
[29] See Gray 2000.

4.1.3 Implicit Definitions

The use of axioms to create a system of objects by specifying the rules for manipu-lating the objects came to be called the "method of implicit definition." It does not so much tell the reader what an object (for example, a geometry, or a line in a geometry) is as what can be said about it. It does not give the meaning of the terms but specifies their use. The collapse of a distinction between meaning and use was a radical philosophical departure, one that considerably predates Wittgenstein's discussions of the much wider implications of that collapse in his *Philosophical Investigations*. The novelty of this move can be measured by Russell's failure to understand what Poincaré was saying in their polemical exchanges about the nature of geometry. As Coffa showed,[30] Russell wanted what he called a philosophical analysis of geometry, one that ultimately rested on undefinable primitive terms that would function as atoms of meaning in the sentences of the theory. Poincaré asked for a geometrical characterization of the terms, knowing much better than Russell that there was an increasing number of distinct theories in which they could be used differently. He objected that Russell had intuitions, such as the equality of two line segments in different places, or of two time intervals, that he, Poincaré, could not have—in other words, that Russell's supposed access to the meaning of the indefinable terms was an illusion, one fostered by a false theory of what the words "distance" or "time" mean. The falsity resided in the idea that distance, for example, was known about inde-pendently of any discussion of how it is determined. To quote from Coffa's con-clusion, which endorses the view of Poincaré (and Hilbert, one might add) (p. 134): "Geometric axioms are definitions disguised as claims, and what they define are the indefinables."

4.1.4 The Nagel-Enriques Thesis

In 1936 Ernest Nagel wrote one of the more interesting papers in the history of mathematics, in which he argued that the reformulation of geometry was an im-portant source of modern logic. His insight was that while duality in plane projective geometry puts points and lines on the plane on a par, intuition always prefers points. So mathematicians were pushed away from regarding intuition as the basis of ge-ometry, and toward formalism and then logic. Nagel's example of the formalist geometer was, inevitably in 1936, Hilbert, but this unfairly slights Enriques. Indeed, the insight of Nagel owes a lot to the original work of Enriques.[31]

Enriques, in his *Problems of Science*, examined the research in projective geometry from the time of Plücker, and noted how various sorts of objects in one space are treated as points in another space. This led him to the paradoxical conclusion that most familiar definitions are no definitions at all, but the more obscure ones are. The usual definitions of a straight line, for example, as the curve of a shortest distance joining any two of its points, or of lying evenly along itself, are rather descriptions or definitions in the psychological sense of being intended to recall certain images. However, more abstract definitions do define. For example, he said, a system of

---

[30] See the discussion in Coffa 1991, 129–134.
[31] This subsection recapitulates arguments I have made in Gray 2006d.

postulates gives an implicit definition of the objects it refers to. The appropriate methods for reasoning about such things being necessarily abstract or logical, Enriques accordingly came to something like the insight that Nagel was to spell out thirty years later, but with a difference. Enriques argued that while the high level of abstraction forced geometers to argue more or less formally, he nonetheless believed that there was a core of meaning in the subject without which it was a sterile activity. Nagel, on the other hand, more aware of the developments inspired by Hilbert's work, but forgetful of Enriques's Italian contemporaries, placed greater emphasis on purely formal reasoning and the counterintuitive (and therefore nonintuitive) nature of the objects.

Non-Euclidean geometry further promoted this tendency. It is the geometry that raises the question of the nature of space, and with it the embarrassing problem of explaining why mathematicians had been so wrong about geometry for so long.

4.1.5 Non-Euclidean Geometry

Beltrami's contribution was taken to be the production of a map, on a disk, of a non-Euclidean plane, with a new metric defined inside the disk so that one can say the intrinsic geometry of the disk (with respect to this new metric) is non-Euclidean geometry. It established the intellectual coherence of the new geometry. It displayed the new geometry in terms that were every bit as good as those of Euclidean geometry. For perhaps thirty years this was taken to be good enough. What more could be asked? With the rise of axiomatic systems for geometry, not all of which were equally plausible or realistic and all of which had to be supported with a suitable model to establish their intellectual coherence, it was perhaps inevitable that someone would finally ask the dangerous question: Was Beltrami's disk model good enough?

That person was Poincaré, and his unsettling reply was tantamount to saying that Euclidean and non-Euclidean geometry stand or fall together. He asked his readers to consider the implications of a self-contradiction in non-Euclidean geometry. Figures in non-Euclidean geometry appear to us as figures in the Beltrami disk and the Poincaré disk. Consider then the supposed argument which shows that something is simultaneously true and false in non-Euclidean geometry. Interpret every stage of that argument in terms of the corresponding Euclidean figure—Poincaré invited his readers to imagine they were using no more than a French-German dictionary. The translated argument shows that something is simultaneously true and false in Euclidean geometry. A self-contradiction in non-Euclidean geometry would also be a self-contradiction in Euclidean geometry. "But," said Poincaré, "no one doubts that ordinary geometry is exempt from contradiction. Whence is the certainty derived, and how far is it justified? That is a question upon which I cannot enter here, but it is a very interesting question, because it requires some further work." Poincaré published these words in 1891, and his mind could first have been set at rest on this question in 1899 when Hilbert spelled out how to establish the consistency of Euclidean geometry by giving it an arithmetical model, thus reducing the question to the consistency of arithmetic.[32] To be sure, Poincaré did not go far enough, but it is

---

[32] These words are usually consulted today in the English translation of his *La Science et l'hypothése*, published in 1905. See p. 43, where the sentence ends "and I think not insoluble," which is something of a free translation from the French "car elle exigerait quelques développements."

patent that one may reverse the roles of Euclidean and non-Euclidean geometry, and then the same argument shows that a self-contradiction in Euclidean geometry is also a self-contradiction in non-Euclidean geometry. Euclidean and non-Euclidean geometry stand or fall together. This remarkable conclusion, which Poincaré was the first even to hint at, implied that all the attempts since the Greeks to establish the absolute truth of Euclidean geometry by finding a self-contradiction in non-Euclidean geometry would, had they ever succeeded, at that moment have destroyed Euclidean geometry as well—like Samson, the authors would have pulled their own house down around them.

The idea that space might be Non-Euclidean was the most mundane novelty entertained in the popular mind, but it was also seen as the one with the most chance of being true. Professional astronomers such as Gauss and Bessel had had no problem accepting the idea, although it was clear to them that space could only be very slightly non-Euclidean. Riemann, in his *Lecture*, had gone as far as to say that all surveyable regions of space were Euclidean, but that still left a theoretical possibility open. Once mathematicians accepted non-Euclidean geometry, the attractive possibility opened up that in some theoretical sense space might be described better by non-Euclidean than by Euclidean geometry.

How might the question be resolved? One could choose a plane, a line, and a point in that plane and look at all the lines through that point to see if they meet the given line, but this is absurd because it involves checking that two lines do not meet, however far they are extended. No conclusive result can reliably be obtained this way. It would be better to investigate the angle-sum of triangles. In non-Euclidean geometry, the angle-sum of a triangle is always less than π, and in principle the accurate measurements of angles in just one triangle would suffice to resolve the matter. Now the problem is just one of precision instrumentation.

Alternatively, one could attempt to draw squares according to the Euclidean recipe: draw a segment of a fixed length, turn through a right angle, draw the segment again, turn through a right angle, draw the segment again, turn through a right angle, draw the segment again, and compare the starting and finishing points. They agree in plane Euclidean geometry, but not in non-Euclidean geometry. Again, it is a matter of precision.

Lobachevskii himself had suggested a third method, which involves measuring the parallax of stars.[33] In a truly Euclidean universe, this can be arbitrarily small, and diminishes with the distance of the star from the Sun, but in a non-Euclidean universe, although it diminishes with the distance of the star from the Sun, it cannot fall short of an amount determined by the radius of the Earth's orbit. Quite apart from the formidable difficulties involved in measuring small angles of parallax, it would be difficult to be sure that one had measured enough sufficiently large triangles to be sure that parallax could not be as small as one wished.

All of these arguments have in common the idea that the question could, at least in principle, be resolved one way or the other. As such, they were all rejected by Poincaré, who wrote a series of papers around 1900 that gave rise to a philosophy of geometry, and indeed of science, called conventionalism. This philosophy was taken up by the Vienna circle, and continues to be of interest to the present day.

---

[33] Lobachevskii had two arguments, one in his first work in 1829 (see Lobatschefskij 1899, §15) and the one referred to here in his last work, the *Pangéométrie* (1856); see §9, 76–78.

FIGURE 4.5. Henri Poincaré.

4.1.6 Poincaré's Geometric Conventionalism

Poincaré's novel answer was that while non-Euclidean geometry made sense, there was no way of telling if space was Euclidean or non-Euclidean. His argument is seductively delightful to quote, as are so many of his:

> If Lobatschewsky's geometry is true, the parallax of a very distant star will be finite. If Riemann's [spherical] geometry is true, it will be negative. These are the results which seem within the reach of experiment, and it is hoped that astronomical observations may enable us to decide between the two geometries. But what we call a straight line in astronomy is simply the path of a ray of light. If, therefore, we were to discover negative parallaxes, or to prove that all parallaxes are higher than a certain limit, we should have a choice between two conclusions: we could give up Euclidean geometry, or modify the laws of optics, and suppose that light is not rigorously propagated in a straight line. It is needless to add that every one would look upon this solution as the more advantageous. Euclidean geometry, therefore, has nothing to fear from fresh experiments.[34]

Thus Poincaré argued that any experiment would involve an interpretation. One could always say that light rays (or whatever played the role of straight lines in the experiment) were indeed straight, and so space was non-Euclidean (if that is what the measurements seemed to indicate). Or, one could say that space was Euclidean, and that rays of light were curved. The key insight Poincaré offered was that there was no possibility, however, of deciding between these alternatives on logical grounds. So the only way forward was an arbitrary choice based on human convenience and maintained by a collective agreement or convention, but nothing was forced.

---

[34] Poincaré 1902c, 95–96; English trans., 72–73.

(Poincaré's reason for advocating that the conventional choice be that space was Euclidean will bear examination, and is discussed below.)

He also offered a related argument, as follows:

> There exists no property which can . . . be an absolute criterion enabling us to recognise the straight line, and to distinguish it from every other line. Shall we say, for instance, "This property will be the following: the straight line is a line such that a figure of which this line is a part can move without the mutual distances of its points varying, and in such a way that all the points in this straight line remain fixed?" Now, this is a property which in either Euclidean or non-Euclidean space belongs to the straight line, and belongs to it alone. But how can we ascertain by experiment if it belongs to any particular concrete object? Distances must be measured, and how shall we know that any concrete magnitude which I have measured with my material instrument really represents the abstract distance? We have only removed the difficulty a little farther off. In reality, the property that I have just enunciated is not a property of the straight line alone; it is a property of the straight line and of distance. For it to serve as an absolute criterion, we must be able to show, not only that it does not also belong to any other line than the straight line and to distance, but also that it does not belong to any other line than the straight line, and to any other magnitude than distance. Now, that is not true, and if we are not convinced by these considerations, I challenge any one to give me a concrete experiment which can be interpreted in the Euclidean system, and which cannot be interpreted in the system of Lobatschewsky. As I am well aware that this challenge will never be accepted, I may conclude that no experiment will ever be in contradiction with Euclid's postulate; but, on the other hand, no experiment will ever be in contradiction with Lobatschewsky's postulate.[35]

It is surely right to see here the hand of someone who had spent many years at the *Bureau des Longitudes* wrestling with questions of what can actually be measured and how[36]—a question we shall pursue further below when measurement is our theme.

This is a dramatic moment. Had not Poincaré earlier proclaimed, after discussing the relative consistency of Euclidean and non-Euclidean geometry, that "Lobachevskii's geometry being susceptible of a concrete interpretation, ceases to be a useless exercise and may be applied"?[37] (Hypothesis, p. 43.) The international expert both on non-Euclidean geometry and measurement, who might well have come up with an ingenious experiment, instead said that it cannot be done. There is, he proclaimed, no logical distinction to be drawn between geometry and physics. Or rather, there are, of course, all sorts of purely mathematical theories about different geometries, but there is no way of granting meaning to their key terms that preserves a separation between geometry and physics. Non-Euclidean geometry had previously been seen to poison fatally the geometers' claim to apodictic reasoning about the world, but the antidote was thought to be but one crucial experiment. Now it was claimed that there was a plurality of geometries, all rigorous intellectual enterprises, and full-blooded science: nothing in between.

---

[35] Poincaré 1902c, 97; English trans., 74–75.
[36] As described in Galison 2003.
[37] Poincaré 1891.

Poincaré's philosophical claims were disputed head-on by Enriques in his *Problemi della scienza* of 1906 which, after some delays, was translated into English as *Problems of Science* and published in 1914. To appreciate it, we need to avail ourselves of Poincaré's anthropomorphized account of non-Euclidean geometry, which he offered his readers so that they may feel comfortable with the subject.

> Suppose, for example, a world enclosed in a large sphere and subject to the following laws:—The temperature is not uniform; it is greatest at the centre, and gradually decreases as we move towards the circumference of the sphere, where it is absolute zero. The law of this temperature is as follows.—If R be the radius of the sphere, and r the distance of the point considered from the centre, the absolute temperature will be proportional to $R^2 - r^2$. Further, I shall suppose that in this world all bodies have the same coefficient of dilatation, so that the linear dilatation of any body is proportional to its absolute temperature. Finally, I shall assume that a body transported from one point to another of different temperature is instantaneously in thermal equilibrium with its new environment. There is nothing in these hypotheses either contradictory or unimaginable. A moving object will become smaller and smaller as it approaches the circumference of the sphere. Let us observe, in the first place, that although from the point of view of our ordinary geometry this world is finite, to its inhabitants it will appear infinite. As they approach the surface of the sphere they become colder, and at the same time smaller and smaller. The steps they take are therefore also smaller and smaller, so that they can never reach the boundary of the sphere. If to us geometry is only the study of the laws according to which invariable solids move, to these imaginary beings it will be the study of the laws of motion of solids deformed by the differences of temperature alluded to.
>
> No doubt, in our world, natural solids also experience variations of form and volume due to differences of temperature. But in laying the foundations of geometry we neglect these variations; for besides being but small they are irregular, and consequently appear to us to be accidental. In our hypothetical world this will no longer be the case, the variations will obey very simple and regular laws. On the other hand, the different solid parts of which the bodies of these inhabitants are composed will undergo the same variations of form and volume.[38]

Poincaré then described how light would behave in such a space, and how sentient beings in the space would build up their sense of geometry exactly as we do. He then concluded:

> If they construct a geometry, it will not be like ours, which is the study of the movements of our invariable solids; it will be the study of the changes of position which they will have thus distinguished, and will be "non-Euclidean displacements," and this will be non-Euclidean geometry. So that beings like ourselves, educated in such a world, will not have the same geometry as ours.[39]

In this passage, Poincaré maintained the traditional idea of a rigid body (one that returns exactly to its position however it is displaced) but broke sharply with the idea that this determines a geometry. Rigid bodies of finite size need no more have their

[38] Poincaré 1902b, 89–90; English trans., 65–66.
[39] Poincaré 1902b, 65–68.

sides straight in some absolute sense than the path of light need be straight: there is an act of interpretation involved.

Enriques objected to Poincaré's account on the grounds that it denies us any way of saying that the changes in length of rulers in the plate is due to changes in temperature. Our experience of heat is that it is a localized phenomenon, and moreover one to which different bodies respond differently. In the Poincaré model, we cannot say these things that characterize what we say about heat, so, said Enriques, heat is not playing the role of a physical concept but of a geometrical one. The same is true of light rays, which demonstrably depart from straightness in inhomogeneous mediums. Therefore, Enriques concluded (p. 178), "in this other world, geometry would be really and not merely apparently different from ours." To think otherwise is to make the contrast between appearance and reality into something transcendental.

If in some strange way, large regions of space were to appear to us as more and more non-Euclidean, as measured by the behavior of light rays, then Enriques would have argued as follows. Either we discover some way in which the light rays are being pulled out of their predicted path by a process we can at least quantify and perhaps even control, or, after exhaustive searching, we do not. On the first alternative, a new physical process has been discovered, and our idea of geometry is left unchanged. On the second alternative, we would have to say that geometry was to be altered.

4.1.6.1 CALINON AND LECHALAS

One measure of the impact of Poincaré's ideas is given by the responses of Auguste Calinon and Georges Lechalas, for whom Poincaré's geometrical conventionalism did not go far enough.[40] Calinon regarded geometry as a purely mathematical construction, which had only to be free of self-contradiction in order to be valid, and which concerned itself, in the first instance, with (precisely defined) lines and planes.[41] Calinon restricted himself to what he called general geometry, which covered the usual three constant curvature geometries. That said, he took a sterner view of the way geometrical descriptions might apply to space than had Poincaré. He agreed that the fundamental geometrical nature of space was forever unknowable, partly on the grounds that we have access to only a very small part of the universe, partly for epistemological reasons. But he disagreed that we would always regard Euclidean geometry as the simplest, and held out the possibility that, for example, long-distance modifications of Newton's inverse square law might suggest that non-Euclidean geometry was in fact simpler.[42] To this Poincaré could have replied that this simplicity was a practical, utilitarian one, whereas his simplicity was rooted in a biologically based epistemology.

Calinon's views were taken up by Georges Lechalas (1896), where he argued that the mind, stimulated by imprecise information, generates infinitely many geometries that are acceptable on intellectual grounds, and that experience then narrows down the choice to within the limits of experimental error. But he seems to have slipped

[40] See Torretti 1984, 272–278.
[41] Calinon 1889.
[42] Calinon 1893.

back toward thinking that experiments could indeed discriminate between Euclidean and non-Euclidean geometry, thus missing the force of Poincaré's original argument.

4.2 Philosophy and Mathematics in Germany

4.2.1 Geometry and Intuition

Geometrical "intuition," as mathematicians came to use the term, is not Kantian intuition—the word carries too many meanings. In the Kantian interpretation, which starts with the passive way in which objects are impressed upon our awareness, an intuition is a representation that gives an object to us. As such, it is necessarily singular; we have an intuition of a specific object. This intuition is brought about through our senses, and Kant accordingly called it "sensible intuition." This sort of intuition is not knowledge. Knowledge requires concepts, and they lack objects (they are the general idea, so to speak). So to have knowledge of an object requires both an intuition, to present the object, and the concept, to convey the knowledge. More importantly, there is the pure intuition, which plays its part in forming synthetic a priori judgments. We routinely perceive appearance as structured. So our intuitions, in the above sense, are structured, and for Kant this meant that they were structured a priori. This form of intuition is independent of experience, and Kant called it "pure intuition." We have pure intuitions of space and time, and the pure intuition of space is what makes it possible for us to have knowledge of geometry. It allows us to grasp in a simple way the necessity of the theorems in geometry. This enabling aspect came to smack of ordinary mental activity, and both schools of neo-Kantians at the end of the nineteenth century rejected the faculty of pure intuition as dualist. They rejected it for being psychological in character, and had therefore to rely on transcendental logic to explain the production of knowledge.

On the other hand, those who were not worried about psychologizing adopted other interpretations of intuition. Herbart had initiated a move toward the study of the individual's acts of intuition, as opposed to the transcendental pure intuition, as part of his study of cognition. From there, and sometimes inspired by his example, others began to worry about different human abilities, and to reconsider the role of intuition in teaching. Different philosophies grew up attaching different weights to learning by rote, learning from the book, and learning from experience. Should geometry be taught as it is presented in Euclid's *Elements*, or in ways that recognized the ways in which concepts are drawn from acquaintance with natural objects?[43] It is not possible here to trace this debate, which was different but important in many European countries. By the end of the nineteenth century, intuition was debated within a rapidly changing spectrum of ideas about what mathematics is, notably the idea that, at its best, it was entirely rigorous. Should this be achieved, if need be, by sacrificing mechanics, even geometry? If it could be achieved, was this because it was in essence logical, even, and simply a branch of logic? Or was this to mistake part of

---

[43] This had been an eighteenth-century tradition in French textbooks, exemplified by Clairaut but reversed by Legendre.

the mathematical enterprise for the whole? No one took up intuition in the face of these concerns more broadly or more seriously than Felix Klein and Henri Poincaré.

4.2.1.1 KLEIN

In 1893 Klein was in Chicago, attending the World's Fair and the Congress of Mathematics as the official representative of Germany.[44] There he gave a speech in which he made a distinction between naive and refined intuition, a distinction he returned to when he gave a series of twelve invited lectures at the nearby Northwestern University in Evanston. In Lecture 6 he explained that refined intuition is found in Euclid's *Elements*—well-formulated axioms, exact proofs, clear distinctions. Naive intuition was exemplified by the carefree way in which Newton had invented the calculus. Klein felt that again at the end of the nineteenth century mathematicians were living through a period of critical analysis of mathematics; but this was not, in his opinion, an unmitigated advantage.

To explain the distinction between naive and refined intuition, Klein made a nice remark: "The naive intuition is not exact, while the refined intuition is not properly intuition at all, but arises through the logical development from axioms considered as perfectly exact."[45] To explain what he meant by the opening part of the aphorism, he returned to views he had expressed twenty years before, when he had written on the general idea of a function and its representation by an arbitrary curve. There he had explained that an arbitrary function need not be differentiable anywhere, but a curve in intuition has a tangent everywhere. Intuition is therefore misleading, and to explain how this can be, Klein argued that our intuition is not of a curve (as mathematically defined) but of a thin winding strip. Such strips have all the intuitive properties one wants, and they offer approximations to the arbitrary plane curves defined by pairs of arbitrary functions. In 1873 Klein had cited Stumpf's recent book for support from the psychological side.[46] In 1893, to quote him from the Evanston lecture: "Now such a strip has of course *always* [emphasis Klein's] a tangent; i.e. we can always imagine a small portion (element) in common with the curved strip. . . . The definitions in this case are regarded as holding only approximately, or as far as may be necessary." These definitions may not have the rigor of definitions in mathematics, but, Klein maintained, we operate with such inexact definitions in real life when we talk about the direction of a river or a road, a line certainly of considerable width.

As for the nature of refined intuition, there were many cases, said Klein, where conclusions drawn by logical reasoning from exact definitions could simply not be verified by intuition, and he gave the example of the almost nowhere differentiable curves discovered by Poincaré in the course of his work on non-Euclidean geometry.

---

[44] For the full story, and Klein's role in shaping American mathematics, see Parshall and Rowe 1994.

[45] Klein 1894, 42.

[46] Stumpf 1873. Stumpf and Klein were lifelong friends. In his autobiography, p. 395, Stumpf recalled: "During vacation I worked on a dissertation about mathematical axioms and at the end of October, 1870, I became instructor in Göttingen. I have never published this dissertation, as the non-Euclidian way of thinking to which Felix Klein had introduced me was, after all, a little beyond me." See http://psychclassics.yorku.ca/Stumpf/murchison.htm. Another mutual friend was William Robertson Smith, who fell foul of the Free Church of Scotland for his modernist theological views and eventually became a pioneer in the analysis of religious life from a sociological perspective.

Such curves are the limit curves that arise from an iterative construction, and, as he put it, "while it is easy to imagine a strip covering the curve, when the width of the strip is reduced beyond a certain limit we find undulations, and it seems impossible to clearly picture to the mind the final outcome."[47]

Klein noted that some other German writers seemed to agree with him: Pasch, and one Köpcke of Hamburg, who advanced the idea that our space intuition is exact as far as it goes, but limited. Where Klein and Pasch disagreed was over the end result of research. Pasch felt that ultimately intuition would be discarded entirely and all mathematics would be based on axioms alone. Klein conceded that this was the traditional view, but felt that for the purposes of research it would always be necessary to combine intuition with the axioms. Intuition, he said, had proved essential in his own case.

Then came a paragraph that became notorious.[48] Klein speculated that the exactness of spatial intuition might vary between individuals along racial lines. Teutons seemed to have strong, naive spatial intuition, while the critical, purely logical sense was more fully developed in the Latin and Hebrew races. A fuller investigation along the lines of Galton's study of heredity might be interesting, he concluded. Given Klein's evident preference for intuition, it is hard not to read this passage as being somewhat anti-Semitic. Moreover, Kronecker, Klein's exemplar of the purely logical mathematician, was Jewish; Peano, a Latin, had already been mentioned by Klein as an axiomatic mathematician. Forty years later, these remarks were brandished by the Nazi Bieberbach as "proof" that "good" German mathematicians such as Klein and Bieberbach were racially superior. Though Bieberbach made poisonous nonsense of it, there is no denying that Klein's remarks partake of the nineteenth-century glibness with which whole races are pigeon-holed. Psychology has ever lent itself to the representation of contemporary prejudice as the latest fruit of science. It should also be remembered that a paragraph, like a kiss, is what it is and not another thing.

Naive intuition was also, Klein suggested, the way in to applied mathematics. Here he had a lighthearted observation not without its charm today: the sciences could be ranked according to the number of significant figures habitually used in each. Astronomy and some branches of physics would come out top, chemistry last, and geometrical drawing somewhere in between. The most accurate science then needed seven significant figures (today physicists can call on up to ten significant figures, good enough to measure the distance from London to New York to within the thickness of a human hair). Because, Klein argued, all science is necessarily approximate, it should be possible to create an abridged system of mathematics, adequate for applications but accessible without having to pass through all of abstract mathematics first.

Intuition was a major theme of Klein's public speeches on mathematics in the 1890s, when he was enjoying his first few years as the leader of German mathematics. There had been a changing of the generations in Berlin: Kronecker had died

---

[47] These are among the famous curves that start the story of fractals. See *Indra's Pearls* (Mumford, Series, and Wright 2002) for an excellent account of fractals and pictures that are so much better than those Klein provided that he could have had no idea what they really looked like.

[48] It is discussed in Rowe 1986 and Mehrtens 1990, 215–218, who also note that elsewhere in his *Entwicklung* Klein is clear that the emancipation of the Jews in Germany in 1812 was quickly fruitful for mathematics, and that Klein was not anti-Semitic in his academic politics, rather the opposite.

in 1891, Kummer in 1893, and Weierstrass had retired in 1894 (he died in 1897). Their successors lacked force and vision, and as the Berlin department settled into comfortable middle age, Klein built Göttingen into a truly international center for mathematics. Intuition was the catch phrase he used to signal what was good about his approach to mathematics and had, by implication, been lacking in Berlin. In 1895 he gave an address to the Royal Academy of Sciences in Göttingen on "The Arithmetizing of Mathematics"—the phrase originated here with Klein.[49] His theme was that mathematics had indeed been made more rigorous by being arithmetized, and indeed increasingly written in symbols to guard against the imprecision of ordinary language. He even granted that more could be done along these lines in many domains of mathematics. But, he said: "I do not grant that the arithmetized science is the essence of mathematics . . . it is not possible to treat mathematics exhaustively by the method of logical deduction alone, but that, even at the present time, intuition has its special province."

Klein then quickly ran through the recent history of mathematics. Arithmetizing mathematics began, he said, with the ousting of space intuition. The problem with geometry was to reconcile the results obtained by arithmetic methods with our conception of space. But he had run a seminar on this sprawling topic for a year, and agreement on what curves and surfaces are could be reached. Different geometries could be studied—that was no longer the difficult point. The question now was to understand why it was justifiable to think of physical space as a number manifold. Why, in other words, could one regard points in space as triples of real numbers? The problem is that space intuition is inexact; the construction of the real numbers is a delicate limiting process. Here he nodded approvingly at the work of the late Benno Kerry and the subtleties of Cantor's theory of point sets. After more of this praise for arithmetization, Klein turned to bury it.

Intuition, as he now defined it, was an instinctive feeling for what is right. It was, he maintained, "always far in advance of logical reasoning and covers a wider field." Not just the origins of the calculus, but Riemann's theory of functions of a complex variable, could be credited to intuition, as could his friend Sophus Lie's work and much of Poincaré's. It was what he called a somewhat refined intuition, to be sure, but especially in applied mathematics it was what enabled the necessary idealization of a real-world problem to be carried out before logical investigation could begin.

Klein now took up the psychological aspect. Plainly, individuals differed widely in their mental abilities, and psychologists distinguished visual, motor, and auditory endowments. Klein hazarded the opinion that mathematical intuition belonged more closely with visual and motor skills, but said that he and many other mathematicians welcomed the recent interest of psychologists in these questions and hoped they would take the matter further. He also alluded to the pedagogical aspects. School teaching, he said, had become too intuitive, and he lodged a vigorous protest. University teaching, on the other hand, was too strict and the merits of intuition were ignored. A number of Klein's later interests surface with these remarks, and cannot be pursued here: his books on elementary mathematics from an advanced standpoint, and his energetic participation in the educational reforms (the Meran Reform) of 1905.

---

[49] Klein 1895. All further references to Klein in this section are to this pages.

4.2.1.2 HÖLDER

Klein's espousing of the cause of intuition drew attention because of his position, but his remarks were nonetheless superficial. Otto Hölder's inaugural lecture in Leipzig in 1899, *Anschauung und Denken in der Geometrie* (*Intuition and Thought in Geometry*), was a more analytical and philosophical critique. He asked, What sort of knowledge is provided by geometry? Is it, for example, an empirical science? Philosophers who have raised this question have not always, he said, mastered the details of the mathematicians' work, but nor have mathematicians adequately inquired into the source of the assumptions upon which they base, as it might be, Euclidean or non-Euclidean geometry.

To bring things together, Hölder reviewed the various ideas with which geometers work. Some, he noted, are fundamental: for example, the idea of a point or a straight line. Others arise from constructions: for example, a square. Alongside the fundamental concepts, which cannot be defined, stand certain axioms, which capture the basic facts about geometric objects and cannot be proved. So the first question for the epistemologist is: Where do the axioms come from? Some reply: "From our intuition of space," others "From experience." The first group are Kantians. The second, Helmholtz being a case in point, regard the intuitive pictures we have in our imagination as memories of our experience. We can manipulate and even change them in our minds, but they derive ultimately from sensation. Hölder inclined strongly to this second, empiricist position, but to advance the congested and often heated debate between the schools, he offered a detailed analysis of a particular case of geometric reasoning.

He took Kant's example, the theorem in Euclidean geometry that the angle sum of any triangle is two right angles. It relies on the existence claim that given any line in a plane, and any point in that plane that is not on a line, there is a (unique) line through the given point that never meets the given line. Now, said Hölder, this claim cannot be supported with school-book logic alone, and he cited Lotze's analysis of syllogistic logic to this end and noted the new theory of the logic of relations (citing Jevons). Rather, the existence claim, which was a statement about infinitely many points and lines, was the result of a thought experiment. So it was necessary to investigate the role of intuition in such thought experiments. Kant believed that it enables the whole thought experiment to be carried out, others that it merely provided the raw ingredients. By way of an illustration, Hölder agreed that intuition might immediately give the result that two angle bisectors of a triangle meet, but not that all three meet in a common point—that requires an argument. This suggests that crude assessments may come about through intuition, but subtler ones require artful deduction.

A further consideration of the way geometric figures enter proofs led Hölder to note that mathematicians also reason by analogy (when they take a figure as typical) and by a form of complete induction (when they vary the elements of a figure so as to satisfy themselves of the truth of a claim). So Hölder proposed that the use one makes of intuition is to bring to light the use of certain rules, tacitly assumed by Euclid, which must be taken as axioms.[50] For example, the proof of the first congruence theorem of Euclid's *Elements*, which asserts that two triangles are congruent if two sides and the angle they include are equal in pairs, makes use of the idea that a

---

[50] Here he cited one of Hilbert's axioms of order.

triangle may be moved from one position to another without altering its shape or size. This idea "comes immediately from intuition or, as we would rather put it, from our experience of the motion of rigid bodies" (p. 14 of his lecture). By proceeding in this way, geometric deduction can unclothe intuition.

When intuition has been thus completely exposed and we can say with certainty what follows from particular assumptions, then we can say when these assumptions contradict intuition. But this form of deduction is certainly not the path to discovery. This is usually through intuitive pictures, sometimes through observation of a series of cases, and therefore by analogy and induction.

Hölder now returned to his opening remarks about figures given immediately in intuition and those that arise by construction, and observed that some ideas, which undoubtedly arise immediately in intuition, also arise as the conclusions of theorems. The theory of proportion was full of such examples; this being a topic dear to his heart, Hölder discussed it at some length. Finally, he concluded that the deductive method has a special form, and therefore the mathematical sciences have a particular method: a long series of intellectual operations are brought together in a characteristic way which ensures that particular conclusions arise. "It is in this sense that one can therefore say that mathematics and the exact natural sciences have their own logic" (p. 23).

Hölder therefore seemed to disagree with Kant and to agree with Helmholtz that experience and logic are the sources of truth in geometry. But this disagreement rests on an avoidance of the concept of pure intuition, which enables us (on Kant's view) to do geometry at all. Rather, it is a preference for empiricism over the views of the neo-Kantians. Intuition has lost its overarching enabling role and become a mere purveyor of truths. These truths are sometimes immediate, and must be taken as axioms, or they may turn out to be consequences of others, or they may be hidden and come to light in the analysis of proofs. But in each case they may be regarded as arising either from experience or deduction, and, for Hölder at least, intuition was left with no foundational role. Its proper place was in the discovery of new mathematics.

Hölder was able to be more explicit about this twenty-five years later, when he published his book *Die Mathematische Methode*.[51] He again argued firmly that logical analysis of an intuitively given insight renders a proof entirely deductive. Against Kant, Hölder now replied that the picture given in intuition is not to be taken as a single picture but as the general one. Berkeley, he noted, had accordingly suggested that, as a result, a process of induction is involved as elements of the picture are varied, but this was not correct. Not only had Kant never allowed objects given in intuition to vary, but the purely logical analysis of proof showed that no such process need be presumed.[52]

Jules Tannery reviewed Hölder's inaugural lecture in the *Bulletin des sciences mathématiques* and he picked up on Hölder's rejection of intuition. He did not

---

[51] This interesting if nowadays neglected book of 563 pages should be rediscovered. Its contents cover most of the examples of modern mathematics discussed here, the consequences for any theory of logic and deduction, and the relationship of modern mathematics to the world of experience. It is particularly interesting on set theory and axiomatics, where Hölder had made original contributions as a younger man.

[52] Hölder here footnoted Leibniz, *Neue Abhandlungen über den menschlichen Verstand* (1904, 380).

dispute that, as Hölder had also suggested, one way to keep intuition out of proofs was to write them out in a purely symbolic way, as Peano had begun to advocate. But he doubted if it was the only way. Could not intuition be the result of a long, even ancestral habituation, and if so, could it not be modified, at least among certain people, and could this acquired intuition not "allow some geometers to proceed boldly with their research in domains subject to other laws than those of ordinary space, while all the time keeping to the customs of geometric reasoning? And to the Kantian who asks what is this 'acquired intuition,' I see no other way to reply that to ask in return what is a 'pure intuition'? Philosophical terms are never more than half understood."[53]

4.2.1.3 BOREL

By comparison with the substantial philosophizing of the German mathematicians, Émile Borel's intervention was more conventional, yet light, precise, and vital.[54] It opened with a familiar complaint: "I always admire the facility with which philosophers speak of the 'method of the mathematical sciences.' It is a curious spectacle to see a candidate for the baccalaureat repeat to his examiner what his professor has told him about the subject, when the examiner, the candidate, and the professor are equally ignorant of what this mathematics is the method of which they discuss."

Borel's ire was directed at a philosophy, lately supported by the logicists, which reduced mathematics to its deductive core. Consider, he said, polynomial identities, such as $(x + 1)(x - 1) = x^2 - 1$. One could imagine these being produced mechanically and entirely correctly. One of these, say number 35,427, could then be $4P^3 = Q^2 + 27R^2$, satisfied when $P = x^2 - x + 1$, $Q = (2x - 1)(x + 1)(x - 2)$, and $R = x(x - 1)$. But this one, unlike the ones on either side, is interesting, not least because it expresses a cube as a sum of two squares.[55] Logic cannot capture the importance and usefulness of this identity, nor is the mechanical production of many results the instructive way to this one. Among other examples, the formula expressing the invariance of the cross-ratio of four points on a line under a perspectivity is easy to find, but it took a Chasles to see in it the key to projective geometry. Numerous "dead formulae" in the calculus were trotted out in Bertrand's textbook of 1868 "without him being able to give them life" because he did not appreciate their significance, unaware as he was of the work in progress of Weierstrass and Méray on complex function theory. Klein's unification of Galois theory with the theory of the symmetries of the icosahedron highlighted another polynomial identity that could only have been found to be valuable if discovered by a nonmechanical route.

He summed up his position as follows: truly fertile invention in mathematics, as in the other sciences, consists of the discovery of a new point of view from which to classify and interpret the facts. This is followed by a search for the necessary proofs, and only in the third and final stage does logic take over, for the purposes of exposition and teaching. Indeed, Borel concluded, it was the needs and methods of exposition and teaching that were responsible for the travesty of mathematics he

---

[53] One wonders if Tannery had in mind an analogy with Lamarckian acquired characteristics.
[54] Borel 1907, 273.
[55] Borel chose not to add that this identity is important in the theory of elliptic functions.

deplored. Echoing Klein's efforts in Germany, he called for reforms to emphasize the importance of intuition in mathematics at every level.

Borel's criticisms are not quite the staples they might seem, and not just because they have a specific resonance in the France of the time. They point quite clearly toward a problem that has not gone away in philosophers' treatments of mathematics: a tendency to reduce it to some essence that not only deprives it of purpose but is false to mathematical practice. The logicist enterprise, even if it had succeeded, would only have been an account of part of mathematics—its deductive skeleton, one might say. Living mathematics, as it is actually done, would remain to be discussed.

4.2.2 Hilbert, Husserl, Frege

Hilbert saw more clearly than anyone the novel importance of the axiomatic method, but, like many a creative mathematician, he saw less well the weaknesses and downright oddities in his presentation of it. His friend Minkowski was quick to tease him with the most salient of these, manifest in his lecture "On the Number Concept" given at the Deutsche Mathematiker-Vereinigung (DMV) meeting in Munich in September 1899, when he wrote to Hilbert: "Your existence is as little to be doubted as in your $18 = 17 + 1$ axioms for arithmetic."[56] The eighteenth axiom differed from the previous seventeen. They described the arithmetical rules that numbers must obey, but this one was about the axiom system itself. It asserted that the axiom system was what he called complete, by which he meant that there was a system of objects obeying the axioms, and there was no larger system of objects also obeying the axioms. It allowed him to claim that the objects really existed. He repeated the claim in the text of his famous lecture on mathematical problems, given at the Paris International Congress of Mathematicians (ICM) in 1900, and again in the second edition of the *Grundlagen der Geometrie* in 1902—the first edition had opened with the claim that the axioms were complete without explaining what that meant.[57]

Trouble came, almost inevitably, from Frege, who wrote to Hilbert to complain reasonably that it was like doing theology with an axiom that says God exists: "Axiom 3, there is at least one God."[58] Frege's entirely sound view was that completeness axioms cannot be used to resolve questions of existence. Hilbert would not back down, and as noted reasserted a completeness axiom for the real numbers, which in his view had the great advantage that it removed the need for explicit constructions in the manner of Cantor (or Dedekind). As the correspondence with Frege proceeded, Hilbert became so carried away that he even claimed that "if the arbitrary given axioms do not contradict one another with all their consequences, then they are true and the things defined by them exist."[59] "True" is too strong. The

---

[56] Rüdenberg and Zassenhaus 1973, 116; Minkowski to Hilbert, June 24, 1899.

[57] The first seventeen axioms are satisfied by many systems of objects, starting with the familiar rational numbers, Q, and including any algebraic extension of them. Hilbert's eighteenth axiom claims that there is a unique maximal collection of objects that obey the first seventeen axioms, and that it coincides with the set of real numbers. Neither claim is obvious, but Hilbert's original and profound view about axioms was that they should organize the theory most effectively, not that they should be indubitable starting points. For many examples of this philosophy at work, see Corry 2004.

[58] Frege to Hilbert, January 6, 1900, in Frege 1980, 46.

[59] Hilbert to Frege, December 29, 1899, in Frege 1980, 39.

FIGURE 4.6. Edmund Husserl. Photo Courtesy of Prof. Ullrich Melle, Director of the Husserl-Archives in Leuven.

example of Euclidean geometry and non-Euclidean geometry should have cautioned Hilbert to say only that theorems are true of the objects defined by the axiom system. "Proved" was the appropriate word in this context, and the one Hilbert preferred in Paris. To say that geometrical axioms are true carries the heavier implication that the objects they define actually exist—exactly the concern that animated Frege, but one that Hilbert was much more openminded and innovative about.[60] Frege was a realist about the existence of geometrical objects, and while this made him sensitive to Hilbert's sweeping claims, it also blinded him to the idea that consistent sets of axioms might create geometrical objects, thus breaking the link with realism in the foundations of mathematics.

Matters went better with Hilbert's Göttingen colleague Edmund Husserl, the former mathematician-turned-philosopher, who heard Hilbert give a lecture at Göttingen on November 5, 1901. Husserl discussed the idea with Hilbert, and then filled pages of his Notebook with a tentative exploration of what it could mean to define or create objects by means of an axiom system, focused, in his case, on defining the integers.[61] He discussed the implications of an axiom system being "closed" (*abgeschlossen*, admitting he had forgotten the word Hilbert used for this concept), but more than a word seems to have slipped and instead the whole conception of

[60] Frege hoped the correspondence would lead either to publication or a joint article, but Hilbert never agreed, and in the end Frege published most of his part of it as an article for the *Jahresbericht der Deutschen Mathematiker-Vereinigung* (Frege 1906).

[61] See Husserl 1970, Hill 1995, and Majer 1997.

completeness was muddled. At this stage, neither Hilbert nor Husserl fully understood it. He also lectured twice to the Göttingen Mathematical Society on "completeness" and "definiteness" at the end of 1901.

To understand the confusion, which Husserl tried to elucidate in his Notebook, we have to broaden the inquiry. When Hilbert gave his famous lecture in Paris in 1900, he spoke of axiomatizing physics in these terms: "The mathematician will have also to take account not only of those theories coming near to reality, but also, as in geometry, of all logically possible theories. He must be always alert to obtain a complete survey of all conclusions derivable from the system of axioms assumed."[62] This is a counsel of perfection. It might be possible to show that a given set of axioms does or does not deliver a specific result, as investigations of the parallel postulate had long shown, and more recently Hilbert's study of Desargues' theorem had established. But how could a mathematician survey all the consequences of an axiom system?

One way to examine this question is to consider sets of axioms that define an essentially unique set of objects, meaning that any other set of objects obeying these axioms is equivalent to the first set in every relevant respect (we say that the sets of objects are isomorphic). This is the property that Hilbert called completeness, but modern writers call *categoric*.[63] It seems best to go with modern usage here.

Another way to examine this question is to consider sets of axioms that cannot be enlarged, meaning that any new statement offered as an additional axiom either follows from the given set or plunges the set into self-contradiction. Therefore, of the two axiom systems obtained by enlarging the given system with the new candidate axiom, one is consistent and the other is not. Confusingly, this property of an axiom system is nowadays called *completeness* or, sometimes, *syntactic completeness*.

Unfortunately, the relationship between complete and categorical axiom systems is a delicate one. It is tempting to argue that a complete axiom system must be categorical, because if it were not categorical there would be two nonisomorphic models satisfying the axioms. But then a statement that is true in one model but false in the other would be independent of the axioms, making them incomplete. This is a contradiction, so the original axiom system must be categorical. The other way around is also plausible. Given an axiom system describing a categorical family of objects, the axiom system must be complete, because if it is not complete, there is a consistent statement *S* that is independent of the axioms. That means there are two nonisomorphic models, one for the original axioms and the statement *S*, and one for the original axioms and the negation of the statement *S*. But the original axioms have a unique model (up to isomorphism). This is a contradiction, so the original axiom system must be complete.

The correct statement is that a categorical axiom system is complete, and the above argument outlines a valid proof (although some comments are in order). The argument that completeness implies categoricity is flawed, and examples will be given below, but it is worth looking at the purported proof in the preceding para-

---

[62] Hilbert 1900, 15.

[63] Following Huntington 1902, and Veblen 1904. For modern discussions, see Awodey and Reck, "Completeness and categoricity," at philsci-archive.pitt.edu/archive/00000544/01/ccI.ps, and Corcoran 1981, 113–119, both for the history and for welcome logical precision on the relationship between completeness and categoricity.

graph. There it was claimed that if a system of axioms A was not categorical, it would have two nonisomorphic models, say M_1 and M_2, and we could write down a statement S that was true of M_1 and whose negation is true of M_2. Were that to be true, then the axiom system $A + S$ would have M_1 as a model, and the axiom system $A + not\ S$ would have M_2 as a model. That implies that S is independent of A, and so the system A is incomplete. The flaw in the argument is that we can write down a statement in the formal system that is true of M_1 and whose negation is true of M_2.

It is just such plausible, but partially false, arguments that Husserl took away from his meeting with Hilbert. Husserl asked himself what it would mean for an axiom system to be categorical (in his word *definit*, or definite), and his answer was: "An axiom system defines a mathematical or 'constructible' manifold, when the sphere of formal objects that it defines, has the property that every object existing in this sphere in virtue of the axioms is uniquely determined, through the operative relations, to a definitely given one, i.e. a uniquely determined and specific object distinguishable in no other way."[64] He then worried away at the idea of defining objects by means of axiom systems for many pages.[65] He asked if what he called a perfect axiom system (one we would call complete) must be definite, and when he restricted his attention to any fixed family of objects, found that it would be.[66] But he left these observations unpublished; plainly he felt that even if Hilbert was right, there was much work to be done to get the necessary clarity.

At this stage in Husserl's career he was making one of the major transitions that mark his intellectual development, and aspects of his journey illustrate very well the concerns that animate mathematical modernism. Husserl had originally studied mathematics at Leipzig, where he also attended some of Wundt's early lectures. He took his PhD from Vienna with a thesis on the calculus of variations, and then worked for a time as Weierstrass's *Assistent*. When Weierstrass fell ill, however, he went back to Vienna with an introduction to Brentano. There he stayed for two years, and then, on Brentano's advice, he moved to Halle where he completed his *Habilitation* in 1887 with a thesis, written under Brentano's pupil Carl Stumpf.[67] This was later reworked and included in Husserl's first significant publication, *Die Philosophie der Arithmetik* (*The Philosophy of Arithmetic*) (1891), by which time he was already dissatisfied with some of the views it expressed.

As a good and loyal student of Bolzano and Stumpf, Husserl, whose major work was *On the Psychological Origin of the Idea of Space* (1873),[68] had initially rooted everything in psychology, but his own training led him to focus on logic. He was also influenced by his reading of Lotze's *Logik*, with its view that mathematics should be grounded in logic. His views were expressly opposed to Schröder's on algebraic logic. Husserl reviewed Schröder's *Vorlesungen über die Algebra der Logik* at length in 1891, and objected to nearly all of it.[69] Logic, for Schröder, was deductive logic, but, said Husserl, deduction was not the only activity even in mathematics. For example,

---

[64] Husserl 1970, 452.

[65] Husserl 1970, 444–488.

[66] Husserl 1970, 433.

[67] In 1904 Stumpf advised Oskar Pfungst, the head of the commission set up to examine the alleged ability of a horse called Clever Hans to do arithmetic; they found it was responding to various subtle clues it had picked up in its training. See Pfungst 1907.

[68] *Ueber den psychologischen Ursprung der Raumvorstellung.*

[69] Original publication, *Göttingische gelehrte Anzeigen* (1891), 243–278.

one also calculates, often extensively. Nor was deduction here analyzed as a philosopher would want to see it, but instead a calculus of deduction was developed. As such, it was very good and generally superior to what had gone before it, said Husserl, but the claims on which it rested were inadequate. As so often with people who emphasised rule-driven calculation, he suggested, Schröder saw mathematics as being about signs, and with this Husserl disagreed. The signs in mathematics have content, as, for example, do the elements of figures in Euclidean geometry. Other substantial disagreements followed, the one of most interest here is over Schröder's confusion between language and algorithm. Schröder advocated replacing words by symbols, but, said Husserl, language is for the symbolic expression of mental phenomena, and it is governed by grammar. A calculus, on the other hand, is for the symbolic derivation of conclusions. So "the logical calculus is *only* a calculus, and absolutely is not a language."[70]

However, it was criticism of his rather dogmatic dismissal of algebraic logic that forced Husserl to rethink positions in his *Philosophy of Arithmetic*. In 1892, Andreas Heinrich Voigt wrote an article on the relation between algebraic and philosophical logic, in which he claimed, not without justification, that Husserl regarded algebraic logic as no logic. Husserl replied, only to find that Voigt, who had just successfully defended a doctorate on symbolic logic with Lüroth and Schröder as examiners, could show that what Husserl claimed was original in his work could be found already in that of Frege and Peirce. Then in 1894 Frege joined in, stung no doubt by Husserl's (correct) claim that Schröder was the first German to write on symbolic logic. But what drew Frege's criticism most was the confusion between treating the laws of logic as natural laws and treating them as normative. Frege had already attacked Husserl's colleague at Halle, Erdmann, on these grounds in his *The Fundamental Laws of Arithmetic* (1893); now he turned on Husserl.

In his review of Husserl's *Philosophie der Arithmetic*, Frege took the opportunity to "gauge the extent of the devastation caused by the irruption of psychology into logic."[71] He found that certain ideas, which are subjective and peculiar to the person having them, were confused with certain objects, such as sets, which are objective. This confusion showed up over zero and one, which Husserl regarded as negative answers to the question "How many?," because number was for him connected to multiplicity. But, said Frege, this is not so. To say zero is the number of predecessors of Romulus is no more negative than to offer the answer "two"; the existence of a number is not denied, merely the existence of a predecessor. Likewise, we cannot have a clear idea of very large numbers, but this does not mean that we cannot think about them. The message was that one should separate "number" from "idea of number" and give an objective account of "number."

Husserl, who became a professor at Halle in 1894, dealt with these criticisms most admirably. He immersed himself in reading everything written on logic, issuing in 1897 two lengthy reports on German work on logic in 1897 and 1903–1904. As a result, he shifted his position over to the normative view of logic, which is the view he expressed at length in his *Logische Untersuchungen*. Husserl was clear that he was after a theory of science along the lines of Leibniz's unsuccessful *mathesis universalis*.

---

[70] Review of Ernst Schröder's *Vorlesungen über die Algebra der Logik* in Husserl 1994, 70.
[71] In Frege 1984, 209.

At Halle, Husserl had gotten to know Cantor, who was fifteen years older than he and had served on Husserl's *Habilitation* committee. Not only did they share a deep interest in the philosophy of mathematics, the course of which has been persuasively analyzed by Hill,[72] they held Weierstrass in the highest esteem. Husserl credited Weierstrass with awakening his interest in seeking radical foundations for mathematics. His *Habilitation* thesis sought to analyze the concept of whole number as a way into the analysis of the philosophy of mathematics, accepting some version of the by then widely accepted thesis that all the other kinds of numbers in mathematics can be produced from the natural numbers. Hill shows that Husserl drew many of his key ideas from this early and inspiring contact with Cantor, notably that the fundamental notion is that of cardinality, which is obtained by a process of abstraction. Husserl proceeded to subject this process of abstraction to a thorough analysis, only to conclude that cardinality was inadequate to ground the concepts of rational and irrational numbers, complex numbers, quaternions, and the like. To give a logical account of these numbers, and indeed to account for our ability to reason even with large natural numbers, Husserl argued that we rely on the use of symbols, an idea he took from Brentano, and it was with the symbol-manipulating idea of mathematics that Husserl was to be increasingly concerned.

Husserl regarded his time at Halle as the unhappiest years of his life,[73] and it is plain from the sprawling and somewhat inconsistent nature of *Logical Investigations* that he regarded it as painful work. He wrote of feeling "tormented by those incredibly strange realms: the world of the purely logical and the world of actual consciousness"[74] Indeed, the *Logical Investigations* arose, he said, as an interruption of his attempts to achieve a philosophical clarification of pure mathematics. He found traditional logic left the rational essence of mathematical theory and method obscure, and peculiar difficulties were raised by formal arithmetic and the theory of manifolds that pushed him toward a universal theory of formal deductive systems. He came to see that quantity did not belong to the "universal essence" of mathematics, but became unsure of the relation between quantitative and nonquantitative mathematics, arithmetic and logical formality. Finally, in 1901, he felt he had reached a certain clarity concerning epistemology and new foundations for pure logic. Even so, the final manuscript apparently had to be wrenched from his hands by Stumpf and sent to the publisher.[75]

The new foundations are explicitly and at length (some eighty pages) antipsychologistic. The conventional view, since Føllesdal's book (1958), is that this shift in Husserl's position is due to Frege's attack on psychologism in his review of the *Philosophy of Arithmetic* (1894). Certainly, as Kusch (1995) has shown in detail, Husserl's antipsychologistic arguments—which we shall not explore—are heavily influenced by Frege. However, more recently a good case has been argued that Husserl had already begun to move away from these psychologistic positions before Frege's attack.[76] What perhaps matters more is that after 1901, with the publication of the *Logical Investigations*, Husserl's arguments against psychologism carried the

---

[72] Hill 2000.
[73] Moran 2000, xxvii.
[74] Husserl 1994, 490–496
[75] Moran 2000, xxxiii.
[76] Haddock 2000, 200.

day in German philosophical circles, while Frege remained intellectually isolated in Jena.[77]

German philosophers had largely been dismissive of the algebraic logic of Boole, Jevons, and others, feeling that it was merely technical and incapable of answering any substantial philosophical question about logic. Some saw a limited value in these techniques as a way of driving away, for example, psychologism.[78] They did not see, as Frege had, that a deep immersion might drive logicians to new and valuable questions about logic. Rather, they could instead agree with Husserl that mathematical logic was not a philosophical enterprise, and that only the philosopher, not the mathematician, could complete the work of the mathematician and ensure that pure and true theoretical knowledge is obtained. Husserl's breadth, his firm but polite criticisms, his occasional closeness to neo-Kantian positions, his pedigree under Weierstrass and Brentano, his position at Halle and, after 1900, in Göttingen, all helped to advance his intellectual position. Frege, isolated and ill-tempered in Jena, suffered under every comparison.[79] All he had to speak for him was the force of his logical analysis, and the hope that it might appeal to the few philosophically minded mathematicians. He was both lucky and unlucky: he was taken up by the imperfectly informed but highly articulate Cambridge mathematician, Bertrand Russell.

4.2.3 Hilbert, Nelson, and the Neo-Friesians

An unusual aspect of the situation in Göttingen was the close connection between the mathematicians and the philosophers, unmatched anywhere else (the closest comparison would be Russell and Whitehead at Cambridge). This was brought about by Hilbert and, for the philosophers, Edmund Husserl. It was strengthened, even as Husserl moved away toward phenomenology, by the arrival of the forceful figure of Leonhard Nelson.[80] As Peckhaus and Kahle (2002, 9) have shown, there was a flurry of activity in the Göttingen Mathematical Society between 1903 and 1907, as the realization sank in, after Russell had torpedoed Frege's work, that arithmetic was not very easily going to be shown to be consistent. The work of Peano was reported on, and Zermelo gave several lectures on topics in mathematical logic.

Nelson had first studied in Berlin and then in Göttingen, where he arrived in the winter semester 1903–1904. There he wrote a paper under Baumann on critical method. It grew into his dissertation and was devoted to expounding the ideas of Jakob Fries. For his doctoral exam he also wrote on theoretical physics (Wiechert was his examiner) and psychology (examined by Müller). Now well prepared, he set about establishing a philosophical school, which became the new Friesian school.[81]

---

[77] See Kusch 1995, 203.

[78] Natorp, cited in Kusch 1995, 204.

[79] Jena was the second smallest university in Germany and the last to be connected to the German rail network; see Kusch 1995, 206.

[80] The best source on this is Peckhaus 1990.

[81] As Peckhaus and point Kahle out, there had been a Friesian school in Fries's lifetime, but this had not been a success.

Nelson brought in Hessenberg, who shortly wrote the first German textbook on set theory,[82] Grelling, and others, and focused the school on problems in the philosophy of mathematics. As the name suggests, they were concerned to revive Kantian, and more specifically Friesian, approaches to these issues. In Christmas 1907 Nelson gave the school a ten-point program, intended not only to keep the school together but to repel boarders (Nelson was not at all averse to the sort of sectarian feuding practiced in neo-Kantian circles). Among other things, members were asked

- To recognize the inviolate nature of the distinction between analytic and synthetic judgments (Peckhaus observes that this excludes most neo-Kantians and Kant historians)
- To find fault with mathematics and scientific empiricism and uphold the synthetic a priori (thus banning the followers of Fichte, Hegel, and Schelling)
- To find fault with mysticism (the idealists again)
- Not to accept the synthetic a priori without demonstrating the grounds for its validity (excluding, therefore, Meinong, Husserl, and Couturat)
- To recognize the distinction between philosophy and psychology.[83]

In fact, the clause about empiricism is much longer than the others and goes on to exclude Frege as well. The program was adopted for the first time at a meeting of the school during the third International Congress of Philosophers in Heidelberg in September 1908, and annually thereafter until 1914.

Nelson also set the tone for the group by his own writings. Hessenberg kept him informed about issues in mathematics, and Nelson naturally took up the topic of how non-Euclidean geometry did not invalidate the synthetic a priori intuition. Nelson (1905) does at least honor the standards of the later program. Mathematical axioms are synthetic (they are not matters of logic); they are not empirical because they are not about empirical objects (they are independent of experience) and so they are pure intuitions; and therefore they are synthetic a priori. Now it had been much disputed among neo-Kantians and beyond if the discovery of non-Euclidean geometry invalidated such arguments. Some felt that the discovery showed that axioms were not pure intuitions, others that non-Euclidean geometry was in some way dubious. Nelson argued that the logical possibility of non-Euclidean geometry has nothing to do with the synthetic characteristic of geometric axioms, and nothing in Kantian philosophy prevents non-Euclidean geometry from being a logical possibility.

Can non-Euclidean geometry be accepted as an intuition, however, and not merely reasoned about (as mathematicians do with spaces of various dimensions, for example)? The answer, for Nelson, was that the thinkable included everything that was not self-contradictory and thus exceeded the domain of the intuitive. But geometry, for Nelson, was not a merely logical game of drawing consequences from arbitrary assumptions, and because the very possibility of non-Euclidean geometry shows that geometric axioms are not entirely logical in nature, it follows, he argued, that Kant's insight was correct: pure a priori intuition was a source of axioms, and

---

[82] Hessenberg 1906; Borel 1898 was the first in any language.

[83] All the comments in parentheses are taken from Peckhaus 1990, 152–153, from which this information about Nelson's program is taken.

Helmholtz and Riemann were wrong to suggest that logic and empirical investigations were the only sources of axioms.

With this distinction in mind, he turned to the question of the origin of geometric axioms in intuition, and Poincaré's views in particular. He recognized that Poincaré had correctly dismissed the idea that geometrical ideas are empirical, and that he agreed with Kant that mathematical knowledge was apodictic without being reducible to logic. But if "this approach of the great French mathematician to the critical standpoint" was an advance, all the more deplorable was the "total misunderstanding" that followed.[84] This was, of course, Poincaré's replacement of the synthetic a priori position with his geometric conventionalism. Poincaré, he said, had asserted that a synthetic a priori judgment would impose itself with such force that a contrary proposition could not be conceived, and therefore non-Euclidean geometry could not exist. In Nelson's view, only an analytic judgment has a negation that cannot be conceived. And what could axioms be, if they are neither a posteriori nor a priori judgments?

Nelson replied to his own question that to distinguish a Euclidean from a non-Euclidean triangle in space could only be a task for someone equipped with the Kantian idea of pure intuition, for someone having only logic and experience to guide them could accept either Euclidean or non-Euclidean geometry, or indeed any geometry, as correct. Poincaré's conventionalist position could always be extended to any new geometry, and so the question of which geometry is valid would have no sense, and one could only ask which geometry is the most appropriate or convenient. However, the origin of geometry is not to be found in logic or experience. Geometry is a different form of knowledge whose judgments are synthetic a priori. Poincaré had failed to recognize this possibility and believed that if a proposition was founded neither in logic nor in experience, then it forfeited any claim to knowledge.

Nelson concluded that the apodictic nature of mathematics together with the possibility of non-Euclidean geometry is a reliable touchstone for any philosophy of mathematics. To hold only one of these views is to see it break on the rock of the other one. The way out, for Nelson and his followers, was to regard non-Euclidean geometry as thinkable, but not as a synthetic a priori judgment such as Euclidean geometry rests upon.

Nelson also discussed the arithmetization of mathematics. He distinguished two forms of it: one he called *syllogistic*, which aimed at a rigorous separation of intuition and thought; the other, due he suggested to Dedekind, aimed at separating analysis from geometry. Each on its own was right, but a confusion of the two promoted the idea that mathematics could and should be reduced to arithmetic, which Nelson denied. Geometry and arithmetic deal with different things.

He then argued that Frege's proposed reduction of arithmetic to logic would necessarily fail, because, given any criterion for recognizing when something is of a purely logical nature (such as the distinction between analytic and synthetic judgments), it is easy to recognize the nonlogical character of the premises of arithmetic. Specifically, the idea that every number has a successor is not at all a necessary idea but arises only in pure intuition.

---

[84] Nelson 1905.

Nelson now gravitated toward more Hilbertian concerns. He distinguished mathematical existence sharply from logical existence (which he took to be freedom from contradiction) on the grounds that one rests on synthetic, the other merely analytic, hypothetical judgments. Logic provides only for statements of the form "If X exists, then . . . ." To guarantee existence of the system of objects described by an axiom system, a further existence axiom is required. Mathematical existence is a matter of pure intuition, not logic (nor, of course, is it an empirical matter). As for the conviction that every mathematical problem has a solution, this has to be answered by looking not at the axiom system that defines a given problem, but at the grounds for those axioms, which form the relevant constitutive principle. If the constitutive principle is self-contained, the problem is solvable, otherwise not. Happily, mathematics has only constitutive principles that rest in pure intuition, so the solvability of every mathematical problem is, apparently, assured.

Prior to this work, Hessenberg had undertaken to show that Hilbert's axiomatic approach to geometry in the *Grundlagen der Geometrie* could be taken as a model for the mathematical component of critical mathematics.[85] He did this in a lecture to the Berlin Mathematical Society in November 1903.[86] He characterized axioms as unprovable statements and endorsed Hilbert's view that all that should be asked of an axiom system is that it be independent, free of contradiction, and complete. That being the case, the first task of critical mathematics was to prove that a given axiom system was free of logical contradiction, and here Hilbert had shown by example how this might be done. The lecture ended with an appeal to philosophers to join in the task.

Wilhelm Meinecke's (1906) response in *Kantstudien* may not have been what Hessenberg wanted, but it presumably was some sort of an orthodox Kantian reply. He conceded that if Kant had thought that the a priori intuition of space makes it Euclidean, then the theory was wrong for the reasons that Helmholtz had given, but he proposed to show that the right slogan was not "Kant or Lobachevskii" but "Kant and Lobachevskii!" To that end he argued that the a priori intuitions of space and time were closely bound together. That of space led to geometry, that of time to the idea of number; but of course we arithmetize geometry, whether as astronomers or, like Hilbert, to demonstrate the consistency of an axiom system. Moreover, the axioms of geometry appeal to motion, a concept that involves time. So axiom systems must respect a close integration of space and time, and what Meinecke called the Lobachevskii-Riemann theory of manifolds is the genus proximum for Euclidean geometry. Meinecke concluded that non-Euclidean geometry does not contradict the Kantian theory of mathematical knowledge, but a derivation on a Kantian basis of the axioms of Euclidean geometry is still allowed.[87]

Considerations of a Kantian, Friesian, or critical kind were, however, to be marginalized for a time by the attention that was paid to the emerging paradoxes of

---

[85] Peckhaus 1990, 159.

[86] Hessenberg 1904.

[87] Meinecke also wrote that when he tried to exercise his powers of non-Euclidean intuition after the manner of Helmholtz and see non-Euclidean parallels approach asymptotically he was prevented by a feeling of dizziness such as one has in dreams. The video *Not Knot*, created by the Geometry Center in 1995, can confirm this impression.

mathematics, a theme to be taken up below when we have further surveyed the advance of modernist mathematics.

4.3 Algebra

The term "modern algebra" is a well-understood phrase in mathematics. It derives from the title of van der Waerden's book *Moderne Algebra* of 1930, and it covers such topics as groups, rings, and fields, Galois theory, and hypercomplex numbers (later known as algebras). The subject of linear algebra, with which it overlaps, is usually not stressed because of the wide range of applications for linear algebra, which gives it a very different feel. Modern algebra is a pure mathematician's subject. As such, and because it is apparently a creation of the period after the First World War, it should be described at length later in this book; but I have decided to omit it for reasons of brevity and instead to describe the origins of some of its separate branches here.

The first appearances of groups, rings, and fields and the subsequent axiomatizations of these subjects have been described at length by historians of mathematics.[88] Undoubtedly, some of that work could be done more carefully[89] but rather than go over that ground it is the acceptance of these ideas that detains us here. Two topics will suffice: the arrivals of abstract group theory, and the axiomatic theory of vector spaces.

4.3.1 Group Theory

Almost forty years ago, Hans Wussing wrote an account of the origins of abstract group theory that has worn well.[90] He traced the concept from its implicit use in geometry and number theory, through the theory of permutation groups in algebra, to the explicit recognition of the role of groups of transformations in geometry, finally to the recognition of the abstract group and its importance as a fundamental concept in algebra. In the geometric setting it is obvious (and usually a simple matter for the doubtful to check) that when one isometry is followed by another the result is itself an isometry, but that fact was long regarded as merely convenient, not in itself interesting. It became interesting with Klein's exploitation of the idea that different geometries have different groups, with Lie's much deeper investigations of the different classes of geometric transformations, and so on. In the algebraic setting of polynomial equations and their solutions known as Galois theory, groups and the different properties they can have were put up front in the French mathematician Camille Jordan's *Traité des substitutions et des équations algébriques* (or, *Treatise on Substitutions and Algebraic Equations*) of 1870. So group theory has its origins in

---

[88] The best single reference is Corry 1996. See also Corry 2007 for an account of how van der Waerden's *Moderne Algebra* changed the shape of algebra.

[89] I have myself contributed an oversimplification to the topic, which I regret.

[90] Wussing 1969.

substantial aspects of mathematical research. To follow the story of its transition to abstract structural mathematics it will be enough to follow the algebraic story, which is layered like an onion.

The decisive novelty of Jordan's presentation, emphasized even more by the succession of papers he then wrote in the 1870s, was to invert the order of ideas that it had been Galois's glory to have discovered. Galois had taken a problem addressed by Lagrange, Abel, and others, and that in turn derived from the common experience of mathematicians down the centuries: one can have an equation and not know how to solve it. Standard methods permit us to solve an equation such as $x^2 + x - 6 = 0$ and find the two solutions $x = 2$ and $x = -3$, and to solve the general quadratic equation $ax^2 + bx + c = 0$, $a \neq 0$ and find the solutions $x = \frac{-b \pm \sqrt{(b^2 - 4ac)}}{2a}$. This last expression is a formula involving the coefficients in the given equation, the arithmetical operations of addition, subtraction, multiplication and division, and the extraction of a square root. As a result, the quadratic equation is said to be solvable by radicals. Mathematicians had found similar but much more complicated expressions for polynomial equations of degrees 3 and 4 (they involve cube and fourth roots, reasonably enough) but had found no formulas for solving the general polynomial equation of degree 5, the so-called quintic equation.

Lagrange had analyzed in 1770 why equations of lower degrees are solvable by radicals and given reasons to believe that the quintic equation might not be. Abel had built on this work to prove, in 1826, that the general quintic was indeed not solvable by radicals. In 1832 Galois began the investigation of which equations or degrees 5 or more are solvable by radicals, as some are, and why most are not. His answer is easy to understand by analogy with, say, using blood samples as a diagnostic tool in the treatment of disease. Healthy people may have such-and-such measurable features in their blood, those with a particular illness measurably different properties. Galois showed that one can associate to a polynomial equation a group, and that equations which are solvable by radicals have groups with a special property.[91] Jordan took the opportunity in presenting Galois's ideas to extend and deepen them and in so doing to show that there was a new subject, the study of groups.[92]

Groups, like prose in the life of Molière's Monsieur Jourdain, turned out to be ubiquitous once they had been identified. For example, the integers form a group and the composition law is addition, as do the rational numbers with the same composition law, and the positive real numbers do when multiplication is the composition law. In the context of polynomial equations, the groups that arise consist of permutations of a finite set of objects, so they form finite groups. Jordan noted that they may have properties other than the one highlighted by Galois, and that in order to investigate Galois's property it was useful to identify certain other properties of groups, so there was a new mathematical object to study—the group—with novel structural properties. In terms of our analogy, it was as if biochemistry intervened in the process of diagnosis and treatment. The German mathematicians Eugen Netto and Walther von Dyck (influenced by Kronecker and Klein, respectively) and others

---

[91] A group is merely a collection of objects and a composition law (here denoted by a ".") such that whenever a and b are in the collection so is their composite $a.b$, and three simple rules are obeyed; see Glossary.

[92] One of Klein's reasons for traveling to France in 1870 was to learn group theory from Jordan.

in the early 1880s then carried out the process of giving a by-now well-motivated theory an abstract setting.

The abstract approach was developed to answer the natural question: How many finite groups are there, and what are they like? The theory divided into an easy part where the groups are commutative (in symbols, $a.b = b.a$) and the much harder part where the groups are not commutative. It proved possible quite early on to classify all finite commutative groups. For the rest, there was some progress on a number of different fronts. For example, it turned out that groups may, in a rather complicated fashion, be built out of groups that merit the title of being simple groups, so the search began for all finite simple groups. Around 1900, the American mathematician L. E. Dickson established the existence of four infinite families of simple groups.[93] Mathematicians who worked in various aspects of group theory other than Jordan were drawn from every country of the mathematical world, and the number of mathematicians who contributed to the enterprise one way or another shows how important group theory had become. The extent of the technical vocabulary that grew up as this work was conducted (normal subgroup, centralizer, normalizer, *p*-group, Sylow subgroup, solvable group, nilpotent group) shows how elaborate it had become. Group theory, as a structural, abstract branch of mathematics, had arrived.

4.3.2 Vector Spaces

The situation with respect to linear algebra is intriguingly different.[94] So many mathematical problems lead to linear algebra—for example, solving a set of linear equations—that mathematicians endlessly repeated, rediscovered, and often simply took for granted this or that piece of what became part of the theory of abstract vector spaces. As with the transition from permutation groups to abstract groups, it is quite easy to see the abstract theory beneath the skin of the earlier approaches, and even to believe that the original mathematicians saw much of it, too. However, in the case of linear algebra this may be reading too much into their work.

If Grassmann's work was too little read to have much influence in his lifetime, one can start by considering what Peano did in 1888 in responding directly to Grassmann's ideas. Much of his book of that year was simply an introduction to Grassmann's ideas that was clearer than the original, but Peano added two chapters, one on vectors and one on vector spaces, which he called linear systems. Here Peano gave a crisp definition of a real vector space without defining the vectors as *n*-tuples of real numbers, and he gave various examples. However, as Moore shows, the book had little effect, perhaps because putting Grassmann's name in the title severely narrowed its readership. On other occasions around this time, Peano gave other accounts in which the vectors were presented less abstractly as *n*-tuples. This shifts attention from the idea of a vector space as an object of interest to the individual vectors and their interpretations in any practical context.

Peano tried again in 1898 in a much more abstract way, but again to little effect. Others, Gaston Darboux and Geong Hamel among them, also put forward similar

---

[93] Simple groups were already known that belonged to none of these families, and the search to find these concluded only in 1982, although even today many details remain to be checked.

[94] See Crowe 1967, Dorier 1995, and Moore 1995.

ideas with similarly meagre results. In fact, matters did not change until after the First World War, when Weyl put forward a set of axioms for finite dimensional vector spaces in his account of the general theory of relativity, and then several people dealt with the much more delicate case of infinite-dimensional vector spaces in the 1920s, when topological questions intervene. Only now did the abstract concept of a vector space emerge with full force.

If group theory emerged quite rapidly and naturally because it had roots in substantial questions in geometry and algebra, it can at least be argued that the other truly elementary concept in algebra, the vector space, was held back both because it is very elementary and because there is almost a surfeit of applications. The axioms for a group are very simple, but the profusion of groups is challenging and requires its own set of definitions and theorems to be understood. (There are, for example, five different groups, each with sixteen elements). Moreover, the uses of group theory, although substantial, remained in the period prior to 1914 largely in areas of most interest to pure mathematicians. But there is only one vector space of any given finite dimension once the ground field is specified.[95] Many of the opening theorems in the theory of vector spaces are so obvious that mathematicians could and did work them out on the spot in an application. On the other hand, the uses of the vector space idea were legion. Among the more important were the theory of linear equations and the theory of linear differential equations, and in each of these it is quite natural to use n-tuples. For all these reasons, it is not surprising that the abstract group idea fared better than the abstract vector space idea, and it points to a general lesson: mathematicians did not modernize their subject from some fascination with abstraction, but because they felt they had to.

4.4 Modern Analysis

4.4.1 The French Modernists

In a few years toward the end of the nineteenth century, a subtle change came over the French mathematical scene, and for the first time the mathematicians who came of age formed something of a coherent group rather than a cluster of individuals, however brilliant. More precisely, they had some of the aspects of a coherent group, notably they thought of themselves as collectively being somewhat in opposition to their elders, if at times as also individually opposed to one another. Their elders reciprocated the feeling and professed to find the new work too abstract and too remote from the proper concerns of mathematicians.[96]

The most prominent of these people were Émile Borel and Henri Lebesgue. Maurice Fréchet, Paul Montel, René-Louis Baire, Jules Drach, and Arnaud Denjoy were also involved. Borel ran the most important seminar and, starting in 1890, organized the publication of results in a series of eighteen monographs, most of them

---

[95] If you think of a vector in an n-dimensional vector space as an n-tuple of coordinates, the ground field is the field from which the coordinates come, and in most applications this will be the real or the complex numbers.

[96] See Gispert 1991.

based on lectures given at the Sorbonne. Lebesgue was to produce the most important reformulation of the concept of the integral; Fréchet and Baire did major work in functional analysis and topology. Borel also founded and ran the journal *Revue du mois*, which he financed out of the money he won in prizes for mathematics. It covered not only mathematical but also scientific, technical, political, philosophical, economic, and other issues.

The main focus of this group was the theory of functions, first complex and later real, which they studied with a vivid appreciation of the power of set theory. They by no means disdained the classical themes of analysis (algebraic functions and differential equations), but they saw themselves as enlarging the range of the subject to include detailed studies of the many different varieties of functions with unexpected properties. To the older generation, this was to lose the plot. They felt that the new generality was artificial, its formalism and degree of abstraction excessive. Nor did the new generation have a distaste for applications—and though they could not know it at the time, they would turn out to be right: functional analysis and the study of functions that are, to speak loosely, integrable would prove to be the mathematics for quantum mechanics. Geometry, too, was acceptable. Lebesgue urged Fréchet in 1904, in terms drawn from some of Klein's remarks about Riemann and modern mathematics, not to be counted among those zealous representatives of hypermodern mathematics who regard geometric intuition as contrary to the spirit of mathematics.[97]

And yet they were abstract. They accepted much more readily than their predecessors the discovery that continuity was not as intuition had presented it. They developed the resources of set theory to describe what properties functions actually possessed, concentrating naturally on the areas in which intuition had proved itself misleading and geometry blind. They had little taste for what they called logic and the axiomatic foundations of set theory, because they cared more for functions. And applications—concrete applications—they left to others. They were, in short, mathematical modernizers.

4.4.1.1 MEASURE THEORY

Ferreirós has usefully pointed out that the naive concept of a set was developed under several distinct headings: as a foundational device, as part of a new theory of the infinite and the transfinite, and within the emerging subjects of topology and Lebesgue's theory of integral and measure.[98] The primary meaning of the integral throughout the nineteenth century was that it stood for an area. Other interpretations, which for example saw integration as an inverse operation to differentiation, were established as theorems, such as the so-called fundamental theorem of the calculus. Multiple integrals stood for the evaluation of volumes, and so on.

What drove the mathematicians onward was a deepening sense that integrals could, or should, be taken of a wide variety of discontinuous functions. This was a natural matter to raise once a function could be compared with its Fourier series representation. Do the function and its representation take the same value at every point, what can be said about the ways in which they do not? The naive theory of Fourier series (see §2.2.1) displays the coefficients as integrals: it is not always

---

[97] Quoted in Gispert 1991, 125; see also Klein 1894.
[98] Ferreirós 1996.

possible to do that, because some functions are not integrable, and this raises the question of whether certain badly behaved functions were going to cause significant amounts of trouble. It is just these questions that led Cantor to discover his theory of transfinite sets and, a generation later, prompted mathematicians to break decisively the link between integral and area.

There are a number of histories of the Lebesgue integral, and it will not be necessary to follow the developments in any detail. In the 1880s and 1890s the most significant and influential theory of the integral was Jordan's. He took an arbitrary bounded set E in a Euclidean space, and supposed both that he could cover it with a set of disjoint regions of known area and that it could be approximated by disjoint regions of known area wholly contained within the set. Then the first set of regions has a known area that presumably overestimates the area of the set E while the second set of regions provides an underestimate. If these approximating regions can be successively refined, the limiting value of the regions that collectively cover the set is what Jordan called the outer content of the set E. Likewise, the limiting value of the underestimates is the inner content, and Jordan said that if the inner and outer contents are equal, then their common value is the measure of the set E.

For the theory to work, delicate topological and measure-theoretic questions must be answered about the set E. These are particularly acute when E lies in a space of dimension 2 or more, which is why the theory of how multiple and repeated integrals are related is so subtle and would become a productive topic for further investigation. But even for single integrals, Jordan's theory amounts to breaking the domain up into measurable sets rather than intervals as Riemann's theory had done.

Lebesgue would later refer to Jordan as a traditionalistic innovator, and Hawkins finds this description to be particularly apt.[99] Jordan's account, which became well known through his *Cours d'analyse* (1892), starts from a naive, intuitive definition of area, proceeds to confront, and generally solve, the problems that arise in extending the naive idea to difficult cases, and along the way proposes some general theorems about the measure of sets, such as that the measure of a union of disjoint sets should be the sum of the measures of the individual sets. The starting point is, one might say, given in nature; the task is to solve the problems that are also given in nature.

By contrast, the modern theory of measure and integral possesses all the characteristic features of mathematical modernism. The central idea concerns measurable sets, so, like Jordan's theory, this theory is an elaboration of naive set theory. But in Borel's presentation of 1895 (*Leçons*) it is firmly axiomatic. As he put it: "A definition of measure could only be useful if it had certain fundamental properties; we have posited these properties a priori and we have used them to define the class of sets which we regard as measurable . . . new elements . . . are introduced with the aid of their essential properties, that is to say those which are strictly indispensable for the reasoning that is to follow."[100] Borel did not connect his theory of measure with contemporary theories of the integral, which may be one reason why Schoenflies was not much attracted to it in his *Report on the Theory of Sets*.[101]

---

[99] Lebesgue, quoted in Hawkins 1975, 9.
[100] Quoted in Hawkins 1970, 104.
[101] Schoenflies 1902.

FIGURE 4.7. Henri Lebesgue, photo from around 1904.

A further impulse to modernism came in 1895 when Borel and Jules Drach wrote up Jules Tannery's lectures at the École Normale on number theory and algebra. Here, as Tannery noted in the preface, Drach treated "numbers as signs or symbols, entirely defined by a small number of properties posited a priori."[102] Drach then took a similar approach to the theory of differential equations in his doctoral thesis of 1898, where, Hawkins notes, solutions were "characterized and classified on the basis of essential properties which they must possess."

Borel's ideas seem to have caught Lebesgue's sympathetic attention, for he took a similarly postulational approach to the definition of integrals, lengths, and areas of surfaces in his doctoral thesis of 1902. The thesis, entitled "Intégrale, longeur, aire,"[103] was published in the Italian journal *Annali di matematica*, apparently because the older generation of French mathematicians thought that it was too abstract.[104] In his thesis, Lebesgue defined a nonnegative measure on sets by some simple axioms, and gave theorems for computing the measure of a set which showed that it coincided with earlier notions due to Jordan and Borel. He then took the significant and original step of connecting his ideas of measurability with the definition of the integral. This led him to introduce the notion of a measurable function, and to define the integral of a measurable function.

A measure, according to Lebesgue's axioms, should be

---

[102] Quoted in Hawkins 1975, 103.
[103] Lesbesgue 1902.
[104] Hawkins 1975, 121–122.

- defined and not identically zero on bounded sets E;
- translation invariant[105] ($m(E + a) = m(E)$); and
- additive on pairwise disjoint sets ($m(\cup_i A_i) = \Sigma_{i=1} m(A_i)$).

Lebesgue then showed that if the measure of the interval [0, 1] is taken to be 1, then the measure of a set on his definition is less than or equal to the outer content of the set on Jordan's definition. He then showed that there is essentially only one definition of the measure of a set that meets the above three requirements, and that all Jordan-measurable sets are Lebesgue-measurable and have the same measure on each definition.

Lebesgue then used his theory of measure to define the integral. It applied to functions he called *summable* or, later, *measurable*. A function is said to be measurable if for all values of c and d the set $x: c < f(x) < d$ is measurable. He then showed that if a measurable function f is defined on a set and has minimal and maximal values m and M, respectively, then the integral of the function f can be defined using partitions of the interval $[m, M]$. The historical importance of using partitions of the range of the function is emphasized by all historians of the subject. Thus Hawkins (1975, 126) writes: "From a historical point of view, however, it seems extremely important that Lebesgue used a partition of $[m, M]$ to determine the nature of the sets e_i into which $[m, M]$ is partitioned. Such an approach necessitated the introduction of the notion of a measurable function; and it was the properties of measurable functions and the structure of the sets e_i that guided Lebesgue's reasoning and led to his major results."

Lebesgue's postulational or axiomatic approach was even clearer in the lectures he gave at the Collège de France in 1902–1903, later published as his *Leçons sur l'intégration et la recherche des fonctions primitives* in 1904. There he wrote down six very simple and natural axioms that anyone would surely subscribe to, and then showed that essentially only his theory met all these requirements. Of course, the skill was in keeping, not making, the six promises.

Lebesgue's theory of the integral rapidly turned out to have many advantages over the old Riemannian theory. Some of these advantages were discovered by Lebesgue and published in his thesis, others soon afterwards. Most famously, the theory is a good theory for taking limits (the sixth of his axioms concerned limits), and, with some further effort by Lebesgue, it was shown to salvage as much as could be expected of the fundamental theorem of the calculus (there are functions whose derivatives are not integrable on Riemann's definition of the integral). Many of these advantages followed from the ingenious idea of partitioning the range interval rather than the domain, as had been the custom hitherto, so it might seem that Lebesgue's success was more a classic demonstration of good mathematics rather than anything modernist. The significance of Hawkins's remark is that it highlights the extent to which Lebesgue's theory works because it is axiomatic and proceeds from an abstract definition of measure that produces, but does not start from, an idea of area. Integral, length, and area are given new definitions, as the title of the thesis suggests, based on the now more fundamental idea of measure that in essence only has to meet the requirements that any theory of the size of sets would have to meet.

---

[105] Here $E + a$ denotes the set of all $e + a$, where e is an element of E.

FIGURE 4.8. Felix Hausdorff in his study at home, Bonn, June 1924.
Universitats-und Landesbibliothek, Bonn, Hss.-Abt. NL Hausdorff.
Kapsel 65: Nr. 29. Photograph by Ludwig Hogrefe, Godesburg.

If the aim of measure theory is to assign a size to every bounded subset of the line, or the plane, or even of n-dimensional space, then against all the many successes of Lebesgue's theory had to be set one irritating problem: it was simply not clear that this could be done. The generalization from one to n-dimensions is the obvious one, and Lebesgue had shown that there was a large class of measurable sets. But he first went on record expecting that his theory would not measure every bounded set, and then constructed examples of sets that were not Lebesgue-measurable sets.[106] This left open the question of whether Lebesgue's definition of "measurable" could not be modified in some way, and this topic caught the interest of Felix Hausdorff.

The significance of the work of Felix Hausdorff for a study of mathematical modernism seems to have been highlighted first by Mehrtens (1990). In recent years a team of German scholars has been working to produce an impressive edition of his works, and as of 2007 four volumes exist of the planned nine.[107] Hausdorff was born in 1868 in what was then Breslau, in Germany, and is now Wroclaw, Poland. With his wife he committed suicide in the little town of Endenich in January 1942 to avoid deportation to a concentration camp, bravely writing the note, "Und vielleicht ist Endenich auch die Ende nicht" ("and perhaps Endenich is also not the end").

Until recently, Hausdorff was remembered only for his work in topology, but, as Mehrtens showed, Hausdorff wrote two books on philosophy and a successful play under the name of Paul Mongré ("Paul to-my-liking" would seem to be an acceptable

---

[106] As by then had Vitali and Van Vleck.

[107] It is therefore premature to speak with full confidence about Hausdorff's work, but I would like to thank Moritz Epple in particular for his help and advice, and also Walter Purkert. What follows is also drawn in part from the unpublished Epple 2004.

gloss) before turning to mathematics in his forties. As a student at Leipzig, he had been impressed by the Kantian Friedrich Paulsen, and Otto Liebmann's *Die Analysis der Wirklichkeit* (1876), whose version of Kantianism was close to the empirical sciences, was another congenial stimulus. As an independent philosopher he saw himself as an upbeat Nietzschean, and his writings are full of stimulating aphorisms. Interestingly, it seems that his philosophical ideas may have influenced him to make a more detailed study of mathematics.

The turn to Nietzsche came about through his involvement in the Akademisch-Philosophischer Verein, which was founded in 1866/67 by some of the Leipzig empiricists (Richard Avenarius, Gustav Theodor Fechner, and Wilhelm Wundt among them).[108] The society became a forum for modernist trends in science, literature, music, and the arts, and all manner of cultural and intellectual issues were fiercely discussed in regular evening meetings.[109]

For our purposes, Mongré's most important work was his *Das Chaos in kosmischer Auslese* of 1898.[110] Chaos in cosmic selection, as it might be called in English, is a Nietzschean theme with a mathematician's twist to it; the term "Auslese" deliberately suggests natural selection. The fundamental idea is that the world in itself is unknowable (a Kantian point about the *Ding an sich*), but everyone produces an intelligible cosmos out of the inaccessible, transcendental chaos. To any kind of metaphysical realism or claim that we could know what the world really is, Mongré opposed what he called a transcendental nihilism. Insofar as any position is advocated, it is a critically refined empiricism. The arguments Mongré used were generally of this type: take a realist conception of time. It presumes to discuss some absolute concept of time and to make a comparison between it and the subjective concept of time. But we can imagine any kind of mathematical transformation of the absolute continuum and a suitable transformation of our minds such that the subjective conception is unaltered. Therefore the absolute concept is meaningless. The same sorts of arguments dealt with realist conceptions of space, and here Mongré ran through the familiar accounts, going back to Riemann, about the significance of non-Euclidean geometry and the nature of geometrical ideas. Indeed, in his insistence that the mathematician can devise all sorts of geometries, Mongré is more Riemannian than many of Riemann's successors in the intervening years. The novelty is that Mongré has a way of squaring empiricism with any kind of claim about the absolute. All this built up to the claims, by the end of the book, that the idea that time is linear, continuous, single valued, does not run backwards, and proceeds uniformly, and the idea that space is three-dimensional, continuous, uniform, and Euclidean, are incapable of proof.

All this is entirely Kantian. Mongré regarded his contribution as producing a strengthened version of this familiar Kantian position. What is more interesting, as Epple writes, is how Hausdorff's thought progressed: "In fact he proceeded in steps, exploring in a sometimes rather poetic fashion stranger and stranger ways in which

---

[108] Avenarius and Wundt were among the founders of the *Vierteljahrsschrift für wissenschaftliche Philosophie*, the first journal devoted to the philosophy of science.

[109] Epple (2004) has pointed out that Otto Erich Hartleben was a member of the Verein, and after 1886, both Hartleben and Hausdorff produced free renderings of Giraud's "Pierrot Lunaire" poems. Hartleben dedicated his poems to several individuals of the Leipzig group, including Hausdorff, and they were later used by Arnold Schönberg for his *Pierrot Lunaire*, op. 21.

[110] Reprinted in Hausdorff 2004, vol. 7.

absolute and empirical time, absolute and empirical space might be related mathematically. It is precisely here, and not in his academic work, that Hausdorff began to rely on the very new mathematical notions that Georg Cantor and others had begun to develop under the label of set theory."

Thus in an unpublished manuscript on "Formalism" of 1904 Hausdorff called upon mathematics to free itself from any kind of intuition, and in a lecture course of 1903–1904 he said:[111] "Mathematics totally disregards the actual significance conveyed to its concepts, the actual validity that one can accord to its theorems. Its indefinable concepts are arbitrarily chosen objects of thought and its axioms are arbitrarily, albeit consistently, chosen relations among these objects. Mathematics is a science of pure thought, exactly like logic."[112] Hausdorff's view quite generally was that no aspect of science can have a unique description, and mathematics should explore all possible descriptions. This is much more radical than Poincaré's conventionalism, which Hausdorff seems to have first heard about sometime after coming to his own position.

After 1911 Hausdorff's interests in mathematics deepened. He worked extensively on set theory, the first to take it up so energetically since Cantor, and his major work, important for both set theory and topology, is his *Grundzüge der Mengenlehre* (1914), very much revised when reprinted in 1927. In 1915 Hausdorff published what became known as his paradox.[113] It concerns any definition of measure that satisfies Lebesgue's four axioms. He assumed, in order to derive a contradiction, that every bounded set is measurable. He took the sphere in three-dimensional space and constructed a division of it into four disjoint sets A, B, C, D, with the property that A, B, C, and $B \cup C$ are all congruent, while D has measure zero. Because the sets A, B, and C are congruent, they have equal measure, and because they cover the sphere except for the set D of measure zero, each must have measure $1/3$. But the sets A and $B \cup C$ are also congruent and cover the sphere except for the set of measure zero, so they must have equal measure, which must therefore be $1/2$. It follows that the set A has measure $1/2$ and measure $1/3$, which is impossible. Accordingly, the assumption that there is some definition of measure according to which every bounded set is measurable is false.

More precisely, one must either agree that that's how it is—there will always be nonmeasurable sets; or one must reject the concept of measure and look for another concept altogether—but Lebesgue's four axioms are very natural ones; or one must scrutinize the proof and hope to find a flaw. This was how Borel reacted. Hausdorff had used the axiom of choice in his proof, and in the second edition of his *Leçons sur la théorie des fonctions* (1914) Borel concluded that the paradox came about not because measure was an inherently flawed concept, but because the set A was not properly defined. To construct it using the axiom of choice was for Borel no construction at all. "If one scorns precision and logic," he wrote, "one arrives at contradictions."[114] Hausdorff, of course, begged to differ. He was not at all bothered

---

[111] I am indebted to Leo Corry (2006) for calling these remarks to my attention; he notes that they come from Purkert 2002.

[112] This remark recalls Poincaré's statement that mathematicians study not objects but the relations between objects; see §3.2.

[113] Hausdorff 1915.

[114] Borel 1914, 256, quoted in Moore 1982, 188, who also notes that a number of Italian mathematicians had already explicitly rejected the idea that one can make infinitely many arbitrary choices: Peano, Bettazzi, and Beppo Levi among them; see Moore 1982, 76–82.

that any definition of the area of a set is inherently imperfect, even though this was a conclusion that could never have been dreamed of by researchers a generation earlier.

At almost the same time, and as part of the same movement among French mathematicians, Maurice Fréchet produced a similarly abstract reformulation of a more technical problem area in mathematics that was to have a decisive influence on the creation of modern topology.[115] Fréchet was interested in the calculus of variations, a topic in which the unknown is usually a function that is required to minimize a certain integral. Examples include finding geodesics in a space, surfaces of least area with given boundaries, and harmonic functions, but many problems in physics can be expressed in the language of the calculus of variations. However, reliable techniques for solving problems of this sort had proved hard to find. It was difficult to tell a minimal value from an extremal value, and therefore a stable solution from an unstable one and the complicated theory that existed was generally agreed to be in need of improvement. Hilbert had raised it as the last of his famous Paris problems, remarking that it was "a branch of mathematics repeatedly mentioned in this lecture which, in spite of the considerable advance Weierstrass has recently given it, does not receive the general appreciation which, in my opinion, is its due."[116]

One technique much employed in contemporary calculus of variations was to find a sequence of successive approximations to the sought-for function. Fréchet proposed a simple analogy. Just as one might approximate π, say, by finding among all numbers a sequence of successive approximations to it, so one might situate the successive approximations to a sought-for function in a space of all plausibly relevant functions. There might indeed be many such spaces, depending on the properties of the functions involved, but a much graver difficulty facing Fréchet was that any such space was surely infinite, indeed infinite-dimensional, whatever that might mean. Undaunted, Fréchet proposed to show that in many contexts one could introduce a sense of distance in a space of functions and speak of a sequence of functions in this space as tending to a limit, exactly as a sequence of approximations to π tends to π. What were functions in one setting became mere points in a space in Fréchet's new view of things.

Fréchet succeeded in showing that there were many problems in the calculus of variations that could profitably be formulated his way. He found spaces of functions of various kinds, and spaces of all curves of certain kinds could be made into metric spaces: spaces in which it made sense to speak of the distance between two points, and where a limiting process could be defined. Most importantly, he showed that these spaces could be complete[117] which means that if a sequence tends to a limit then the limiting value is also in the space.[118] Thus in Fréchet's vision, the concept of distance is greatly generalized away from any sense of distance in any Euclidean space. Moreover, the drive was completely axiomatic in spirit. Fréchet was clear that

---

[115] Bearing in mind later discussions of Esperanto, I note that Fréchet was a lifelong enthusiast for the language.

[116] Hilbert 1901 a, in 1976, 29.

[117] Strictly, sequentially compact, but the details need not concern us.

[118] This need not be true: any sequence of rational numbers tending to π has a limit (π, of course), but π is not a rational number, so it is not an element in the set of rational numbers.

what was wanted was the ability to talk of distance and to take limits, in the abstract senses he had in mind, and one had this ability whenever a space satisfied certain axioms. Initially he talked of spaces of class D (or distance spaces) where it made sense to talk of distance. These spaces were later called *metric spaces*, admitting a distance, by Hausdorff. Then Fréchet introduced spaces of class V (for *voisinage* or neighborhood in French) where it made sense to talk of points being close, which he soon realized were equivalent to spaces of class D. Finally, Fréchet spoke of normal spaces of class V when it is possible to talk of limiting processes converging to an element of the space.

4.4.2 Dimension

We may start the discussion of the dimension of a space with another essay by Poincaré(1912b), "Why Space Has Three Dimensions," which was one of the last essays he wrote. After briefly indicating that the question is a topological one, he turned to consider the results obtained by Cantor and others. Cantor had shown that there is a one-to-one correspondence between the line and the plane, and more generally between the continuum of p and the continuum of n dimensions. But this map is not continuous. Nor, said Poincaré, is the analytic definition of n dimensions satisfactory, which asserts that an n-dimensional continuum is a set if it has n independent coordinates. Such a definition minimizes the importance of the intuitive origin of the notion of a continuum. At best it would satisfy a mathematician of the arithmetizing kind, but not a philosopher. Indeed, Poincaré should have gone further: it has to be shown that the analytic definition is a definition at all, and that it makes sense for a continuum to have a unique dimension.

It would be better, said Poincaré, to define dimension inductively. A closed curve is a continuum of one dimension. If a pair of points is removed from the curve, it necessarily falls into two pieces. But we may remove arbitrarily many points from a continuum of two dimensions without it falling into pieces. This suggests that if we regard one, two, or finitely many points as having zero dimension, we may distinguish the one-dimensional continuum from higher dimensional continua by saying that they, and only they, fall into two pieces if a zero-dimensional continuum is removed from them. Similarly, a two-dimensional continuum may fall into two or more pieces when some one-dimensional continua are removed from it, but three-dimensional continua do not. So one may say a continuum of n dimensions is one that can be cut into pieces by $n - 1$-dimensional continua.

It is hard to know what to make of Poincaré's rather superficial account. Certainly, it captures something of our geometric sense of what a continuum is, but the concept of a continuum is not defined. We must know, for example, that a closed curve without self-intersections is a continuum, but a closed curve that crosses itself is not. There needs to be an argument, of which Poincaré offers not a trace, to establish that what seems to be true for continua as we expect to see them is true for topological copies of them. There is such an argument, but the matter is not obvious. Indeed, it was not rigorously proved that a closed curve with no self-intersections shares with the circle the property of dividing the plane into two regions, the inside and the outside, until Veblen (1905) showed this. But the most decisive criticisms of the argument are these: it does not so much define dimension as capture a property of it that we are happy to recognize; and, as a result, the argument does not even hint at how one might define dimension for more complicated spaces. Indeed, if you unthinkingly apply Poincaré's iterative procedure, you may get the wrong answer: Brouwer pointed out that the double cone surely has dimension 2—it is a surface (except perhaps at the vertex)—but it falls into two pieces on removing the vertex, so on Poincaré's "definition" it has dimension 1. Poincaré was operating firmly in the old world of intuitive concepts abstracted naively from familiar objects and their most immediate generalizations.

To Brouwer is due the first proof that the concept of dimension can be applied consistently and sensibly to a wide class of spaces. He showed in 1910 that if two manifolds are homeomorphic, then they have the same dimension. Unlike Poincaré's naive argument, the proof of this theorem is both rigorous and difficult. Freudenthal, in editing Brouwer's *Collected Works*, said that on its first appearance it must have looked like witchcraft. Intuition has been replaced by definition and proof, in this case at a forbidding price in abstractness, and even then only for a class of spaces that have a clear intuitive concept of dimension, because they are "locally," as mathematicians say, like continua. It remained unclear if there was a large class of topological spaces to which a generalized concept of dimension would apply.

Also in 1910, Fréchet made progress on defining dimension for a large class of spaces. He wrote down a set of axioms that he asserted any concept of "dimension" would have to satisfy, and constructed examples of countable sets of points with a "dimension" a rational number between 0 and 1. He then showed how to construct sets with arbitrary rational dimensions.

Fréchet's definition was largely eclipsed by the one Hausdorff offered in a paper of 1919, in which he gave another axiomatic definition of dimension and exhibited spaces having any prescribed real number as their dimension.[119] The definition is inevitably complicated, but the basic idea can usefully be explained. Consider a rectangle in the plane. We believe it has an area because we can argue as follows. Let the rectangle be covered by whatever number it takes of squares of a standard size (allowing for some overlaps). If we decide to cover the rectangle with smaller squares, say squares whose sides are half the size of the first ones, we expect to use about four times the number of squares (the overlaps may be smaller, too). Each time we halve the sides of the squares, we expect to increase the number we need by a factor of four. So the product (number of squares)×(side length of a square)^2 is roughly constant, and gives us better and better estimates of the area of the rectangle as the squares get smaller. If we think the rectangle has a length, we would try to cover it with line segments, but when we do we have to use infinitely many (a mathematician might prefer to say we fail). And if we think the rectangle has a volume, we would try to cover it with little cubes, and when we do, because we need about $8 = 2^3$ times as many cubes when the side length halves, we wind up looking at the product (number of cubes)×(side length of a cube)^3, and this tends to zero as the cubes get smaller. In fact, it turns out that for rectangles, the only product of the form (number of squares)×(side length of a square)^k that tends to a finite amount as the squares get smaller is k = 2. If k is smaller than 2, this ratio tends to ∞ as the cubes get smaller, and if k is greater than 2 this ratio tends to zero. So we say that

---

[119] This concept is currently central to many branches of mathematics where repeated iteration is studied.

the rectangle has dimension 2. Hausdorff refined this definition until it worked for a large class of spaces, and he gave examples to show that there are spaces of every dimension greater than or equal to zero. Once again an intuitive mathematical concept has been redefined into something axiomatic and almost counterintuitive.

4.4.3 Continuous Curves

The importance of the naive idea of continuity has already been discussed. People believe that objects do not jump from place to place but move in a continuous fashion, if, on occasion, very fast. Some might even say that it is precisely because shadows can jump in this fashion that we recognize that they are not "proper" objects. And if position can change only continuously, never abruptly, then it seemed natural to say the same of velocity, acceleration, and all the other quantities involved in the description of motion. Thus the belief that the calculus always applied and functions were, to recall the technical term, analytic. It seemed entirely possible to live with a "creative tension" between such smoothness and sudden change: blows, impulses, shocks. However, by the end of the nineteenth century it was recognized that this profound intuition was almost wholly misplaced as a source of confidence in geometry, mathematical physics, and even mathematical analysis itself. The damage was done by the acceptance of Cauchy's definition of continuity, coupled to that of the Cantor or Dedekind definitions of the real numbers. Far from completing a program of rigorizing the calculus of Newton and Leibniz, the arithmetized calculus had to be seen as answering quite different questions altogether.

It had always been allowed that a curve may have a few sharp corners. As early as 1834 Bernhard Bolzano had constructed a continuous function that has no tangent anywhere, but he never published it, and it did not disturb the mathematical world until the twentieth century.[120] However, when Riemann announced a similar thing in his lectures in 1861, and when Weierstrass in 1872, pursuing a related line of enquiry and always with a view to being more accurate than Riemann, came up with a whole family of such functions, the trouble was plain to see.[121] Two consequences flowed from these discoveries. One was that there can be no retreat to the concept of differentiable. These examples showed that a point could move in a continuous fashion without the moving point having a tangent direction at any stage, and they showed just as surely that velocity might be continuous but not differentiable, and so on. Every rung of the "analytic" ladder broke. The other consequence was that the only naive intuition underpinning the application of the calculus to geometry and mechanics also collapsed. As Pierpont, the leading analyst at Yale, put it in 1899: "Our intuition is utterly helpless to give us any information in regard to such curves. Indeed our intuition would rather say that such curves do not exist."[122] Needless to add, proofs that depended at crucial points on appeal to intuition might, at that very moment, be in fatal error. It was small comfort when Jean Perrin confirmed in 1908 what Einstein had assumed in 1905 in order to give an accurate estimate of the

---

[120] See Kowalewski 1923.
[121] Weierstrass 1872.
[122] Pierpont 1899, 398.

amount of energy involved in Brownian motion: the general path of a particle in the air is continuous but never differentiable. For this work, which was decisive in establishing the existence of atoms, he was to be awarded the Nobel Prize, but mathematicians had long since ceased to base their work on appeals to physics.

The proper mathematical response to these discoveries was to accept them, and to admit that there was a realm of analytic functions, outside which lay an unexplored wilderness. Of course, it would now be necessary to check the provenance of each new function thrown up by research, but the analytic realm was rich. It was a matter of taste whether the other functions should be studied or not. Hermite wrote to a fellow mathematician to say that he turned in horror from these pathological functions.[123] He did not dispute that they were there, but he thought mathematicians had much better things to do than investigate them. Even Poincaré, who had been among the first to study such functions, conceded that "logic sometimes engenders monsters. For half a century one has seen a slew of bizarre functions arise that seem to strive to resemble as little as possible honest functions that do something. No longer continuous, or if continuous without derivatives, etc. Moreover, from the logical point of view it is these strange functions that are the most general, those that one seeks without having looked for them appear as no more than a special case."[124]

Just as the English Lake District challenged the boundaries of eighteenth-century English taste, and the American Grand Canyon challenged European sensibilities, the nonintuitive functions were in some sense an affront.

However, this response, although many mathematicians took it, was inadequate, rather as Poincaré hinted. There was and is a very short list of things that mathematics is taken to be about. One of them is geometry, and since Archimedes this has as its key component continuous shapes. It is therefore important for mathematicians to be able to say what a continuous curve is. Consider then a curve traced out in the plane by a moving point. We may suppose the point starts at some moment of time, and at each subsequent moment until it stops it has a position and therefore an x and a y coordinate. The obvious definition of a continuous curve is exactly that the coordinates of a point on the curve are given by two continuous functions.

The question now arises: Does this definition isolate a class of curves having more or less the expected properties? The answer is dramatically that it does not. Peano in 1890 was the first to show what Hilbert and then E. H. Moore made more visible: there are space-filling curves—continuous curves (on the above definition) that pass through every point (inside and on the boundary of) a square.[125] A solid square is a most unacceptable picture of a curve.

Nor need a curve have a length. To measure length, one lays off successive unit lengths along a curve until one reaches the end. While the process will be very inaccurate if the unit is too large or the curve is very crinkly, one supposes that by choosing smaller and smaller units, better and better approximations will be found that settle down and converge to a value for the length of the curve. It was discovered by a number of mathematicians in the 1880s that this need not be so:

---

[123] Quaded in Hawkins 1975, 122.
[124] Poincaré 1905a.
[125] Peano 1890. Incidentally, the reviewer for *Fortschritte* was Hilbert.

there are continuous curves that have no length, or, if you prefer, have infinite length.[126]

Faced with such a problem, mathematicians looked for a retreat. Can one add to the definition some sensible requirement so that the new class of continuous curves does have the "right" members? It was noted, for example, that the Peano space-filling curve crosses itself infinitely often. Could one do better with curves that may not cross themselves? To be sure, such curves may still fail to have tangents, and fail to have lengths, but surely they do not fill whole regions of the plane? They do not, but an ingenious example of the Harvard mathematician Osgood in 1903 showed that they may still have an area.[127]

What about this simple question: Does the continuous image of a circle that never crosses itself enclose a region of the plane, just as the circle does? In other words, do continuous closed curves that do not cross themselves have an inside and an outside? It is a measure of how hard this question is that it was raised but not solved by Camille Jordan in the 1890s, and only solved for the first time some years later, by Veblen. The answer is "yes," but there is still no simple proof.

Pierpont's concluding remarks in an address to the American Mathematical Society are indicative of the shock mathematicians felt on meeting the consequences of rigorous, arithmetized continuity:

> The opinion the intuitionist holds in regard to the value of our intuition is untenable. [These] notions are vague and incomplete and . . . it is impossible to show the coextension of these notions and their arithmetical equivalents. The practice of intuitionists of supplementing the analytical reasoning at any moment by arguments drawn from intuition cannot be justified. . . .
>
> We have two worlds, the world of our senses and of intuition, and the world of number. Objects in the first world give occasion to form certain objects in the world of number which we strive to make as close to the original as possible. How close the copy is we can never know. Doubtless they are sufficiently approximate. . . .
>
> The analysis of today is indeed a transparent science. Built upon the simple notion of number, its truths are the most solidly established in the whole range of human knowledge. It is, however, not to be overlooked that the price paid for this is appalling, it is the total separation from the world of our senses.[128]

Schoenflies, when he surveyed the implications of set theory for geometry and function theory, was of a similar opinion.[129] His opening words were that the idea that a continuous curve is the image of an interval by means of two continuous functions had long since forfeited its naive geometric content. Then he noted that other definitions of a curve are possible. For example, one might define a curve in the plane as a particular kind of subset of the plane. The nascent subject of topology offered a way to do that, but on that definition a piece of the curve $y = \sin(1/x)$ (see

---

[126] The space-filling curves are indeed examples of curves of infinite length, but the reader may feel that they are already ripe for exclusion on other grounds. See Sagan 1994.

[127] For the wise, in the sense of Lebesgue measure. See Osgood 1903.

[128] Pierpont 1899, 405–406.

[129] Schoenflies 1906b.

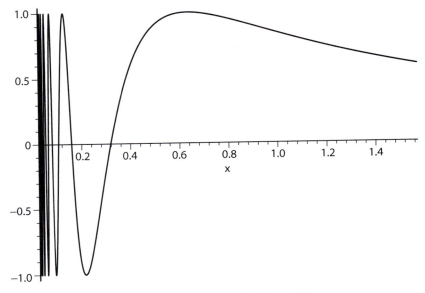

FIGURE 4.9. The graph of $\sin(1/x)$.

fig. 4.10) can be joined to a piece of the y-axis through the origin so as to make a curve.

The graph of $y = \sin(1/x)$ oscillates infinitely often between $+1$ and -1 as x tends toward the origin, the curve cannot be described as the image of an interval under a pair of continuous functions, and its shape and behavior inside any circle enclosing the origin make it hard to say that it enclosed a region, although it does. The reader may readily and correctly imagine that topologists could easily come up with much worse.

Nor were these examples artificial, and only intended to demolish poor definitions. Schoenflies noted that Poincaré, in the course of his remarkable technical work on function theory and non-Euclidean geometry, had constructed a set of points that met the topological definition of being a curve but was far from being a curve in the usual sense. Not only did it have many points where it had no tangents, there were infinitely many points on it around which the curve winds infinitely often. No wonder Poincaré raised the question of the extent to which this curve could be considered as a curve at all.[130] Schoenflies rightly observed that "the unlimited slew and manifold of geometric forms that is contained in this concept of a curve, scarcely lets itself be recognised by any hints."[131]

It is not so much the curves, as the mathematicians' shrewd yet anxious remarks about them, that attract our interest. Intuition might remain as a guide to research, but as a source of knowledge it had to be abandoned. This was, as Pierpont said, a high price to pay, and giving up intuition in this way was not at all the answer envisaged by anyone who had questioned the foundations of the calculus. Such

[130] Poincaré 1883.
[131] Schoenflies 1906b, 563.

questions had been answered, but the answers were couched in the language of number and set theory. By then Hilbert was already asking mathematicians to dig deeper: Could that be done, and would set theory suffice?

4.4.4 Riesz on Space and Topology

A good measure of how far mathematics had gone in setting out its own position on scientific questions, and in what the difference between mathematical and scientific formulations had become, is given by a paper the young and distinguished Hungarian analyst Frigyes (Friedrich) Riesz presented to the Hungarian Academy of Sciences in 1906. The paper (Riesz 1906) is called "The Genesis of the Concept of Space."[132] He began by saying that, as an exact science, geometry was built on a system of consistent and complete axioms. Then, to be a natural science, it must be the case that its axioms agree with our intuitions and representations of space. But, he said, there is a gap right at the first step, because the mathematical continuum is a problematic concept. Riemann had introduced, rather obscurely, the concept of an n-manifold, which Hilbert had recently defined more precisely. (As noted earlier, Hilbert had given a complicated topological characterization of the plane in 1902, but deep mathematical difficulties had prevented him from studying the case of three dimensions.) Unfortunately, Riesz continued, many nonmathematicians had offered opinions on the philosophy of continuity who were at least as mystical as the oracle at Delos, and Riesz quoted with approval Russell's remark[133] as applying to more than Hegelians: ". . . the Hegelian dictum that everything discrete is also continuous and vice versa. This remark . . . has been tamely repeated by all his followers. But as to what they mean by continuity and discreteness, they preserved a discrete and continuous silence." However, he said, there is also a deeper question: grant that the mathematical question of the continuum can be solved, and a geometric system constructed accordingly, which agrees without intuitions. The question of the development of the concept of space is this: to what extent is this system determined by our ideas, whether by the nature of our thinking (*Denkarbeit*), or psychological life, or different systems having the same purpose. The answer will be derived from essentially psychological hypotheses built up inductively from observations and that control our representations of space and time.

This, he said, has nothing to do with Kantian theory. Rather, it is a matter of abstracting from the organization of our sensations of space and time the sort of physical continua we take space (and time) to be, and comparing them with the mathematical continua we can construct. For example, Poincaré had addressed the question of why our space has three dimensions. He had considered the way we create our concept of space from various different sensations (touch, sight, the motion of our bodies) and how one can define the concept of dimension. But the double cone (Brouwer's example) shows that Poincaré's definition was too naive, and unfortunately, said Riesz, it is very difficult indeed to define dimension for

---

[132] Much more information about Riesz and his study of geometry will be found in the PhD thesis of Laura Regina Rodriguez Hernandez (2006). I am very grateful to the author for sending me this work, which I hope will be fully published; it reached me as I was putting the final touches to this book.

[133] Russell 1903, vol. 1, 287.

arbitrary point sets. So the idea of dimension cannot presently be taken as fundamental, for it lacks clarity. And even if we grant, or can show, that space is a mathematical continuum of some sort that, locally, has three dimensions, and can indeed be described by real numbers, we still have to consider the topology of space and investigate the idea of a straight line before we can discuss the geometry of space.

Riesz, in agreement with Poincaré, did find it possible to say some things about the physical continuum. It is made of points that are distinct but, if too close together, indistinguishable (that is, cannot be told apart by the senses). It is connected, meaning that any two points can be joined by a chain of distinct but indistinguishable points. Riesz now defined two points a and b to be logically indistinguishable if there is no third point, c distinct from both of them and distinguishable from only one of them, and asserted that a physical continuum has at least two indistinguishable points that are not logically indistinguishable. It remained to refine this idea and to compare it with mathematical continua.

So far, one might say it is a traditional program but with a vastly enlarged repertoire. One can imagine Aristotle having some views about space that were encapsulated by the (later) axioms of Euclid. But Riesz knew that a vast amount of mathematics can be built up without any note being taken of our experience of space. Not merely Euclidean and non-Euclidean geometry, or even the different geometries of Hilbert. Not even, indeed, the plurality offered back in the 1860s by Riemann, all those metrical geometries with different curvatures. For Riesz, the starting point has to be a topological analysis, prior to any metrical one. This in turn determined the sorts of experiences he was willing to say affected the physical continuum that one is then to compare with the mathematical ones on offer. He was not looking at sundry global properties: It space were infinite, what properties would straight lines have? Riesz only considered the fact that points of space can be so close as to be indistinguishable. It is true that Riemann, and after him Clifford, had been prepared to consider that space might not be continuous but somehow granular on very small scales, but that idea had lain abandoned as a mathematico-physical jest from the moment it was written down. Riesz was prepared to take it seriously, if only to rule it out and take as his starting point for the analysis of the physical continuum that it contained points arbitrarily close together.

Accordingly, he argued that what we apprehend at any moment is a tiny smear of space and of time composed of the place and time and all the places and times indistinguishable from that one. The mathematical continua worth discussing must have the corresponding topological properties: every mathematical point must be surrounded by a continuum of arbitrarily nearby points, where "nearby" is to be understood in a Cantorian topological sense, not a metrical one. Riesz therefore elaborated a theory of suitable point sets, which he noted had some similarities to Fréchet's recent work on topology. Some theorems can be proved about such spaces. For example, they obey the Bolzano-Weierstrass theorem, which may be rendered vaguely but well enough for present purposes as saying: every such set which is bounded and infinite has at least one point arbitrarily close to the rest of the set. Space is connected—it comes in one piece, and that means that the mathematically relevant continua must be connected in the mathematical (rather more precise) sense. Space has certain order properties (this point is to the left of that one, or perhaps indistinguishable from it, as far as one of its coordinates is concerned). These too must be matched by the mathematical continua.

All this was difficult, as Riesz had said it would be, and ultimately inconclusive. But the philosophical point was for that reason all the clearer. He concluded that our hypotheses are not unconditional truths, but assumptions that arise from experience and whose strongest support is necessity. But what can we say if there is no necessity? What are the assumptions we can make without falling into contradiction, and what are their consequences? These are the typical questions in axiomatic research, he said, but if pursued the axiomatic study of geometry as a science of experience would for the first time attain a certain completeness. How complicated the relation of mathematics to experience has become! What had once seemed to have been settled by Euclid was, in 1906, the goal of difficult researches in topology.

4.4.5 Modernism and Modern Analysis

It is now possible to examine the claim that analysis was always abstract and became irrevocably so in the 1820s with Cauchy's work. This claim points to the new definitions of limit, continuity, differentiability, and integrability Cauchy introduced, and the elaborate theory he produced that wove these concepts together and delivered all the important results of the old calculus. To be sure, there were things left to be done, notably clarifying the concept of real number and therefore of limit, but much of what Cauchy did survives to this day in rigorous presentations of mathematical analysis.

There is no doubt that this work was a formidable piece of abstract reasoning. Geometric and mechanical analogies were removed from the foundations of the calculus, thereby eliminating the risk that mathematical science might all rest on a vicious circle: the calculus supporting dynamics and geometry which in turn support the calculus, which . . . Instead, reliance was placed on the arithmetic of the real numbers. However, the case for modernism does not rest on refuting these claims, which are more or less unassailable, but on calibrating them. The challenge to the modernist case is that mathematics did not become significantly more abstract after Cauchy, and that in the history of analysis continuity greatly outweighs change. To deal with this challenge, let us first consider an analogous claim. Let us suppose some historians of music assert that between Bach's creation of the classical theory of musical harmony and Schoenberg's twelve-tone theory stand, most importantly, the architects of the classical style (Haydn, Mozart, Beethoven) and the romantics (say, Schubert, Schumann, Chopin). To sustain this position, these historians of music have to give criteria for recognizing these four episodes, for appreciating their significance (regarding them as roughly comparable and more noteworthy than other changes of style) and of course for ranking Schoenberg's alongside the first three. The position is threatened by an argument that relegates any of these to a footnote on the others because it was an inevitable development. The modernist case, by analogy, can recognize (to name only names) Newton and Leibniz, Euler, Cauchy, and finally Cantor and Dedekind.

There is no dispute that a line runs from Bach to Haydn, Mozart, and Beethoven and from there to the Romantics. There is some dispute that it runs to Schoenberg, and there are many who would suggest that music outside Vienna shaped the musical world as much as the second Viennese School. There is little dispute, however, that a line runs from Bach to Schoenberg. The details of the case that it is the main line, or a direct line, may be left to music historians, but we should note the outline. Increasing levels of chromaticism eroded the concept of governing keys and the inner logic of the coherence and development of a piece of music. Responding to this need, Schoenberg proposed a new organizing principle, one that was apparent to the composer but perhaps not apparent to the audience or, indeed, likely to produce music that agreed with "natural" ideas of what sounded harmonious.[134]

Similarly, there is no dispute that a line runs from Newton and Leibniz to Cauchy, or that it runs onward, past Dirichlet and Weierstrass. The analogy breaks down in that, as with all artistic matters, it seemingly becomes impossible for one generation to write the music the previous generation did not, whereas it is not hard to imagine mathematicians continuing to churn out arguments in the style of Cauchy. Let us grant that they did. They also found, as Cantor did, that the limit process needed a much better theory of the real numbers, that differentiation and integration were not, after all, inverse processes, that complex analysis required deep results from the novel field of topology. Point set topology and, more fundamentally, set theory itself was needed if concrete results of an entirely technical kind were to be obtained. In many areas, mathematicians turned to a form of reasoning largely missing from Cauchy's work: abstract axiomatics.

Cauchy's work proceeds, tacitly and naively, on the prevailing assumption that the calculus applies to all functions, and on his firm conviction that power series are to be rejected when they do not make arithmetic sense. By the end of the century, there was no such belief. Functions, it was recognized, may be analytic, smooth, differentiable, continuous, integrable, measurable, or come in a variety of steadily more awkward forms (not given by a power series, not differentiable, . . .). Mathematical analysis was to be extended at each stage to find what it could say about such functions. Differentiation and integration were no more the touchstone of analysis than tunes were to the contemporary composer.

In this connection, what might seem a technical matter of interest only to a select group of mathematicians in fact illustrates very clearly the arrival of the modernist point of view. This is Ernesto Cesàro's work in the 1880s on the summation of series. Whereas Cauchy had taken the naive realist view that a series $a_0 + a_1 + \ldots + a_n + \ldots$ has a sum if and only if the sequence of partial sums $S_n = a_0 + a_1 + \ldots + a_n$ converges to a limit, Cesàro took the view that the sum of series was something a mathematician defines. Confronted with series that did not have a sum on Cauchy's approach, he argued that one may define its sum in any one of a number of ways and see if it is then summable. The new definition should satisfy certain rules if it is to be properly called a summation; for example, the new value for the sum should agree with Cauchy's for series that are summable on Cauchy's approach, but it is otherwise arbitrary. Indeed, Cesàro came to feel that even Cauchy's definition of a sum was merely a convention, and not a necessary matter. The details of Cesàro's definitions need not concern us, but the attitude should, because it is axiomatic and because it initiated a large modern theory devoted to the summability of divergent series, in which Borel, Poincaré and later Hardy were prominent figures.[135]

Cauchy's work was about functions of one or more real or complex variables. So was late nineteenth-century analysis, in the main, but the advent of topology changed

---

[134] Rosen 1996.
[135] See Ferraro 1999.

the picture. The subtleties of point sets cannot be regarded as a footnote to Cauchy's work, nor can the global topological considerations of Riemann surfaces in complex analysis. These raised wholly new types of questions.

Insofar as it rested on the concept of a real number, one might argue that Cauchy's work just needed the amplification and precision of later theories. But Cauchy did not possess, and did not hint at, a reduction of the real numbers to the natural numbers, and such a theory turned out to bring in too much set theory to count as a mere amplification.[136]

The late nineteenth-century cannot, therefore, be written down as a footnote to Cauchy. The work of Cantor, Dedekind, and others is, for solid internal reasons, in the direct line from Cauchy. As this book proceeds, we shall deal with the claim that it is of at least comparable significance.

4.5 Modernist Objects

Modernism in mathematics did not only mean new foundations for mathematics, although they, naturally, attracted the attention of philosophers. It also meant new mathematical objects inaccessible in any other way. Some of these we have already encountered. The axiomatic method in geometry produced a number of novelties, including non-Archimedean geometries. Two further ones shall be mentioned here, because the vertiginous journey through their description will make three valuable points:

1. The objects were introduced for substantial reasons.
2. They were then transformed into objects that could only be studied by abstract methods, and so
3. They became of interest to pure mathematicians almost exclusively.

4.5.1 Hensel's New Numbers

Kurt Hensel, a student of Kronecker's, pioneered the introduction of a new type of number that was to prove particularly useful in the more advanced reaches of number theory.

We can get a good sense of what these numbers are, and their significance for the modernist stance in mathematics, without getting lost in the details by looking at the claims and motivations on offer in Hensel's book, *Theorie der algebraischen Zahlen* (or, *Theory of Algebraic Numbers*) of 1908. He began by observing that modern complex function theory had successfully pushed the relation of rational to algebraic functions very far, and he now proposed to do the same for the domains of rational and algebraic numbers. Although he did not stress the point, this put him quite close to ideas raised by Dedekind and Weber in a major paper of 1882. We can next

---

[136] See Moore 1982 for accounts of many pieces of work that turned out to involve tacit, or illicit, appeals to contentious principles as the axiom of choice, some of which will be discussed below.

observe that he reviewed the concepts of addition and multiplication for numbers and noted that the inverse operations, subtraction and division, did not apply unreservedly to integers. They required the definition of negative numbers and fractions. But, he said, "These classes of number quantities are only calculation symbols whose meanings initially are entirely arbitrary. Their introduction is governed solely by the rule that the fundamental laws of calculation apply to them as they do to the integers."[137] It is the arbitrariness that Hensel observed, and which allowed him to define a new kind of "number," that is characteristically modernist. The analogy he was pursuing then led him to introduce new expressions that obeyed this rule and so had to be considered numbers.

These new numbers resembled the familiar decimals. They had the form of infinite series in powers of an arbitrary prime number p, perhaps multiplied by a negative power of p. So just as the decimal expansion of $659/3 = 219 + 2/3$ is $219.6666\ldots$, which is $2.10^2 + 1.10 + 9.10^0 + 6.10^{-1} + 6.10^{-2} + \ldots$, which is an infinite power series in negative powers of ten, a p-adic number is something of the form

$$c_{-m}p^{-m} + \ldots + c_{-1}p^{-1} + c_0 + c_1p + c_2p^2 + \ldots, \qquad (4.1)$$

which is an infinite power series in positive powers of a prime p. (The coefficients c_i are chosen from the remainders, called residue classes, $0, 1, \ldots p-1$, so if $p=5$ from among 0, 1, 2, 3, 4. Accordingly $2.5^1 + 4 + 3.5^2 + 3.5^3$ is a 5-adic number.) This in turn forces the counterintuitive but coherent idea that p must be considered to be small, p^2 even smaller, and so on, so that the series (4.1) converges.

Hensel found that he could say when two p-adic numbers were close together, and indeed that the p-adic numbers resembled the decimals in another way: they filled in all the gaps in the rational numbers. But they did so in a new way, one that played havoc with the usual sense of when one number is smaller than another. This might seem a high price to pay, although Hensel put his new numbers to good use; but in the next generation Hasse incorporated them even more powerfully into the study of algebraic number theory.

Hensel's p-adic numbers were discussed by Perron in his *Habilitation* lecture, entitled, with more than a nod to Dedekind, "Was sind und sollen die irrational Zahlen?" (or, "What Are the Irrational Numbers and What Are They For?").[138] He first discussed the different theories of the familiar irrational numbers due to Dedekind and Cantor and observed that whether the resulting numbers were defined as "cuts" (with Dedekind) or signs (with Cantor) or in some other way is ultimately a matter of taste. But, he went on in impeccably modernist fashion, "it is essential that in each case that they are our own creations and there is nothing mystical about them" (pp. 148–149). That, however, only invited the question of what the irrational numbers are for—what do we want to do with them? Numbers, said Perron, are completely independent of the idea of measurable quantity, but we can make them useful in that context, and that is their proper purpose. Poincaré, he supposed, would say we made it this way not because we must but because it is convenient and, please note, convenient for our quite definite purpose.

[137] Hensel 1908, 12.
[138] Perron 1907

Perron now turned to Hensel's *p*-adic numbers, another valid, arbitrary creation. What, he asked, are they for? He noted that they cannot play a role in daily life because they cannot represent quantities. Rather, their purpose is to extend the methods of number theory from the domain of the integer and rational numbers, where they have been so fruitful to the study of transcendental numbers. This task would be difficult, Perron suggested, for what one lacked was a method of analytic continuation, as he called it, that would allow one to compare a *p*-adic and a *q*-adic number for different primes *p* and *q*. But once such a principle was found, he concluded, the way would be open to exploit an analogy between number theory and complex function theory in which the prime numbers corresponded to points of the complex plane. It is a dramatic answer that suggests how strong was the appeal of the structural approach to mathematics—even to Perron, who grew into a markedly traditional creative mathematician.

4.5.2 Knots and Topology

The nascent field of topology displays the most obviously modernist objects, even if they can also be presented very informally. This has been well studied by Epple, who shows how modernist the new topology was.[139] His account focuses on the topic of knot theory, which he shows has significant roots in the work of Gauss on electromagnetism and Helmholtz on hydrodynamics, and is therefore not simply an idle question about bits of string. Even the Scottish applied mathematician and physicist Peter Guthrie Tait, who made a very thorough study of the different kinds of knots with no more than nine crossings, had in mind that atoms might be knotted pieces of ether ("vortex atoms"). As anyone knows who has ever fiddled with string, it is not clear when one knot can be rearranged into another, and in particular when a given tangle is actually not a knot. But recreational questions of this kind were rescued from the margins of mathematics by the work of Poincaré on celestial mechanics.

In a series of works on the three-body problem, Poincaré was led to consider the orbit of a satellite moving in the gravitational field of two much larger bodies that move in circles around their common center of gravity and influence the satellite but are not affected by it. Remarkably, even the motion of the Moon can profitably be studied this way, and Poincaré's name first reached the general public when, in the late 1880s, he won the prize awarded by King Oscar II of Sweden for an essay on celestial mechanics and the stability of the solar system.

The motion of the satellite is determined once its initial position and velocity are known, so one might say that the motion is determined by these four numbers. But Poincaré was able to show that these four numbers cannot evolve in an entirely arbitrary way (analogous to the fact that in many problems in mechanics energy is conserved) and so the future evolution of the system takes place in a way that requires only three numbers to describe.[140] The possible values of these coordinates fill out a three-dimensional space that is not the physical one. What then is it?

---

[139] See Epple 1999.
[140] See Barrow-Green 1997 for a definitive account of how the essay came to be written and revised, and Poincaré, *Oeuvres*, vol 7 for the final essay.

The question is certainly confusing. It confused Weierstrass, who felt that people would confuse Poincaré's "space" of position and velocity coordinates with the "real" thing. It also raised acutely the question of how such an intractable object could be studied. This is where knots come in. A reasonable way to compare two knots is to compare the spaces surrounding the knots. One supposes the knots are made out of rope or solid tubes, as if they are art objects in exhibition cases, and compares the spaces around them. If the knots cannot be deformed one into the other, then presumably the three-dimensional surrounding spaces cannot, and vice versa. A mathematician taking this route has turned the problem of classifying knots into the problem of classifying three-dimensional spaces.

In 1895 Poincaré began to publish a series of papers on topology—*analysis situs*, as he called it—which were intended to open up the field. The opening paper began, "The geometry of *n* dimensions has a real object; no-one doubts this today. The objects of hyperspace can be given precise definitions just like those of ordinary space, and if we cannot represent them, we can conceive of them and study them." In fact, so great were the difficulties in the way of anyone studying three- or *n*-dimensional topology that precise definitions proved hard to give, and even harder to work with, so many alternatives were pursued. It was well known, for example, that surfaces may be made out of polygons by gluing pairs of edges together. An analogous procedure can be followed one dimension up: take a polyhedron, and glue pairs of faces together. This was one of the ways in which Poincaré constructed three-dimensional "spaces" that then had to be classified.[141] I put quotation marks around the word because even the choice of word is a problem. What are these multidimensional objects? What aspect of the study of curves and surfaces in ordinary space do they generalize, what features are entirely new?

A productive way forward was to take an idea that had been proved useful in the classification of surfaces, and to consider loops in the three-dimensional space. Two loops are considered equivalent if one can deform one into the other. There are fundamentally two kinds of inequivalent loops that one can draw when dealing with an unknotted solid tube: ones that, when pulled tight, wrap around the tube, and ones that do not. There is an obvious sense in which the former loops go around the tube once, or twice, or *n* times. But there are more ways a loop can go around a solid tube knotted into the shape of a trefoil knot, a fact that is not so easy to prove, obvious though may seem. Poincaré was not interested in the mathematical study of knots, but took a similar approach to the three-dimensional "spaces" he was considering and studied the different types of loops one can draw in them. He used the fact that if all the loops start and finish at the same point in the "space" then one can follow one loop by another and obtain a third loop. In this way he was able to regard his loops as elements of a group.[142] For the unknotted solid tube, the group obtained in this way is the integers; for the solid tube knotted like a torus, the group is more complicated. So this approach distinguishes some different spaces.

---

[141] Here is the origin of a famous remark of his that has since become known as the Poincaré conjecture, and was only solved, over a hundred years after his comments, in work for which the Russian mathematician Grigory Perelman was awarded a Fields Medal in 2006. Such are the difficulties of this branch of mathematics.

[142] I omit some nontrivial technical points that Poincaré dealt with. For example, equivalent loops define the same group element.

The modernist shift here is from the world of solid tubes, even string—genuine three-dimensional objects manipulated in the natural way—to artificially defined "spaces" studied via the groups that describe some of their cruder geometric features. It would be easy to dismiss it as a frivolity, but the fundamental question here is how to extend the methods of geometry to problems involving more than three variables, and thus was born algebraic topology.

One novel feature of the new topology is worth stressing. Whatever a three-dimensional "space" might be, small pieces of it will look like small pieces of Euclidean three-dimensional space. This condition was built in deliberately as a restriction on the sorts of objects that were to be studied. Nonetheless, the objects taken as a whole can be very different. It is the global differences between locally similar objects that were at issue. So local, piecemeal methods would not work unless they could be made into global tools. This was a further challenge to mathematicians hoping to work in this area, and by and large one for which existing tools proved useless.

4.6 American Philosophers and Logicians

4.6.1 Peirce

By 1886 Charles Sanders Peirce had been dismissed from both the Coast survey and Johns Hopkins University. He was never again to enjoy permanent employment or the opportunity to focus on research without financial worry. The rest of his long life was spent being dependent on friends for opportunities, which, in the long run, he invariably squandered. He was paid to work on the *Monist* by its founder, the zinc millionaire Edward C. Hegeler; Baldwin's *Dictionary* retained him to write on logic, philosophy, mathematics, mechanics, astronomy, and weights and measures; William James and Josiah Royce enabled him to lecture occasionally at Harvard; Wendell Phillips Garrison regularly paid him to write for *The Nation*; the New York Public Library paid him to advise on the purchase of scientific books. None of this lifted him out of poverty for long, and when it was clear that he could no longer support himself, William James organized a fund in 1907 to support Peirce, lasting until his death in 1914.

He therefore poses a notorious problem for historians. On the one hand, he was a man of remarkable brilliance and originality; on the other hand, his difficult personality and troubled life stunted the appreciation of many of his ideas. He might nonetheless have earned a secure place posthumously, but the original four-volume edition of his work in 1931–1935 did little to establish his reputation, which only began its rise with more recent investigations of his extensive legacy. He has long been deprived, therefore, even of the status of being a major figure neglected in his lifetime, and one may presume that further research will bring to light more important aspects of his life and work. That said, there has been a valuable biography,[143] a good symposium on his work on logic (broadly defined),[144] an analysis of

---

[143] Brent 1993.
[144] Houser, Roberts, and Van Evra 1997.

his place in American intellectual life,[145] as well as numerous more specialized studies. He is an inescapable figure in any study of mathematical modernism, and a way in to the work of other, lesser figures. As a minor indication of his charm, take this quotation from Fiske. Peirce was addressing a meeting of the New York Mathematical Society in the early 1890s, before it joined with Chicago to become the American Mathematical Society, and he proclaimed "that the intellectual powers essential to the mathematician were 'concentration, imagination, and generalisation.' Then, after a dramatic pause, he cried: 'Did I hear someone say demonstration? Why, my friends,' he continued, 'demonstration is merely the pavement on which the chariot of the mathematician rolls.' "[146]

The predictable results of Peirce's profligacy were a number of articles by him on a variety of topics, and an even greater pile of unfinished papers and ideas. But his poor reputation and lack of a permanent position weakened the impact of anything he had to say, and he was deprived of the beneficial effects of colleagues interested in what he had to say and able to follow it, which led in turn to his sounding more and more like an autodidact. Neologisms abound in his work, new terms are often divided into triples, distinction is piled upon distinction before the reader can be sure that anything works. There are a number of undoubted achievements in Peirce's work: he is the founding father of pragmatism ("pragmaticism" was his term, and he intended something more elaborate than William James preferred); he was the first to proclaim the existence of truly random, causeless events, as opposed to events that are merely too difficult to predict or of which the causes lie hidden; he made significant advances in the logic of relatives and in semiotics; the term "abduction" in the philosophy of science is due to him; and there is a larger number of smaller achievements to his credit.[147] The conventional view of the reception of Peirce's logical work is that Peirce belonged to a tradition begun by Boole and continued after Peirce by Schröder, which may be called the tradition of algebraic logic. But when the independent quantification-theory tradition associated with Frege and Peano came along and was decisively taken up by Russell and Whitehead, it swallowed up what was valuable in the algebraic tradition. This view, spread by Russell himself and echoed by many historians since then, down to van Heijenoort, has always had its dissenters, but their numbers have only grown recently with the growth of scholarship about Peirce. Peirce was in a poor position himself to contest the influence of Russell and Whitehead's *Principia Mathematica*; by 1912 he was very ill and he died in 1914. After the First World War, there were eloquent spokesmen for the dominant viewpoint, only few for that of Peirce. Their voices, and that of Peirce himself, were raised by Anellis (1995), who also offers a decidedly sharp view of Russell as a historian.

4.6.1.1 RUSSELL VERSUS PEIRCE

Bertrand Russell came to feel strongly that all of mathematics could be produced from logic. This is the celebrated logicist position, which he took over from Frege and adapted some of Peano's ideas and notation to promote. Russell and Peirce exemplify the distinction drawn by Jean van Heijenoort and Jaakko Hintikka between those

---

[145] Menand 2001.
[146] In Fiske 1988, 16.
[147] See Houser 1997, 3.

who view language as a universal medium that is impossible to escape, and those who view language as a calculus. The former find themselves trapped within language, and semantics is opaque, even inaccessible to them. The latter propose semantical theories of various kinds, often model theoretical ones, and allow for multiple interpretations of syntactical propositions. Frege, Russell, and, later, Wittgenstein belong to the universalists, Schröder and Peirce to the calculators.

Since the later history of logic in its relation to mathematics was for a long time dominated by the claims made about Russell and Whitehead's *Principia Mathematica* of 1910—and often couched in its language—this may be the place to note Anellis's (1995) attempt to redraw the map the way it stood at that time. Russell argued throughout the first decade of the twentieth century that his work surpassed that of Schröder and Peirce, and his sincere admiration for the work of Peano and his successful run-in with Frege seem to have convinced historians down to van Heijenoort that indeed this was the case. This was not the view at the time, nor indeed does Whitehead seem to have held it with the same intensity; but writing in notes for a lecture class at Columbia, Christine Ladd-Franklin observed that Whitehead and Russell "plainly imply that P[eirce] and S[chröder] were absolutely non-existent!"[148] Royce too noted that Russell acknowledged Peirce's work inadequately. Anellis goes on to suggest that Russell's guilty awareness of this lapse may explain his hostile reaction to Norbert Wiener, who arrived in Cambridge, England, when not quite nineteen years old with a PhD from Harvard in which he had made a comparison of the work of Schröder and Russell that did not coincide with Russell's own view.

Peirce, too, was aware that Russell had minimized the debt. He had a better philosophical eye than Schröder had had for the significance of his own work and the relationship it established between logic and mathematics. He knew that he was one of the founders of the theory of quantifiers (with which, as Anellis points out, he was only to be praised much later, by Quine) and that as far back as 1867 he had had the idea of how to define addition and multiplication that later surfaced in *Principia Mathematica*. But the tradition of algebraic logic did not survive the First World War and the many years during which Russell was left to propagandize for the merits of Russell and Whitehead.[149]

The points to determine, if the conventional view is to be significantly modified, are the extent to which Peirce's work either contains more of Russell's than one might suppose (and so it was not simply swallowed up), or if it indeed surpasses it, or even if it is doing the same sort of thing. Could more have been made of the connections between the two schools of logic? Here we may turn to Houser (1997) and other essays in Houser, Roberts, and van Evra (1997).

It is not to be doubted that Peirce placed mathematics at the foundation of the sciences, and not logic. His logic of relatives, which his admirers seem to suggest may be Peirce's finest work, certainly looks more like mathematics than logic[150] and belongs with mathematics. Only after phenomenology, aesthetics, and ethics do we

---

[148] Quoted in Anellis 1995, 291.

[149] It is tempting to see a reevaluation of this history and to connect it to the steady rise in esteem of Tarski, who always was sympathetic to algebraic logic but who is left out of van Heijenoort's otherwise highly valuable *Source Book on Logic*.

[150] Houser 1997, 15.

come to logic in Peirce's scheme of things. What Peirce took mathematics to be, other than, as his father had said, the science of drawing necessary conclusions, is less clear, and not only to us. In 1885 Peirce submitted the second part of his paper on the algebra of logic to the *American Journal of Mathematics*, edited by Simon New-comb. Newcomb wished to know if the paper, which discussed quantifiers, was mathematics or logic, because it would only be accepted if it was mathematics. Peirce replied that it was logic, and Newcomb rejected it. It would seem that, in 1885 at least, not only was mathematics not part of logic, being more fundamental, but neither was a work on logic ipso facto a work of mathematics.

But logic, as Houser observes, had a broad definition in Peirce's scheme of things. He divided logic as semiotic into three branches: speculative grammar (the role of signs), critic (the classification of arguments, and truth conditions), and speculative rhetoric (communication between minds, the relation of representation to thought). It might seem that this formulation would have generated a philosophy of mathematics, in which mathematics was taken as a special case of semiotic, but Peirce persistently ruled that out. "Mathematics," he wrote, "is not subject to logic. Logic depends on mathematics. The recognition of mathematical necessity is performed in a perfectly satisfactory manner antecedent to any study of logic. . . . The only concern that logic has with [mathematical reasoning] is to describe it."[151] The only use mathematicians had for logic, in Peirce's opinion, was for the most elementary and intuitive kind of logic, not the science of logic as defined by Peirce.

Peirce argued that all modern mathematicians agree that "mathematics deals exclusively with hypothetical states of things, and asserts no matters of fact whatever; that it is thus alone that the necessity of its conclusions is to be explained."[152] Consistent with this, he also asserted: "The certainty of pure mathematics . . . is due to the circumstance that it relates to objects which are the creations of our own minds."[153] The truth of a mathematical statement is therefore a matter of its consistency with other mathematical statements, and not a matter of referring it to the real world. The formation of hypotheses, as well as deductions from them, should therefore have been prominent in his philosophy of mathematics, but as Levy has observed it was not,[154] and the reason for this, Levy suggests, is that hypothesis formation casts doubt on the independence Peirce felt existed between mathematics and logic. Levy suggests that awareness of this problem accounts for some of the contortions in Peirce's position as it evolved over the years. For example, Cantorian set theory struck Peirce as much more of a logical than a mathematical problem or topic, especially when the "set" was taken as the fundamental mathematical concept. This was partly because of the paradoxes of set theory, some of which Peirce learned about in the late 1890s.

Peirce's position emphasized the hypothetical nature of mathematical reasoning. This, too, led him back to logic as a useful tool in the analysis of the exploitation of axiom systems, to which the hypothetical view of mathematics is naturally inclined. In this connection his account of logical truth in Baldwin's *Dictionary* is striking: "Logical truth is a phrase used in three senses, rendering it almost useless," he wrote,

[151] Hawkins 1997, 121.
[152] Peirce 1960–66, vol. 4, 232. See Houser et al. 1997, 81.
[153] Peirce 1960–66, vol. 5, 166. See Houser et al. 1997, 81.
[154] See Houser et al. 1997, 86.

of which the second is: "The conformity of a thought to the laws of logic; in particular, in a concept, consistency; in an inference, validity; in a proposition, agreement with assumptions. This would better be called mathematical truth, since mathematics is the only science which aims at nothing more" (and he cited Kant in his support at this point, that is, in support of a distinctly non-Kantian point of view!). The laws of logic, one must presume, are the intuitive ones; logic is not taken here in the exalted, philosophical sense of logic as semiotic as defined above.

The objects of mathematics, in Peirce's view, are intelligible to the mind, which manipulates them and perhaps has access to them, only through the rules governing deductions about them. Multiple interpretations are therefore possible, and even when the ostensible interpretation is logic, the result can look much more like mathematics.

4.6.2 Royce

What then of the influence of this view on those most impressed by Peirce? Among these was the American philosopher Josiah Royce, who taught for many years at Harvard. Royce was born in Grass Valley, California, in 1855.[155] The town had been founded not many years before, during the California Gold Rush. The Royces were a devout family who believed in the immediacy of the divine, and religion was to be a major theme of Josiah's philosophy throughout his life. He studied at the University of California in Oakland from 1870 to 1875, graduating with a degree in classics, and then spent a year in Germany, where his favorite course, he later recalled, was Wundt's course on logic. He first came to Harvard in 1882 as a temporary replacement for William James, with whom he became good friends. He stayed on, and ten years later he became professor of the history of philosophy there, writing on philosophy, psychophysics, ethics, psychology, idealism, logic, and, inevitably, pragmatism. He died in 1916.

In 1905 Royce wrote a paper responding to two papers by the British mathematician A. B. Kempe, published in 1886 and 1890 but "have remained almost unnoticed."[156] In the second of these papers Kempe showed how to take a set and impose on it a relation he denoted $ab.p$ satisfying five axioms ("laws," in his terminology) and showed that a possible interpretation of this system was the geometry of lines in a linear space of some dimension. The theorems of the system of triads $ab.p$ were deduced from the axioms and could be interpreted as theorems of geometry (concerning which he commented rather interestingly that "we need not trouble ourselves about metrical geometry; that is now fully understood to be included in the more general descriptive geometry."[157] The interpretation was that a, b, and p were three collinear points, and p lies between a and b (and may coincide with one of them). Peirce was very impressed by Kempe's essay, and went so far as to call Kempe the greatest living logician.[158]

---

[155] See Clendenning 1999.

[156] Alfred Bray Kempe (1849–1922) graduated 22nd Wrangler from Cambridge in 1872, and became a successful ecclesiatical lawyer. He was elected an FRS in 1881 and knighted in 1912.

[157] Kempe 1890, 178.

[158] In Murphey 1961, 139–140.

Royce picked up on the fact that Kempe's geometrical relation was a triadic (i.e., three-term) relation of betweenness, a very similar relation being fundamental to Veblen's (1904) recent axiomatization of geometry, and the fact that the algebra of logic, as presented by Peirce and Schröder and refined by Royce's Harvard colleague Edward Huntington, was a dyadic (i.e., two-term) relation. Royce noted that the dyadic relation may be extended in various ways to a triadic one, and so geometry appeared as a special case of logic. As he put it: "One may view the points of a space as a select set of logical elements, chosen, for instance, from a given 'universe of discourse,' " a view he modestly regarded as "the essential thought at the basis of Mr. Kempe's paper of 1890."[159]

Royce's axioms applied to any set Σ of objects, with some distinguished subsets called O-collections. He was able to show that the set Σ must be infinite, and that it admitted a two-term asymmetrical relation F (which may be regarded as an abstraction from such concepts as "on the same line and to the left of "). The relation F allowed Royce to define sequences and chains of elements, which admitted the geometrical interpretation as lines. Royce followed Christine Ladd-Franklin's original idea, that in the presence of negation a symmetrical relation can be made to define an unsymmetrical one,[160] and Kempe in introducing a relation they called "obversion," the meaning of which need not detain us except to note that if the obverse of a is denoted a' then $(a')' = a$, which goes some way to explaining the name. Obversion shares this property with negation (the proposition "not not A" means the same as A). He then showed how his symmetrical relation of obversion could be made to define Kempe's unsymmetrical relation of betweenness. He ended his paper by showing how Veblen's axioms for geometry could be expressed in terms of his relation of obversion.[161]

The point of all of this is not to present Royce as a neglected mathematician, but to illustrate how boldly he had grasped the idea that the algebra of relations could be made to yield a system of geometry, and thus to produce geometry out of logic. More precisely, on the view that an axiomatically defined relation is, qua relation, a logical entity, then the betweenness aspect of axiomatic geometry can be derived from logic. This boldness impressed his former student C. I. Lewis, who wrote at the end of the preface to the first volume of his *A Survey of Symbolic Logic* (1918) that he was most "indebted to my friend and teacher, Josiah Royce, who first aroused my interest in this subject, and never failed to give me encouragement and wise counsel." Lewis singled out the work of Kempe and Royce, in contradistinction to that of Russell in *Principia Mathematica*, as exemplifying the possibility of an entirely formal, syntactic view of mathematics, a "mathematics without meaning," which, if successful—and Lewis admitted it needed more work—offered a formulation of mathematics that was largely free of metaphysics and risked no confusion of form and content.

What one might call the American axiomatic or structuralist view of mathematics of Royce and Veblen was also on view elsewhere, for in this respect, the definition or definitions of mathematics in Baldwin's *Dictionary* are interesting. They were mostly written by Henry Burchard Fine, the professor of Mathematics at Princeton and one of the small group of men whom Oswald Veblen credited with carrying "American

---

[159] Royce 1905, 355.
[160] Peirce 1883.
[161] For a discussion and interpretation of Royce's paper, see Kuclick 1985.

Mathematics forward from a state of approximate nullity to one verging on parity with the European nations." He had originally studied Latin at Princeton but was converted to mathematics by Halsted's enthusiasm for the subject. In 1884 he went to Leipzig to study with Klein, surely on Halsted's recommendation, and took his PhD there in little more than a year. It seems that Eduard Study may have been more help to him than Klein, and soon afterwards Fine gravitated more toward Kronecker and his style of mathematics.[162] Fine was a friend of Woodrow Wilson from his student days, and when Wilson became president of Princeton in 1903 he made Fine the dean of the faculty. Thereafter, Fine's career lay in administration, although he wrote *College Algebra*, which students found too hard, and so successful was he in directing Princeton toward research that the mathematics building at Princeton was named after him.[163]

As a definition of mathematics, Baldwin himself and Fine jointly offered "a science of abstract relationships," noting that "these are by no means exclusively quantitative. The projective properties of curves, for instance, . . . , are purely positional. And the whole of analysis may be presented in the form of an abstract calculus of symbols to which no meaning of any kind need be assigned, the operations themselves being defined by certain formal laws." Then they quoted Chrystal: "Any conception which is definitely and completely determined by means of a finite number of specifications, say by assigning a finite number of elements, is a mathematical conception. Mathematics has for its function to develop the consequences involved in the definitions of a group of mathematical conceptions," adding that these conceptions are combined with certain fundamental principles, or axioms and postulates. Fine then went on to remark that the more fundamental conceptions correspond immediately to things and relations among things in the external world, from which they have been derived by a process of abstraction, for example, cardinal numbers. In his opinion, the function of mathematics is to make the simplest independent selections of these primary conceptions, and to combine and generalize them to create a body of more complex conceptions having intrinsic interest and beauty and a value for the furtherance of the science itself or for the study of other positive sciences. Then, in a passage which would have delighted Dedekind had he ever seen it, Fine wrote:

> Very characteristic of the modern mathematics is the prominence of the notions: *manifoldnesses* or *assemblages* of actual or ideal elements of some kind, e.g. of points, lines, or surfaces; the *transformations* or *substitutions* by which the elements of two assemblages may be brought into definite relations of correspondence; and the *invariantive properties* of assemblages, i.e. the properties which remain unchanged by given *groups of transformations*. In fact, to such an extent do these notions, in some form or other underlie the various branches of mathematics, that it is almost admissible to define mathematics as the science of assemblages.[164]

This strikingly captures the modern, structural view of mathematics. So much so that it is also interesting to note that the sources for this view that Fine adduced were precisely these: Clifford's *Essays and Addresses*; Klein's *Lectures on Mathematics*—

---

[162] Parshall and Rowe 1994, 194.
[163] The information in this paragraph comes from Leitch 1978.
[164] Baldwin and Fine, "Mathematics," in *Dictionary of Philosophy and Psychology*.

the *Evanston Colloquium*; Russell's *The Foundation of Geometry*; and recent papers by Mach and Poincaré in the *Monist* and the *Revue de Métaphysique et de Morale*.

4.6.3 American Axiomatizers

American mathematicians in general also proved receptive to the idea of axiomatizing not just geometry but other branches of mathematics. Huntington at Harvard gave axiomatic treatments of the real numbers as well as the positive natural numbers, culminating in his *The Continuum and Other Types of Serial Order*, originally published in the journal *Annals of Mathematics* in 1905 and then as a book (shortly translated into Esperanto). The phrase "serial order" shows the importance Huntington attached to Russell's ideas; he saw his own work as a contribution to the logic of mathematics, and was as much interested in the elegance of his axiom systems as the objects they axiomatized. Axiom systems, he said, should be consistent, sufficient, and independent.

E. H. Moore in Chicago led another and larger school of mathematicians, with strong ties to Göttingen. He first ran a seminar in Chicago on Hilbert's *Grundlagen der Geometrie* in 1901, Moulton was there the next year, and most significantly Oswald Veblen took his PhD there in 1903. Veblen's thesis based Euclidean geometry on twelve axioms for which the primitive notions were point and order. His book *Projective Geometry*, written in two volumes with J. W. Young in 1910 and 1917, rapidly became the definitive axiomatic treatment of geometry in the English-speaking world, and it was oriented about equally to Italy and Göttingen.

The Texan Robert Lee Moore was also at the University of Chicago from 1903 to 1905. He had been captivated by Halsted's advocacy of non-Euclidean geometry—Halsted had translated both J. Bolyai (1832) and Lobachevskii (1840) into English—while at the University of Texas, and had already studied Hilbert's *Grundlagen der Geometrie* at Halsted's suggestion. On his eventual return to Texas, Moore wrote an unpublished attempt to characterize the positive integers and their arithmetic (a topic connected to Hilbert's second problem). He then published a set of axioms for plane Euclidean geometry, and then sought a topological characterization, using the primitive, undefined notions of point, region, and motion. This work directly shaped his teaching. He pioneered a teaching method, known today as the Moore method, which prohibited the use of textbooks and required students to work obtain for themselves a structured set of results starting from some small set of axioms, the aim being to encourage students to prove results for themselves. Discussion was allowed, firm criticism the order of the day.[165]

Halsted himself wrote a book, *Rational Geometry*, which is almost unadulterated Hilbert, being an axiomatic treatment of plane and solid geometry that, like Hilbert's, is based on axioms for five types of geometric property. Unlike Hilbert, however, Halsted was concerned to derive all of elementary Euclidean geometry from the axioms, and not to compare and contrast different geometries. Rather, it is the book Euclid would surely have written had he lived after Hilbert.

Everything so far has been naive about sets and logic. That comes to an end with the following topic.

---

[165] For a recent biography of Moore, see Parker 2005.

4.7 The Paradoxes of Set Theory

4.7.1 Thinking about Sets

It is hard to imagine a more elementary foundation for mathematics than naive set theory. The mental ability to recognize distinct objects and to form them into collections that may again be regarded as objects, seems to be all that is required. So if it is agreed that mathematics is not to be founded on primitive intuitions of number, shape, and function or transformation, then constructing these things out of sets would seem to be the natural way forward. All the more disturbing, then, when it began to seem that naive set theory was fraught with self-contradictions.

The problems arose not at the naive level but in the higher reaches of set theory, and some, indeed, in the uses of set theory in other domains of mathematics, such as topology. Recognition that very large sets, or the cavalier formation of sets, could cause problems was at first a hesitant business. Georg Cantor seems to have been aware that the set of all cardinal numbers was an untenable object, because it too would have a cardinality, which would be greater than any cardinal in the set, and so the set could not be the set of all cardinals. He also knew of the very similar problem caused by the set of all ordinal numbers. But these were not problems for him, because his deep theological convictions allowed him to distinguish between various orders of infinity and the absolute infinity of the divine. Even the transfinite, he believed, cannot bring mere mortals close to the Absolute, which can only be acknowledged but never known.[166] Cantor divided sets into two types, the "definite" and the "absolutely infinite" or "inconsistent" and clearly thought that it would only be possible to prove results about definite sets, which were legitimate objects of thought, while the other sort would lead to contradictions. But he gave no criteria for recognizing a set to be definite, so his way of heading off disaster was not likely to help other people, although Schröder independently found such a distinction helpful and introduced it in his *Algebra der Logik* (1890).[167] He even used the distinction to prove theorems, as Cantor was to do later and more profoundly, so the distinction, however open it is to criticism on the grounds of imprecision, initially did not seem threatening.[168]

In 1897 Burali-Forti published the argument that leads to the paradox named after him.[169] It proclaims that the set of ordinal numbers is well ordered[170] and therefore has an ordinal number, but this ordinal is both a member of the set of all ordinal numbers and greater than any ordinal number. This is an impossibility. What does it show? Burali-Forti himself hoped that refining the concept of a well-ordered set would yield a way out, as he implied in a letter to Couturat in 1905. To modern eyes

---

[166] Cantor 1883, 205, quoted in Ferreirós 1999, 291, who notes that Cantor was a Platonist about mathematical existence, so it was no problem for him that collections beyond the reach of set theory might be part of the Absolute, or in some other way unknowable.

[167] See Schröder 1890, 213 and 343.

[168] What Cantor meant by "finished" or "definite" (*definit*) has remained insufficiently clear to this day.

[169] See Ferreirós 1999, 205–207.

[170] A well-ordered set is a totally ordered set such that every set has a least element; a set is totally ordered if any two elements can be compared.

it shows that the collection of all ordinal numbers cannot be said to form a set, and most people who picked up his result did suspect that the paradox shows that something deep was wrong with elementary set theory.

The year 1897 is also when Cantor discovered the very similar paradox about the "set" of all cardinals. Cantor wrote to Hilbert about this, in his capacity as editor of the *Mathematische Annalen*, and suggested that the way out was to regard the "set" of all cardinals as unfinished, in which case the notion of cardinality did not apply.[171] He responded to the Burali-Forti paradox similarly, writing in a letter of 1899 to Dedekind that it showed that the collection of all ordinals was an inconsistent set.[172] But Cantor also noticed that if one cannot consider the totality of all thinkable things, then his friend Dedekind's "proof" that infinite sets exist became invalid, and he asked Dedekind to send his student Bernstein over to work on the problem. Bernstein later recalled that Dedekind began to doubt if human thought is completely rational. Hilbert's first reaction was one of simple disbelief, but he came to see the force of the paradox and to discover one of his own which, he said, had for a time caused him to think that it would cause "invincible problems for set theory that would eventually lead to the latter's eventual failure."[173]

What is at stake? It might seem that mathematicians have followed a line of inquiry into a very recondite area, possibly necessary for the treatment of some difficult questions in mathematical analysis, but otherwise irrelevant. If this leads them to paradoxical situations, why not leave them to find their own way out? So it might have been, had Cantor's transfinite sets only been a way of raising, and possibly answering, questions about continuous functions. But matters had changed since the 1870s, and naive set theory was increasingly being seen as the foundations of mathematics. This view, originally due more to Dedekind than Cantor, became his view, too, in the 1880s, as he turned more toward philosophy (as his letter to Mittag-Leffler shows).

It may seem odd that the paradoxes did not immediately signal danger ahead, but there are two reasons why this sort of retrospective history is wrong in this case. The first is that quite generally in their research mathematicians find there are deep problems with their overall position and have no choice but to go forward and hope that solutions can be found, or at least that the problems can be avoided. The opening chapter of this book presented many such problems in other domains of mathematics. Initially, there was no reason for the paradoxes to be any more threatening than the paradoxical conclusions to hand in non-Euclidean geometry or infinite series.[174]

The second reason is that set theory itself generated many conundrums, many of which snared those who entered into the field.[175] It may be helpful to indicate a few,

---

[171] Cantor to Hilbert, September 26, 1897, in Cantor 1991, 388–389.

[172] Cantor to Dedekind, August 3, 1899, in Cantor 1991, 407–411.

[173] Peckhaus and Kahle 2002, 163.

[174] In the epilogue to the reedition of his *Labyrinth*, published in 2007, José Ferreirós sheds new light on the ways sets were defined or regarded in this period and on to the 1950s, one that points up a helpful distinction. He argues that Dedekind was willing to think of a universe of sets and pick out subsets by a defining clause, whereas Cantor was to some extent aware that paradoxes lie in that direction. Ferreirós's comments on the path from Zermelo to Gödel and the modern theory are particularly insightful. The reader is referred to those helpful pages, and I am happy to thank José for sending me an advance copy.

[175] The guide here is Moore 1982.

although the implications between them are tricky and took a long time to elucidate. Consider the following statements:

1. Given any nonempty collection of sets, it is possible to define a function that assigns to each set in the collection an element of that set (so, if A is a set in the collection, and we denote the function by f, then $f(A) \in A$). This is the famous Axiom of Choice (AC).
2. The Axiom of Choice holds for denumerable sets (a set is denumerable if it can be put in a one-to-one correspondence with the natural numbers). This is the denumerable axiom of choice (DAC).

So tempting are both of these statements that they were often used implicitly in arguments. Nor is it clear whether the choice of an element from each set in a collection is to be made simultaneously or sequentially, or indeed if that is a distinction without a difference. The axiom DAC turns up, for example, when it is shown that a countable union of countable sets is countable. What of the assertion that every infinite set has a denumerable subset? Cantor, Russell, and Borel all separately gave proofs of this assertion using DAC implicitly.[176]

These examples are typical, and there are many more.[177] They are enough to establish that for those who took set theory seriously, there were many problems. There were problematic proofs, there were unproven claims, there were tempting steps to take that really required proof and might not even turn out to be true. None of this was a reason to give up the project.

Two more must be cited because they came to dominate research in the field. Cantor first came to think of the trichotomy of cardinals in 1878. This is the claim that, given two sets A and B, only three things could happen. Either the cardinality of A was greater than the cardinality of B, or the cardinalities were equal, or the cardinality of A was less than the cardinality of B. It shows that the cardinalities of any two sets are comparable, and so there is an ordering on the cardinalities. The other is the well-ordering principle, the claim that every set can be well ordered. This is much more complicated. An ordering on a set S is a relation imposed on the set (let us denote it suggestively by $<$) with the following properties: given any two distinct elements of the set S, say a and b, either $a < b$ or $b < a$; if $a < b$ and $b < c$ then $a < c$; and for no element a of the set is it the case that $a < a$. The set S is said to be well ordered if every nonempty subset T of the set has a least element, t_0, one that is with the property that if t is any element of the set T, then $t_0 < t$. Notice straight away that the usual order relation on the real numbers does not make them well ordered (there is no least element in the set of real numbers x such that $0 < x < 1$, for example). This does not mean that the set of real numbers cannot be well ordered, but an explicit well-ordering of the real numbers has never been given.

Cantor knew, as he said in a letter to Dedekind in 1882, that the continuum hypothesis implies that the real numbers can be well ordered (see §3.2.2). Recall that the continuum hypothesis is the assertion that the cardinality of the real numbers is the smallest cardinal after $\aleph_0$, which is $\aleph_1$. It is straightforward that the cardinality of

---

[176] It later turned out that any proof of this assertion must use DAC, for there are models of set theory without DAC in which the set of real numbers has an infinite set lacking a denumerable subset (and another model in which the set of real numbers is both uncountable and a countable union of countable sets).

[177] See Moore 1982.

the continuum is $2^{\aleph_0}$, for this is clearly the cardinality of the numbers between 0 and 1 in binary notation, so the continuum hypothesis may be written as the claim that $2^{\aleph_0} = \aleph_1$. If the continuum hypothesis fails, therefore, it would be because $2^{\aleph_0}$ was equal to some cardinal greater than $\aleph_1$.[178]

Cantor's faith in the continuum hypothesis translated into a faith that every set can be well ordered, but never into a published proof. Eventually, in his last and most thorough paper on set theory, the *Beiträge* (1895), Cantor admitted explicitly that a number of statements required proof, and that he had been unable to prove them. These were trichotomy and the continuum hypothesis. At this stage Cantor was still confident of the well-ordering principle, but, as Moore notes, the *Beiträge* "left unresolved a number of the most fundamental and inter-related questions: the Trichotomy of Cardinals, the Equivalence Theorem, the Well-Ordering Principle, and the Continuum Hypothesis."[179]

4.7.2 Paradox

Cantor sent a draft of a proof of the well-ordering principle to Hilbert in late 1896 or 1897, where it caused quite a stir. But Hilbert was presumably intrigued and unconvinced, which is why he proposed the construction of a well ordering of the real numbers as part of the first of the Hilbert Problems. He also raised the continuum hypothesis as part of the same problem (the Hilbert Problems are often multiples), and this seems to have directed some attention to these problems in set theory. Schoenflies's long survey article in the *Jahresbericht der Deutschen Mathematiker-Vereinigung* cast a more skeptical eye on these issues, noting that the trichotomy lacked all objective validity.

Among the French mathematicians, Émile Borel, who had read Cantor's *Beiträge*, met Cantor at the first International Congress of Mathematicians in Zurich in 1897 and asked him about the theory of cardinals. Cantor told him that his student Bernstein had proved the Equivalence theorem. Borel included the proof in his book on set theory and complex function theory.[180]

Outside Germany, it seems that English mathematicians were most strongly drawn to these difficult issues in set theory. Russell, Hardy, and Jourdain all took them up, to indifferent effect. Of these, Hardy emerges as the (over)confident mathematician, Jourdain as the set theorist, and Russell as the most effective critic. As Moore (1982, 58–59) noted, Russell's early Kantian period, with its antinomies, may well have disposed Russell to look for contradictions in set theory, and Peano's symbolic logic may merely have sharpened his tools to that end. At all events, in 1902 Bertrand Russell discovered the paradox that destroyed so much of Frege's work. This concerned the set of all sets that are not members of themselves. If this set is a

[178] Machover remarks that $\aleph$ is not only the first letter of the Hebrew alphabet but the first letter of the Hebrew word "*einsoph*" meaning infinity and as such a cabbalistic appellation of the deity, and notes that Cantor was deeply interested in mysticism; see Machover 1996, 95.

[179] Moore 1982, 46. The Equivalence theorem is the assertion that if $A \subseteq B \subseteq C$ and A and C have the same cardinality, then B has the same cardinality as A and C.

[180] Borel 1898. Here is where he also proved that every infinite set has a denumerable subset. As noted above, without him realizing it, his proof assumed the axiom of choice—as indeed it must, but he was not to know that.

FIGURE 4.10. Émile Borel.

member of itself, then it is not a member of itself, and if it is not a member of itself, then it is. Russell drew attention to the fact that this paradox made no reference to anything other than the ability to form sets and determine membership thereof. It was therefore more fundamental, and more worrying, than paradoxes about ordinals or cardinals.[181] More attention was drawn to this paradox when he admitted in his *Principles of Mathematics* (1903) that the problem posed by considering the set of all predicates that are not predicable of themselves led to difficulties he could not resolve except by something like a theory of types.

Soon there were other paradoxes. The paradox of the liar, which went back to the Greeks and runs: "Epimenedes the Cretan says, 'Cretans only tell lies." If he is lying, then his statement is true, if he is telling the truth then it is false; either way we have a contradiction. Richard's paradox, which introduces the smallest number not definable in English in less that fifty syllables, and which would seem to define a number in twenty-one syllables.[182] We shall encounter more below.

Historians of mathematics at one time suggested that the discovery of the paradoxes was immediately the cause of a crisis. We have seen, however, that their initial discovery did not plunge everyone into a profound state of dread. Rather, as in many a novel branch of mathematics, the first occurrence of the paradoxes in the study of very large sets was nothing more than a warning to be cautious. Even the word "paradox" may signal as much. Indeed, when Zermelo went on to write up his

[181] Coffa 1991, 115–116, has a nice explanation of how Russell was led to his paradox by his disbelief in Cantor's claim that the power set of a set is always larger than the original set. Russell believed that this must be wrong because it seemed paradoxical: surely there cannot be a larger set than the set that contains everything? Cantor had proved his theorem by a reductio argument that used sets which are not members of themselves, and Russell located the source of the paradoxical conclusion precisely in sets of that kind—thus his own celebrated paradox.

[182] See Richard 1905.

resolution of the paradoxes, he objected to the word "paradox" as merely expressing a contradiction with common opinion, whereas the term "antinomy" expresses an inner or self-contradiction.

The paradoxes gave their sting when they cropped up in the logicist program. If one cannot be allowed to form sets at will, only two alternatives would seem to be present. Either there is a genuine flaw in the whole business of analyzing rational thought, or some ad hoc rules have to be brought in to constrain mathematics, and these ad hoc rules will not be self-evident truths of logic. The first alternative is disaster, the second the likely end of logicism. It is notable that the paradoxes only became serious when they were construed as threatening the very foundations of, or the possibility of giving foundations to, mathematics. This is a concern of one kind of those committed to anchoring mathematics in set theory, and of course a concern of another kind to those for whom the set-theoretic enterprise is unattractive or ill-founded.

4.7.3 Hilbert's First Thoughts

Hilbert seems to have recognized the gravity of the situation quite quickly. Indeed, there is evidence that he had known of such difficulties for some time. Recent scholarship has established the details behind Hilbert's response to Russell's paradox. Hilbert learned of it from Frege when Frege sent him a copy of his book, with its brave recognition that much of this, his life's work, was called into question by the paradox. Hilbert replied that this example was already well known in Göttingen and had been discovered by Zermelo some three or four years earlier. He added that he himself had "found other even more convincing contradictions as long as four or five years ago."[183] The example, although technical, is all the more telling precisely because its construction is technical, and for that reason suggests that no amount of clever talk will obviate the difficulty.

Hilbert had not had to enter these vexatious matters when he gave his celebrated foundations for geometry in his *Grundlagen der Geometrie* in 1899, because his analysis of geometry rested on an unanalyzed use of number. But even then he seems to have known that it would not be easy to give an analysis of number, for the obvious reason that it was not clear upon what it could rest. As he put it when he spoke at the International Congress of Mathematicians in Heidelberg in 1904: "In establishing the foundations of geometry, certain difficulties of a purely arithmetic nature could be put on one side; but it seemed impermissible to establish the foundations of arithmetic on another discipline." Hilbert therefore did what he often did on public occasions: he surveyed the field.

Kronecker, he said, had taken the idea of an integer as a fundamental one and did not see that it could be analyzed; this was a dogmatic position. Helmholtz was an empiricist and had attempted to derive numbers from experience. But, said Hilbert, this foundered on the fact that experience established neither the existence nor the possibility of very large numbers. Christoffel was among the "opportunists" who, sensing that without the idea of irrational numbers mathematical analysis would be

---

[183] Peckhaus and Kahle 2002, 499.

unfruitful, opposed Kronecker by seeking to find "positive" properties of the idea that would lead to a proof that they exist, but Hilbert thought that this would not succeed.

Frege was among those who had gone farther, as had Dedekind in a different way, but both their enterprises now ran into the paradox first discovered by Cantor: the uncritical use of the word "all" when forming a set. To avoid these paradoxes, Hilbert now advocated an appeal to the axiomatic method. Rather than think of arithmetic as a part of logic, he suggested it would be better to develop logic and arithmetic simultaneously. To this end, Hilbert sketched out a program that began with the idea of an "object of thought" (*Gedankending*) such as could be denoted by the sign "—." This sign could be repeated indefinitely. Another sign, "=," could be introduced with the obvious rules for when one collection of signs equals or differs from another. One then separates the combinations of simple things ("—" and "=") into two classes, beings and nonbeings (or existing and nonexisting, *Seienden* and *Nichtseienden*) and introduces an expression (a bar) for negation, so that if a collection a exists, the collection $\bar{a}$ does not. The point here is to show how to have a class of statements that do follow from some axioms, and a class of statements that do not. A consistent axiom system will not mix up these two classes of statement.

Signs could also be introduced for such useful matters as the Peano axioms, which define the natural numbers and provide for the first infinite set $\{0, 1, 2, \ldots\}$. Hilbert indicated that the Peano axioms could be shown now to be defined in a way free from contradiction. The proof consisted of showing that a contradiction would be an expression of a certain form, and that the rules governing what signs can be written did not permit such an expression to be obtained. The basic idea, he said, that showed the validity of his approach was that it reduced to finite considerations and certain ideas about when objects are equinumerous which can be established without effort. He was also pleased to note that his argument was the first that showed that a collection of axioms could be free of contradiction directly and did not proceed by specializing and constructing examples.

Hilbert now indicated how the paradoxes can be avoided. New statements can always be introduced into a theory at any stage, provided they do not generate contradictions with what has already been established. Expressions can only be introduced which are, so to speak, properly expressed, and in particular these expressions are necessarily finite. So it should be possible to show that a contradiction cannot arise in an axiom system by arguing that any contradiction that arises at whatever stage could always have arisen earlier (an argument known in number theory as the *principle of infinite descent*). He then ended his brief presentation with a new defense of his notorious eighteenth axiom, the axiom of completeness, which gave him the real numbers and a refutation of Kronecker's dogmatism.

Hilbert's lecture was necessarily superficial, a mere hint of what might be done and of what Hilbert was to do with the help of his assistants Wilhelm Ackermann and Paul Bernays in the 1920s.[184] It neither provided a resolution of Russell's

---

[184] Bernays had taken his doctorate at Göttingen with a thesis on number theory, and attended Nelson's lectures in philosophy. He then taught at the University of Zurich for some years until Hilbert brought him back to Göttingen in 1917. He remained close to Hilbert, editing the eighth and ninth editions of Hilbert's *Grundlagen der Geometrie*. He edited the tenth in 1968, upon which the second English translation, by L. Unger, is based (Hilbert 1971).

paradox, nor did it hint at what changes might be involved on the logical side if his ambitious project would succeed.

The implicit logicism of Hilbert's approach meant that the paradoxes loomed larger rather than smaller, and this seems to have galvanized Nelson's neo-Friesians. In 1906 Nelson sent Hessenberg a copy of Russell's *Principles of Mathematics*, which Hessenberg found a poor thing indeed, not least for its logicism. Hessenberg was more inclined to improve set theory than defend Russell's variety of logic. All the same, an impressive list of the more philosphically minded members of Nelson's school did respond to the paradoxes. Among these was Kurt Grelling, who came up with a paradox of his own, now named after him:

> Let $\phi(M)$ be the word that denotes the concept defining M. This word is either an element of M or not. In the first case we will call it "autological" in the other "heterological." Now the word "heterological" is itself either autological or heterological. Suppose it to be autological; then it is an element of the set defined by the concept that is denoted by itself, hence it is heterological, contrary to the supposition. Suppose, however, that it is heterological; then it is not element of the set defined by the concept that is denoted by itself, hence it is not heterological, again against the supposition.[185]

This paradox, mistakenly attributed to Weyl by Frank Ramsey, and dismissed as "scholasticism of the worst kind" by Weyl himself, was presented as a variant of Russell's paradox, but it is different in that it strikes at our ability to classify at all, and thus to use language.

The flurry of activity notwithstanding, no satisfactory solutions were proposed to any of these paradoxes. But if mathematicians, inspired by Hilbert's characteristic optimism, were waiting for him to step forward and take over the enterprise, they were to be disappointed, for Hilbert shortly found he had other, more attractive things to do, as the burgeoning theory of integral equations took over his research life. It was left to a young colleague of his to take it up: Ernst Zermelo.

4.7.4 Zermelo and Well-Ordering

Zermelo had begun his mathematical life studying the calculus of variations and then turned to thermodynamics. He worked as an assistant to Max Planck and discovered a paradoxical result of his own. Thermodynamics is based on the law of increasing entropy, but Poincaré's recurrence theorem for dynamical systems shows that every closed system will return infinitely often arbitrarily close to its initial configuration, which would seem to suggest that the entropy must be constant. In 1899 he transferred from Berlin to Göttingen and came under the influence of Hilbert. He shortly discovered the Frege-Russell paradox, some two years before Russell; but having a different approach to these issues, he did not publish it. For Zermelo it was merely one of those difficulties at the frontier that would eventually have to be sorted out.

In September 1904 Zermelo had a paper ready for publication, which Hilbert quickly ushered into *Mathematische Annalen*. In three pages, it offered a proof that every set can be well ordered. From this he deduced the trichotomy of cardinals and that the cardinal of every infinite set is an aleph (that is, is one of a sequence of

---

[185] In Grelling and Nelson 1908, quoted from Peckhaus and Kahle 2002, 13.

FIGURE 4.11. Ernst Zermelo during his time in Zurich.

transfinite cardinals Cantor had already in some sense characterized). It is appropriate to quote our guide at this point, because the appearance of quasi-philosophical questions at a time of crisis is precisely what this book is about:

All of these questions echoed a broader problem which had rarely been enunciated explicitly: What methods were permissible in mathematics? Must such methods be constructive? If so, what constituted a construction? What did it mean to say that a mathematical object existed? Normally mathematicians avoided such quasi-philosophical questions, and addressed them only when they felt their discipline to face a crisis. That Zermelo's proof precipitated such a crisis was shown by the extent of the resulting controversy. From 1905 to 1908 eminent mathematicians in England, France, Germany, Holland, Hungary, Italy, and the United States debated the validity of his demonstration. Never in modern times have mathematicians argued so publicly and so vehemently over a proof.[186]

Moore's analysis of the remarkable argument is focused with great clarity on the mathematical issues it engendered, their resolution by Zermelo, and the subsequent work that defined the landscape of theorems, axioms, and philosophical preferences in the whole area. It does not venture far into a discussion of why the topic was so vexing, finding, reasonably enough, that the underlying issues are so momentous that

---

[186] Moore 1982, 85.

it is worth disentangling them. But more might be said. It is reasonable to imagine that a different history might have happened in which a few mathematicians with an atypical interest in very large sets found it necessary to reformulate naive set theory in an axiomatic way that ruled out the paradoxes. They did so by clearing up which formulations of which statements in this area entailed which other implications, but, working as they did in an area cut off from much of the mathematics of the day, their work had, by and large, for the rest of the mathematical community all the interest of any other recondite branch of mathematics. Which is to say: very little. In many ways, this sad story recounts what actually happened, only a generation or so later, when mathematicians started to lose interest in logic. In some domains of mathematics logic or set theory was found to be important for its applications (group theory in some places, some branches of topology), and it was always recognized to be of interest in its own right, but many departments of mathematics found they could manage perfectly well without a logician.[187] So the reasons for the intensity of the debate around the axiom of choice can be analyzed a little further.[188]

As for the reception of Zermelo's argument, Moore helpfully notes the objections that were raised to this proof almost paragraph by paragraph. Zermelo assumed the axiom of choice: the French mathematicians Baire, Borel, and Lebesgue objected, as did Peano and Russell. The proof made ingenious use of the construction of a certain subset of the set of all sets of a certain kind—this subset was then shown to be well ordered: Poincaré objected to its impredicative definition. The proof ended with an argument by contradiction: Bernstein and Schoenflies feared that the Burali-Forti paradox had crept in.

When Zermelo's proof was published, Hilbert, as editor of *Mathematische Annalen*, sought the opinion of various authors as to what had been accomplished, itself evidence of Hilbert's keen interest in the question. Among these was Émile Borel. He agreed that Zermelo had shown that to prove the well-ordering principle it was enough to assume the axiom of choice.[189] But he disputed that this resolved the question of whether every set can be well ordered. He did this not on the banal grounds that the axiom of choice required proof, but on the altogether harder ground that no such proof could be given, or accepted if offered. It was no better, he said, than using transfinite induction to well-order an arbitrary set, which, he said, no mathematician would accept. He left open the possibility that a merely denumerable set could be well ordered.

4.7.5 The "Five Letters"

Borel's reply elicited a vigorous French debate, published in the form of five letters between Baire, Borel, Lebesgue, and Hadamard.[190] Hadamard led off. He disputed the analogy between the axiom of choice and transfinite induction. Transfinite induction involves making a sequence of choices, each depending on the previous. This

---

[187] Some forms of theoretical computer science have changed this picture, but only a little.
[188] See also Ferreirós 1999, 311–315.
[189] The converse is obvious, thus making the two axioms equivalent, as Borel noted.
[190] Reprinted in Borel 1898, 2nd ed., 1914, 150–159; English translation in Ewald 1996, vol. 2, 1077–1086.

might well be illegitimate, but the axiom of choice made those choices independently. Nor, he said, could he understand the force of the distinction between denumerable and nondenumerable sets that Borel found so natural. What Zermelo had done, in Hadamard's opinion, was to show that any set can be well ordered. He had not shown how to do so effectively—but that was a different question.

Baire replied to Hadamard by siding with Borel, but from a more extreme position. He disputed that even when one is given a set one is *given* all the subsets of that set. This is an epistemological position, which asserts that we simply do not know about such sets and cannot therefore reason with them.[191] Lebesgue's position was intermediate between Borel and Hadamard: He wondered if it was legitimate in mathematics to say objects exist when they cannot be specified (or "named," as Lebesgue put it). He was inclined to the position he associated with Kronecker, that a nonconstructive existence proof that asserts that such-and-such an object exists but gives no way of identifying it is no proof at all. What is lacking is a construction (in Zermelo's case, a construction of the function that picks out a distinguished element from each set of a family of sets). So Zermelo's argument was "too little Kroneckerian to have meaning."

Hadamard, in his next letter, replied that naming "is the issue for you; it is not the issue for me." His underlying position was antipsychological. The mathematical question is, "Can an arbitrary set be well ordered?" not "Can we well-order a set?" Even Hilbert-style talk of an argument being noncontradictory seemed to him to be too subjective. As he went on:

> I believe that in essence the debate is the same as the one which arose between Riemann and his predecessors over the notion of function. The rule that Lebesgue demands appears to me to resemble closely the analytic expression on which Riemann's adversaries insisted so strongly. (Here Hadamard inserted a footnote: I believe it necessary to reiterate this point, which, if I were to express myself fully, appears to me to be the essence of the debate. From the invention of the infinitesimal calculus to the present, it seems to me, the essential progress in mathematics has resulted from successively annexing notions which, for the Greeks or the Renaissance geometers or the predecessors of Riemann, were "outside mathematics" because it was impossible to describe them.) And even an analytic expression that is not too unusual. Not only does the cardinality of the choices fail to alter the question, but, it seems to me, their uniqueness does not alter it either. I do not see how we have the right to say, "For each value of x there exists a number satisfying. . . . Let y be this number . . . ," whereas, since "the bride is too beautiful," we cannot say, "For each value of x there exists an infinity of numbers satisfying. . . . Let y be one of the numbers. . . ."[192]

Borel was, as is the way in such debates, not persuaded, writing:

> One may wonder what the real value of these arguments that I do not regard as absolutely valid but that still lead ultimately to effective results. In fact, it seems that if they

---

[191] Readers with a taste for paradox may savor another of Borel's positions. Consider an arbitrary infinite decimal, $0.a_1 a_2 \ldots a_n \ldots$. It is fairly clear that the number of sentences in any given language that define such a sequence is countable. So at most countably many infinite decimals can be defined, but there are uncountably many infinite decimals. What does it mean to talk about an arbitrary undefined infinite decimal?

[192] In Borel 1914, 156–157.

were completely devoid of value, they could not lead to anything, since they would be meaningless collections of words. This, I believe, would be too harsh. They have a value analogous to certain theories in mathematical physics, through which we do not claim to express reality but rather to have a guide that aids us, by analogy, in predicting phenomena, which must then be verified. It would require considerable research to learn what is the real and precise sense that can be attributed to arguments of this sort. Such research would be useless, or at least it would require more effort than it would be worth. How these overly abstract arguments are related to the concrete becomes clear when the need is felt.[193]

The analogy to mathematical physics, with its strikingly contemporary ring, will detain us later. It is also interesting to see what, for Borel, was concrete no-nonsense mathematics and what was almost meaningless.

The French debate did not end—it was not resolved—but it revealed a number of different positions. There is a strange paradox about it, in that the whole topic only became controversial when a major result was rigorously proved. Where previously there had been conflicting claims and a number of arguments of uncertain merit, not all of them published, Zermelo brought a clear result. As if propelled by a strong electric charge, these French mathematicians immediately took up strong positions, often contradicting those they had held before. Borel was not the only one; Baire too moved away from his earlier, implicit, use of the axiom of choice. What had at one time been an entirely natural activity of the human mind suddenly became off limits.

Baire seems to have been the first to jump. He wrote to Borel on seeing Zermelo's paper that the paper was a real vicious circle: "At most, he proves the equivalence of two equally hypothetical facts. . . . I am astonished that Hadamard believes in it. If some *Revue Rose*, or one of any other colour, opened an investigation, I should not hesitate to make my opinion known. Between ourselves, these are our N-rays."[194] After asking what Hilbert thinks about this, Baire went on: "It is surprising that it is the Germanic races, with a heavy and correct spirit, that throw up these rash assertions; it would perhaps be more comprehensible among the hare-brained Latins."[195]

The vigor of Baire's opinion is striking.[196] His work had involved him in a deep study of discontinuous real functions and point sets in the plane, where he had found solid ground in the study of denumerable sets. Why did he object so violently to the work of Zermelo? The answer usually given is that he had gravitated to the constructivist or Kroneckerian end of the mathematical spectrum, which disliked abstract existence proofs and preferred explicit constructions, and while this is true, as an answer it invites further questions of the same kind. Baire was not interested in the foundations of mathematics. In common with Borel and Lebesgue, he had a poor

---

[193] In Borel 1914, 159.

[194] N-rays were the celebrated mistaken case of a discovery in physics. The more counterevidence against them appeared, the more adamantly their "discoverer," Blondel, believed in them, until he too had to admit his error.

[195] In Dugac 1990, 79.

[196] Mathematicians do have strong opinions, but Baire seems to have repeated bouts of depression that made it difficult for him to work, and to have had prolonged fallings-out with people, Lebesgue among them.

opinion of Couturat's work and was not attracted to basing mathematics on some form of logic. He saw himself as a mathematician interested in the rightful business of the mathematician, functions, and the point sets they engender. In this connection, he raised questions about infinite nondenumerable sets and the well-ordering principle.

In January 1903 he had written to Borel that "my opinion is not that it is impossible to define a well-ordered non-denumerable set. On the contrary, I think that the latest memoirs of Cantor . . . are excellent from the point of view of logic and rigour. . . . I consider that transfinite numbers of the second class [non-denumerably infinite] are as well and *as badly* defined as the integers."[197] Later that month, in his reply to Borel, he wrote that

> "we are in agreement about the transfinite, but I consider that one may not say that "one gives names to the transfinite numbers." No, one only does that at the start, and the question is important. I believe that this difficulty already exists, in a certain sense, for the ordinary integer numbers, and has never been noticed, neither by Drach nor by Tannery, and I do not speak of Couturat. . . . Crudely, it's this: one does not know the very large numbers, because, to write them, to name them, it is already necessary to know the very large numbers (whatever the system of numeration employed) ((defined necessarily by a finite number of conventions)).

The point about very large numbers is a shrewd one, but the fact remains that in 1903 Baire was comfortable with the well-ordering principle, and we have no evidence to help us understand what drove him, in the face of Zermelo's clear, short paper of 1904, to such a strong rejection.

4.7.6 Zermelo's Axiomatization

Faced with such criticisms, Zermelo knew he had to defend his proof of the well-ordering principle. He had by now obtained a lectureship in mathematical logic and related fields, and he went on to give the first university lectures on mathematical logic in Germany. His defense took the form of an axiomatization of set theory, which he published in 1908. Van Heijenoort, in his introduction to Zermelo's paper axiomatizing set theory, distinguished between Russell's theory of types and Zermelo's contribution: "The former is a far-reaching event of great significance for logic and even ontology, while the latter is an immediate answer to the pressing needs of the mathematician." One hesitates to argue, but the reply is surely possible that while both attempts created more problems, the ramified theory of types is one of such burdensome complexity that Russell tried to abandon it, whereas Zermelo's axiomatization could be repaired and in its improved version serves to this day as a basis for mathematics and a doorway to more philosophical questions.

Moore rightly notes that Zermelo was not the first to attempt to axiomatize set theory. Burali-Forti had earlier indicated the desirability of such an attempt, but Zermelo offered a complete axiomatization, ostensibly driven by the need to obviate the paradoxes. As Zermelo put it:

---

[197] Dugac 1990, 59.

Set theory is that branch of mathematics whose task is to investigate mathematically the fundamental notions of "number," "order," and "function," taking them in their pristine, simple form, and thereby develop the logical foundations of all of arithmetic and analysis. At present, however, the very existence of this discipline seems to be threatened by certain contradictions, or "antinomies," that can be derived from its principles—principles necessarily governing our thinking it seems—and to which no entirely satisfactory solution has yet been found.

His attempt relies on refusing to allow collections that are "too large" to be sets.[198] In true axiomatic style Zermelo began by stating that there is to be a domain **B** of individuals, called objects, among which are sets, and an object is said to exist if it belongs to the domain **B**. There is a fundamental relation, written $a\varepsilon b$, which expresses that a is a member of b, and similarly terms like subset are defined.[199] The weakness of Zermelo's system came next. He said that an assertion was "definite" (*definit*) if "the fundamental relations of the domain, by means of the axioms and the universally valid laws of logic, determine without arbitrariness whether it holds or not." He did not discuss these universally valid laws of logic, and later writers had to do so, which they did in various ways. One wonders if he simply felt nothing more needed to be said, of if he could only gesture in that direction. As he wrote, a gap was opening between the laws of thought and the appropriate rules for dealing with mathematics.

Then came his seven axioms, which he interspersed with elementary deductions, and which he claimed are mutually independent. They are as follows:

- Axiom (i) (Axiom of extensionality) If every element of a set M is also an element of N and vice versa, if therefore $M \subseteq N$ and $N \subseteq M$, then $M = N$; or, more briefly, every set is determined by its elements.
- Axiom (ii) (Axiom of elementary sets) There exists a (fictitious) set, the null set, 0, that contains no element at all. If a is any object of the domain, there exists a set $\{a\}$ containing a and only a as element. If a and b are any two objects of the domain, there always exists a set $\{a, b\}$ containing as elements a and b but no object x distinct from them both.
- Axiom (iii) (Axiom of separation) Whenever the propositional function $\sigma(x)$ is definite for all elements of a set M, M possesses a subset M_σ containing as elements precisely those elements x of M for which $\sigma(x)$ is true.
- Axiom (iv) (Axiom of the power set) To every set T there corresponds a set PT, the power set of T, that contains as elements precisely all subsets of T.
- Axiom (v) (Axiom of the union) To every set T there corresponds a set UT, the union of T, that contains as elements precisely all elements of the elements of T.
- Axiom (vi) (Axiom of choice) If T is a set whose elements all are sets that are different from 0 and mutually disjoint, its union UT includes at least one subset S_1 having one and only one element in common with each element of T.

---

[198] There is a long history of distinguishing between collections that can be considered as sets in an axiomatized theory of sets and other collections which cannot, on the ground that the latter are somehow "too large." See Hallett 1984.

[199] Zermelo used Schröder's sign for a subset; it will be replaced here by the sign $\subseteq$.

- Axiom (vii) (Axiom of infinity) There exists in the domain at least one set Z that contains the null set as an element and is so constituted that to each of its elements a there corresponds a further element of the form $\{a\}$, in other words, that with each of its elements a it also contains the corresponding set $\{a\}$ as element.[200]

Zermelo admitted that he had not been able to show that his axioms were consistent, only that the known antinomies were banished. Crucial in this is Axiom (iii), which only creates a "set of all x such that $\sigma(x)$" out of members of an existing set. So, as Zermelo showed, one is not allowed to create the set of all sets, or the set of all ordinal numbers, and thus the corresponding antinomies are banished. The Russell paradox is turned into the theorem that every set M possesses at least one subset M_0 that is not an element of M. The proof is that the proposition $x \notin x$ is definite for every $x \in M$, and defines a subset M_0 of M. The original argument for the paradox now shows harmlessly that M_0 is not an element of M, and "this disposes of the Russell antinomy as far as we are concerned." Any defining expression must also be definite, and so Richard's paradox is struck down. So too, however, was the Frege-Russell definition of cardinal numbers, and so Zermelo devoted several pages to showing how the essentials of that theory could be developed from the idea of a one-to-one correspondence between sets. In the course of this he proved a result that says, once cardinality was patched back in, as it was by later writers, that the cardinality of a set M is strictly less than the cardinality of the power set of the set M.[201]

Moore (1982, 159) has an interesting argument that the paradoxes did not motivate Zermelo to anything like the degree that they did Russell. He points out that they are banished rather perfunctorily, and that Zermelo's whole attitude toward them in the opening decade of the twentieth century is that they are problems, not fatal threats. Rather, says Moore, what drove Zermelo was the need to give a proof of his well-ordering theorem that was beyond cavil. This is likely to be the case, not only for the detailed reasons to do with the reconstruction of Zermelo's route to his axiomatization that Moore supplies, but because the antinomies really are only a threat to logicism, and Zermelo was never a logicist.

4.7.7 Poincaré: Impredicativity

Poincaré's responses to set theory changed during his life. As noted above, his response to the topological implications of Cantor's set theory was wholly positive, but his feelings about making set theory fundamental to mathematics were more complicated, and the paradoxes, when they emerged, did not strike him as fatal. In 1908 Darboux presented a speech to the International Congress of Mathematicians in Rome on Poincaré's behalf (Poincaré had fallen ill) entitled "The Future of Mathematics"[202] in which Poincaré surveyed many new developments in the subject. Turning to what he called Cantorism, he lauded it for being one of the two

---

[200] Taken from the translation in van Heijenoort 1967a, 201–204 of Zermelo 1908.

[201] A fine account of Zermelo's set theory, together with Fraenkel's later additions to it, that shows just how natural it is, is in Boolos, "The Iterative Conception of a Set," reprinted in Boolos 1998, 13–29.

[202] Poincaré 1908, chap. 2.

developments in recent mathematics that have won a central place, noting that it proceeded by defining only by the method of the nearest genus (genus proximum) and specific differences. He remarked that some—he cited Hermite—had been appalled by this departure from the method of the natural sciences, although "most of the rest of us," as he put it, have overcome these prejudices; but now certain paradoxes have appeared that would have delighted Zeno of Elea, and everyone who looks for a remedy. His own, he said in words that will detain us shortly, was to restrict mathematics to objects that can be defined completely in a finite number of words. But, he went on, "whatever be the remedy adopted, we may look forward to the joy of a doctor called to a beautiful pathological case."[203]

Poincaré did not say that "later generations will regard *Mengenlehre* as a disease from which one has recovered," but this has become one of the most widely quoted remarks about mathematics. It is found in otherwise reputable sources,[204] and since it is a travesty of his views it may be worth amplifying the point. What he wrote in the published address and in *Science et Méthode* was that set theory has an interesting disease. He did not say that mathematics has a disease, namely, set theory. For that matter, nor did he so confidently predict a recovery. What then seems to have happened is that Hölder (1924, 556) wrote that Poincaré said that one would look back on set theory as a disease one has overcome. But Hölder was not at the Rome congress in 1908, nor did he give a precise reference to the speech. From there the misquote passed to Pierpont, a mathematician and historian of mathematics at Yale; from there to E. T. Bell—at which point the damage was done—and from there to Kline and many others.[205]

Poincaré's response, not only to Zermelo's work but to the presence of the paradoxes, was to single out for attack what he called *impredicative definitions*, such as, "The least integer not definable in less than one hundred French words," because of Richard's paradox to which it leads.[206] Poincaré argued that the danger to be headed off is that the classification of an object will be altered by objects arriving later—indefinitely later, in the case of infinite sets. In thinking about how the integers may be defined without impredicativity creeping in, he was led to consider Russell's theory of types, but he was not attracted to it because he was not persuaded of the axiom of reducibility. This left him with Zermelo's axiomatization of set theory.

He argued that Zermelo had offered a set of seven rules concerning what you could do with a *Menge* (he deliberately used Zermelo's German word). Poincaré was willing to allow that such axioms could provide genuine foundations provided they were self-evident, there being no other way to show that they were free of contradiction (plainly, no mathematical models can be adduced for axioms intended to found all of mathematics). The first six he could agree were self-evident if "Menge" meant what it usually did: a set—and the sets were finite. However, Zermelo's seventh axiom admitted the possibility of infinite Mengen, so the theory was plainly intended to cover the case of infinite sets. Might we than consider, says Poincaré, an eighth axiom: any collection of objects forms a Menge? This is explicitly rejected by Zermelo, who spurned Cantor's naive definition because it would lead to a contra-

[203] Poincaré 1908, 41.
[204] For example, Kline 1972, 1003.
[205] For a full account, see Gray 1991.
[206] See Poincaré 1913, 7–31, first published in 1909; English translation in Poincaré 1963, 45–64.

diction. But on what basis, asked Poincaré, since Zermelo has provided no definition of what it is not to be a Menge. Plainly, with some idea of what a set might be, so we are returned to the idea that there is some intuition of what a Menge is, to which the seven legitimate axioms conform.

This Poincaré believed he had found in Zermelo's use of the word *definit* (again Poincaré kept the German word). An expression is "definit," Zermelo had said, if its domain of validity can be deduced logically from the axioms in an unarbitrary way. This looks like predicative, but Poincaré argued that it was not identical, and asked his readers to consider an infinite Menge and the question: Does this element of the Menge possess such-and-such a relation to all the other elements of the Menge? Zermelo will not permit himself to consider the set of all objects with such-and-such a property, but he will consider the Menge of all objects with that property. However, said Poincaré, this is difficult to do if the Menge is infinite, because not all the elements can be conceived of as existing beforehand all at once, and it is possible for new ones to arise constantly. For Poincaré an infinite set consists of a finite but growing set of elements along with all the objects specified by any method that can be invented, "And it is this '*that can*' which is the infinity" (italics in original).

We come, accordingly, to Poincaré's famous pieces of advice:

1. Never consider any objects but those capable of being defined in a finite number of words;
2. Never lose sight of the fact that every proposition concerning infinity must be the translation, the precise statement of propositions concerning the finite;
3. Avoid non-predicative classifications and definitions.[207]

As a small piece of theater, we note that Poincaré came to Göttingen in 1909 on money from the Wolfskehl fund that had been set aside for a solution to Fermat's Last Theorem but, of course, was left unspent. He gave six lectures, one of which was on set theory, where he rehearsed problems with impredicative definitions and Richard's paradox, and remained unconvinced that the second transfinite ordinal existed. According to Courant's later reminiscence, "the lecture left Zermelo in despair and fury, and at the dinner the same day he would have shot Poincaré, if he had been a little bit more skilful, but he was a very clumsy person."[208]

The publication of the paper generated much debate, which Poincaré alluded to in a later essay, without much change of mind.[209] Now Poincaré distinguished between what he called pragmatists and Cantorians, admitting that these were not perhaps ideal names. Pragmatists heed his advice, they adopt the point of view of extension. Cantorians adopt the point of view of comprehension. They would accept Zermelo's proof that every set can be well ordered, even though no process of well-ordering, say, the reals, has been exhibited and none is known. Poincaré seems to have shared Zermelo's feeling that the well-ordering principle and the axiomatization of set theory stood or fell as a package. In fact, as Poincaré pointed out, such a procedure would involve more than aleph-zero operations, and the pragmatists could not proceed past aleph-zero, and do not think it makes sense to talk of larger sets (ones

---

[207] Poincaré 1913, 31; 1963, 63.
[208] See Courant 1981, 162, from which we learn that Weyl was also in the audience.
[209] Poincaré 1912a.

neither finite nor in a 1-1 correspondence with the set of cardinality aleph-zero) as having a cardinality at all.

The main characteristic of Cantorians, as opposed to pragmatists, is that given a set they believe that they know *all* its members,[210] something the pragmatists believe requires a construction for each object. Only after its construction does an object exist.

Cantorians, Poincaré suggested, spoke often of epistemology, an epistemology independent of psychology, a science of science that taught what science would be even if there were no scientists. For them Nature offers a reality independent of the physicist, and indeed physics is a reality that would exist even if there were no physicists. So they are realists (object realists and theory realists, too, in later terminology). They are realists also about mathematical objects, which for them have an independent existence; they are committed to infinite sets. The pragmatists, on the other hand, are a species of idealist, prepared to discuss objects only when conceived by a mind, and so they only admit arbitrarily large finite sets.

Real mathematicians, he immediately went on, are more complicated, and he gave the example of Hermite, who he said, often claimed "I am anti-Cantorian because I am a realist." Cantor, in his view, had created, not discovered. For Hermite, as a realist, there was only one reality, and it is exterior to and independent of us. However, "all that we can known of it depends on us, so it is merely a becoming, a stratification of successive conquests. The rest is real but eternally unknowable."

4.7.8 The Schoenflies-Korselt Exchange

A further indication of the way foundational questions were regarded comes from an exchange in the *Jahresbericht der Deutschen Mathematiker-Vereinigung* for 1911 between Arthur Schoenflies and Alwin Korselt. Schoenflies had come to prominence by enumerating the 230 crystallographic groups, which he did independently of Fedorov in 1891.[211] On the strength of this work Klein had him appointed a professor extraordinarius in applied mathematics at Göttingen, and in 1899 he went to Königsberg to become a professor ordinarius. In the 1890s he turned to the new subject of point-set topology, on which he wrote extensive reports, subtle errors in which led Brouwer to some of his best work. Alwin Korselt is remembered, if at all, for being the target of one of Frege's longer critical attacks, which he wrote in opposition to Korselt's "Über die Logik der Geometrie."[212] There, Korselt had argued as a Bolzano loyalist that Hilbert's *Grundlagen der Geometrie* should be interpreted as a purely formal system, and on that account found wanting on occasion, and that Hilbert's axioms were not axioms but theorems. This was because he believed that axioms really should be unprovable, true assertions.

Schoenflies led off his "On the Role of Definition in Axiomatics" with the stark position that mathematics had lost its cutting edge by falling too much under the influence of philosophy.[213] There needed to be a distinction between mathematics

---

[210] Poincaré 1912a, his italics.
[211] See Scholz 1989.
[212] Korselt 1908, 98–124.
[213] Schoenflies 1911.

and discussions of the theory of knowledge, even though Hilbert denied this in his Heidelberg lecture (Hilbert 1904). Korselt replied that mathematics was not in a position to pronounce upon its own validity, only the logic of mathematics could ratify mathematics itself.[214] Schoenflies emphasized the logical foundations of mathematical conclusions and the contradictory nature of mathematics, by which he meant that in mathematics a statement was either true or false. Korselt observed that this confused Leibniz's law of contradiction with the law of excluded middle. Schoenflies then elaborated in the particular nature of the axiomatic method, exemplified, as he saw it, by Pasch's *Neuere Geometrie* (1882) and Dedekind's *Was sind* (1888), and consummated by Hilbert's *Grundlagen der Geometrie* (1899) with which, as he put it, the axiomatic method conquered the world. Korselt objected to Schoenflies's account of what Hilbert had done, because he spoke of mathematical objects. Schoenflies had said that a definition in mathematics was a name for a mathematical object. For Korselt, in agreement with what Paul du Bois-Reymond had argued just the previous year, mathematics has no objects. At most, it has objects in an extended sense of the term. Only science has objects—real things in the real world.

Schoenflies had surveyed other types of definition, reserving his scorn for what he called "word definitions." He regarded these as makeshift even in the sciences, which aim to clarify completely what presently is conveyed only vaguely by the defined word, and he was worried that they still survived in mathematics, for example in Hilbert's definition of a set in his Heidelberg lecture. The sort of "definition" that gave rise to the Richard paradox should be thrown out as inapplicable to mathematics. Korselt would also reject such formulations, because they do not define, but for him, as he noted, a definition is a theorem!

Schoenflies believed that mathematics could and should advance through the free creation of new objects and relations that are eventually to be axiomatized, and was critical of the restrictions Zermelo, for example, had resorted to. Rather, mathematicians should axiomatize everything, and the only restriction should be to throw out contradictions. He felt that Zermelo's foundations of set theory, although axiomatic, had a flaw, which he also showed how to remove. The flaw had the effect of smuggling a meaning into the word *Begriff* and as it stands spoils the transformation of Russell's paradox into a theorem. Here he announced we see the bad influence of philosophy on mathematics. In philosophy, he said, the complete harmony of laws and purpose is not to be found. Language and the world of ideas are not completely logical, unlike mathematics—thus his plea for a separation. And he ended his essay with the rallying cry: "Up with Cantorism—down with Russellism!"[215] To which Korselt objected that only logic properly grasped could rescue mathematics from the paralogisms of Russell, and if not language then at least the world of representations is as logically constructed as mathematics. Less stridently than his opponent, Korselt asked his readers to consult Bolzano's *Wissenschaftslehre*.

Schoenflies, one might say, was a mathematician without the means to analyze his mathematics, and Korselt, who had identified the need for an independent analytical-process mathematics, was still wedded to a realist philosophy of mathematics. Or, one might say that Schoenflies saw mathematics entirely as autonomous syntax, and Korselt wished to ground it in an enfeebled semantics. Schoenflies was insufficiently

---

[214] Korselt 1911.
[215] "Für den Cantorismus, aber gegen den Russellismus!"

reflective, Korselt reflective but out-of-date. Mathematics was surely no longer as either of them sought to portray it.

4.8 Anxiety

4.8.1 The Appreciation of Error

Most productive mathematicians in general do not have much inclination to engage with such philosophical speculations about the very earliest foundations of their science, any more than a real musician feels the need to give an account of how the musical logic that his ear has learned, is properly based.

—Friedrich Engel, in review of Russell's *Principles of Mathematics* in *Fortschritte* 34, 34.0062.14

The growth of mathematics in the nineteenth century is generally portrayed as a success story. A long list of remarkable achievements is recorded: the rigorization of the calculus,[216] the rediscovery of projective geometry,[217] the discovery of non-Euclidean geometry, the emergence of group theory,[218] the development of Fourier theory from its inception to the discovery of infinite sets of different cardinalities,[219] complex function theory from Cauchy to Riemann and Weierstrass.[220]

These achievements may be thought of in many ways. They were of two overlapping kinds: some enlarged the sphere of mathematics, some performed existing tasks better than before. Multidimensional geometries were only created in the nineteenth century; the theory of real functions was an eighteenth-century invention, but rigorous function theory with good proofs of the existence of solutions to differential equations happened only in the nineteenth century. Group theory is a nineteenth-century branch of mathematics; the misnamed fundamental theorem of algebra finds proofs to support its by then well established role in the nineteenth century.

Mathematicians in the nineteenth century knew very well that they often surpassed their predecessors. They knew they were doing new mathematics in areas hitherto unsuspected, as well as doing mathematics their predecessors had not done with sufficient precision. They knew that they were intimately, if not always satisfactorily, tied into the new sciences. What has not been sufficiently discussed by historians of mathematics is the note, hesitant at first but growing to a crescendo around 1900–1914, of anxiety. The mathematics of the nineteenth century is marked by a growing appreciation of error. For although mathematicians—with some notable exceptions—have traditionally had a low tolerance of errors, during the nineteenth century the awareness of errors grew and became a source of anxiety in

---

[216] Grabiner 1981 is a good example, but there are many other texts.

[217] We still lack a good modern historical account of projective geometry, but Coolidge 1940 is informative, and the algebraic side is covered in Gray 2006c.

[218] Wussing 1969.

[219] Hawkins 1975.

[220] Another topic that lacks a good, full-length modern historical account, but for the moment see Bottazzini 1981, chap. 4, and Bottazzini 2003.

mathematics, for reasons it is worth analyzing. Success, of course, can easily coexist with anxiety. Indeed, some authors have characterized the start of the twentieth century culturally as an age of anxiety, despite (even because of) the immense achievements of the time and the profound changes they unleashed. To cite only one author: "[T]he perception that [popular sovereignty] was now the reality, or at least the destiny, of the modern state, . . . , was a powerful conditioner of middle-class anxiety, pessimism and fastidiousness."[221] Indeed, once the safe havens of traditional mathematical assumptions were found to be inadequate, mathematicians began a journey that was not to end in security, but in exhaustion, and a new prudence about what mathematics is and can provide.

The failure of mathematicians in the eighteenth century to come up with foundations of the calculus that were generally acceptable is one kind of failure—one might call it a failure in the philosophy of mathematics, or a failure in foundational studies.[222] Early in the nineteenth century standard methods in the calculus failed to produce convincing evaluations of integrals, a failure highlighted by the existence of conflicting values.[223] The failure of Euclidean geometry is a third kind of failure: against almost all expectations it was shown that there were two mathematically sound descriptions of physical space.

None of these failures could be called routine failures, those occasions when, simply, a problem is too difficult. They formed, and form, the sour parts of daily mathematical life. The failures that came to animate whole aspects of nineteenth-century mathematics are of a deeper kind, because they pointed to deeper difficulties with the whole enterprise of mathematics. It is characteristic of them that they did not go away; resolving the original difficulty typically led to others. To establish this point, let me recall briefly what was said earlier (see §2.2.2). Cauchy showed how some questions about integrals could be tackled by thinking more carefully about the path of integration. Related questions arose in the eponymous work of Green and Stokes.[224] But the delicate questions about paths, and about the continuity of the functions involved not only remained unsolved all century, but came to seem more elusive and more profound.[225] Cauchy's definitions of differentiability and integrability survive to the present day. But his definition of continuity was less satisfactory, and later mathematicians were to realize that he was presuming more about the nature of the underlying system of real numbers than he had thought to question. Controversy about the nature of the real number system grew as the nineteenth century came to an end, even if workable (or rather, rigorous) versions of infinitesimals had to wait until the 1950s.[226] Attempts to shift the foundations of mathematics onto set theory foundered with the recognition that naive set theory led to irresolvable paradoxes, but axiomatic set theory has not led to unanimity yet concerning its deepest assumptions (the axiom of choice).

Frege was famously passionate about error in what should be rigorous and sound. To quote just one remark: "Mathematics should properly be a paradigm of logical

---

[221] Burrow 2000, xi.
[222] See the discussion in Grabiner 1981.
[223] See Bottazzini 1981, 137.
[224] See the discussion in Bottazzini 1981, 137.
[225] See Gray 2006c.
[226] See Epple 1999 and Ehrlich 2006.

rigour. In reality one can perhaps find in the writings of no science more crooked expressions and consequently more crooked thoughts, than in mathematics."[227] Frege is notorious for his provocative style, but it is interesting that after quoting this remark of Frege's, Voss elected not to argue the charge (either for or against) but to examine the sources of error that can arise, as he put it, despite an absolutely certain method.

Pervasive error would challenge the reliability of mathematics that is one of its most cherished features. Mathematics, in the opinion of many who use it, could no longer be regarded as among the most certain and true things we know. Many philosophers have claimed that mathematics is a particularly clear kind of knowledge, even paradigmatic; few, by contrast, have sought to deny the claim or limit it severely. Mathematical physicists happily consume and put their trust in a great deal of mathematics, much of which they produce. No mathematician can get very far in any ambitious program of research without believing that he or she stands on the firm ground of proven results. This widespread confidence is based on the fact that mathematicians habitually prove their assertions. But the nature of proof, what it is for an argument to be a proof, became just as unclear during the nineteenth century as any of the specific failures of important arguments to be a proof.

This anxiety about the very nature of proof coexisted with successful theorem proving for a variety of reasons and in a variety of ways. Nineteenth-century mathematicians were robust people in the main, and quite capable of taking the commonsense position when faced with seemingly philosophical dilemmas, of saying, "I may not know what a proof is, but I know one when I see one." The social aspect of proving theorems, the shared agreement that such-and-such an argument was indeed a proof, was also reassuring to them, as was the concomitant ability some had of detecting false "proofs." Jacobi wrote: "If Gauss says he has proved something, it seems very probable to me; if Cauchy says so, it is about as likely as not; if Dirichlet says so, it is certain. I would gladly not get involved in such delicacies."[228] And Dirichlet's wife tells us that Jacobi would spend hours with Dirichlet "Being silent about mathematics. They never spared each other, and Dirichlet often told him the bitterest truths, but Jacobi understood this well and he made his great mind bend before Dirichlet's great character."[229] The ability to recognize fallacious arguments tends to strengthen one's confidence that one can also recognize valid arguments.

Mathematicians also developed their understanding of the subject by rederiving results that others have found. Poincaré was certainly not the only mathematician who, on hearing of a new result, would immediately set about seeing how to prove it, and would only laboriously follow the original proof when his own intuition failed him.[230] Lesser, but still very good mathematicians would do the same, quite possibly with higher standards of rigor. The ability to find a new, and perhaps different proof also strengthens one's confidence, not only in the original result, but in the activity of proving. Even the up and down business of finding a proof, then finding a flaw in it, then a new proof, and perhaps a new flaw, often converged on a satisfactory argument; which was further evidence that mathematics is based on proof.

[227] Quoted in Voss 1914, 26.
[228] Jacobi to A. von Humboldt, 1846, quoted in Pieper 1980, 23.
[229] Quoted in Scharlau and Opolka 1985, 148.
[230] See the account of his way of working below, §6.2.1.

The existence of proofs of often spectacular results was nonetheless not as secure as it might seem. Numerous mathematicians accepted arguments that were shoddy or fanciful. Mathematical journals filled up with poor arguments as well as sound ones. As Jacobi's remark indicates, confidence in proof was likely to be confidence in idealized proofs, or those produced only rarely and by a select few mathematicians, not confidence in real existing proofs published in contemporary journals. Mathematicians had simply to accept that the standards in the field were variable. Routine failures in analysis included inadequate convergence arguments (which were sometimes wholly lacking).[231] In geometry, arguments were often restricted to the generic case.[232] In algebra, the slack and incorrect assumption was often made that all roots of a polynomial are distinct.[233] Faced with such mistaken arguments, the better mathematician would usually be able to check that the given argument could be repaired or supplemented, or they could see how to restrict the proof to the situation it really did cover and to employ it with greater care. Such work, often important, often interesting, was and is nonetheless routine. It is mentioned here only to establish that standards in the field were not uniformly high. Mathematicians in the nineteenth century could not and did not believe that they and their contemporaries lived up to the highest standards of proof. Daily mathematical life was error prone. Once disquiet arose, it was not going to go away with a reassuring immersion in what one's colleagues do; such reassurance was simply not there.

This might explain why some topics that were not pursued that might have been. After Darboux's study of uniform convergence in the 1870s, which included his account of a continuous, nowhere differentiable function, French mathematicians waited over a decade to respond.[234] From lack of interest, or from a distrust of one's abilities in such a novel, and unintuitive, area? In the later nineteenth century there was a curious lack of interest in differential geometry in dimensions greater than two, despite the pioneering work of Riemann. Some work was done, much of it by Bianchi (as noted in §3.1.4), but less than one might imagine before Einstein's theory of general relativity.

There are also significant results that people refused to accept. French textbooks discussed continuous functions as if they were differentiable[235] and, as noted above, no less a mathematician than Bertrand endorsed defenses of the parallel postulate well after non-Euclidean geometry was established—although he was speedily criticized for this by Darboux and others.[236]

As has been noted, historians of mathematics often deal with specific failures of mathematicians. The dreadful reception of Bolyai and Lobachevskii's non-Euclidean geometry is a case in point, as are the controversies surrounding the discovery and use of uniform convergence is another and the use of infinite sets is a fourth. These were not isolated cases, however, but part of a larger picture.

The discovery of non-Euclidean geometry could not be regarded, even by its supporters, as a simple success. Although recognition of this fact took time, the new

---

[231] Lützen 1990, 456.
[232] Hawkins 2000, 109.
[233] Hawkins 1977.
[234] Darboux 1875.
[235] Laugwitz 2000.
[236] See Pont 1986, 638.

geometry established that the old, Euclidean geometry, was not necessarily true, and so the claim that mathematics was true had either to be restricted to arithmetic or reformulated in some way. The discovery of non-Euclidean geometry is a classic case of a discovery that raised more questions than it answered and more problems than it solved.

Projective geometry is a subject that was revived from a moribund state in 1800 to become one of the hegemonic subjects in 1900. But the synthetic approach to the subject that was a great appeal of the subject for many proved difficult to use. Steiner, its greatest exponent, was so obscure that Cremona called him "the celebrated sphinx."[237] It also fitted uneasily with the algebraic approach that offered rigor and flexibility, and the use of imaginary points remained, burdening the research and pedagogic literature until it contributed to the demise of the synthetic branch of the subject in the twentieth century.[238] To quote Coolidge: "It is hard to escape the conclusion that the field of synthetic projective geometry, except perhaps in the matter of studying the postulates, is pretty much worked out."[239] Yet the algebraic style of projective geometry as initiated by Möbius and Plücker had its problems. It relied too much on the naive idea that one could count the number of configurations of a particular kind by counting the number of solutions to a system of equations, and this sometimes led mathematicians into error. Plücker himself slipped up this way in his groundbreaking study of bitangents to a quartic curve,[240] but his mode of reasoning remained popular throughout the nineteenth century, and bedeviled the study of complicated singular points and their resolutions. As late as 1923 Bliss was to record, in a presidential address to the American Mathematical Society, that the matter was not satisfactorily resolved.[241]

In axiomatic geometry, the process, which had started with Pasch in 1882 and reached by 1900 to Hilbert's axiomatic geometry and the even more radical accounts of several Italian mathematicians, sought foundational clarity quite openly at the expense of intuition. In a journey of barely twenty years this broke entirely with Pasch's own views on the nature of geometry. We must conclude that geometry, for all its successes, did not even provide geometers with an entirely reliable body of knowledge.

Real analysis was considered above (§2.2), and the vigorous debates described there, and between Kronecker and Cantor, do not suggest that the line from Cauchy to Cantor and Dedekind and beyond was a simple case of progress. Hermite's strong language concerning continuous nowhere differentiable functions is more consistent with an active dislike of the topic than merely finding it trivial, or likely to be a dead end, or otherwise irrelevant to serious mathematics. And since it concerns a crucial point in real analysis, one may feel that Hermite's strong dislike for an existing topic contained an element of anxiety.

The realization spread steadily among mathematicians that their intuitions about curves applied only to differentiable curves and not, as they had always thought, to continuous ones. So the American mathematician E. B. Wilson in 1905 could write

---

[237] Cremona 1868, 3.
[238] Samuel 1988, v.
[239] Coolidge 1940, 104.
[240] Plücker 1839; see Klein 1926–27 and Gray 2000, 151.
[241] Bliss 1923.

plausibly and at length on the collapse of this intuition, observing in effect not only that mathematicians had been wrong all along but that with their newfound clarity on the issue there was no chance of restoring the status quo ante. At the very least, rigorous mathematics had to be conducted according to much more stringent rules. The foundations of real analysis were no longer certain, and the capacity to match intuitive and formal reasoning had been thrown into doubt.

Complex analysis is one of the major success stories of the century, but it too was not without its problems. Fundamental to any approach derived from Cauchy or Riemann was the Cauchy integral theorem, relating the sum of residues inside a contour to the value of an integral taken around the contour. But as Jordan observed, this requires that one can distinguish the inside and outside of a contour, a task he admitted he was not able to do precisely. A long debate then ensued concerning the meaning and validity of the Cauchy integral theorem for arbitrary contours. The highly counterintuitive space-filling curve of Peano, and its refinements due to E. H. Moore and Hilbert, have an extra significance when placed in this context. The Cauchy integral theorem was not a problem for close adherents of the Weierstrass school, because Weierstrass eschewed the use of integral methods whenever possible in favor of power series ones. But Weierstrass's strictures against Riemann's ideas, in particular his view that defining a complex function by the Cauchy-Riemann equations, was dangerously imprecise and did so much to polarize the community of German function theorists that a unified approach to the subject largely free of error had to wait until the twentieth century.

One reason mathematicians' soul-searching became acute as the century drew to a close is surely that by 1900, the nature of mathematical objects was contentious and obscure. The arithmetization of analysis broke the naive identification of mathematical with physical objects. Geometry was no longer true in any simple sense, and was increasingly abstract in its formulation. To these ontological shifts can be added a related epistemological shift, as the nature of proof and the relation of mathematics to logic became a matter of animated research.[242]

Another set of reasons is of a sociological nature. The growing separation of mathematics from physics was quite marked by 1900.[243] In many universities across Europe there were separate departments for mathematics and physics, there were separate journals, and recognizable divisions existed between the subjects and between their practitioners. This weakened mathematicians' ability to rest their subject on the security (such as it was!) of science. The more critical mathematicians were aware that they therefore had to base their claims for the quality and value of mathematics on more intrinsic grounds. This raised the stakes. One traditional argument for mathematics that it is a species of pure, abstract, science was weakened, and the new discipline of physics could complete very favorably with mathematics on utilitarian grounds. When academic turf wars were to be fought, as they were when Klein sought to enlarge Göttingen in competition with Berlin, or when American universities expanded, or when times were harder, as they were in the 1890s in Germany, mathematicians knew that their arguments for their subject had to be recast.

[242] Most thoroughly described in Grattan-Guinness 2000.
[243] There is a large literature on this; see the introductory essay to Gray 1999 and the papers cited there.

In addition, the professionalization of mathematics also forced mathematicians to be more self-critical. A profession has a declared set of standards, to which its members are held, and which is open to public scrutiny. Anyone who meets the criteria, perhaps by passing a public examination, and who continues to remain competent, is a member of a profession, all others are out. In the case of mathematics it is only other mathematicians who are capable of making such judgments, and the various national mathematical societies that came into existence in the nineteenth century did not police their members. But the fact remained that a research mathematician was only as good as his (or, sometimes, her) arguments as they were set out in journals and books. The failure of mathematicians to live up to these standards was in theory open and public, however much a matter of likely public indifference. But part of the burden of being a professional is to care about these standards even without the threat of public sanction, and some of the publicly fought debates were provoked by just such a sense of responsibility. In addition to which, the great popular literature on mathematics around 1900 is evidence that in fact the educated public was interested in some of the problems that also animated mathematicians.[244]

4.8.2 Anxiety: Kronecker and Enriques

Two examples may help make more precise the way in which this anxiety was manifested. In 1891, the distinguished German mathematician Leopold Kronecker gave what was to prove to be his last lecture course at the University of Berlin.[245] The course was entitled "On the Concept of Number in Mathematics," deliberately, Kronecker explained, so as not to engage with matters of philosophy.[246] He distinguished three mathematical domains: mechanics, geometry, and the "so-called pure mathematics," which he preferred to call arithmetic, noting that mechanics involved time, geometry involved space but not time, and that arithmetic was free of both these concepts and was more rigorous.[247] The language of anxiety then crept in, when he said:

> If one wants to define the idea of magnitude quite generally, so that it is also valid for geometry and mechanics, it becomes more and more blurred. . . . I am not of the opinion that the mixing of disciplines, and also in the expressions, will bring trouble; I only believe that if one goes down to foundations an absolute separation is necessary. . . . But if one wants, for example, to discuss the idea of number, one must define it in its most precise sense, namely as number [Anzahl], and may not mix it with what originally does not lie within it. . . . If this is not the case then number becomes a battered coin whose impression can no longer be correctly recognised. . . .[248]

The message is clear: unless mathematicians retreat to the secure heartland of arithmetic, what they do will become contaminated and devalued; the core concepts of pure mathematics must be separated absolutely from the rest, even though they

---

[244] See Gray 1999, 58–83.
[245] See Boniface and Schappacher 2001.
[246] Boniface and Schappacher 2001, 223.
[247] Boniface and Schappacher 2001, 227.
[248] Boniface and Schappacher 2001, 231.

have been traditional domains of mathematics. To speak of rigorous separation and a willingness to abandon part of what was hitherto one's own but has become dangerous is to use the classic language of anxiety.

Another example is offered by the distinguished Italian algebraic geometer and writer on scientific and philosophical issues, Federigo Enriques.[249] He was originally close in his opinions to his friend G. Vailati, but they drifted apart over the issue of foundations. In 1911, Vailati had summed up the role of postulates in the modern axiomatic conception with an appropriate metaphor: "Postulates had to give up that sort of divine right which their assumed obviousness entitled them to, and they had to become the lowliest of the low, the mere servants of the associations of propositions which characterise the various branches of mathematics."[250]

Enriques was hostile to this point of view, and in his address to the International Congress of Mathematicians held in Cambridge, England, in 1912 he replied:

> An Aristophanes could object that the availability of unlimited choice risks turning democracy into demagogy; too often dishonest functions replace the simple but honest functions satisfying the theorems of infinitesimal calculus, some bizarre Geometries (justified initially as a means of investigating some relations of subordination) affirm the freedom of the inspiring idea in the same way as the governments which follow one from the other in the Principality of Monaco under the auspices of Rabagas.[251]

What Avellone et al. charitably call "baroque exaggerations" is highly charged, emotional language: democracy will decay into demagogy, simple honesty give way to dishonesty, good government collapse into parody. Enriques continued:

> The history of mathematics has shown the work of logical criticism in the reworking of notions over the centuries. Contrary to the consequences that would follow from a view of postulates as implicit definitions, history shows that the definitions of mathematical entities are not arbitrary because they are the outcome of a long process of acquisition and an effort revealing of some of the general aims of research. There is a tradition of problems and there is an order on the extensive and intensive progress of science, therefore there is a subject matter of mathematics that definitions aim to reflect. Hence the decisions of the mathematician are not different from those of the architect who lays the stones of a building according to a carefully arranged project The act of free will that the mathematician embarks upon in the formulation of problems, in the definition of concepts, or in the assumption of a hypothesis, is not arbitrary. It is the opportunity to get closer—from more than one perspective, and by *continuous approximations* to some ideal of human thought, i.e. an *order and an harmony* that reflect its intimate laws. If this is the view that emerges from an historical view of science, mathematical logical pragmatism will have given research a better understanding of its aims rather than a proliferation of bizarre constructions. Moreover, by purifying Logic it will have made

---

[249] This information is taken from Avellone et al. 2002, 409–11.

[250] Vailati 1911, 690.

[251] Rabagas was the central figure in Sardou's play of that name, a satirical attack on Gambetta first performed in 1872. He was a radical demagogue won over to the Court's party by the simple expedient of being invited to dine at the palace. His name became synonymous with lack of principle. Interestingly, one of the terms hurled at Mussolini in 1914, when he was expelled from the Italian Socialist Party, was "Rabagas."

clear its inadequacy and the need to investigate more in depth *the other psychological elements which give meaning and value to mathematics.*[252]

Threatened by chaos and collapse, mathematicians may yet find security in tradition if they subject the arbitrary exercise of free will to natural ideals of order and harmony; purifying logic will show what needs to be done to give meaning and value to mathematics. Again we hear the language of anxiety urging a strategic withdrawal, keeping (redefining?) the correct traditional values and rejecting new ones. Kronecker's objections are those of a profound mathematician drawn, despite his prudent disclaimer, to a philosophical position of his own that acutely prefigured intuitionism. It would be wrong to see it as the nervous, even prejudiced, stance of an old man opposed to so much of what was new around him. But it would also be wrong to just read it as another statement of Kroneckerian finitism; the language is too resonant of a desire for purity and protection for that. The case of Enriques is more clearly that of someone who, despite an equal enthusiasm for philosophy, is simply opposed as a mathematician to some trends in contemporary mathematics. They do not strike him as harmless and self-limiting; as can be seen from his choice of dramatic political language, they worry him. Kronecker and Enriques shared a fundamental attachment to meanings in mathematics, and in very different ways and for different reasons saw mathematics as closely related to natural science. Kronecker explicitly compared mathematics to science in the fourth lecture.[253] Enriques defended the pursuit of mathematics because of its utility throughout his major philosophy work, his *Problemi della scienza* of 1906. By holding onto meaning, they belong to the group of mathematicians Mehrtens called the *Gegenmoderne* (anti-moderns), and I think that it is indeed likely that mathematicians prone to the anxiety I am calling attention to in this book will lean to *Gegenmodernismus*; but there is no reason why anxiety should not strike the *Moderne* as well.[254]

4.8.3 Perron's Inaugural Address

A remarkable expression of this anxiety is to be found in Oskar Perron's *Antrittsrede* (inaugural lecture), given when he became a professor at Tübingen in 1911.[255] He began by observing that few nonmathematicians know what is of interest to mathematicians, whereas the public knows more about science. Moreover, what the public knows about mathematics is often of no interest to the mathematician: he cited the problems of trisecting the angle and squaring the circle, and observed that Fermat's Last Theorem had recently joined the ranks of popular problems because of the prize of 100,000 marks offered for its solution (it is not clear what importance Perron attached to the problem itself). But, he went on, even if one can debate the utility of mathematics, one cannot doubt its truth: "Of all the sciences this is the one whose results are the most securely grounded" (p. 197). Or rather, this perception of the methods and results of mathematics was exactly what he proposed to deny: "This

---

[252] Quoted in Avellone et al. 2002, who added the italics, 410–11. Enriques 1912, 77.
[253] Boniface and Schappacher 2001, 232.
[254] In keeping with the German focus of Mehrtens 1990, Kronecker is discussed, but not Enriques.
[255] Perron 1911. This section is adapted from Gray 2004b.

complete reliability of mathematics is an illusion, it does not exist, at least not unconditionally."

Perron began his attack by observing that confidence in old mathematics was misplaced. Euclid's *Elements* was now beset by critics finding fault with it; Newton, Leibniz, and other early workers might have reached correct conclusions in the calculus, but only by unconvincing methods. Nor was it any better with contemporary mathematics. "Indeed, there is one branch of mathematics today over which opinion is divided, and some consider right what others reject. This is the so-called set theory, in which the certainty of mathematical deduction seems to be becoming completely lost" (p. 198).

He then asked rhetorically how mathematics had fallen into such a state of error and doubt, and answered his question by considering the sorts of errors that could be found. Apart from straightforward errors of calculation, there were errors in which ideas were poorly expressed. Often, mistakes crept in where the matter was thought to be true "in general" (a term Perron thought was quite useless). Words, of course, can have multiple meanings, and it is necessary to be sure there is no slippage between them. This could be done by using a language from which multiple meanings were excluded, such as formulas, and here Perron noted the efforts of Peano to enrich mathematical notation. This can be done advantageously not only in the statement of results but in definitions, for example in the ε—δ formulation of key concepts in analysis. Such steps had led, he said, led to the disentangling of convergence and uniform convergence.[256]

Another class of mistakes arises from statements that are thought to be self-evident but may indeed be false, however attractive. Multiple limit processes were adduced as a case in point, where the exchange of their order is a common source of error. Another kind of error is often to be found in Steiner's work on the isoperimetric problem, where he assumed that there is a curve of a given length that contains the greatest area. Perron might have noted at this point that a similar error was often taken to be the fatal flaw in Riemann's use of the Dirichlet principle.[257]

Spatial intuition is a very frequent source of error, especially when it is used to supplant a proof, as, for example, in proofs of the intermediate value theorem.[258] "Intuition is a crude instrument that lets us make out true relationships only imprecisely" (p. 204), and this is particularly so of our understanding of curves, which may fail in all sorts of ways to have the intuitive properties one suspects.

Perron then gave two lengthy examples I omit, and went on that it was not, however, his intention to pronounce the death sentence on intuition, far from it. "Intuition and Imagination are the tools with which one *finds* new mathematical truths" (p. 207, italics in original). However, after potential truths have been found in this way it is absolutely necessary that they be exposed to the light of logic. The way forward will be difficult and dangerous, and yet, he said, it is very simple to avoid all the errors just mentioned. One has merely to keep clearly in mind all the

---

[256] For a historical account of how entangled these were, see Lützen 2003.

[257] A well-known refutation of this method of argument is this: if there is a largest integer, it is the integer 1. Proof: any integer $n > 1$ has the property that $n^2 > n$, and it so cannot be the largest integer!

[258] The intermediate value theorem states that a continuous function on an interval that is positive at some point and negative at another is zero somewhere in between.

definitions one is using and all the hypotheses one is making. In this way, in fact, "a degree of reliability has been reached in many partial domains of mathematics" (p. 208), but not in all, and certainly not in set theory.

The problem is that every mathematical theory must begin with undefined terms and results that cannot be proved but are taken as axioms and postulates. The question then arises as to the consistency of a system of axioms, for only a system of axioms free of contradiction can be taken as definitions. What can be done in this way for geometry, however, is not so easy for set theory. There is not, presently, a system of axioms known to be free of contradiction, but there is an idea (the set of all sets) that is known to engender contradictions. So one cannot be sure that the other basic ideas of set theory are not also flawed. The result is that at present "we do not always know if the methods of mathematics, the most exact of the exact sciences, are really exact" (p. 211).

To conclude his talk, and in keeping with the generally up-beat occasion of an inaugural address, Perron urged his audience not to despair. Some exactness has been obtained, the situation is much better than a hundred years ago, we can hope that in the near future many problems will be solved, and we can console ourselves with the words (p. 211), "Better than a known truth is the search for truth."

This rhetorical exercise is not difficult to understand. Perron was not going to stand up and say, what in any case he did not believe, that mathematics was in a total mess. The purpose of the occasion was to demonstrate to a wide audience in the university and beyond, that he was worthy of his appointment, and that he had a sufficient command of his subject to lead it forward. As he delicately reminded his audience at the start of his talk, this is not as easy for a mathematician to do as a scientist. His own specialty, Diophantine approximation, might have served from the standpoint of its results, but certainly not from its methods, and indeed even the results can be seen obscure and pointless to those outside the field. What then was he to do? It was quite usual on these occasions to find a philosophical perspective on mathematics, but what is striking is the critical tone of the lecture. It is unrelenting in its catalog of errors. The different types of error are carefully described, but the exactness of some domains is merely asserted. Perron did not assert that all mathematics reduces to set theory, but he gave the impression that it was not merely new but somehow fundamental, and very much in trouble.

It is striking that someone in Perron's position should stand up and point to the profusion of error in his subject. An inaugural lecture is not the occasion for mere pedantry; plainly Perron felt that the matter was so grave that the scale of the problem was exactly what could and should interest his audience. And while he had a standard of mathematical practice that he felt mathematicians should be kept to, he offered no remedy, and his remarks about set theory, prominently placed near the start and finish of his talk, were such as to suggest the problems there might be insoluble. It is reminiscent of the situation that developed in quantum mechanics, where it became common to talk about the failure of classical physics. But even then there was a more precise description of the way forward. What Perron felt the need to do was to spell out the pervasiveness of error in mathematics, which went far beyond simple slips of the pen to loose reasoning, ambiguous terms, failure to be aware of one's presuppositions, and undue reliance on intuition as a source of proof. In setting out his view of the present state of mathematics in this way, he plainly thought it was a matter of some concern that the subject was as he found it to be.

Enough has now been said to indicate that mathematicians who heard Perron's talk or read it in the pages of the *Jahresbericht der Deutschen Mathematiker-Vereinigung* would have had to agree with its criticisms in the main. That there were many mistakes of the kinds he described, from the routine lapses in calculations to the more damaging failures to be scrupulous about one's assumptions and the uncritical but misplaced reliance on intuition, was common enough. It might be that Perron, by taking the side of logic against intuition, was at least implicitly opposing Felix Klein, who had contrasted the logical, formal, and intuitive styles of mathematics in his Evanston colloquium lectures in 1893, to the advantage of intuition, and had sought to steer the arithmetization of analysis away from a total adherence to Weierstrassian rigor in his lecture in Vienna in 1895. Klein's own mathematics always inclined to the sort of imprecision that Perron was adamantly against. Even if Perron had not had Klein in mind, Klein's example was the kind of thing he was against. No mathematician tolerates a mistake once it is pointed out, but they differ over the extent to which they are prepared to tolerate mistakes in pursuit of discoveries, and the extent to which they think mistakes are harmless and will be corrected. Klein was undoubtedly more prone to mistakes than Perron, and more likely to regard them as, ultimately, harmless. Perron, in making this public attack on the many errors of mathematicians, was much more concerned about them.

4.9 Coming to Terms with Kant

4.9.1 The Leibnizian Revival

A striking feature of the debates in philosophy and the foundations of mathematics is the role played by Leibniz. Interest in Leibniz's ideas had been revived as the nineteenth century progressed, first by Erdmann and then by Gerhardt's edition of much of his unpublished work, but a decisive moment came with Couturat's *La logique de Leibniz d'après des documents inédits* (Paris, 1901). In that book, Louis Couturat set out, as it seemed for the first time, a coherent yet critical interpretation of all of Leibniz's philosophy, and this gave the forces opposed to Kantianism a focus. In particular, it offered a rallying point for those who thought that proper attention to logic could unlock questions about the foundations of mathematics.

Couturat aimed to show that "Leibniz's metaphysics rests solely on the principles of his logic, and proceeds entirely from them." In this he noted that he was in agreement with Bertrand Russell, who had arrived at the same conclusion but from a completely different interpretation.[259] As part of showing what this claim should be taken to mean, Couturat argued that Leibniz's principle of sufficient reason meant nothing more than the more intelligible, if by no means obviously acceptable, claim that all truths are analytic. As Leibniz put it, "No proposition can be true unless there is a sufficient reason why it should be thus and not otherwise."[260] Or, in his second letter to Clarke:

---

[259] Russell 1900.
[260] Leibniz n.d., 8–9.

The great foundation of mathematics is the principle of contradiction or of identity, that is to say, that a statement cannot be true and false at the same time and that thus *A* is *A*, and cannot be not *A*. And this single principle is enough to prove the whole of arithmetic and the whole of geometry, that is to say all mathematical principles. But in order to proceed from mathematics to physics another principle is necessary. As I have observed in my *Theodicy*, that is, the principle of a sufficient reason, that nothing happens without there being a reason why it should be thus rather than otherwise.[261]

The contentious case is that of contingent truths, ones that are the way they are but could surely have been otherwise. Leibniz sought to finesse the point by arguing that the lack of analyticity is the fault of our finite minds, and better ones would find the explanation of events seemingly random to us. To God, all would be governed by the law of sufficient reason, nothing would be contingent. Mere mortals are condemned to the sort of imperfect approximate grasp of things that rational approximations provide for irrational numbers.

Couturat drew out the remarkable consequence of Leibniz's claim that every truth is analytic: "As a consequence, everything in the world must be intelligible and logically demonstrable by means of pure concepts, and the only method of the sciences is deduction. This can be called the postulate of universal intelligibility. The philosophy of Leibniz thus appears as the most complete and systematic expression of intellectualistic rationalism."[262] He had been engaged, he said, in a study of Leibniz as a precursor of modern algorithmic logic, but on seeking the philosophical principles of these theories he "saw on the one hand that they proceeded from Leibniz's seminal conception of a universal mathematics [. . . , and] on the other hand, that they were tightly bound up with his attempts at a universal language." These became Couturat's intellectual concerns for the rest of his life, until his death at the age of forty-six at the start of the First World War.

Leibnizian ideas met with a modern movement already actively reengaged on some of Leibniz's abandoned projects, including most significantly the derivation of mathematics from logic, or at least the profitable clarification of the relationship between the two. Indeed, it was this prior enthusiasm that guided Couturat through the confused mass of Leibniz's unpublished papers and led him to identify a logicist core. The point here is not whether such an interpretation is fair to Leibniz, but that it was emblematic of what its exponents wished to bring about more fully, even completely. In particular, it was a radical standpoint with which to drive out the various forms of Kantianism.

Kantianism was widely agreed to be in trouble, not least by the various stripes of neo-Kantians who disputed among themselves for the honor of being the ones to rescue Kant from this or that criticism with an astute reinterpretation or return to the "real meaning" of the texts; Nelson was by no means the only one. As the flood of German idealism and Hegelianism subsided, Kantian ideas once again seemed to be the right direction to take for anyone concerned, as Kant had been, to locate philosophy's place in the growing empires of knowledge. Once again, the philosophical enterprise seemed to be threatened by the successes of contemporary science. By the end of the nineteenth century, German philosophers were locked in difficult struggles

---

[261] Leibniz, n.d., 193–194.
[262] This and the next quote come from Couturat 1901, ii.

with psychology on the one side and the growing reach of the physical sciences. But if Kant's sophistication was to be of any help, much graver problems faced it from within science and mathematics than had been the case in Kant's own time. Foremost among these was the nature of geometry and of physical space, but another was contemporary formulations of the nature of mathematics itself, which seemed to have no need for any kind of Kantian intuition.

Leibnizians were encouraged not only by Leibniz's emphasis on the primacy of logic, but by his attempts to create a symbolic calculus capable of analyzing any problem according to formal rules. No matter that the attempt seems hopeless, or at best quixotic, given the sheer complexity of the world, the idea that formal system should be capable of giving such results was once again popular. In part, this was because of the much greater degree of success in mathematical physics since Leibniz's time, and in part this was because mathematics now rested on a much more formal reduction to abstract entities tantalizingly close to the purely logical. Finally, in part this was because Boole's work seemed to have successfully initiated just the sort of symbolic, logical analysis Leibniz might have had in mind. So Leibnizians looked to Boole's work as a most fortunate repair, done without knowledge of Leibniz's own work, of the defects they detected in Leibniz's original system. Kantians, on the other hand, found Boole's contribution to be superficial, at best an efficacious rewriting of simple logical arguments, but not an insight into the nature of logic and reason. They could not so easily dismiss Leibniz, but they could and did disagree with the logicist or rationalist interpretation placed on him by Couturat and Russell.

4.9.1.1 RUSSELL

Russell warmly welcomed Couturat's book when he reviewed it in *Mind*, writing that "it may almost be said it constitutes a new book by Leibniz. For those who have not read this book it will be impossible henceforth to speak with authority on any part of Leibniz's philosophy."[263] In the same review in *Mind*, Russell then turned to an interpretation of Leibniz from the Kantian camp, Cassirer's *Leibniz's System in seinen wissenschaftlichen Grundlagen* (Marburg, 1902). Poor Cassirer, it seemed, had not grasped "the very modern discovery of the importance of symbolic logic." There was little realization of the arithmetizing of mathematics, "and none at all of the still more recent 'logicizing' if such a word be permissible. Mathematics, for Dr. Cassirer, is not synonymous with Symbolic Logic, and Logic is synonymous with the theory of knowledge. In both these respects," Russell concluded fairly, "the work is Kantian," going on immediately to make it clear that in his opinion it was, on those grounds, also incorrect.

For Russell the Leibnizian, the Kantian view espoused by Cassirer that geometry and algebra offer a form of knowledge distinct from that of logic was known "with all the certainty of the multiplication table" to be wrong and

the connexion of mathematics with the sciences of experience is the same as that of logic, i.e. they cannot violate mathematics, which is concerned wholly and solely with logical implications, but also they all of them, including the geometry of actual space, require premises which mathematics cannot supply. This conclusion, originally suggested by

---

[263] Russell 1903, 177–201.

non-Euclidean geometry, . . . , has been wholly removed from the realm of dubitable hypotheses.

In terms of Russell's growing identification with the logicist philosophy, this is several claims: that geometry, algebra, and presumably any other form of mathematics are identical with logic in the nature of the knowledge they offer (nothing more nor less than chains of logical inference); that indeed mathematics is synonymous with symbolic logic; that scientific knowledge is logical inference applied to various extralogical premises; and that these conclusions are now known with certainty.

The neo-Kantian response to these ambitious, programmatic, and far from well-established claims was, as Russell noted, to dispute the meanings of the key ideas. They did not dispute that mathematics was logical, but they meant by this claim that knowledge is rooted in thought. For the neo-Kantians, especially those of the Marburg school to whom we briefly now turn, Kantian intuition in any of its forms was suspect because it was known to open the door to subjective, psychologistic accounts of how knowledge is acquired. The neo-Kantians took the philosopher's task to be to explain how scientific knowledge was possible, and, since science was evidently in flux, it was its success as a method that had to be explained. This knowledge was expressed in the mathematical laws that had been discovered, not in the empirical facts of science, and so the philosopher had to explain how lawful accounts were possible. This ability was grounded in the nature of thought, and the demonstration of this insight was provided by a transcendental critique.[264] The scientific method is logical and mathematical par excellence, but the logical analysis of science (of the conditions that make science possible) is the contribution of philosophy, and therefore logic (on the neo-Kantian sense of the term) cannot be identified with mathematics. Logic is, rather, what enables the transcendental critique.[265]

It is generally agreed that contemporary disputes about number establish the extent of the disagreement about logic most clearly. The approach of Russell and Frege to the definition of number rests crucially on the idea of a one-to-one correspondence. To use Russell's language from 1903, in his *Principles of Mathematics*, two sets are similar if they can be put into a one-to-one correspondence with each other, and a number is the set of all sets similar to a given set. To neo-Kantians, this approach presumed that one knew what the number "one" meant—one had to know what a single individual of a set was before one could pair off elements of two sets—and so the definition was illusory; knowledge of numbers was prior to being able to define sets in this fashion. To the logicists, the definition was fine, because a one-to-one correspondence does not require that the number "one" be known, only that it can be defined purely logically. This is not difficult: a set has precisely one element when it is not empty, and if x and y are in it then $x = y$. The harder task is to separate the loose talk of an individual object being unique, being a "one," from the novel, precise talk of a set having exactly one member. Cassirer replied that the mental act

---

[264] See Alan Kim, "Paul Natorp," in *The Stanford Encyclopedia of Philosophy* (Fall 2003 Edition), ed. Edward N. Zalta; Mark Balaguer, "Platonism in Metaphysics," in *The Stanford Encyclopedia of Philosophy* (Summer 2004 Edition), ed. Edward N. Zalta, http://plato.stanford.edu/archives/sum2004/entries/platonism/.

[265] I regret I have not been able to take on board Lydia Patton's recent thesis on Hermann Cohen, the leader of the Marburg school: "Hermann Cohen's History and Philosophy of Science," PhD diss., McGill University, 2004, http://home.uchicago.edu/~patton/Dissertation.pdf.

of identifying the x and the y presumed that it was possible to identify them as individuals, and so they are candidates for the question, "Are these elements of the set the same?" and that this ability presumed that the x and the y were each one thing (whether the same or different does not matter).[266]

Disagreements of this kind are not to be resolved. Each side held an awkward position about the relation of mathematics to the world: the Leibnizians that mathematics is reducible to logic and all truths are analytic, the neo-Kantians that the possibility of science depends on a transcendental argument. Each side's argument was implausible, but each preferred to attack the other for not agreeing with their fundamental presuppositions rather than attend to where they were most exposed. These disagreements also have resonances at several levels. Anglo-Saxon analytic philosophy goes one way, continental European philosophy another in the interwar years, and these arguments are one indication of that incipient split.[267] Another is the place of philosophy vis-à-vis science and mathematics. Around 1900 philosophy was engaged in one of its recurrent examinations of this question. Kantians and neo-Kantians alike sought to locate philosophy just ahead of any apparently more successful field (as it might be, physics or mathematics) by establishing the preconditions that made such a field possible. Leibnizians located all the philosophical problems of mathematics within mathematics, which they then claimed to be reducible to logic. That claim was to prove indefensible, but it is the direct ancestor of those philosophies of mathematics that locate the significant problems within set theory.

4.9.2 Poincaré Replies

Why mathematics has a particular claim on our attention was exceptionally unclear around 1900, and so it is not surprising to find mathematicians engaging with it at a significant philosophical level, and one of those who did so most influentially was, of course, Poincaré. He was an eloquent opponent of Russell and Couturat, and showed himself to be almost Kantian in his attitude to number. In his essay "Les mathématiques et la logique" (1904/5) Poincaré sought to oppose Couturat's reduction of mathematics to logic, inspired perhaps by Couturat's speech on the occasion of Kant's jubilee, at which a neighbor of Poincaré had said in a stage whisper, "One sees very clearly that this is the centenary of the death of Kant."[268]

---

[266] Cassirer 1910.

[267] A split well described in Friedman 2000b and also in Kusch 1995.

[268] One source for Poincaré's Kantianism, if that is not too strong a term, is the work of Emile Boutroux and the group of people around him, which included the brothers Jules and Paul Tannery. Boutroux, who married Poincaré's sister Aline, had studied in Heidelberg before the Franco-Prussian War, where he got to know both Zeller and Helmholtz and acquired a taste for broad intellectual thinking in contrast to what he saw as narrow French specialism. In 1874 he defended his doctoral thesis, "De la contingence des lois de la nature," which takes a move toward Kant and opposes Comtean positivism. Among his friends in the audience for his defense of his thesis was Jules Tannery, who later became the influential director of the science curriculum at the École Normale Supérieure, and who regarded scientific theories as symbolic creations. Boutroux found much that was arbitrary in science, including scientific laws, which he argued are themselves evolving—a view Poincaré would explicitly deny. What Boutroux and the Tannery brothers may well have helped bring about was a shift in French intellectual life away from scientism and positivism, and the creation once again of a space for philosophy in discussions of science. See Nye 1979.

Poincaré began by noting the markedly formal character of contemporary mathematics, exemplified by Hilbert's approach to geometry. Poincaré had no quarrel with this, but others had attempted to do the same for analysis and arithmetic. Now, he said, even if the logicians succeeded in these attempts, a Kantian would not be condemned to silence, because this new mathematics would start from axioms, and a philosopher reserved the right to contest the choice of axioms. Nor would the practice of mathematics be reduced to the mechanical deduction from the axioms, but would be guided by the mathematician's instinct to search ever deeper. That said, Poincaré took his stand on his former position that one cannot establish all the truths of mathematics without appeal to intuition.

He had already defended the view that the principle of complete induction is both necessary to the mathematician and not reducible to logic. Given the advances in logic since then, he needed to check if his arguments still held up, or if the new definitions of number somehow entailed the principle of induction. He found that Burali-Forti (in the course of his paper in which he set forth his famous paradox) had defined the number using Peano's notation in this way:

$$1 = \iota T'\{Ko \cap (u,b) \in (u \in Un)\},$$

a definition, he remarked, "eminently suitable to give an idea of the number 1 to people who have never heard it spoken of before" (p. 168). He did not, he said, understand Peanian well enough to risk a critique, but he did suspect that this definition contained a vicious circle, because the "1" that is to be defined seems to occur inside the definition as "Un." Be that as it may, he noted that Burali-Forti then showed that, as he put it, $1 \in No$, and so Un is a number.

Couturat, Poincaré now observed, had defined both 1 and 0. Zero is the number of elements of the null class, and the null class is that class which contains no element. "To define zero by null, and null by nothing," he said, "is truly to abuse the riches of the French language," so Couturat had introduced an improvement to the definition by writing

$$0 = \iota \Lambda : \varphi x = \Lambda.\supset.\Lambda = (x \in \varphi x),$$

"which is to say . . . : zero is the number of objects which satisfy a condition that is never satisfied. But, as "never" signifies "in no case," I do not see that the progress is considerable." Poincaré concluded that the logicians definitions of number were inadequate, and that his original defense of intuition still stood.

In his next essay, "Les logiques nouvelles,"[269] Poincaré took issue with Russell. Russell based mathematics on certain indefinable notions and certain indemonstrable principles that Poincaré said were intuitive and generally found, more or less explicitly, in mathematical treatises. But these principles were intuitions, synthetic a priori judgments. They did not cease to be such just because the new logic vastly exceeded logic in the Aristotelian sense; they just changed their place. Could these principles be disguised definitions? For this to be the case, they would have to be shown to be incapable of introducing a contradiction. But no such proof could be

---

[269] Poincaré 1908, 152–71.

given, because it would have to proceed by complete induction, and that, Poincaré reminded his readers, was not established. Thus the Russellian enterprise rested upon nine indefinable notions and twenty indemonstrable principles (Russell's own estimate—Poincaré claimed he would have counted more).

There were other criticisms. How could a two-term relation be understood without a prior understanding of the number 2? But it is more instructive to see how Poincaré responded to Hilbert's 1904 lecture at Heidelberg. There Hilbert noted that certain arithmetic notions seemed entangled with logical ones, and so the only way forward was to develop the principles of logic and arithmetic simultaneously. Poincaré took this as an implied criticism of the work of Couturat and Russell, but still felt that Hilbert had not overcome the difficulties he faced. By the end of the memoir, he wrote: "The contradictions accumulate; one feels that the author is vaguely aware of the petitio principe he has committed and that he searches in vain to cover up the fissures in his reasoning."

But if such distinguished authors are failing to define correctly, what is a definition? To help his readers, Poincaré ended his paper with some examples. The parallel postulate is, he agreed, a disguised definition of a straight line. It can be shown to be free of contradiction. But complete induction is not a disguised definition of number for the reasons Poincaré had by now given several times and which he now gave again. First of all, a definition must be provably not a consequence of the law of contradiction; complete induction fails this test. For, argued Poincaré, there are two definitions of the natural numbers. One asserts that they are all obtained by successive addition from 1, the other that one may reason with numbers by recurrence. In Poincaré's opinion, the two definitions are not identical (they do not say the same thing)—they are equivalent, but the equivalence is a synthetic a priori judgment and not a matter of logic.

Russell and Hilbert, he concluded, had written thought-provoking works that were often profound, original, and contained many results that were destined to last. "But to say that they have definitively decided the debate between Kant and Leibniz and destroyed the Kantian theory of mathematics is evidently inexact. I do not know if they really thought they had done so, but if they so believed, they have deceived themselves."

Couturat naturally replied. Poincaré had defined existence as freedom from contradiction, just as Hilbert had done. Couturat replied that existence was rather a proof of absence of contradiction. Then, he went on, admitting that what he had to say would seem paradoxical, individuals always exist; what has to be shown is that they exist in a class. Poincaré unwrapped the paradox by saying that it was presumably meant to mean that a single object cannot be in contradiction with itself, it is only a system of objects and postulates about them that can be in self-contradiction. That said, Couturat's enterprise still foundered on the rock of complete induction. Nor did Poincaré see any reason to change his opinion of Hilbert's work, although he noted that Couturat had excommunicated him from the community of logicians. Then Poincaré switched tack, and discussed Russell's account of the paradoxes of set theory, as has already been discussed.

Hadamard followed this debate and did not agree completely with either Poincaré or Couturat in Hadamard (1906). He agreed with Poincaré that Couturat's account of number was inadequate; the definitions of "one" and "two" were circular. These attempts to define number drew from Hadamard a distinction between nominal definitions, which implied the existence of the object thus defined, and definition by postulates, which he proposed to call the characterization of an indefinable idea. Couturat, he said, had tried to define number by postulates, thus conceding that the concept cannot be defined. However, said Hadamard, he sided with Frege and against Poincaré and many other mathematicians, including Hilbert, in doubting that existence was implied by freedom from contradiction. At the very least, this freedom from contradiction (he used the German word *Widerspruchlosigkeit*, showing how un-French the idea was) must be established rigorously and not merely asserted. What was even stranger in Hadamard's eyes was Couturat's apparent opposition to such proofs, which can often be given by the process of complete induction (provided the definition of number is not at stake). Couturat, said Hadamard, was preaching against his own saint.

But his deepest objection to logistic was that it did not make errors of reasoning impossible to make. Logisticians could not escape the Burali-Forti paradox, said Hadamard, any more than mathematicians could, but at least the mathematicians could explain the error. Indeed, he said, Poincaré had explained it precisely and the flaw lay in the incautious use of the word "all," exactly as Richard's paradox showed. Logistic, however, was gravely challenged by these paradoxes, and Russell had sought to demolish the system he had created by modifying the fundamental principle of reasoning, the notion of "class" as commonly understood. Hadamard suggested that logisticians could make this sacrifice only because it destroyed the logic of the mathematicians at the same time as their own. Safer, and better, he concluded to stay with mathematics.

4.9.3 Russell and Whitehead

Russell famously praised Frege in later life for his remarkable honesty in responding as he did to the postcard Russell had sent him demolishing his major work just as it reached its end. In a more modest way, Russell was also honest about his intellectual failures, skipping from one position to another, first as he learned about the philosophy of mathematics, then as a leading figure in that rapidly advancing field. As the editors of his *Collected Papers* make clear, once he was through his idealist neo-Hegelian phase (well described in Griffin 1991) he was at times a Leibnizian, at others not. He was remarkably responsive to criticism, while at the same time remarkably confident that whatever his current project, he had swept away all the outstanding problems, or else he soon would.

Philosophers of mathematics speak with great respect of Russell and Whitehead's *Principia Mathematica*, so much so that it can be hard to discover how it is flawed, while mathematicians find so little mathematics in it that they are perhaps too dismissive of Russell (and simply ignorant of Whitehead). The confidence with which Russell wrote sits uneasily with his imperfect grasp of mathematics—the product of a Cambridge education he himself found unsatisfactory—and may contribute to a feeling that he was overambitious and ultimately unconvincing. For all these reasons he has become one of the most well-studied figures in the history of philosophy, although his place in the history of mathematics is less secure. He may have known this himself, as his delightful nightmare relates: A librarian's assistant is working through the books in the University Library at Cambridge in 2100, selecting a few meritorious books for eternity, casting most into oblivion. Eventually he comes to the letter R, and to the last surviving copy of *Principia Mathematica*. He holds the book in his hand, pauses in thought . . . and Russell woke up![270]

Russell's first work of philosophy is his revised Cambridge thesis, published in 1897 as *An Essay on the Foundations of Geometry*. The topic is to what extent should Kantian views be modified in the light of non-Euclidean geometry. Russell was later to be very dismissive of the book, and it does reveal the limitations of a young Cambridge-educated mathematician whose postgraduate training was not in mathematics but philosophy. Nonetheless, it gave Russell, who spoke and read German well, a thorough grounding in the neo-Kantian philosophy of the day. It also brought him a rapturous review from Couturat, which proved to be the start of a deep and valuable intellectual friendship. Dissatisfaction with the *Essay* took him to a neo-Hegelian idealism, then popular in Cambridge, before acquaintance with the work of Frege, and still more that of Peano, changed him thoroughly, and he came to the view that mathematics was identical with logic. He was prepared for this influence by the ideas of A. N. Whitehead, especially his *Universal Algebra* (1898). As Griffin (1991) writes: "The need to do justice to Whitehead's algebra strained Russell's fundamentally Kantian approach to pure mathematics to breaking point" (p. 272).

Even the preface to Whitehead's book alerts the reader to all sorts of themes that would occupy Russell in the years after the book came out. It investigates "various systems of Symbolic Reasoning allied to ordinary Algebra" that will also be considered as "engines for the investigation of the possibilities of thought and reasoning connected with the abstract general idea of space." "Mathematics," wrote Whitehead, "in its widest significance is the development of all types of formal, necessary, deductive reasoning. . . . The sole concern of mathematics is the inference of proposition from proposition." And a page later: "Historically, mathematics has, till recently been confined to the theories of Number, of Quantity (strictly so-called), and of the Space of common experience." One final quotation, from the opening page of the book proper: "Words, spoken or written, and the symbols of mathematics are alike signs. Signs have been analysed [here Whitehead gave a footnote to a paper by Stout from 1891 and to 'a more obscure analysis' in Peirce 1867] into (α) suggestive signs, (β) expressive signs, and (γ) substitutive signs. . . . The signs of a Mathematical Calculus are substitutive signs."

The first strain was adjusting to the idea of mathematics as not being wholly or essentially about quantity, but being heavily symbolic. The "possibilities of thought and reasoning" are not, of course, Kantian conditions for the possibility, but explorations of what is possible; but that they are so heavily symbol-driven would have been a novel perspective. That the sole concern of mathematics is inference must have struck home, because in 1903 Russell opened his *Principles of Mathematics* with his famous definition of the subject, which begins, "Pure mathematics is the class of all propositions of the form 'p implies q.'" That idea, combined with the view that written mathematics is made up of signs (whose semiotic analysis will detain us later) governed by symbolic logic, took the philosophy of mathematics away from idealism and neo-Kantianism and pushed it into the arms of Leibniz. Russell would join in the embrace.

---

[270] In Hardy 1941, 23.

We do not know how much of the *Universal Algebra* Russell read (he is particularly thanked for reading the parts on non-Euclidean geometry), but we do know that in 1899 he deputized for McTaggart at Cambridge and lectured on Leibniz. At the same time, as his preface to the *Principles of Mathematics* acknowledges, he came under the influence of the Cambridge philosopher G. E. Moore.[271] This was decisive in moving him to a hard-line antipsychologistic position. Among the fruits of his intensive labors was another book, *A Critical Exposition of the Philosophy of Leibniz* (1900), with which Couturat was happy to find himself in general agreement.

It must have been deeply reassuring for Russell to find that the *Universal Algebra*, his first proper acquaintance with state of the art mathematics as opposed to the undergraduate version, was so richly suggestive philosophically. His *Principles of Mathematics* more or less opens, as does Whitehead's, with a lengthy discussion of symbolic logic because (p. 8) "all mathematics follows from symbolic logic." It is then, naturally and rightly given its theme, heavily philosophical, more so than Whitehead's book, which itself went some way in that direction. Upon this philosophical basis—with its 20 premises—Russell erected accounts of number (based on a theory of classes and on part-whole theory), quantity, and order or series. Order, it emerges, after a long discussion of awkward cases, is an asymmetric transitive relation between terms (as, for example "less than," $a < b$). The inspiration for this material was Dedekind's account of the integers. The correlation of series leads to a Cantorian theory of the real numbers and the infinite. An account of geometry follows, after which the book concludes with an account of matter and motion.[272]

The geometrical part proceeds under the provocative idea that "*Geometry is the study of series of two or more dimensions*" (p. 372, italics in original). The order idea is invoked to generate from a single term, say (*a,b*) first a series of *a*'s and so the terms (*a_i,b*), and then a series of *b*'s and so the doubly infinite series of terms (*a_i,b_j*). Multidimensional terms are defined analogously. For all Russell's claim that his definition of geometry "represents correctly the present usage of mathematicians," it does not so much describe what contemporary geometers did so much as mimic it in a form amenable to his logicist thesis. In particular, it shows that geometry needs no new indefinable terms. It is closest to those contemporary accounts that also had a strong formalist urge, and farthest from those mathematicians who just wanted to get on and do geometry. And this is typical of Russell's mature writings on mathematics. His account of mathematics is adapted for a philosophical purpose and can alienate mathematicians who do not recognize their subject and the way they do it in his account. As befits someone in the grip of Peano's ideas, it is well informed about, and sympathetic to, Italians in his school, such as Pieri, but Hilbert's *Grundlagen der Geometrie* barely gets a look in, being confined to three footnotes.

The part on geometry ends with Russell's defense of opinions that he cheerfully admitted were diametrically opposed to those of Kant. This disagreement arose from Kant's view that the truths of geometry were synthetic, while those of logic were analytic, and therefore geometry could not be reducible to logic. To this Russell

---

[271] See the discussion in Griffin 1991, 297–309.

[272] Ronny Desmet has pointed out in a paper presented to the Perspectives on Mathematical Practices conference, Brussels, 2007, that Whitehead did not seek to reduce mathematics to its foundations, as Russell did, but was a unifier and a generalizer, as is demonstrated by the range of topics in *Universal Algebra*.

blithely replied that it now appeared (and he had shown in his book on Leibniz) that logic "is just as synthetic as all other kinds of truth" (p. 457). No matter that he had recently agreed with Couturat that all truths are analytic. Kant had been driven to explain the veridical quality of mathematics by the vexed synthetic a priori propositions, intuitions about which supply the characteristic methods of mathematical reasoning. But, Russell proclaimed, these extralogical arguments were simply unsound. The only sound arguments in mathematics were the newer purely logical ones, which needed the synthetic a priori not at all, and so the concept expired of inanition. As for the a priori recognition that physical space was or is Euclidean, Russell held that this was not a question for the philosopher of mathematics, who does not care if mathematical terms correspond to things that actually exist.[273] The book then ends with an account of applied mathematics, matter and motion, and an appendix connects it to the work of Frege, which Russell had only recently picked up.

The *Principles of Mathematics* is a halfway house, and not only because it was to be eclipsed by the joint work of Whitehead and Russell a few years later. It proposes that mathematics is nothing more than symbolic logic, and to defend this successfully it must show that symbolic logic provides all the raw ingredients for mathematicians: the terms and the methods of reasoning. It recognizes that classical logic has been greatly extended. What then of the additions to logic that make symbolic logic? What is their logical, mathematical, philosophical status? What of the twenty indefinable items? Does the book deliver the raw ingredients, or is it in some way muddled? As Russell openly admitted in the preface and again in chapter 10, its whole argument is imperiled by the paradox (the "singular contradiction," p. 101) of predicates not predicable of themselves. For this he hinted at a theory of types, and publication only proceeded, he said (p. xix), when there seemed to be no prospect of adequately resolving the contradiction, "or of acquiring a better insight into the nature of classes. . . . It seemed better, therefore, merely to state the difficulties, than to wait until I had become persuaded of the truth of some almost certainly erroneous doctrine."

Mathematicians not persuaded of the prior claims of logicism took refuge in the paradoxes. For example, Jacques Hadamard, in two essays in volume 17 of the *Revue générale des sciences* largely agreed with Poincaré and opposed Couturat on the grounds that logicism collapsed under the paradoxes and whatever small virtues it might possess were worthless if it could not guard against such failures of reasoning. The unspoken premise was that some other defense for the practice of mathematics could be found.

4.9.3.1 HAUSDORFF

Felix Hausdorff, in his review of Russell's book,[274] found it scholastic in its acuteness, and an orgy of subtlety. It was exhausting, both for the author and the reader, a continual climb to no summit. In a few words, "acute, and still not clear."[275] This

---

[273] Russell also dealt with the Kantian antinomies, with which we have not been concerned.

[274] In *Vierteljahrsschrift für wissenschaftliche Philosophie und Sociologie* 29: 119–124, to be reprinted and discussed in vol. 1 of the Hausdorff *Gesammelte Werke*.

[275] "spitz und doch nicht klar," 119.

was the more to be regretted, because at many points it was a valuable philosophy of modern mathematics. Its delineation of the problems was as radical as only a mathematician could want—no psychology, no science, no empirical investigation. Here there was a philosophy of the formalism that was an ever-growing force in modern mathematics, free of talk about intuition, and largely correct in its account of mathematical topics.

But, Hausdorff went on, Russell went farther than any formalist could when he entered the unclear realm of the primitive ideas and assumptions. No idea, said Hausdorff, was indefinable, no axiom unprovable per se, but only in the context of other axioms and ideas. But when it comes to the necessarily somewhat arbitrary business of the fundamental ideas and axioms of logic, Russell is unclear. This was because Russell had based his study of formal implication, with its insistence that variables vary arbitrarily and without restriction, on the work of Peano but in so doing had made it worse. For Russell, material implication was the deduction of one proposition about particulars from another, and formal implication involved variables. Crucially this was exemplified not by the deduction "Socrates is mortal" from "all men are mortal" but by the deduction [x is mortal] from [x is a man], where the variable x is entirely arbitrary. So a formal implication embraces a whole class of material implications. But, said Hausdorff, no sensible person would deduce that a triangle is mortal from the admittedly false proposition that a triangle is a man, although this was no problem for Russell (because for him anything can be deduced from a false proposition).

The problem was the Burali-Forti paradox, mentioned but not resolved by Russell. It was, observed Hausdorff, simply not possible to talk loosely about all objects. This meant, he said, that "formal implication is a class of material implications endowed with a contradiction!" (p. 123). Not, he noted ominously, that this need be a problem for Russell, who, in a discussion with Couturat, had noted (with awe-inspiring perversity, as Hausdorff put it) that if it could be shown that any theory of space was necessarily contradictory, then any noncontradictory theory would have to be rejected. Hausdorff sided with those who, "doubtless with rash optimism," hoped for foundations of thought that were free of contradiction, and would find a geometry with antinomies scarcely more bearable than an alogical logic.

Hausdorff concluded his review with his judgments on the individual chapters of the book. The problem posed by the Burali-Forti paradox also made the definitions Russell gave of cardinal and ordinal number unacceptable. But the account of order, the infinite, and continuity could be recommended as a clarification of the modern idea of the infinite and formed the best part of the book, while that on quantity was the weakest. The part on space was an advance on the *Essay on the Foundations of Geometry*, although the account of how distance could be introduced into projective geometry repeated too many of the earlier mistakes.

As is well known, Russell and Whitehead, who had each planned second volumes of their individual works, realized that they agreed on enough for a fruitful collaboration to be possible, and they set to work. To quote from the *Stanford Encyclopedia*:

> By agreement, Russell worked primarily on the philosophical parts of the project (including the philosophically rich Introduction, the theory of descriptions, and the no-class theory), while the two men collaborated on the technical derivations. Initially, it

was thought that the project might take a year to complete. Unfortunately, after almost a decade of difficult work on the part of both men, Cambridge University Press concluded that publishing *Principia* would result in an estimated loss of approximately 600 pounds. Although the press agreed to assume half this amount and the Royal Society agreed to donate another 200 pounds, that still left a 100-pound deficit. Only by each contributing 50 pounds were the authors able to see their work through to publication.[276]

The essential novelty of Whitehead and Russell's *Principia Mathematica* was the ramified theory of types, which carried the burden of obviating the paradoxes. It is hardly possible in a work such as this one to do justice to the theory of types, but fortunately it is not necessary. There are many good accounts,[277] and in any case we have been led to describe it as part of our inquiry into Russell's claim that mathematics could be reduced to logic. The situation is, in the end, quite simple. *Principia Mathematica* is not hopelessly flawed on its own terms, which is to say that where it is unclear (for example, in matters of syntax versus semantics) it can probably be put right. It does, therefore, deliver something like mathematics out of an elaborate version of symbolic logic. There are problems at each end, however, and they are bad side-effects of the theory of types.

To oversimplify, Russell and Whitehead regarded sets as having types, reflecting the complexity of their construction. A set of objects has a lower type than a set or sets of objects, and so on. To bar the paradoxes, a set could only be made up of sets of lower types, but this impoverishes mathematics. For example, consider the least upper bound of a set of numbers. This is a number, but of a higher type than the set of numbers with which you began. It cannot be used, therefore, to exhibit a zero of a continuous function defined on the unit interval which takes a negative value at 0 and a positive value at 1, and with that failure goes one of the most vital theorems in real analysis.

Russell and Whitehead attempted to save the situation by licensing a way in which types could be reduced, the so-called axiom of reducibility. Indeed, it was to turn out that without this axiom the whole program of *Principia Mathematica* was in serious trouble, for it was shown that without it the natural numbers cannot be defined but must be accepted as an indefinable primitive idea.[278] The axiom does what it is intended to, but mathematicians, and Russell himself in later life, found the philosophical arguments in favor of it somewhere between downright mysterious, unintelligible, and an attempt to fudge the issue. The issue of the nature of mathematics was not resolved by all the three dense volumes of Russell and Whitehead's *Principia Mathematica*; it was merely shown to be unexpectedly hard.

To put the point another way, the axiom of reducibility made it very clear that there was no easy way to claim that the principles from which mathematics can be derived are simply logical. At least the set of principles set forth in that book did not seem any better than the better claims made by mathematicians for the raw ingredients of their subject.

---

[276] Irvine 2006. To give an impression of these figures, the EH.NET Web site estimates the value of £100 in 1910 as over £7,000 or $14,500 in 2006 using the retail price index and more by other methods.
[277] For example, Potter 2000.
[278] Myhill 1974 cited in Potter 2000, 154.

On the other hand, what emerged well from *Principia Mathematica* was the symbolic logic itself. Suddenly it looked like the way a number of deep and delicate issues in the foundations of mathematics could be resolved. That is why, when Gödel came to his celebrated incompleteness theorem, he couched it in the language of *Principia Mathematica*. The great claims for logic may have been blocked, but the engine was there ready for the mathematicians to use for their own purposes.

4.9.4 Around 1910: Weyl, Winter, Study, and Cassirer

A perspective on the philosophical state of mathematics in 1910 that is interesting both because it is indicative of the view from Göttingen, and because of its author—Hermann Weyl—is offered by an essay published when its author was twenty-five. The essay is the first of his many forays into philosophy, which later became rather different, as we shall see in chapter 7. Weyl, like Hilbert, was at that time open both to the possibilities of logicism and to the significance of its role in science. As logicism ebbed, mathematicians could more easily espouse other philosophies of mathematics. Two, Maximilien Winter and Eduard Study, chose to argue the case for mathematics as part of science, and to defend science using what they took to be the methodology of science rather than metaphysical grounds. They were both, therefore, anti-Kantian, and both are interesting for their views on modern mathematics and on the philosophical positions of Poincaré. Winter was one of the small group who assisted Xavier Léon to found the *Revue de métaphysique et du morale* in 1893; after the First World War he ran the *Supplément* to that journal until his death in 1935. In 1914 Eduard Study was a professor of mathematics at Bonn with a particular interest in geometry. Cassirer's neo-Kantian book of 1910 will be discussed last because it has proved to be altogether more substantial and lasting than the other works.[279] Indeed, it may be argued that it provided the most satisfactory resolution of the contest between the ghosts of Leibniz and Kant.

4.9.4.1 WEYL

Weyl's essay "On the Definitions of the Fundamental Concepts of Mathematics" is an intriguingly well-constructed piece.[280] It opens with a nod to the idea that Euclidean geometry is, or was, a synthetic subject in which ideas derived by a process of abstraction and idealisation are subjected to logical analysis and reorganized axiomatically. But, said Weyl, it was not at all necessary to start with Euclid's axioms. Pieri had indeed shown, he said, that one could start with a few undefined terms and the primitive concept of a point being equidistant from two given points.[281] As an indication of what Pieri had done, Weyl briefly sketched how this concept had to be worked with so that one could give a criterion for three points to lie on a straight line. But, said Weyl, the whole idea that one idea be definable in terms of another required one to be clear when something was defined, and Richard's paradox (Weyl called it

---

[279] A fuller appreciation of Cassirer than can be attempted here will be found in chapter 6 of Friedman 2000b.
[280] Weyl 1910.
[281] See Pieri 1899b.

an antinomy) made that seem altogether problematic. He suggested, however, that it would be enough to work with only finitely many basic ideas and principles of definition, and that when this was done one could then handle whole infinities of objects.

Once it is clear what it is to define something, said Weyl, one must recognize that the objects that satisfy the definitions will not be unique. Just as there are synthetically defined objects that obey Euclid's axioms for plane geometry (on either the traditional or Pieri's approach) so too there is the Cartesian plane. The best that can be said is that the mathematical definitions and relations that constitute geometry on one approach are completely isomorphic to the definitions and relations that constitute geometry on the other; there is no meaningful way to tell them apart mathematically. They have the same sense but different meanings (*Sinn* and *Bedeutung*, Frege's terms, though he makes no mention of the irascible sage of Jena). This led Weyl to discuss the new mathematical technique of implicit definition by means of systems of axioms, which had the advantage, he said, of putting the key properties of the objects so defined up front, but gave no sense to them. That was not a problem for Weyl; instead, he acknowledged that one must also check that a system of axioms is free of contradiction.

Weyl now turned to the attempts to logicize mathematics (*Logisierung*). He nodded toward Cantor, Dedekind, and Weierstrass for their work reducing the real numbers to the natural numbers, and raised the idea that the logicizing might go on and cover all of mathematics. For this to happen, he said, a distrust of set theory that had been generated by the paradoxes would have to be overcome. Here, logical and mathematical inquiries were mixed up with psychology. Small finite numbers were understood, he said, citing Kerry and Husserl, by association with definite psychic acts involved in counting. True symbolic representations arrived with Cantor's extension of ordinal arithmetic to the transfinite, and for mathematical analysis one required full-blown set theory. He noted, without mentioning names, that some had taken refuge in the psychologically monstrous idea that we understand !!! as 3 by reference to the set of all sets that can be put in 1-1 correspondence with the set {!,!,!}, but it had recently be shown, he observed, that it could be incoherent to speak of the set of all sets having such-and-such a property. Rather, a better attempt at explaining the concept of number had been initiated by Dedekind and continued by Zermelo, who defined 0 as the cardinality of the empty set, ∅, 1 as the cardinality of {∅}, and so on.

To handle all of mathematics, Weyl said, turning to his conclusion, the leading candidate was Zermelo's set theory, with its use of the symbols " = " for equality and "∈" for membership. He noted that Zermelo's definition of the term *definit* was a problem, and he repeated his idea that the problems inherent in Richard's paradox could be avoided if one used only finitely many concepts and principles of definition, and used the symbols = and ∈ only finitely often. But he admitted that this approach would be difficult to make work in the absence at this fundamental level of a concept of "finite."[282] Moreover, even if this logicizing succeeded, he felt that mathematics would still need to be comprehended intuitively, and that its applicability was a symptom of its rootedness, not its applicability. The proud tree of mathematics, he concluded, with its broad crown unfolding in the ether and its thousand roots

---

[282] He tried to do this later, but abandoned the attempt when it became too complicated.

drawing strength from the soil of intuition and representation, should not be cut back disastrously with the scissors of a petty utilitarianism.

These are lofty words, indicative of Weyl's justly earned later reputation as one of the great writers of mathematics, but nineteenth-century piety all the same. Indeed, the whole essay is much less radical and original than the same author's views after the encounter with Brouwer and the dramas of the First World War. It is, however, a curious mixture of the views of the later Weyl—sympathetic once again to Hilbert's attachment to Cantor's paradise of set theory and more convinced than ever that the meaning of mathematics resided in its place in science—and the ideas of the day. Implicit definition and the power of axiomatization is highlighted, set theory is the way to ground a fully logical analysis of mathematics, there are problems with set theory that are serious but not, apparently, fatal, and a remedy is suggested. What will be produced, if the whole enterprise is successful, will be a system of relationships between objects known only up to isomorphism: mathematics will be the study of sets and structures. It is a wholly modernist perspective, down to the names of the leading figures in the enterprise, that is by no means disdainful of science but meets it as an independent equal.

4.9.4.2 WINTER

In 1911, Maximilien Winter published his *La méthode dans la philosophie des mathématiques* (Method in the Philosophy of Mathematics). He began by noting that Hilbert and Peano had presented theories of geometry and of number without entering into philosophical discussion at all, but he wanted to go beyond such simple positivism and examine the metaphysical principles that could serve as foundations for science. He did not want to ground science in vague and indefinable notions, such as Couturat had done when speaking of "rational" ideas of unity, plurality, and quantity. Instead, he advocated a careful attention to the nature of definition (here he discussed Burali-Forti's work on nominal and abstract definitions), a modest use of logic and grammar, and augmenting a static science of principles with an account of change (thus, as he put it, secularizing transcendental doctrines). Each of these suggestions was then put to work in the three chapters of the book, with the overall hope of understanding Science by using only the methods of science. The third of these, however, on "a science of which they seemed almost completely ignorant: the higher theory of numbers," followed by an account of the theory of equations and Galois theory, does not escape the technicalities and can be ignored.

Winter immediately rejected Kant's ideas, which he claimed were no longer capable of supporting contemporary science, logic in particular. But this, said Winter, had long since been recognized by the neo-Kantians, Cohen and Natorp. However, Natorp hung his philosophy from a peg called the "original idea," something Winter found altogether too Hegelian, and which in any case delivered not a single scientific property. Faced with all the advances in non-Euclidean geometry and the geometry of any number of dimensions, the original Kantian position that the three dimensions of space were apodictic and a necessity of thought was plainly untenable. While Natorp accepted this, he distinguished these truths of transcendental logic from a logic of existence. The logic of existence provides for only one space, which is the a priori three-dimensional Euclidean one that is a precondition of experience. Worse, Natorp had presumed to quote Poincaré in his support and in opposition to the idea that geometrical empiricism made any sense. But, said Winter, quoting Poincaré, experience is a useful guide, which it could not be if a priori form imposed itself on us. To cut short this interminable quarrel, as Winter called it, it was best to note that Gauss, Riemann, and Hilbert had not sought to rejuvenate this metaphysics, but to present mathematical methods. These methods created many geometries, and so one had to investigate the relation of the postulates of geometry to experience. Here Winter counseled caution: experience is only approximate and will only provide us with alternatives to choose between, never a unique answer, so we must resign ourselves to some uncertainty.

Winter now turned to logistic, as he called the enterprise of Russell and Couturat and which had culminated in *Principia Mathematica*. Having rejected Kant, he was no hard-line Leibnizian. He found logistic, when taken in modest doses, a valuable method for the critical analysis of science, but taken as something that aspired to absorb all human thought it became as sterile a scholasticism as anything from the Middle Ages. The paradoxes into which it then led were pathological cases, and to take them too seriously was to reverse the importance of the healthy and the un-healthy. It is one thing to analyze the forms of language that enter mathematics, and another to pretend to reduce all mathematics to some elementary forms. This chapter struck Abel Rey, in his review of the book for the *Revue générale de philosophie des sciences* (vol. 73, pp. 629–50) as the most incisive.

4.9.4.3 STUDY

Eduard Study was a more substantial mathematician, and one not afraid of controversy. A rigorous mathematician in his own right, he did not oppose methodological purity, but preferred to advocate a wide variety of methods; for him it was better to be fruitful than pure. His *Die realistische Weltansicht und die Lehre vom Raume; Geometrie Anschauung und Erfarhrung* (or, The Realist Worldview and the Study of Space; Geometry, Intuition, and Experience) of 1914 likewise wished to defend mathematicians (Gauss, Riemann, Helmholtz) and their achievements against the misunderstandings of philosophers. In this case the hostile philosophers were idealists (for Study, Kant, and the neo-Kantians), positivists (Ostwald the chemist, Hume, and Comte), and pragmatists (Poincaré among them), who consider any science only insofar as it is useful; and Study compared his enterprise to that of Erdmann (1877), which he said was now out of date.

Study's realism asserted that the space of physical objects exists independently of the knowing subject. Realists, he said, know about it by abstraction, their knowledge of this as certain as anything can be that is not mathematics (p. 7). It emerges that hypotheses play a crucial role in Study's realism, and the criterion for truth and falsehood of these hypotheses should be presented as separated as far as possible from the idea that there are absolute criteria for truth. Rather, the realist speaks of hypotheses as truthful or realistic, and means that the hypotheses coincide with the things in the world (p. 19). Indeed, Study's distinction between realism and pragmatism would have been more interesting, because of his emphasis on the role of hypotheses, had he not painted it in such utilitarian terms.

Study's opposition to the neo-Kantians, Cohen and Natorp, followed Winter's very closely and shall not be discussed further. Where he said something new was in his discussion of what he called "natural geometry," by which he meant something that is simultaneously n-dimensional geometry, a branch of pure mathematics and as such a free creation of the mind, and also a true picture of the space in which we live, and as such is preexistent and an object of our knowledge. This is described as a system of abstract ideas and theorems that possess not only the "reality" of mathematics (Study's quotation marks) but the reality of physical bodies. This is offered as an unproven but very plausible assumption (p. 58). We arrive at this system in the same way we arrive at any body of science: through experience and reflection.

Study explained the marriage of imprecise experience and precise mathematics by arguing that items of experience are connected by hypotheses, whose purpose he said, echoing Wundt, was to bring out the logical connections between facts. Such hypotheses include the possibility of causal laws, the assumptions needed to proceed with geology, and so forth. There should be no restriction placed on hypotheses in advance, but science preferred appropriate and simple ones. A hypothesis may be found to be consistent or inconsistent with the current state of knowledge, but it is a daring act to say that something is not the case and still bolder to say that something *is* the case.

The geometry of space was a key issue for Study. He defined a geometry as Klein had done in his Erlangen Program, which he felt was insufficiently appreciated.[283] This generated many systems of geometries, from which, using the standard argument about rigid bodies, he selected the constant curvature geometries, and which he now tested against empirical observations. Schwarzschild's discussion of parallax (mentioned above, §3.1.4) led him to the conclusion that the curvature of space was within 6.10^{-24} of zero. Mathematics per se, however, was not yet tightly constrained, but in line with his views on hypothesis formation Study set his face firmly against purposeless speculation. He was left with spherical geometry, elliptic geometry, Euclidean geometry, and non-Euclidean geometry. He concluded that the curvature estimate allowed us to say that space is Euclidean to all intents and purposes, certainly in all but very large regions.

While Study found some of the objections to his position to be mathematically worthless, he could not dismiss Poincaré's geometric conventionalism on such grounds. He opposed Poincaré's conclusions by noting, first, that in geodesy one has no problem in saying that the geodesics on the (spherical) Earth are arcs of great circles, and that the assumption of a spherical Earth is the simplest we can usefully make. But, he said, it is the same with the geometry of space and unreasonable to assume that the Euclidean hypothesis is the simplest. Poincaré, he observed, had opposed no less than Gauss, Riemann, and Helmholtz when he claimed that geometry was not an experimental science. What is at stake is not the role of empiricism in geometry but the role of geometry in empiricism. Alas, Poincaré's authority had helped spread his error—"how beautiful it would be if true thoughts would spread even half as fast!" (p. 119n)—to Natorp, Müller, Cohn, and Cassirer. But Cassirer's *Substance and Function* was almost correct, although idealist (Study suggested eclectic) in noting that the role of experience does not lie in founding a single system but in choosing between them.

Study concluded his book with an unusual and, one might say, reactionary account of axiomatics in geometry. Hypotheses, not axioms, are appropriate in re-

---

[283] A footnote directed the reader to Enriques's article in the *Encyclopädie der Mathematischen Wissenschaften* and its account of the Helmholtz-Lie space problem.

search. The natural philosopher need not consider the recent outbreak of non-Archimedean, non-Pascalian, and other geometries constructed axiomatically. While the pure mathematician may marvel at them, the fruitful path remained that of analysis, and Study preferred the arithmetization of analysis and the arithmetic models of Euclidean and non-Euclidean geometry. Hilbert, he said, and Schur in his *Grundlagen der Geometrie* (1909), had proposed a logical analysis of spatial intuition, which they claimed was geometrical when it agreed with the simplest results of observation. But Study found only a system of abstract theorems. His sympathies were more with Pasch and his formulation of geometry back in 1882. It could seem, he said, as if Pasch's *Grundsätze* lacked mathematical content these days, but they should be seen as hypotheses. To be sure, modern geometry was not geometry as Gauss, Lobachevskii, Poncelet, and others had practiced it, but it seemed to Study that a purely axiomatic geometry was a utopia. When such a geometry said, "Let there be objects that have this or that property," one also had to ask: "Are there such things?" The axiomatizers knew very well that intuition and experience could not say, and they turned to analysis for their answers. When the question of the freedom of analysis from contradiction arose and analysis was treated as a branch of pure logic, there was no alternative but to assume it.

Study's independent position was offered as that of the working mathematician, more interested in solving problems than in advancing an ideology. It points up by contrast the view of the mathematical modernist, who asserted a higher degree of autonomy for the pure subject. However, this point also makes clear the opportunity Study missed to discuss how the modern view can enrich the analysis of experience. Study's hypotheses work underground, in a spirit of anything that works, and he presented no critical analysis of this procedure. Hilbert, by contrast, had a much clearer idea of the way an abstract, axiomatic analysis of the mathematical formulation of a science could simply result in better science. Study's realism might have been an account of the production and use of several theories, but he believed in it too strongly to want, or see the need, to take it apart and put it back together again. The old-fashioned viewpoint is also clear in the complete lack of mention of Einstein, and the brief mention of Minkowski's space-time geometry lacks any discussion of its significance for physics.

Einstein, however, read Study's book in September 1918 and wrote to him to say that he liked it. Study replied to ask about points where Einstein disagreed with him, and Einstein replied on September 25:

> I am supposed to explain to you my doubts? By laying stress on these it will appear that I want to pick holes in you everywhere. But things are not so bad, because I do not feel comfortable and at home in any of the "isms." It always seems to me as though such an ism were strong only so long as it nourishes itself on the weakness of its counter-ism; but if the latter is struck dead, and it is alone on an open field, then it also turns out to be unsteady on its feet. *So, away with the squabbling.* "The physical world is real." That is supposed to be the fundamental hypothesis. What does "hypothesis" mean here? For me, a hypothesis is a statement, whose *truth* must be assumed for the moment, *but whose meaning must be raised above all ambiguity.* The above statement appears to me, however, to be, in itself, meaningless, as if one said: "The physical world is cock-a-doodle-doo." It appears to me that the "real" is an intrinsically empty, meaningless category (pigeon hole), whose monstrous importance lies only in the fact that I can do

certain things in it and not certain others. . . . I concede that the natural sciences concern the "real," but I am still not a realist.[284]

Einstein was torn, here as elsewhere, between the physicist's natural desire to describe the world accurately, even correctly, and his sense that theory building was a creative act involving many choices, an activity that was very far from being determined uniquely and easily by the experimental evidence. So one can do certain things but not others, but one is not describing the universe as it really is. One is constructing a theory, and that is a different matter. His use for philosophy was that of the reflective scientist who finds that philosophers have spelled out the consequences of this or that position, without ever reaching secure, fundamental conclusions. Rather, philosophy is at its best when it deals with disagreements and conflicts. Thus, those Study opposed also had some merit, as Einstein continued:

> The positivist or pragmatist is strong as long as he battles against the opinion that there [are] concepts that are anchored in the "A priori." When, in his enthusiasm, [he] forgets that all knowledge consists [in] concepts and judgments, then that is a weakness that lies not in the nature of things but in his personal disposition just as with the senseless battle against hypotheses, cf. the clear book by Duhem. In any case, the railing against atoms rests upon this weakness. Oh, how hard things are for man in this world; the path to originality leads through unreason (in the sciences), through ugliness (in the arts)—at least the path that many find passable.[285]

4.9.4.4 CASSIRER

Cassirer had first replied to the logicists in his paper "Kant und die moderne Mathematik," published in *Kantstudien* in 1907. He did not dispute that modern accounts of logic had gone a long way to making the reduction of mathematics to logic plausible. Rather, his disagreement with Couturat and Russell was that even a successful reduction would not show that mathematics was analytic. The issue is what mathematical knowledge is knowledge of. As a Kantian, Cassirer argued for the existence of synthetic a priori knowledge, that is, knowledge not analytic and yet necessary.[286] Even if all mathematics were deducible from a suitably enriched logic, it remains the case that mathematics plays a major role in science, and scientific knowledge is empirical, it increases, and it cannot therefore be analytic. Or, as Cassirer put it: "Only when we have understood that the same fundamental syntheses on which logic and mathematics rest also govern the scientific construction of empirical knowledge, that they first make it possible for us to speak of a fixed lawful order among appearances and thus of their objective meaning—only then is the true justification of the principles [of logic and mathematics] achieved."[287]

One might say that the logicists hoped for a reduction of mathematics to logic, but Cassirer wanted an account of mathematical knowledge. Or, that mathematics was

---

[284] Einstein 1997, vol. 8, doc. 624; all emphases in original.

[285] Einstein 1997, doc. 624.

[286] Recall that synthetic for Kant simply meant not analytic, and analytic that the predicate is contained in the subject.

[287] Cassirer 1907, 44–45, quoted in Friedman 2000b, 92–93.

pure mathematics for the logicists, but for Cassirer (and others) it was also applied mathematics, including empirical geometry.

Cassirer greatly amplified these views, and drew out their implications for epistemology, in his *Substanzbegriff und Funktionsbegriff* (or, *Substance and Function*) of 1910, which is arguably the most lasting and cogent of all the works of this kind of its period. He emerges here as an exceptionally well read and sensitive thinker. Even to list the people and ideas this book discusses in its opening one hundred or so pages is to delineate much of the history and philosophy of mathematics discussed in later works (this one being no exception): Mill, Frege's foundations of arithmetic, and those of Dedekind, Helmholtz and Kronecker, and of Russell; reductionist concepts of negative and irrational numbers; projective geometry, including the work of Cayley and Klein; pure geometry (Hilbert), and meta-geometry (Pasch). Then come a hundred pages on the philosophy of science (from Plato to Rickert) and as many pages again on relational concepts and the concept of reality. There are gaps: Riemann and Poincaré are mentioned only in footnotes, and Einstein is not discussed— but amends were made with a second book in 1921.[288] One might even say that the gaps have been transmitted quite often, as well.

The purpose of *Substance and Function* is to respond fully to the new developments in logic that, in Cassirer's opinion, had forced a reconsideration of Kantian philosophy. The problem is to make sense of conceptual thinking, especially in mathematics and the sciences, and, therefore, first of all, to be clear about what a concept is. The answer in mathematics will be different from the case of science, because, as is often remarked, there are no actual points, straight lines, and mathematical planes in nature from which one can abstract the relevant concepts, whereas there are many natural objects of various kinds. Taking his cue from the eighteenth-century philosopher and polymath Johann Heinrich Lambert, Cassirer argued that the crucial feature of a mathematical concept is that the particular is contained in, and can be recovered from, the general. In his example, the equation of a circle can be recovered from the general curve of the second degree by the right choice of parameters. There is no process of abstraction from a variety of examples, as there is in science, but what Cassirer rather confusingly called a principle of serial order that generates individual members from a general expression. It seems to me that nothing is lost and something gained in our understanding of Cassirer's work on replacing the terms "serial order" and "serial form" by the term "law defining a set," and "manifold" by "set," so I shall do so. A law, and even a set determined by a law, is not the same thing as its elements, and can only be determined, defined, and distinguished from others (p. 26), "by a synthetic act of definition, and not by a simple sensuous intuition." The reference to synthesis is not only to make clear that in forming a concept we are not engaging in an act of intuition, but to show that the nature of the necessity in reasoning about sets will be appropriate to the a priori, not the analytic.

We may then "quote" Cassirer as describing the purpose of his book in these terms (p. 26):

> The totality and order of pure laws defining sets lies before us in the system of the sciences, especially in the structure of exact science. Here, therefore, the theory finds a

---

[288] *Zur Einstein'schen Relativistätstheorie*, translated as *Einstein's Theory of Relativity* and published with *Substance and Function* in one volume in 1923.

rich and fruitful field, which can be investigated with respect to its logical import independently of any metaphysical or psychological presuppositions as to the "nature" of the concept. This independence of pure logic, however, does not mean its isolation within the system of philosophy. Even a hasty glance at the evolution of "formal" logic would show how the dogmatic inflexibility of the traditional forms begins to yield. And the new form that is beginning to take shape, is also a form for a new content. Psychology and criticism of knowledge, the problem of consciousness and the problem of reality, both take part in this process. For in fundamental problems there are no absolute divisions and limits; every transformation of the genuinely "formal" concept produces a new interpretation of the whole field that is characterized and ordered by it.

Plainly, Cassirer's Kantianism is to be free of the legalistic maneuvers that have made the work of his contemporary neo-Kantians seem so dated.

His presentation of the neo-Kantian critical theory was less contrived than those of his peers because it was not so preoccupied with abandoning intuition while dogmatically keeping the rest of the Kantian edifice. Rather, it was an attempt to rethink the whole of the theory of knowledge in the expectation that many of Kant's fundamental ideas would survive. What Cassirer regarded as the "copy" theory of knowledge—the view that our correct ideas are true because they copy a relationship between some (inaccessible) substances—was in his view mistaken and based on inadequate, Aristotelian foundations. Critical theory sought a way of anchoring empirical knowledge in ideal, formal mathematics. This explains the prominence Cassirer gave to mathematics in his theory of knowledge. To this end, the logical elucidation of mathematics was welcome, as was Helmholtz's theory of signs, which Cassirer regarded as more formal, and less pictorial, than some other writers. Critical theory's provisional, evolving character explains his further interest in the history of mathematics.

He presented geometry in a truly Kleinian spirit, through its evolution in the nineteenth-century, as having taken a "final and decisive step" (p. 88) with the addition of the theory of groups. Pure geometry (that is, projective geometry) has realized Leibniz's ideal as being about the order of possible coexistences (p. 91) in which meaning is conferred by axioms, and indeed a further Leibnizian ideal, whereby mathematics is a science of quality not quantity, is approached. This is clearest in Hilbert's procedures by which "the nature of the original geometrical objects is here exclusively defined by the conditions to which they are subordinated" (p. 93). Cassirer noted approvingly Wellstein's characterization of Hilbert's geometry as "a pure theory of relations,"[289] and went on to say that "the peculiarity of the [mathematical] method is bound and limited by no particular class of objects" (p. 95).

All this led, after some pages on Grassmann and Cohen, to "the problem of metageometry." Cassirer insisted that mathematicians and philosophers agreed that there had been a justified extension of the field of geometry; the question was whether "the new content breaks through the logical form of geometry or confirms it" (p. 101). Cassirer argued that it was confirmed. He noted the old view that metageometrical considerations determined the empirical content of geometry (meaning that Euclidean geometry had to be true). But now, he said, citing Veronese's *Grundzüge der Geometrie von mehreren Dimensionen*, which he hailed as "the first complete historical survey of all critical attempts to reform the theory of the principles of

---

[289] Wellstein 1905.

geometry," the consensus was that at least ordinary geometry is founded merely on experience.

However, this consensus would not stand up to scrutiny. The best attempt to found geometry entirely on empirical considerations was Pasch's (1882), and it foundered because the price of rigor was the introduction of nonempirical ideal elements (geometric points and lines). In responding to this deference of approximate geometry to pure geometry, Veronese had sought to base geometry on "mental facts," or postulates. This also allows for higher-dimensional geometries and other novelties. Cassirer now insisted that this logical freedom also apply to the concepts and postulates of ordinary Euclidean geometry.

This led him to the Kantian albatross: the supposed special, intuitive space of ordinary geometry. Cassirer argued that we have a collection of geometries to choose among. However, there can be no decisive experiment to choose between them, for reasons that (unnamed) philosophers had long argued but which had first gained wide acceptance with the writings of Poincaré. Other rational criteria must therefore be sought, and logical consistency is possessed by all the candidate systems so it will not serve. What did, for Cassirer, was uniformity. The fact that Euclidean space is uniform, by which he meant that it held infinitesimally, whereas non-Euclidean space is not (in the infinitesimal it is Euclidean), allowed him to proclaim Euclidean geometry as the simplest, and therefore that it had "a peculiar advantage in value and significance" (p. 109).

This concedes more of Kant's original position than Cassirer was willing to state. Kant had allowed the mathematician, but not the philosopher, to proceed immediately by a simple construction in pure intuition to deduce a theorem that is only true of Euclidean space. Cassirer allowed his much more richly equipped mathematician any number of geometries, from which the meta-geometer, or indeed the philosopher, is free to advocate one on grounds of simplicity that, ultimately, is an irresolvable matter of taste.[290]

Cassirer returned to these themes as part of his final work, specifically in the book *The Problem of Knowledge: Philosophy, Science and History since Hegel*, which he wrote in Sweden in 1940. It was first published in English in 1950, after his death. The first part of this book deals with exact science and largely recapitulates Cassirer's earlier ideas, with a greater openness to Poincaré's geometric conventionalism, and makes some remarks about intuition as it is expressed by Kant and Klein, and about the intuitionism of Brouwer and Weyl. It is quite noticeable how much his choice of mathematicians is a roundup not only of the suspects from his earlier writings but how familiar they have become, right down to Mehrtens's book and beyond.

To the end of his life, Cassirer saw the discovery of non-Euclidean geometry as leading inevitably to the destruction of the idea that space was a substance that geometry had as its subject matter. He placed Klein's Erlangen Program at the epicenter of this transformation, with Hilbert's *Grundlagen der Geometrie* and Poincaré's geometric conventionalism as the epitomes of the purely relational view of geometry and of its application to the study of space. This made Cassirer an avowed idealist, but he avoided nominalism by bringing space back, not as it had been seen as the ground of knowledge, but as a problem and as the goal of knowledge. He noted

---

[290] Cassirer only a few pages later speaks quite warmly of du Bois-Reymond's dualistic philosophy, so it seems reasonable to insert such a note here, too.

that the freedom and ability of mathematicians to create continuous but nowhere differentiable and space-filling curves further emphasized the new nature of mathematics, but his philosophical focus led him to seek a tradition that also gave this mathematics a history. He found it by seeing the hints and fragmentary remarks made by Leibniz as indicative of the way mathematics was to go, perhaps even *had* to. As Friedman has rightly noted, Cassirer interpreted mathematics as exemplifying Leibniz's universal characteristic and came close to the logicist idea that mathematics is an outgrowth of logic.[291] But he argued explicitly that the construction of the various types of number was a synthesis, not a process of reduction, and so on this occasion he judged Kant's views to have prevailed.[292] As for intuition, as a specifically Kantian faculty, that seems to have diminished in Cassirer's opinion to Klein's much more psychologically enabling kind, despite some cautionary remarks from Leibniz that Cassirer quoted near the start of the book.

Coffa has argued that Cassirer's views on the role of geometry in the study of space were outdated by the 1920s, and that Einsteinian relativists can be as absolutist about space-time as Newton ever was about space, seeing it as a substance to be understood.[293] This criticism is true in proportion to the amount of Kant one still detects in Cassirer's philosophy of geometry, and that may depend on the latent degree of realism in the reader. This reader, at least, finds a gap between the philosophy of mathematics Cassirer presented when he was interpreting the views of mathematicians, and the philosophy of mathematics he presented when trying to connect them to the philosophies of Leibniz and Kant. This also shows up in his priorities: Riemann's importance is underestimated, and for all the range and subtlety of Cassirer's writing a unity of vision among the mathematicians is suggested that exaggerates what was there.

That said, Cassirer can be appreciated as one of the last philosophers to take the transformations of mathematics that he saw happening around him as significant for the development of philosophy as a whole. After him, the philosophy of mathematics took its deep but highly technical turn to mathematical logic, and philosophers, following Wittgenstein, found their intellectual problems in the very possibility of (elementary) mathematics. Cassirer's identification with the problems raised by non-Euclidean geometry reflects his origins among the neo-Kantians, just as his writing about Einstein reflect his later position with respect to the Vienna circle. But he saw very clearly that leading mathematicians around 1900 knew that their subject had become markedly more abstract and conceptual, he read everything relevant that a well-educated German would read, and he drew out its philosophical significance in ways that addressed the questions he took to be fundamental. Seen historically, he may serve as the last attempt of the Kantians to match the insights of their master to the advances of mathematics and science, while those in Vienna who came after him were able to start to one side of the long-running debate between Kant and Leibniz.[294]

---

[291] See Friedman 2000b, 108.
[292] Cassirer 1950, 76.
[293] Coffa 1991, 193–220, esp. 196–197.
[294] Schlick and Carnap are discussed briefly in the final chapter; much more can be found in Coffa 1991 and Friedman 1999.

4.9.5 Brouwer

Few have an importance for the development of mathematics and its philosophy in the twentieth century to match that of the Dutchman Luitzen Egbertus Jan Brouwer, and perhaps none have his complexity. He was a strong, independent character, quite capable of defending his various positions yet liable to emotional collapse. He was a powerful, original mathematician with many lasting achievements in topology to his name, yet driven by deep, often mystical, philosophical convictions. He lived a life so consistent with these convictions that it excited the admiration of Hermann Weyl, yet he was investigated for suspected Nazi sympathies after the Second World War. He developed an alternative formulation of mathematics to the standard one, but this intuitionism contradicted many of his best-known mathematical results.

Brouwer was born in 1881 and studied mathematics at the University of Amsterdam, where he was fortunate to attract the support of the mathematics professor D. J. Korteweg, and to meet one of Korteweg's protégés, Gerrit Mannoury, who in 1903 became a *privaat docent* in the logical foundations of mathematics at the university. Mannoury sustained Brouwer's interests in philosophy and language, he was a prominent leftist and a writer and pamphleteer who, Brouwer's biographer has said, "was, in a sense, in an extracurricular way, the mentor of Brouwer."[295] In 1904 Brouwer married and settled in Blaricum, a village not far from Amsterdam that was home to a number of artistic and spiritually inclined communities, including one that called itself Walden Two and lasted from 1899 to 1904; the painter Mondriaan lived in nearby Laren. Also in 1904 Brouwer began his doctoral thesis, and in 1905 he published his philosophical booklet *Leven, Kunst en Mystiek* (*Life, Art, and Mysticism*). So controversial was this work that it was to be omitted from Brouwer's *Collected Works*, despite its undoubted importance for understanding his ideas; an English translation was published for the first time not until 1996.[296]

Life, Art, and Mysticism does indeed expound a number of views generally unacceptable in academic circles a century later:[297] the subordinate role of women, an anguished Protestant religious sensibility, a cultivated obscurity of language (not, alas, foreign to academic prose today), romantic historical inventions, and a mystical hostility toward and renunciation of, the world.[298] Such views have their historical roots and their psychological resonances. But the booklet also sets out a position on the intellect and on language that would animate Brouwer all his life. It begins with his observation that the construction of the canals was the cause of Holland's problems with the sea, because the canals interfered with Nature. Chapter 3, entitled "Man's Downfall, Caused by the Intellect," argues that the intellect forces man to live in desire and fear. Chapter 5 called language the "immediate companion of the intellect" and observed its inadequate nature. A modest degree of communication is possible in mathematics and logic—"a sharp distinction between the two is hardly possible," he said (p. 401)—"but even then two different people will never feel them

---

[295] Van Dalen 1999, 260.

[296] See *Notre Dame Journal of Formal Logic* 37.3: 391–430.

[297] See the discussion in van Dalen 1999, chapter 2, and Hesseling, *Gnomes in the Fog*, chapter 2.

[298] We have been denied a New Age author with anything like Brouwer's mathematical talent, but his booklet does bring to mind the ideas that animated Grothendieck in the 1970s.

in exactly the same way." For other purposes, language, in Brouwer's view, is ridiculous.[299]

Brouwer plunged into the work for his doctoral dissertation, and therefore into original mathematical research, with a remarkable eye for important topics. He selected for his attention no less than three of Hilbert's problems, the first, second, and fifth. To take them in reverse order, the fifth is a problem about the topology of transformation groups, and Brouwer's account, amounting to a partial solution, was good enough to be published. The second asked for a proof of the consistency of arithmetic, and the first was the continuum hypothesis; almost incredibly, Brouwer claimed to have solved them. To understand what he meant, we must turn to his interpretations of them, which were unusual but indicative of the future direction of his work. He also wished to discuss a number of philosophical issues, but Korteweg refused to let him have his way completely, and the tone of Brouwer's remarks is considerably more moderate than it had been in *Life, Art, and Mysticism*. Here we learn that Brouwer now endorsed Kant's idea of time as a synthetic a priori intuition, but not that of space, which seemed to Brouwer to be no longer tenable after the elaboration of non-Euclidean geometry. Instead, for Brouwer, the image of space is a free act of the intellect.

Brouwer was moving to a hard-line position that said that a mathematical object exists if and only if an explicit construction for it is given. At this stage in his work, Brouwer felt that certain mathematical operations having to do with infinite sets were unintelligible. Mathematics was therefore restricted to the realm of the constructible, and being constructible was therefore consistent. As for the continuum hypothesis, Brouwer's continuum was a simpler thing than Hilbert's, and it was not clear to Brouwer that the sets Hilbert talked about existed (one might even say that it was clear to him that they did not). Brouwer denied the existence of sets larger than the denumerably infinite. Brouwer's position was closer to that of Poincaré, whose philosophical and popular works he read and mostly agreed with, but he took a harder line on constructibility and existence. Brouwer also had a very low opinion of logic, and felt strongly that logicians talk about mathematical construction whereas the real mathematical work of construction is wordless.

Shortly after his thesis was successfully defended, Brouwer began to work out in detail what mathematical arguments he was prepared to accept and what, in his view, made no sense. He came to believe that the principle of the excluded middle was invalid for infinite sets. This principle asserts that for any given statement S either it or its negation is true. A statement about all the elements of a finite set is in principle capable of being checked element by element until the set is exhausted, so the principle of the excluded middle necessarily applies to it. Hilbert had built a career on the use of the principle with infinite sets, analyzing them along these lines. Does such-and-such a set have an element of such-and-such a kind? Suppose that it does not. If a contradiction follows, then the set does have at least one element of the required type, even though no construction to exhibit the element may be at hand. Brouwer disagreed. He felt that some mathematical statements might be neither true nor false. And he did so without regard for the consequent damage to mathematics.

---

[299] Brouwer's attitude to language is discussed further in §6.1.6.

For example, the intermediate value theorem asserts that a continuous function defined on the real numbers between, say 0 and 1, that takes a negative value at 0 and a positive value at 1, must take the value zero somewhere in between. The proof is not constructive, and Brouwer therefore rejected it in favor of a constructive statement (given any $\varepsilon > 0$ there is an x in the interval such that $f(x)$ is within ε of 0). But this constructive statement does not permit one to talk of an x such that $f(x) = 0$, because such an x has not been shown to exist, and this is a deep hole to punch into the body of mathematics.

The branch of mathematics upon which Brouwer first worked intensively was topology, and one notes that the first paper on topology published in a Dutch journal was written by Mannoury. Paradoxically, Brouwer's best results in topology, the ones for which he is remembered to this day, are not constructive. To give but two examples, Brouwer generalized this result, where I denotes the set of real numbers between 0 and 1 (the set of real numbers x such that $0 \leq x \leq 1$): for any continuous map f of the set I to itself there is a number x that is mapped to itself: $f(x) = x$. This result is immediate from the intermediate value theorem applied to the function $f(x) - x$. But now consider the unit ball B_n (all the vectors in Euclidean n-dimensional space of length 1 or less. Brouwer showed that any map f from B_n to itself also has a fixed point.

The second example illustrates the power of topology to rescue mathematicians' intuition. The results of Cantor and Peano described above (§§3.2.2 and 4.4.3) had made it seem possible that there was no deep distinction between the line, the plane, and indeed any Euclidean n-dimensional space. Brouwer was able to define the dimension of a space and show that if there is a continuous map of one space to another that has a continuous inverse, then the spaces have the same dimension. This result, called rather obscurely the *invariance of domain*, was one of Brouwer's finest achievements, but it too is nonconstructive.[300] Brouwer proved a number of profound results like this, on the way finding numerous errors in then-standard work on topology (the report by Schoenflies). All of them are much more powerful in a nonconstructive setting than in a constructive one.

Brouwer's philosophical position on mathematics is known as intuitionism. It seems that the first time Brouwer publicly used the word to describe his own position was in his review, published in 1911, of his friend Mannoury's book about elementary mathematics.[301] Mannoury there distinguished between symbolism, as he called the style of Russell, Hilbert, and Zermelo, and Kantianism, the style, he said, of Poincaré and Borel.[302] Brouwer changed the names, and reviewed the debate as being between formalists and intuitionists. Hesseling, in his study of the reception of Brouwer's intuitionism,[303] suggests three ways in which Brouwer might have hit upon this name. One is from his reading of Poincaré, Kant, and even Schopenhauer. Another is directly from Klein or Bergson, and the third is from ethics where, intriguingly, intuitionism is described and contrasted with formal and logicist positions.

---

[300] In symbols, if f is a continuous map from a space X to a space Y and g is a continuous map from the space Y to the space X with the property that whenever $f(x) = y$, it is the case that $g(y) = x$, then the spaces X and Y have the same dimension.

[301] Mannoury 1909.

[302] Mannoury sided with the symbolists.

[303] Hesseling 2003, 53.

Brouwer called the symbolists "formalists" because the word already had negative overtones in some circles and alleged that they incorrectly identified mathematics with mathematical language, and, following Poincaré, that they needed the intuitive use of the principle of complete induction, so were surrendering to the intuitionists, anyway. But, as Hesseling points out (p. 56), even a year later, on the occasion of his election to the Dutch Academy of Sciences, Brouwer would still speak of these alternative philosophies of mathematics in almost neutral terms.

Brouwer, by 1912, was the leading point set topologist in the world. Had he confined his attention to mathematics, his fame and influence would have been assured. But he was driven by philosophy, and after the war he would to turn more and more to criticisms of mathematics that excited and appalled mathematicians in equal measure, until he came into a conflict with the aging Hilbert in the late 1920s that would prove disastrous for him.

5

FACES OF MATHEMATICS

5.1 Introduction

As mathematics advanced in the late nineteenth and early twentieth centuries, mathematicians fashioned for themselves a new image of the subject: autonomous, abstract, largely axiomatic, and unconstrained by applications even to physics. At the same time, they often valued and cherished the link to physics, whether from a genuine appreciation of, and interest in, science, or from a shrewd sense that the utilitarian arguments for science and technology could only benefit mathematicians if they were somehow conspicuously tied to those subjects. On the other side of the divide that was growing up, albeit in rather different ways in different countries, physics too was turning more theoretical and abstract, so much so that in 1915 the Nobel Prize-winner Wilhelm Wien could rejoice that "the great masters of theoretical physics have also been the greatest scientists" and claim that theoretical physics had now become a separate science that was joined to experimental science in a higher unity, so that he could speak of "the now mighty theoretical physics."[1] Physicists generally shared Boltzmann's rough and ready distinctions between mathematical and theoretical physics. The former was really mathematics, the latter had more physical content and stronger connections with experimental physics. As ever, physicists knew very well that for many of the problems they would like to understand, mathematics had as yet nothing to offer, nor did they think that resolution of such problems would necessarily require new, advanced mathematics. In this, when it came to the motion of the electron, Einstein was to prove them right. Nor did physicists believe that much more was required of mathematics than that it accommodate the much-needed, but lacking, physical insight; the theory of differential equations would most likely prove equal to all that could be asked of it.

Modernized mathematics, however, offered much more. It offered ways of creating and organizing ideas that were open to a variety of new interpretations, it

---

[1] Quoted in Jungnickel and McCormmach 1986, vol. 2, xv, from which much of the information below about Boltzmann and others is taken.

allowed for novel forms of mathematics, and generally it allowed for novel applications—even within physics. In the comparisons that follow in section 5.2, Maxwell, Hertz, and finally Einstein show how these possibilities were recognized by physicists; Riemann, Hilbert, and Minkowski show how mathematicians tried to reintegrate their subject with physics. Poincaré interestingly straddles the divide. A permeable boundary was created, based in the professional academic separation of mathematics and physics and reflecting the new constitutions of those disciplines.

Physics would not be possible without measurement, and the way numbers are assigned to physical quantities can seem either banal or a matter for the technician. But once mathematics was no longer, in any simple way, the science of quantity, mathematicians discovered whole realms of quantities with novel properties, as section 5.3 shows. Debates about what could be measured, and the kinds of "numbers" that the measurements could yield, provide a vivid illustration of the ways in which the emancipation of mathematics proceeded and of the deepening sophistication of ideas of measurement.

All this renegotiation of the boundaries between mathematics and related disciplines fed into popular debates about the subjects, and after a brief look at the field in general, section 5.4 offers an account of some of Poincaré's ideas and the response they drew from his most important interlocutor, the Italian Federigo Enriques. If nothing else, this examination should dispel the idea that Poincaré's essays swept all before them philosophically, save perhaps for the acute and critical reading Einstein gave them as a young man.[2]

5.2 Mathematics and Physics

5.2.1 On the Roles of Mathematics in Physics

Newton's *Principia* set a standard for the use of mathematics in questions of physics (or natural science) that in many ways still persists. Most famously, it showed, or rather it claimed very vigorously to have shown, that only an inverse square law of force could produce orbits consistent with observations of the planets, the satellites of Jupiter, and, less certainly, the Moon. In particular, the prevailing vortex theory could not simultaneously obtain all three of Kepler's laws. Thus was gravity defended, although Newton cheerfully admitted that he could not explain it, and while defenders of Descartes' vortex theory put up a vigorous defense for a time, the balance of the argument shifted decisively in Newton's favor, so that by the mid-eighteenth century the leading Continental mathematicians were winning prizes from the major scientific academies of the day for their confirmation of one after another of the merits of the theory of gravity. Indeed, one has the impression that so sweeping was the victory of Newton's physics and the advancing calculus that by the start of the nineteenth century the very idea of gravity had become only too well established, and the genuinely mysterious nature of a force acting over such considerable distances had been forgotten.

---

[2] Scott Walter has pointed out to me that acceptance of Poincaré's conventionalism seems to have been confined to France, most recently in his "The Cash Value of Conventionalism: Henri Poincaré on Geometry and Physical Relativity," International Workshop, History of Modern and Contemporary Geometry, Centre International de Recherches Mathématiques, Luminy, 28.08–02.09.2005 (30.08.2005).

What Newton showed so successfully was that hypotheses about physics can be tested in the laboratory of mathematics, and that when this is done the victorious idea may lack intuitive, physical sense. Yet Newtonian mechanics is straightforward in its assumptions about what are the physical quantities that enter the mathematics—position, velocity, and mass among them. The situation in any form of nineteenth-century electrodynamics was much more complicated, because the crucial physical quantities had to be discovered. The forces involved were more complicated, the question of their relationship to matter harder to decide, the connection to light difficult to elucidate, and, partly as a result, mathematicians loosened their grip on the idea that all quantities that enter the mathematical description should have a physical interpretation. A proliferation of theories resulted, and with the proliferating theories came a proliferation of the ways in which mathematics and physics were related.

It is unlikely there will ever be a convincing, rich description of the roles mathematics played in the development of physical theory in the nineteenth century. There is simply too much to say, and much that we would like to know now lies buried in archives if it is not lost altogether. Warwick's *Masters of Theory* stands as an example of what can be done for one context (nineteenth-century Cambridge); comparable studies are needed, and generally lacking, for the other major centers of research. Warwick's account makes clear that mathematical techniques were drilled into generations of students and were instinctive and habitual among the best of them when it came to research. Almost any page of Buchwald's earlier *From Maxwell to Microphysics* or Darrigol's *Electrodynamics* shows that the same must be true of practitioners in continental Europe and the United States. The studies by Jungnickel and McCormmach show how the profession of physics itself evolved, and how, in Germany, theoretical physics came to dominate the experimental side.

It seems clear from these and other studies, as indeed it does from the original works themselves, that by the nineteenth century physics was impossible without mathematics, and it is tempting to argue that mathematics entered in a naive, descriptive way. Certain quantities were naturally represented or described as functions or vectors, the physical interaction of this with that was naturally to be described by such-and-such an equation; formalism followed function. This is not to say that any of this was easy, routine, lacking in conceptual innovation, or that physics rather than mathematics was in the driving seat, whatever that might mean. Physicists could readily make such a separation if pedagogic need required it, so Thomson and Tait could announce that the object of their *Treatise of Natural Philosophy* (1879) was to give a tolerably complete account of the subject in language adapted to a non-mathematical reader, and consign material to mathematical appendices. But they nonetheless opened their two-volume work with two hundred pages of difficult mathematics, the better to meet their other objective, which was to give a connected outline of the analytical processes by which knowledge has been extended into regions as yet unexplored by experiment.

5.2.2 Maxwell

Mathematics, for all its sophistication, and for all the years of training needed to master it, was used in physics in a more or less straightforward way. For the British, this was often via mechanical analogies, as can be seen very readily in Thomson's case by looking at his famous Baltimore Lectures (as Lord Kelvin).[3] When treated in such a way, the physics leads directly to its mathematical account, not just because the language of mechanics is transparent in this way, but because the mathematics of potential theory was a suggestive way to formulate ideas to do with the ether and the wave theory of light. One of Lord Kelvin's targets by this time, 1884, was the much more elaborate methodology of Maxwell. As Harman describes in his essay appended to the reedition of the Baltimore Lectures, Maxwell had been impressed by an early reading of Fourier's *Chaleur* (1822) as well as by Kelvin's study of the partial differential equations that arise in that work. He was first led to model electricity and magnetism on fluid flow and later to a theory of molecular vortices and idle wheels. The molecular vortices were there to express his belief that there must be a genuine rotation in a magnetic field to explain its effect on light. The idle wheels moved around and corresponded to the flow of an electric current. So far, so mechanical, but the mathematics gradually took over, and when all this was brought to a high state of finish in his *Treatise on Electricity and Magnetism* (1873), more sophisticated methods were in play.

In this famous, and difficult, work, the forces of nature were equipped with the full force of contemporary mathematics—more precisely, with Lagrangian dynamics. Lagrangian dynamics was much appreciated, at Cambridge as elsewhere, because it was a very general and flexible method.[4] In keeping with the freedom built into that theory, the connection between the mathematics and the physics is now more attenuated, although as a physicist Maxwell always directed his readers to look again at the work of Faraday. Even so, no specific mechanical model is invoked to explain what the mathematics describes, and Kelvin was unhappy as a result and proclaimed himself unable to get "electromagnetics," as he said in Baltimore, Lecture 20. Maxwell's theory is a pure field theory: current is a transfer of polarization, charge a discontinuity in polarization, but polarization is a primitive concept incapable of reduction to the microscopic reduction of electric charge.[5]

Indeed, Maxwell argued forcefully that there can be no simple transcription of physics into mathematics. He invoked the metaphor of bell ringers pulling bell ropes that are connected not to bells but to a machine hidden in a room above their heads. The bell ringers may make careful study of what happens when they pull the ropes, couched in terms of potential and kinetic energy in the ropes, and as a result they may work out what they are doing to the machine and express their answer in terms of how it in turn pulls the ropes. But they cannot infer much about how the machine works; in particular, "If the machinery above has more degrees of freedom than there are ropes, the coordinates which express these degrees of freedom must be ignored. There is no help for it."[6] This is a position quite unlike Kelvin's. It is agnostic about fundamental physical processes—those not accessible to the bell ringers—and it invites the mathematical physicist to employ terms and concepts in the mathematical analysis that have no counterparts in the physical theory of observable or measurable quantities. These might correspond to hypothetical entities in the machine and de-

[3] See Kargon and Achinstein 1987.
[4] See, for example, Harman 1982.
[5] See Darrigol 2000, 173.
[6] Maxwell 1879, 784.

scribe the way the extra degrees of freedom work, but they must, ultimately, be ignored. They are arbitrary, and accordingly complicate any account of the relation of mathematics to physics.

This comparison between Kelvin and Maxwell indicates that mechanical models of electromagnetic phenomena inclined to the naive application of mathematics, but that existing mathematics, taken off the shelf by skillful hands, could also generate more elaborate explanations of physical processes. To see that mathematics could do even much more than that, and could disturb the ways in which mathematics can inform physics, it is necessary to go back some years, to the mid-1850s.

5.2.3 Riemann

Maxwell's electromagnetic theory was not the only one to distrust action at a distance as an explanation, and to seek to supplant it with some local process. Most famously, once the idea was accepted that light was a wave phenomenon, there was a reasonable amount of agreement that it was transmitted as a wave in an all-pervading substance called the ether. The properties of this ether were to prove ever more remarkable, but the ether gave substance to space. It became possible again to imagine mechanisms that carried the gravitational influence of one body on another. One who was strongly drawn to such questions for a while was the young Bernhard Riemann. When he was finishing his *Habilitation* in Göttingen in the early 1850s he was also working in the laboratory of Gauss's friend and colleague Wilhelm Weber on, he tells us, "the connection between electricity, light, and magnetism."[7] He began to sketch out a paper, which he intended for publication, drawing on his reading of the work of Newton, Euler, and, more surprisingly, Herbart. He transformed Herbart's plenum into a universe filled with a substance (*Stoff*) that flowed through atoms and out of the material world. He sketched out a geometric theory of how this Stoff behaved as if in, or perhaps as if it was, an elastic ether, according to which stresses and strains in the ether show up as deformations of the local metric. This variation in the metric would in turn be felt by a particle as a force, and by resisting this force the particle might move through space.[8]

To proceed further with this idea, Riemann distinguished two cases, one in which the particle resists changing its volume, the other in which a physical line element resists changing its length. He assigned gravity and electrostatic attraction to the first case, and light, heat diffusion, electrodynamic and magnetic attraction and repulsion to the second, but he found he could go no further and the paper was abandoned. Later, he returned to the theme of gravitation and light and tried to explain them via the motion of a substance spread continually through space. The substance would be physical space, moving in geometric space. In so doing, he followed both Newton, in his General Scholium, and Euler who, in his various papers and in his *Letters to a German Princess* (1770), had also hypothesized a "subtle spirit" or ether and charged it with conveying these physical forces. It is very likely that Riemann knew these sources, and that would explain why they are, with Herbart, the inspiration for this work. This time he made more progress and obtained a plausible theory by which

---

[7] Riemann 1990, 580.
[8] See the account in Bottazzini and Tazzioli 1995.

motion through finite regions of space could be explained purely in terms of relations defined infinitesimally, but again nothing was published.

Although Riemann's published papers on these topics tell a different story, in that they do not invoke an ether, Riemann did discuss the properties of the ether in his lectures of 1861, published by Hattendorff in 1876. Various ways in which the density of the ether varied would be responsible for electrostatic and electrodynamic effects. That said, nothing in Riemann's work on these topics was to prove decisive for physics. But his thinking on these matters was surely decisive for the views on geometry that he set out in his Habilitation lecture.[9] The idea of local distortions of space leads very naturally to the idea of a variable metric on a mathematically defined space. The result is curiously Newtonian and anti-Newtonian at the same time. On the one hand, Riemann proposed a vast new mathematical laboratory for the formation and testing of hypotheses of a physical kind. On the other, he let the local properties of space create the global phenomena, whereas Newton had assumed the global properties of space were those of Euclidean geometry. Riemann's work therefore raised in a dramatic way the possibility that there might be many forms of mathematics, all potentially valid or at least useful, in physics.

5.2.4 Poincaré contra Duhem

Poincaré also thought deeply about the relation between mathematics and physics.[10] His philosophical writings drew heavily on his work as a theoretical physicist in the 1890s, starting with his *Électricité et optique* (first edition 1890). In this book he considered the ideas of Maxwell, Helmholtz, and Hertz. It will be convenient here to discuss Hertz's work after Poincaré's.[11] Poincaré observed that there were several theories of electromagnetism, and he set himself the task of making mathematical sense of them and adjudicating between them by setting out their respective merits and problems. He noted that the experts managed very well with a variety of ontological commitments and remained open to ontologically incompatible theories. Rather, following Maxwell, he insisted that there can be many different physical hypotheses that lead to equivalent mathematical conclusions. The only requirement upon a mathematics analysis was that it be correct, but its terms need carry no ontological weight. Hypotheses of a physical kind should only be introduced when the mathematics demanded it, but this was seldom. For example, he argued that there could be no mathematical way of discriminating between one- and two-fluid theories of electricity, and the only atoms he would admit were those "purely geometrical points which obey only the laws of dynamics." These are the mathematician's atoms, not the physicist's (which he came around to only much later, after Perrin's work on Brownian motion). Faced with conflicting explanations, he said that a choice could perhaps be made one day if physicists were to take up questions that were inaccessible to positive methods and had been abandoned to the metaphysicians.

---

[9] Hermann Weyl, it should be noted, thought differently, but he seems to have underestimated the weight of the connection between Herbartian ideas and physical ideas in Riemann's mind.

[10] This section overlaps with Gray 2006b and Gray 2006d.

[11] A further account of Hertz's ideas was published in Poincaré 1894.

Valid mathematics and established experimental results were the pillars of Poincaré's theoretical edifice. Experiments determined the laws of physics, correct mathematics determined their consequences, and anything logically prior to the terms involved in the statements of the laws had only a notional, fictional existence. Poincaré denied that there was any logical way to choose between incompatible physical theories incapable of further analysis. Indeed, he took this to be Maxwell's fundamental idea: "In order to show that a mechanical explanation of electromagnetism was possible, it was not necessary to find the explanation itself, but it would be enough to find expressions for the two parts T and U of the energy function, to write down the corresponding Lagrangian, and to compare the results with experiment."[12] He then suggested that any choice was arbitrary, thus opening the way to his later conventionalism.[13] When, however, a convention had ample physical support, Poincaré came to refer to it as a principle, as in his the paper "Space and Time," which he presented in London in May 1912, where he wrote, "We therefore realize the meaning of the principle of physical relativity; it is no longer a simple convention. It is verifiable, and consequently it might not be verified."[14]

This did not prevent Poincaré from doing physics. Indeed, he was generally accepted as France's leading theoretical physicist, and it has been suggested that it was his work, along with Hertz's discovery of electromagnetic waves, that made the ether theory widely accepted.[15] His book was criticized by the Maxwellians in Britain, but the Germans, who shared Poincaré's misunderstandings of Maxwell, saw that it was quickly translated into German. It was there, nonetheless, that the real difficulties began.

In his accounts of the various electromagnetic theories, Poincaré took certain conservation laws as fundamental. Along with a small number of other laws of physics (Newton's laws, the relativity of motion, and the principle of least action), they make the theory comprehensible, as Poincaré explained in 1897. However, Poincaré found that in Helmholtz's theory Rowland's experiments on electric convection were incompatible with the law of action and reaction.[16] Poincaré deduced that there was a contradiction between experimental facts in optics and Newton's third law. Poincaré's decided preference was for Newton's law, but he could not resolve the contradiction, and in 1904, in his address to the St. Louis Congress (1904), Poincaré abandoned the law of action and reaction.[17]

5.2.4.1 LE ROY AND DUHEM

Poincaré's conventionalism was not the only kind in town, and he found he had to sharpen it around 1900 when he was outflanked by a curious alliance of French neo-Kantians and Catholics, including a bastion of the Catholic modernist movement and

---

[12] Poincaré 1890, viii.

[13] For Poincaré's interpretation of Maxwell, and his misunderstandings, see Darrigol 2000.

[14] Poincaré 1912b, 105; English trans., 21.

[15] Miller 1981, 166.

[16] See Darrigol 1995, 12–13, and 2000, 351–360, where he notes that these criticisms were not reproduced in the second edition of *Électricité et optique*.

[17] Poincaré's problems evaporated with the ether and his assumptions about its interactions with mass and energy. The problem that most concerned him has to do with radiation pressure and is resolved once the equivalence of mass and energy was established.

a formidable conservative. He distinguished his geometric conventionalism (see above, §4.1.6) from other arbitrary acts involved in theorizing, such as making hypotheses that are natural and necessary (for example, that the influence of distant bodies can be ignored on such-and-such an occasion) and making indifferent hypotheses (when the same conclusion is reached on either assumption, for example, that matter is continuous and that matter is discrete). There remained real generalizations, to be confirmed or refuted by experiment, and, said Poincaré, there are such nonconventional hypotheses in optics and electrodynamics.

The challenge he faced revived an old issue Poincaré might well have thought was thoroughly resolved: the rotation of the Earth. It was raised by Edouard Le Roy in an essay of 1899 called "Science et philosophie" and again in the paper he presented to the International Congress of Philosophers in Paris in 1900. Le Roy was then teaching mathematics in a Parisian *lycée*, which he did for many years until he succeeded his friend Henri Bergson as professor of modern philosophy at the Collège de France in 1921.[18] In his book *Dogme et critique* of 1907, he adapted Bergsonian vitalism to a modernist philosophy of Catholicism, arguing that dogma could be a source of moral values without being either inscrutable or in contradiction with rational knowledge. For this he was attacked by Pope Pius X in his encyclical of 1907, when the pope moved to shut down the Catholic modernist movement.

Le Roy's vitalism led him to claim that the only true knowledge lies in an authentic and immediate relationship with one's surroundings, and all theoretical knowledge is a matter of invention. This is not far from Boutroux's neo-Kantianism, as he admitted, but the article went further in advocating a radical form of conventionalism, according to which there are no facts in science, only inventions, which are entirely arbitrary even though they may be necessary on pragmatic grounds. And as examples of scientific "facts" that were in reality only inventions Le Roy cited the atom, the phenomenon of eclipses, and the rotation of the Earth. Furthermore, Le Roy argued that because this rotation was only an invention with no *truth* in it all, not only had the Catholic Church done nothing wrong in condemning Galileo, but Protestant and anticlerical criticisms of the Church seeking to accuse it of bigotry and hostility to science were profoundly misplaced.

This made it a highly charged matter. Did Poincaré's conventionalism commit him to such a position? It could well seem so, for he had written in *Science et hypothèse* (p. 117) that the "affirmation: 'the Earth turns round,' has no meaning, since it cannot be verified by experiment, [. . .] or, in other words, these two propositions 'the earth turns round,' and 'it is more convenient to suppose that the Earth turns round,' have one and the same meaning."

Poincaré naturally thought he was not thus committed. Where Le Roy had seen a sharp distinction between brute facts and all the scientific facts that are really inventions, in his essays "La science est-elle artificielle?" and "La science et la réalité," Poincaré saw a succession of gradations, each with their own claim to compelling assent.[19] From a state of ignorance one could progress to astronomical predictions,

---

[18] See Don Howard, "Le Roy, Édouard Louis Emmanuel Julien," in *Routledge Encyclopedia of Philosophy*, ed. E. Craig (London: Routledge, 1998), http://www.rep.routledge.com/article/Q057?ssid=187996079&n=7#.

[19] In Poincaré 1905b, 213–247 and 248–276. All the following page references are to the English translation of this book.

to accepting Newton's laws, and finally to the deduction of the rotation of the Earth (and a defense of Galileo). The role of convention was restricted to the choice of units of length and time in physics, of definitions and postulates in mathematics. Once those were established, much else followed inevitably, and in particular scientific facts were merely the translation of brute facts into the language of science. Newton's laws follow from some simple assumptions, or may indeed be taken as assumptions and their consequences may be tested—that, for Poincaré, is all a matter of facts. The rotation of the Earth, however, he allowed (p. 231), should not be spoken of as a simple fact because it is not comparable to the statement that at such-and-such a time the sky will darken and an eclipse will occur.

After a lengthy discussion of the nature of scientific laws, Poincaré then addressed the question of the objectivity of science, which he grounded in communication. Pure sensations (*my* experience of this patch of red, etc.) cannot be communicated, but relations can. What is objective is what is the same for all of us, once the translation between different conventions is allowed for. Science is a classification of apparently different appearances, so it is a system of relations (p. 266). Therefore, to say that science cannot be objective because it can speak only of relations and never of things "in themselves" or "as they really are" is to have the problem entirely backwards. Science does not reveal the true nature of things, but that is because nothing can. Indeed, said Poincaré (p. 267), if some god did know the true state of things not only could he not find the words to express them, but we could not understand them if he did. Indeed, it can seem that a theory is born one day, fashionable the next, classical the day after, then antiquated, and on the fifth day forgotten; ruins accumulate on ruins. But a closer look shows that it is the theories that presume to tell us what reality is that perish, and more relational theories persist (p. 268).

Now Poincaré could return to the claim that the Earth rotates. It was a misinterpretation of his earlier work, he said, to say that it put the rotation of the Earth on the same footing as the parallel postulate (p. 272). Rather, it belonged with claims about the existence of the external world. And indeed, the two claims, that the Earth rotates and that it does not, cannot be told apart kinematically because there is no absolute space. But the claim of rotation is attached to a much richer dynamical theory, which covers the apparent motion of the stars, Foucault's pendulum, and much else that would be disparate phenomena on a Ptolemaic theory. So the distinction, as Poincaré saw it, between brute facts and scientific facts was merely that scientific facts were brute facts translated into the language of science by being incorporated in a theory. The choice of theory was arbitrary, in much the way that one may speak French or German, the facts were intertranslatable.

This did not impress Le Roy's ally, the staunch Catholic and French nationalist Pierre Duhem. He had graduated in mathematics from the Sorbonne in 1888 with a thesis on the theory of magnetization by induction, and became a prolific author of articles on various topics not only in mathematical physics, but also on the history and philosophy of physics.[20] However, his conservative Catholic views as well as technical disputes about thermodynamics had not endeared him to the influential anticlerical Berthelot, and Duhem was exiled to the provinces, where he made a

---

[20] See Don Howard, "Duhem, Pierre Maurice Marie," in *Routledge Encyclopedia of Philosophy*, ed. E. Craig (London: Routledge, 1998), http://www.rep.routledge.com/article/Q027SECT1. Maz'ya and Shaposhnikova 1998 speak of four hundred articles and twenty-five books.

successful career for himself at the University of Bordeaux. Between 1894 and 1906 he wrote a number of articles on the philosophy of physics, many in the neo-Thomist journals the *Revue de philosophie* and the *Revue des questions scientifiques*, the organ of the Société scientifique de Bruxelles, which had obeyed Pope Leo XIII's instructions to espouse Thomism in 1890.[21] This work culminated in his book *La théorie physique*: *Son objet et sa structure* (1908), which had a considerable and lasting impact on the field.

The book (p. 149) makes clear that Duhem was scornful of Poincaré's idea about scientific languages: "It is therefore clear that the language in which a physicist expresses the results of his experiments is not a technical language similar to that employed in the diverse arts and trades. It resembles a technical language in that the initiated can translate it into facts, but differs in that a given sentence of a technical

FIGURE 5.1. Pierre Duhem in his study (c. 1905).

[21] Paul 1979, 171.

language expresses a specific operation performed on very specific objects whereas a sentence in the physicist's language may be translated into facts in an infinity of different ways."

Duhem's own philosophy of science emphasized that a network of physical theories is present at every stage in scientific work, and which generally permits different interpretations of any given result. This network rules out any idea of there being a "crucial experiment" in physics, the outcome of which is decisive for a theory. Rather, the outcome of an experiment is a challenge to the whole network; this is Duhem's contribution to what has become known as Duhem-Quine holism. It implies that refuting a belief such as the nonrotation of the Earth would be a very complicated matter indeed, going well beyond the observation that assuming it fits into a broad general theory. It goes much farther than Poincaré's mixture of geometric conventionalism and principles in regarding much more of the web of theory and experiment as, in principle, arbitrary. However, Duhem located the ambiguities and rival interpretations entirely within the physics. Duhem never acknowledged the modernist shift except to say, as he does of his friend Hadamard's work and that of Poincaré on celestial mechanics, that it is useless to the physicist.[22]

Duhem's remarks about mathematics were entirely conservative. As befitted both its author's education and his nationalist sympathies, the book has a vigorous defense of French mathematical physics against British mechanical models. The trouble, apparently, is that the English (a word that generously includes Maxwell and Lord Kelvin) wish to reduce physical theory to nothing but models, something "French or German physicists" would never have done "of their own free will." That said, no less a figure than Hertz had already reduced mathematical physics to a collection of algebraic models. This weakness was pandered to by Poincaré and has spread on a fashion for all things English, especially among the engineering schools, producing piles of faulty reasoning and false calculation, a confusion of science and industry, and the rejection of abstract and deductive theories in favor of the concrete and inductive.[23]

The path to redemption would be, and had been all along, geometric, formal, and even axiomatic. As Duhem (1908) put it: "A physical theory will then be a system of logically linked propositions and not an incoherent series of mechanical or algebraic models" (p. 107). To produce an error-free account of a substantial theory is so difficult that only mathematics (specifically, arithmetic with its extension to algebra) is adequate to the task. "It owes this perfection to an extremely abbreviated symbolic language in which each idea is represented by an unambiguously defined sign," a language he said was created in the sixteenth and seventeenth centuries. Moreover, it emerges, after a discussion of measurement and of the distinction between quantity and quality, that the traditional and valid aim of a physical theory is to work with measured quantities and measured intensities of qualities, where measurements are simply numbers.

However, it should be noted that Duhem's intention was to show that science was an exercise in classification, and that it was independent of any metaphysics. Scientific laws were not capable of being true because they were only representations,

[22] Duhem 1908, 138–143.
[23] Duhem 1908, 89–93.

and because they could not be true science was not capable of conflicting with religion. This was not the Thomist position, and Duhem would be criticized from that quarter, even though he had offered a spirited defense of scholastic philosophers as being much more akin to modern scientists than secularists liked to claim they were. It would be the Thomist claim that reason (properly guided) led to truth that would fare much less well.

5.2.5 Hertz

Heinrich Hertz, Helmholtz's finest pupil and the famous discoverer of radio waves, is also known to philosophers of science for his theory of images. His ideas could hardly be better for demonstrating how a distinguished experimental physicist had come to see the relation of mathematics to physics, even if one should be careful about drawing conclusions from so exceptional a figure. Hertz was an unregenerate thinker in some ways: he subscribed to the Kantian a priori, he was entirely happy that space was Euclidean, and he was a naive realist about the existence of external objects. For him, the real business of thinking began elsewhere.

In principle, said Hertz in his *Principles of Mechanics* (1895), experience and experiment should produce images (*Bilder*) in the mind, which might be images of external objects or of the relations between these objects. Specific items in an image might be called (following Helmholtz) signs. These images he came to think of as having to be correct, but not necessarily true. To be correct they had to satisfy the requirement that any necessary relation between objects had to appear in the image as a necessary relation between the corresponding signs. As he put it: "We make mental images (*innere Scheinbilder*) of outer circumstances, and indeed we make these of such a kind that the necessary consequences of the images shall again be images of the necessary consequences of the represented circumstances."[24] But the signs were otherwise arbitrary and the image might also contain signs that did not correspond to objects (which is how, or why, a useful image might not be true). The arbitrariness of the sign was an idea he took over from Helmholtz, but he dropped Helmholtz's condition that the correspondence between objects and signs must be one-to-one and allowed that there may be many different signs for the same object. An image must also be permissible, that is, logically permissible and amenable to the laws of thought (which were given a priori for Hertz), and to underline this point Hertz adopted an axiomatic style of reasoning in his *Mechanics*. Finally, Hertz demanded of his images that they be appropriate, that is, that they capture more of the essential features of the matter at hand and fewer of the superfluous ones.[25]

In practice, the same Hertz in the same book was understandably more tentative. Different images might be permissible to various degrees, and correct to various degrees, although in theory it seemed as if these were absolute, yes-or-no properties. Even in principle the judgment that an image was correct was held to be context dependent and open to revision. And on the other hand, the relative property of appropriateness was sometimes given absolute status. None of this would matter had Hertz not put his mind to constructing a remarkable theory of matter in accordance

---

[24] See p. 1 of Hertz 1895, English translation, quoted in Toepell 1986, 64.
[25] This discussion follows Lützen 2005, Corry 1997, and Corry 2004.

with his principles, and it is this theory that shows just how imaginative the use of mathematics could be for him in physics.

Hertz wanted to make the best possible sense of mechanics as the simplest branch of physics. He found that it was based on concepts such as mass and energy, with which he had no quarrel, and force, to which he objected strongly. To remove force from the list of fundamental concepts, Hertz more than made up for any philosophical naivety about the nature of space by a virtuoso display of differential geometry in the spirit of Riemann.

Hertz began, as was customary in mechanics, with the consideration of a system of n particles, each of a certain mass and having positions and velocities. Each particle has 3 coordinates, so the position of all n of them is described by a system of $3n$ coordinates, which Hertz thought of as defining the position of a new "point" in a $3n$-dimensional space, $\mathbb{R}^{3n}$. He then reexpressed the equations of motion for the n masses in ordinary space so as to give equations for the motion of the new "point," and in this way introduced a Riemannian metric on this space. The idea of thinking of the positions of the particles in physical space as defined by a point in a different space, $\mathbb{R}^{3n}$, was not new—it goes back at least to Hamilton and Lagrange—but the novel metric was. It implies that the intrinsic geometry of the space need not be Euclidean. Similar ideas had been expressed by Darboux in his *Leçons sur la théorie générale des surfaces* of 1887, where he used n independent variables as coordinates and spoke of trajectories in the new space as geodesics and of orthogonal families of curves; but he did not use the word "space" or the term "n-dimensional space."[26] One can almost hear the concept being—politely—stretched.

As ever, the study of systems of points was intended as but preliminary to systems of large rigid bodies, so Hertz now moved toward systems of infinitely many points. We need not worry about the mathematical rigor needed to carry this out, because Hertz did not, but note that a rather remarkable space is in process of construction that will be an image of an otherwise mundane mechanical system. In this space, Hertz defined notions of length and angle, velocity and momentum. He could now talk geometrically about the time-evolution of a mechanical system as it appeared in his artificial space. This path might well qualify as straight—a concept more or less available in contemporary differential geometry—but it is a path in an intrinsically curved space. If, moreover, the motion of the particles in Euclidean three-space is constrained in some way, then the path in the artificial space will also be constrained to lie on some subspace of $\mathbb{R}^{3n}$.

In familiar three-space, the particles are often supposed to interact with one another by their mutual gravitational attraction, but this is just the sort of action at a distance that Hertz was keen to remove from the theory. In the artificial space, these forces appear as what Hertz called "connections," and he gave these connections a fundamental role. He could show that for unconstrained systems the time-evolution appeared as a straightest path in the artificial space, determined by its initial point and velocity only and satisfying the law of conservation of energy. If the system was constrained, then Hertz regarded it as but a part of a larger, unconstrained system, and the forces he wished to eliminate he sought to describe as connections between visible and new, hidden masses (all these masses are, of course, in the usual Euclidean three-space).

---

[26] See Darboux 1887, vol. 2, 496–527.

Present readers may take on trust that much shrewd mathematics underpins Hertz's arguments, and in particular that a connection is a different object from a force. They may form their own opinion about hidden masses and the merits of replacing the concept of force with one that invokes hidden masses instead. No less an authority than H. A. Lorentz was charmed by the use of higher dimensions, writing that "it seems hardly possible to doubt the great advantage as to conciseness and clearness of expression that is gained by the mathematical form," but he did not endorse Hertz's connections.[27] The truth is that, with Hertz's early death, these particular ideas never had much of a life. What is undoubted is the extraordinary sophistication of a Hertzian image. Not only is a great deal of advanced mathematics required to make the image visible, the image itself is in a number of ways remote from what it purports to be about. It is multidimensional, it has a geometric structure quite unlike familiar three-space, and it is accessible only through the intellect.

Mathematicians had been here before—in a simpler fashion but one that was destined to make much greater immediate headway—when Poincaré drew out from his study of celestial mechanics the need to study three-dimensional topology (see § 4.5). Hertz's ideas were different in that they were grounded in differential geometry rather than topology, and because, as a result, they retained much more detail of the mechanical problem. Hertz's work shows how abstract geometrical ideas were now entering the thoughts of at least some of the best theoretical physicists.

5.2.6 Hilbert

Hertz's ideas may not have captured the minds of most physicists, but quite properly they did capture the mind of one mathematician with a serious interest in mathematical physics: David Hilbert. So firmly is Hilbert remembered as a great pure mathematician, perhaps the greatest and most influential of his generation, that it is worth stressing that this is not how he seemed to those who knew him in Göttingen. There, between 1905, 1909 (when Minkowski died), and 1913 he lectured often on topics in physics and reflected extensively on the role mathematics should play in constructing a physical theory. With Klein, Hilbert was the person Einstein talked to most productively when Einstein was thinking his way toward the general theory of relativity, and there is a famous story that in the end these two were engaged in a race to produce that theory first.[28] The fifth of the problems in Hilbert's celebrated list of twenty-three problems calls for the axiomatization of physics, and this was no idle nod toward mathematics' roots in science. It may be that much of Hilbert's best work is in pure mathematics, and even the creation of the basics of the mathematics used in quantum mechanics was done before the new physics was discovered, but Hilbert was not a narrow pure mathematician. It would be more useful to say he was a pure mathematician actively interested, as a mathematician, in contemporary physics.

Numerous historians have recently traced a debt of Hilbert's to Hertz's idea of an image. As early as 1894, when Hilbert lectured on geometry in Königsberg, he called it "a science in which the essentials have been so well developed that all their im-

---

[27] Lorentz's remark of 1902 is quoted in Lützen 2005, 286.

[28] See in this connection Corry 2004 for what actually happened, and Sauer 2005 for the refutation of an unsavory recent twist to the story.

plications (facts, *Thatsachen*) can already be derived logically from those derived earlier."[29] This was, he said, quite unlike the situation in electricity or optics. But, he continued, since not all the ideas are to be derived by pure logic and some come from experience, it is important to isolate the fundamentals, to be called the axioms, that are sufficient for all geometry. The central problem of the lecture course was therefore to determine what are the necessary and sufficient independent conditions that one must impose on a system of things so that to every property of these things there corresponds a geometric fact and conversely, so that by means of this system of things it is possible to give a complete description and ordering of all geometric facts.

"Axioms," Hilbert continued in a close paraphrase of Hertz, "are what Hertz would call images or symbols in our minds, such that consequences of images are again images of consequences, i.e. what we logically derive from the images is found again in nature."[30] As Toepell (1986, 114) points out Hilbert drew from this the idea that spatial intuition should only be employed as one possible intuitive analogy.

Hilbert's axiomatic presentations of geometry have already been discussed. But Hilbert also took a cautious axiomatic approach to physics.[31] He argued that geometry was a branch of physics already sufficiently well understood to be axiomatizable, that mechanics would surely be the next to reach that state, and that in any case it was always good to know what the logical relations were between the various parts of any particular theory. In a lecture course in 1899 he said:

> Geometry also [like mechanics] emerges from the observation of nature, from experience. To this extent, it is an experimental science. . . . But its experimental foundations are so irrefutably and so generally acknowledged, they have been confirmed to such a degree, that no further proof of them is deemed necessary. Moreover, all that is needed is to derive these foundations from a minimal set of independent axioms and thus to construct the whole edifice of geometry by purely logical means. In this way [i.e., by means of the axiomatic treatment] geometry is turned into a pure mathematical science. In mechanics it is also the case that all physicists recognize its most basic facts. But the arrangement of the basic concepts is still subject to a change in perception . . . and therefore mechanics cannot yet be described today as a pure mathematical discipline, at least to the same extent that geometry is.[32]

Hilbert first set out his own set of axioms for mechanics in 1905, and they are significantly different from the sorts of things he presented as axioms in geometry. From a rather obscure discussion of time, intended to convey a sense in which it flows uniformly and only in one direction—a view Poincaré was already on record as finding meaningless—it emerged that Gauss's principle of least constraint must also be assumed, and then the Lagrangian equations of motion can be derived from this principle. But, said Hilbert, one could also start by taking the Lagrangian equations as axioms, in which case many of the previous axioms now became theorems. Or, indeed, one could take the law of conservation of energy, or Hertz's own (the principle of the straightest path, which reduces to Gauss's principle in the force-free case).

---

[29] Toepell 1986, 58.

[30] Hilbert repeated this Hertzian idea in his course in 1899 on Euclidean geometry; see Toepell 1986, 204.

[31] See the full accounts in Corry 1997 and Corry 2004.

[32] Quoted in Corry 2004, 90.

One axiom Hilbert was inclined to insist on was that, as the famous phrase has it, "Nature does not make jumps." In other words, the functions of mathematical physics are continuous. After 1905 he could point to a short paper by Hamel in his support.[33] In that paper Hamel took the simplest functional equation of them all, $f(x+y) = f(x) + f(y)$, and asked for what functions f does this equation hold. When f is assumed to be continuous, the answer is easily seen to be just what you expect: $f(x) = \alpha x$, for any choice of constant α. The equation had been used, he noted, by Darboux in 1880 to analyze the fundamental transformations of projective geometry, and the law of composition of forces in statics. But if the function f is not continuous, then, as Hamel showed, it follows from Zermelo's well-ordering theorem that there are other remarkably different, totally discontinuous, solutions.[34] It follows that if the laws of physics are to turn up in the form of functional equations, then physically meaningful solutions will only arise if the functions are assumed a priori to be continuous. Contrariwise, it would be a thoroughly modernist thing to study functional equations for discontinuous functions.

There are two crucial points here. The first is that Hilbert was prepared to take certain dynamical principles as axioms, thus respecting what the practitioners of the discipline had to say, and the second is that Hilbert was prepared to consider different principles as axioms. In this second connection, his approach was interesting. He could come to various opinions about the axiom systems he generated: this one might not be the simplest, this might need to be supplemented with further axioms, and so on. In his next topic, thermodynamics, Hilbert began to suggest axioms in which physics was initially subordinate to mathematical elegance. An axiom might be preferred not because of its more elementary nature, but because it opens the way to a better-organized account. This is a long way from the usual approach to axioms, which was that axioms were primitive statements that could not be doubted or further analyzed.

It was to turn out that this way of thinking was not much appreciated by physicists—an indication of the growing gap between them. In 1909 Constantin Carathéodory proposed an axiomatization of thermodynamics very much in the Hilbertian spirit.[35] But his friend, the physicist Max Born, who was very interested in thermodynamics at that time, noted some years later that Carathéodory's axiomatization had not caught on, probably because it was too general and abstract, and because it was published in *Mathematische Annalen*, a journal physicists did not consult. Born then offered a popularization of Carathéodory's work for physicists, but toward the end of his life he noted that his attempt had also fallen on stony ground.

5.2.7 Minkowski

Minkowski's contributions to physics are chiefly of two kinds. First, he brought geometry to Einstein's much more computationally and metrologically focused

---

[33] Hamel 1905.

[34] They have the disconcerting property that the "curve" of points $(x, f(x))$ gets arbitrarily close to every point of the plane.

[35] Carathéodory was part of the Hilbert circle. He had taken his doctorate at Göttingen in 1904 and habilitated in 1905.

original presentation of the special theory of relativity. Although initially Einstein dismissed the presentation as burdensome, he came to approve of it. Second, he introduced an axiomatic formulation of the ideas that Hilbert, at least, liked very much.[36]

It is indeed due to Minkowski that we speak of space-time. As he put it in his celebrated address to the German Society of Natural Scientists and Physicians in Cologne in September 1908: "The views of space and time which I wish to lay before you have sprung from the soil of experimental physics, and therein lies their strength. They are radical. Henceforth space by itself, and time by itself, are doomed to fade away into mere shadows, and only a kind of union of the two will preserve an independent reality."[37]

After this opening peroration, Minkowski settled down to show how geometry and mechanics could profitably be brought together. In coordinate geometry, we do not regard different positions for the coordinate origin as having any significance (nor, he might have said, is there a privileged orientation for the axes); in mechanics we have no preference for one system of axes over another if both are in uniform motion relative to the other. By bringing space and time together in a four-dimensional space-time, these two beliefs are unified. One merely requires, said Minkowski, that the coordinate transformations that express this indifference are such that what is given by $x^2 + y^2 + z^2 - c^2 t^2$ is given by $x'^2 + y'^2 + z'^2 - c^2 t'^2$ in any other allowable coordinate system, where c is the velocity of light in a vacuum. Minkowski graciously allowed that physicists had found this first, and while mathematicians could have done so they could only claim staircase wit, but in any case, in this way, "three-dimensional geometry becomes a chapter in four-dimensional physics." Minkowski then gave a few reasons for accepting the new ideas that would appeal to physicists: the coordinate transformations did not alter the equation of a path of light, and they explained away the so-called Lorentz contraction. He had set these arguments out at greater length in a previous paper, which had provoked Einstein to say that four-dimensional geometry made unusual demands on physicists.[38]

Minkowski noted that Einstein's interpretation of the proper time of an electron in its motion as equivalent to the proper time of any other electron deposed time from its high seat as an unequivocally determined phenomenon, but that Einstein and Lorentz had made no attack on space. Once this was done—"another act of audacity on the part of the higher mathematicians," he joked—he felt that calling the postulated invariance of measurements with respect to the new group of transformations the relativity principle was to give it a very feeble name. Instead, he proposed to call it the "postulate of the absolute world"—a clever choice of name that did much to promote his ideas.

Minkowski's proposal did more than suggest that the coordinate transformations of special relativity would sort out the kinematics of the electron; that had been Einstein's contribution. The underlying geometry he introduced made it clear that these transformations were not that different from the familiar rotations of Euclidean geometry, and he suggested a formalism that, to a mathematician or a mathematically

---

[36] A detailed comparison of Minkowski's work with that of Lorentz, Poincaré, and Einstein will not be necessary here. See, e.g., Corry 1997, 1999, and 2004; Walter 1999, and the literature discussed therein.

[37] Minkowski 1908a.

[38] See Minkowski 1908b.

trained physicist, was little more than a change of sign in a formula. Mathematicians would already know that non-Euclidean geometry was not far away.

T. Hirosige and other commentators have noted that Minkowski's address played a crucial role in the acceptance and agreed centrality of Einstein's special theory of relativity.[39] Scott Walter's analysis shows that indeed the number of articles on the subject rose after 1909, peaked in 1911, and did not fall back to their 1908 levels until 1915, by which time the First World War was underway.[40] Despite Einstein's misgivings, Planck and Wien, the editors of *Annalen der Physik*, welcomed the four-dimensional geometry, although they did not approve of the attempt to base the special theory of relativity on mathematical, rather than physical, arguments. Arnold Sommerfeld, however, a former pupil of Hurwitz and of Hilbert, a one-time protégé of Klein's, and a well-regarded physicist, liked Minkowski's mix of mathematics and physics. In lectures in Munich in 1909–10 he showed how the geometric formulation facilitated the derivation of the formula for the addition of velocities in special relativity and made it seem quite natural.

Sommerfeld's connections made him the obvious person for the mathematicians at Göttingen to turn to, especially after Minkowski's early death in 1909.[41] Naturally enough, Klein could not resist returning to mathematics to show how well Minkowski's geometrized theory of special relativity fitted into his Erlangen Program—one can almost hear his feet on the stairs. A number of other mathematicians then joined in. It would be too technical a digression to explain here how close the topic of non-Euclidean geometry is to that of special relativity, but the key points to note are that the addition formula for velocities in the special theory of relativity and the addition formula for lengths in non-Euclidean geometry can be made to appear identical, and the transformation groups of special relativity and non-Euclidean geometry are almost identical. These facts were happily seized upon by a number of mathematicians, and they show how far non-Euclidean geometry had, inadvertently, prepared the ground for the special theory of relativity.

To what extent, then, is the Minkowskian version of Einstein's ideas a geometric theory of physics? And to what extent is it, in that way, novel? Certainly Einstein, as he deepened his grasp of gravity and his appreciation of geometry, came to recognize that Minkowski's presentation of the theory had helped him come to his general theory of relativity. But the clearest way to see the role of geometry in Minkowski's theory is to bring out its axiomatic formulation. This was more fully expressed in the *Grundgleichung* paper. There Minkowski observed the mathematical fact that Maxwell's equations for electrodynamics are invariant under Lorentz transformations; he called this the "theorem of relativity." He called the postulate of relativity the idea that all physical processes concerning massive bodies (including processes as yet unknown) are Lorentz covariant.[42] And what he called the principle of relativity was the claim that Lorentz covariance is true at the level of the observable magnitudes associated to a moving body.[43] He then set himself the task of deriving the equations of motion of a moving body from the principle of relativity.

[39] See Hirosige 1976, 78.
[40] Walter 1999.
[41] Sommerfeld also helped the physicists by introducing vector methods into the subject.
[42] This is in part a claim about gravity that neither Minkowski nor Hilbert was to pursue.
[43] See Minkowski 1908b, 353.

To do this he first treated the case of a space containing only electric and magnetic fields, then space with matter at rest in it, and finally matter moving in space. The principle of relativity enters as an axiom at the third stage. Minkowski then examined other equations of motion in the literature and showed that some were incompatible with his axioms. These included the usual equations for Newtonian mechanics, which Minkowski explained by saying that Newtonian mechanics is a special case of his theory when the value of c is allowed to become infinite. Therefore, Newtonian mechanics should be seen only as a low-velocity approximation to the correct theory, and the correct theory of mechanics needed to be spelled out. On the other hand, the relativity of simultaneity so basic to Einstein's approach emerges as a theorem in Minkowski's.

Minkowski's theory avoided any commitment to the fundamental nature of matter (should it be mass, or electrical in nature, or some mix of the two?). It said: accept a certain set of equations and a certain group of transformations, and deduce "everything" from them. It suggested rejecting mathematical formulations of physics that were inconsistent with this approach on the explicitly stated belief that nothing in physics was or would be inconsistent with it. It was much clearer that the best-known sets of equations in electrodynamics were Lorentz covariant (and therefore acceptable) than that the natural world would comply, and Minkowski did not even attempt an analysis of physical processes to show that it would. So it was, as Planck and Wien had said, a mathematician's theory, not a physicist's. That said, it exemplified what Minkowski's friend Hilbert wanted from an axiomatic theory: it was well organized and consistent. By contrast, Einstein's original 1905 paper was just what Hilbert was worried about: resolving physical difficulties by adding further postulates.

Minkowski's axiom was also geometric in exactly Klein's sense. It proposed a certain space, and the allowable properties of "things" in this space were those of a certain group naturally associated with this space. One might say that Klein had promoted a one-person view of geometry in which the mathematician, by means of a group, transforms figures in some mathematical space and recognizes as geometric just the properties of those figures that remain invariant under those transformations. In this spirit, Minkowski proposed a two-person view, according to which different observers agree to regard as "absolute" just those properties of space that are invariant under the Lorentz group. The underlying logic of Minkowski's approach was that if you think about what doing physics involves, then you are led to a Kleinian position. There are many such approaches, but so much of physics is based on Maxwell's equations that it seems best to put your trust in them. If you agree that the transformations of physics should preserve straight lines (that two observers in a state of uniform relative motion agree about straight lines), then the Lorentz group is the correct group. His theory therefore stands in a mathematical tradition going back to Riemann, but notably reformulated by Hilbert, which says that it is for the mathematician to formulate theories, from which the physicist selects according to what, it is presumed, will be simple but fundamental requirements.

Such a view presents a division of labor entirely in accord with the modernist separation of mathematics and physics. The modern mathematician stocks the shop, perhaps with a whole variety of garments. The physicist consumer has a difficult job knowing what to choose, because the transaction does not involve taking the clothes back repeatedly for minor alterations. Rather, the expert physicist must decide, on whatever grounds and after as much or as little experimental work as is needed, precisely what is wanted. With luck, the sale is made and the newly attired physicist moves successfully onward. Even Minkowski, whose interest and immersion in physics was deep, and Hilbert, equally interested, could not tell physicists what to choose without becoming physicists, as Minkowski did more nearly than Hilbert.

As is well known, Lorentz covariance was not the exclusive property of Minkowski. It was Poincaré who had first suggested it. Minkowski's originality lay in the axiomatic character of his account and the degree of organization it brought to the whole body of ideas. Its acceptance by mathematicians was enhanced by the way it captured the drama of recent developments in physics while making them wholly over into an almost familiar branch of mathematics. Physicists, on the other hand, accepted it because it delivered what Minkowski had cleverly called an absolute world. It gave the puzzling, and still imperfectly understood domain of electrodynamics a setting that was as reassuring, and objective, as Newtonian mechanics.

5.2.8 Einstein

As already noted, Einstein's arguments in his paper of 1905 that ushered in the theory of special relativity were entirely about physical quantities and their measurement.[44] Einstein was not impressed with Minkowski's geometrical reformulation until he tried to incorporate gravity into his relativistic scheme.

In 1909 he began to look at the case of the geometry on a rotating disk, and he first wrote about the rotating disk in two papers published in 1912.[45] The rotating disk is one of Einstein's thought experiments—he never seriously maintained that space might be rotating. Special relativity theory predicts that if one object is moving with respect to another, it will appear contracted along its direction of relative motion, and this effect increases with the relative velocity. So to an observer at the center of the disk, the further away a meter stick lies, the more it will appear contracted, and this contraction will be greatest along the direction at right angles to the radius joining the observer to an end of the stick. The situation is very similar to the cooled sphere that Poincaré had described (and Einstein had studied Poincaré's popular essays with great care). Einstein showed that the observer at the center will indeed think that the geometry on the disk is a non-Euclidean geometry—not, to be precise, the non-Euclidean geometry of Bolyai and Lobachevskii, but a geometry different from Euclid's.

Even Einstein found the subject matter difficult.[46] He was bothered on and off for many years by the relationship between coordinates on the one hand, and measurements with rods and clocks on the other. John Stachel argues that by 1912 Einstein was ready to deal with what are called stationary gravitational fields (ones that vary from point to point but not over time) and of which the simplest example is the rotating disk. From it, he learned that coordinates have no simple meaning in such contexts, and so his famous equivalence principle can only apply infinitesimally. This led him to recall the way Gauss had introduced coordinates on curved surfaces,

---

[44] See, for a persuasive account of Einstein's route to special relativity in 1905, Norton 2004.
[45] Einstein to Sommerfeld, September 29, 1909, in Einstein 1993, doc. 179.
[46] This suggestion is due to Stachel 1989, where he discusses two papers by Einstein, 1912a and 1912b.

a topic he had learned when a student at the Federal Polytechnic Institute in Zurich as a student in 1900, in a course he regarded as a "masterpiece of the pedagogical art."[47] Now he began to appreciate Minkowski's four-dimensional geometric account of special relativity, and could see, in 1912, that he needed a four-dimensional generalization of Gauss's ideas. This he obtained with the help of his friend Marcel Grossmann.[48] As Einstein put in in his book *On the Special and the General Theory of Relativity* (1917), discussing the simpler example of the non-Euclidean geometry on the surface of an ellipsoid: "Gauss indicated the principles according to which we can treat the geometrical relationships in the surface, and thus pointed out the way to the method of Riemann of treating multi-dimensional, non-Euclidean *continua*. Thus it is that the mathematicians long ago solved the formal problems to which we are led by the general postulate of relativity."[49]

The best of the Einstein literature describes how useful Einstein found all this differential geometry when creating the general theory of relativity, a long, complicated process that took him the next three years, to 1915. The result was a theory of gravity that treated it as a nonuniform force, varying from point to point, but affecting all bodies equally, that changes lengths and so is geometric. Accordingly, Einstein's theory of general relativity is a geometric theory of physics, but this is a point of view that requires some careful articulation. Fortunately, the experience of producing the general theory of relativity made Einstein an authority on the relationship of mathematics to physics, and he took this as the subject of one of his most famous expository papers, "*Geometrie und Erfahrung*" ("Geometry and Experience," 1921).

In this paper, Einstein began by posing the problem of how mathematics can contribute so strikingly to the success of science in these terms:

> How can it be that mathematics, being after all a product of human thought which is independent of experience, is so admirably appropriate to the objects of reality? Is human reason, then, without experience, merely by taking thought, able to fathom the properties of real things? In my opinion the answer to this question is, briefly, this: As far as the laws of mathematics refer to reality, they are not certain; and as far as they are certain, they do not refer to reality. It seems to me that complete clearness as to this state of things first became common property through that new departure in mathematics which is known by the name of mathematical logic or "Axiomatics." The progress achieved by axiomatics consists in its having neatly separated the logical-formal from its objective or intuitive content; according to axiomatics the logical-formal alone forms the subject matter of mathematics, which is not concerned with the intuitive or other content associated with the logical-formal.[50]

The possibly misleading reference to axiomatics will detain us in a moment, but the thrust of the argument is that mathematics in its own right proposes; it out-

---

[47] In 1912 Einstein knew only Gauss's work, but not that of Riemann (and still less that of Ricci or Levi-Civita).

[48] As Stachel notes, if this account of Einstein's route to his famous discovery is broadly correct, then so is Einstein's own theoretical account in his relativity book of 1917 broadly fair to the historical development. There, the rotating disk is discussed in chapter 23.

[49] See the footnote at the end of Chap. 24 of Einstein 1917.

[50] Einstein 1921, 209.

lines the possibilities, and this it does quite formally, and with full logical force. Its conclusions are, therefore, certain. As for which of these possibilities is to be selected as correct as the one that describes reality, that is another matter, and certainty is not to be had. Einstein's argument recalls Riemann's own position on these issues, perhaps more than he was aware.

The role of axiomatics for Einstein was to ensure the logical force of mathematics. He did not write his own, increasingly mathematical, work in the style of Euclidean geometry, of course. But he did write in the way that Hilbert would have regarded as axiomatical, and for much of the time between 1912 and 1921 Einstein had enjoyed a fruitful and often amicable relationship with Hilbert. Indeed, he found he got a better hearing from Hilbert and Klein in Göttingen than he did from his colleagues in Berlin, although to be sure the business of the so-called race to general relativity soured things for a while.[51] Axiomatic mathematics, for Einstein, is to be mathematics governed by clearly stated principles, such as a set of equations.

Equations, per se, are meaningless on this formulation. They acquire a meaning, Einstein went on to say, "by the co-ordination of real objects of experience with the empty conceptual frame-work of axiomatic geometry." The usual naive commonsense theory of rigid bodies yields Euclidean geometry—note again the echoes of Poincaré's ideas, to which Einstein immediately appended a Hilbertian gloss: "Geometry thus completed is evidently a natural science; we may in fact regard it as the most ancient branch of physics."

Einstein now made a distinction between what he called "purely axiomatic geometry" and "practical geometry," and remarked, "I attach special importance to the view of geometry which I have just set forth, because without it I should have been unable to formulate the theory of relativity." This was because of his work on rotating systems.

Then he went on to recall the ideas of that "acute and profound thinker, H. Poincaré" that Euclidean geometry is distinguished above all other imaginable axiomatic geometries by its simplicity. Certainly, said Einstein, we should retain Euclidean geometry whatever may be the nature of reality, because when contradictions arise it is easier to change physics than axiomatic Euclidean geometry. Why then, he asked, is the naive commonsense theory of rigid bodies nowadays denied? Because real solid bodies are less than ideal, they vary with temperature and the action of forces.[52] Better therefore to express Poincaré's conventionalism this way:

> Geometry (G) predicates nothing about the relations of real things, but only geometry together with the purport (P) of physical laws can do so. Using symbols, we may say that only the sum of (G) + (P) is subject to the control of experience. Thus (G) may be chosen arbitrarily, and also parts of (P); all these laws are conventions. All that is necessary to avoid contradictions is to choose the remainder of (P) so that (G) and the whole of (P) are together in accord with experience. Envisaged in this way, axiomatic geometry and the part of natural law which has been given a conventional status appear as epistemologically equivalent.

[51] See again Corry 2004, and for further detail Sauer 2005.
[52] Nor, indeed, are they rigid, but Einstein felt that they were sufficiently so for this observation to hold less force than might seem appropriate at first glance.

So, Einstein concluded, "Sub specie aeterni Poincaré, in my opinion, is right." This raises a problem, because it is hard to see by what token Einstein can argue for the correctness, rather than merely the simplicity of his general theory of relativity, and the role non-Euclidean geometries play in formulating it. The way out is to retreat from eternal things to the realm of the practical. Einstein argued as follows. Call a "tract" a region between two boundaries on a practically rigid body, and say two tracts are equal to one another if the boundaries of the one tract can be brought to coincide permanently with the boundaries of the other. Assume that if two tracts are found to be equal once and anywhere, then they are equal always and everywhere. As Einstein noted, the practical geometries of Euclid, Riemann, and the general theory of relativity all rest upon this assumption. The propagation of light in a vacuum provides examples of tracts, and so two clocks that agree once and anywhere allow us to measure tracts. Treating atoms as clocks, the existence of sharp spectral lines is therefore a convincing experimental proof of the above-mentioned principle of practical geometry.

By means of tracts, the nature of the space-time continuum becomes a physical question that must be answered by experience, and not a question of a mere convention to be selected on practical grounds. Accordingly, "This is the ultimate foundation in fact which enables us to speak with meaning of the mensuration, in Riemann's sense of the word, of the four-dimensional continuum of space-time." It may be that conventionalism holds, said Einstein, but nature seems to offer only one way in which measurement in physics can be formalized, and when this way is taken, gravity becomes a geometrical property of space.

Einstein's sharp separation between mathematics and physics, and the characterization of mathematics as logical and formal, show how much he accepted the modernist transformation of mathematics, which was not only eloquently on display in Göttingen, but was congenial to his own temperament, which naturally made such a distinction. As Howard has argued, Einstein was concerned to defend the autonomy of theory choice against his supporters among the logical positivists, who seemed to wish theory to be a logical consequence of experience.[53] One may say more (still following Howard's analyses):[54] Einstein's position is in some ways closer to Duhem's than Poincaré's. All of physics, including the nature of clocks and rods, is but imperfectly understood by Einstein, whereas some of physics was accepted without argument by Poincaré. So Einstein's idea of what stands or falls with what when geometry is applied in physics, or physical observations are interpreted theoretically, is much broader than Poincaré's.

Einstein's friend Moritz Schlick held similar views, indeed they may have influenced Einstein when he came to write "Geometry and Experience." In his *Allgemeine Erkenntnistheorie (General Theory of Knowledge)* of 1918, Schlick advocated a Duhemian position on measurement together with a Hilbertian position on implicit definitions in mathematics.[55] But Schlick was opposed to the neo-Kantians, even in

[53] Howard, "Albert Einstein, Philosophy of Science," *Stanford on-line encyclopedia*, http://plato.stanford.edu/entries/einstein-philscience.

[54] Howard 2004.

[55] Schlick's ideas are discussed briefly below in §7.3.2. The book was translated as *General Theory of Knowledge* by A. E. Blumberg and published by Springer-Verlag in 1974, then republished by Open Court in 1985.

the sophisticated form of Cassirer, and so he argued that what Cassirer saw as constitutive principles are more like conventions than either contingent or apodictic a priori components of perception. Einstein was sympathetic to this position, but some years later Schlick wished to push his position into a form where it conspicuously clashed with Cassirer's idea of incremental convergence to the best account. So in 1924 he argued that there was an empirical element to the determination of the metrical structure of space-time (thus making measurement much less problematic and moving toward Poincaré's geometric conventionalism). The scope of convention was now restricted to operationalizing of primitive concepts, and all others acquire their meanings from these acts. Einstein apparently shared Schlick's feeling that Cassirer's ideas were too accommodating of change to be adequately precise, but he would not endorse Schlick's new position because he could not see a principled distinction between the statements one chooses to regard as conventions and the rest.[56]

5.3 Measurement

5.3.1 Classical and Representational Theories

There are a number of ways one can think about measurement. Up close the question might be one of instrumentation (how can I get a thermometer there?) or lack of access (how big is this star?). Or it might be one of comparability (is this length here the same as that length there?). One can take a step or two back and ask what sense it makes to attempt such comparisons (is it meaningful to ask if an hour today is as long as it was a year ago?). One can wonder, as a mathematician might, about what sorts of numbers or quantities can be offered as measurements. A philosopher might ask what sorts of things can be measured, and if measuring a length and measuring, say, happiness is in any useful way the same sort of activity. All of these questions were raised in the late nineteenth and early twentieth centuries.

What Michell (1992) usefully called the classical theory of measurement is the idea that measurement establishes a ratio between some aspect of the object being measured and a convenient unit.[57] Length along a line is the simplest case. The object to be measured (the side of a field, let us say, is a quantity, meaning that it can be infinitely divided and is measurable (if it is not infinitely divisible it can be counted, rather than measured). Moreover, it is a quantity of the same kind as the quantity with which it is being measured (the yardstick), and the result of the appropriate comparison is measurement, which is a ratio between the side of the field and the yardstick. This kind of idea goes back at least as far, in a sophisticated form, as Newton in his *Arithmetica universalis* of 1707, and in the nineteenth century it became widely agreed that the concept of ratio could be applied to incommensurable pairs and can yield a number. That is to say, a comparison of the diagonal of a square with the side of the square (taken as a unit) now yielded a number ($\sqrt{2}$). In general,

---

[56] See Howard 2004, 15–19.

[57] In his book (1999) Michell seems much more critical of the representational theory and its application in psychology. There he seems to suggest that measurement must by definition be as the classical theory has it, and anything else, such as scales of happiness, is merely qualitative.

the classical theory of measurement discusses quantities that may be added, divided indefinitely (or else they would be countable), and admit an arbitrary, conventional choice of unit, so that the magnitude of every quantity may be expressed as a ratio of the quantity to the chosen unit. Quantities naturally come in sundry incomparable types, such as lengths, areas, and volumes.

The representational theory of measurement, which Michell attributes to Russell, does not assume that the result of a measurement is a number, even if the outcome is quantitative. The example to think of here is happiness. It can make good sense to say someone is happier today than he or she was yesterday, perhaps much happier, but very little sense, if any, to imagine these estimates can be turned into numbers. Evidently, the representational theory includes the classical one.

It may readily be granted that the nineteenth-century physicist dealt whenever possible with quantities capable of measurement, and that the measurements were numbers, so that physical quantities could be described as mathematical variables. As discussed above, Helmholtz asked himself what must be the case before some combination of two objects is such that some attribute of the objects can be considered a quantity and so expressed by a number. He argued that it was necessary that if two quantities are each equal to a third, then they are equal to each other, and also that the sum of similar quantities must be similar to the summands. The measurements must also obey the commutative and associative laws. Examples are: weighing objects in a balance, and adding lengths by joining the objects in a straight line. Curiously, this account of measurement leaves out one conspicuous success of nineteenth-century physics: Friedrich Mohs's elaboration of a scale of hardness (see below, §5.3.9), which is entirely qualitative.

5.3.2 Poincaré

Helmholtz's view was that there are certain properties of objects that exist independently of us and can be measured. By abstracting from familiar cases, one can specify what measurement involves and so be in a position to measure new properties of objects. Poincaré came close to denying this for one of the most fundamental quantities in physics: time.

Poincaré had developed a deep interest in measurement, but from a novel perspective. In February 1897 he was appointed the secretary of a commission of the Bureau des Longitudes to examine the question of whether France should scrap the twenty-four-hour day and the 360 degree circle in favor of a truly decimal system.[58] The background here is that France was losing the international argument in favor of the zero meridian of longitude passing through Greenwich, which the British and Americans preferred and which suited the maps used by 75 percent of the world's shipping. Strict decimalization was either an attempt to install a rational metric system (French in origin) or, at best, a delaying tactic. Despite recognizing that the metric system of mensuration had succeeded but attempts to decimalize time during the French Revolution had failed, the commission voted to decimalize time. Poincaré's own preference was for a twenty-four-hour day but a degree circle divided

---

[58] For a full account of these developments, set in the context of a need to globalize time, see Galison 2003.

into 240 parts. The same commission, with Poincaré's support, even revived an old proposal of the French Revolution to divide the circle into a new "degree," with 400 "degrees" in a circumference. It is best simply to note that France finally adopted the Greenwich meridian on March 9, 1911.

All this comic nationalism had a serious side. Both railway timetables and accurate maps of the globe required the determination of longitude, and the recently invented and rapidly spreading electric telegraph was crucial to bringing this matter about.[59] For the first time it was easy to see how one could precisely determine the time in New York when it was noon in London. Poincaré was heavily involved with all the work on the telegraphic determination of longitude, and it forms the background to the paper that concerns us here, his essay "La mesure du temps" (or, "The Measure of Time"), published in 1898.[60]

One of his concerns in the essay was psychological time, which, we all feel, can pass painfully slowly or only too quickly. Can it be quantified? And what about time in different worlds? Poincaré based his analysis on the observation that we do not have a direct intuition of two intervals of time. There is no objective sense in which an hour today is as long as an hour yesterday, unless we make some further conventions. We might, perhaps, count the beats of a pendulum; but they depend on temperature, air resistance, and many other factors. These can be allowed for, of course, but an element of imprecision irretrievably creeps in. Let us leap over it, he said, and imagine a perfect clock. We suppose that the duration of two identical processes is the same. Could our experience overthrow this idea?

Poincaré answered his own question affirmatively. We can imagine that the processes in Paris and Quito (the place names are not in Poincaré's essay) once kept the same time but now do not, and the process in Paris happens in less time than that in Quito. This is because all that we can say is that one complex of causes and another almost identical complex produce almost identical outcomes in almost identical times. Time can only be measured with respect to some agreed-on sort of clock. Over time, or when great distances are involved, the near identity may not be enough to measure time properly. Consider, said Poincaré, a current issue: astronomers claim that the daily rotation of the Earth is slowing down. They note the effect of the tides and cite the law of conservation of energy; they observe the motion of the Moon and compare it with what Newton's laws predict, and they explain the discrepancies by proclaiming that the Earth is slowing down. Now, if the Earth is the clock, one may say either that time is slowing down or switch to a new clock of some sort, but then Newton's laws would only hold in an altered sense.

The astronomers, at least, would hold out for a definition of time that made the laws of mechanics as simple as possible, and so one would choose a clock on the only possible grounds: convenience. There is not, he said, a way of measuring time that is more true than any other, only one that is more convenient.

Nor was this all. Suppose, said Poincaré, that you hear thunder and recognize that it has been caused by an electric discharge. This physical phenomenon is prior to the sound, because it is the cause of it. But very often when two phenomena are constantly associated, we say that the earlier one is the cause of the other. So sometimes we allow our sense of causality to determine our sense of time order, and sometimes

---

[59] See Kern 1983.
[60] Reprinted in Poincaré 1905b, 35–58.

we allow our sense of time order to determine our sense of causation. We cannot do both without creating a vicious circle. Once again we must conclude that the way out is to rely on the best available science, and, therefore, he concluded: "The simultaneity of two events, or the order of their succession, the equality of two durations, must be defined in such a way that the statement of the laws of nature is as simple as possible. In other words, all these rules, all these definitions are only the fruit of an unconscious opportunism."

Poincaré's view of the concept of time was therefore a deeply operationalized one, very much indeed what one might expect from someone engaged in establishing a global network of synchronized clocks. He did not agree that time was what Newton or Kant had taken it to be. It was something that only made sense with respect to a measuring instrument. That said, his theory did operationalize time as Newton had regarded it: the sense in which separated observers could agree on the simultaneity of an event took no account of their relative motion, as Einstein was to consider not many years later.

The measurement of time differs from the measurement of length in one crucial respect: whereas one can place one brick alongside another and compare their lengths, one cannot position a minute yesterday alongside a minute today. There is no way in which one can make such a comparison directly; one can only do so by comparing this clock and that clock in the same place and at the same time. It might seem that this is the same "truism" that attends discussion of length. If all it means for two rods to have the same length is that they coincide when brought together, then there is no sense in the question, "Are they changing their lengths as they move?" However, there is a crucial difference when space (unlike time) is allowed to have two or more dimensions.

In space, one may start with a square (a figure with four equal sides, four equal angles) and attempt to construct a chessboard. If one succeeds, one is inclined to say space is Euclidean, otherwise non-Euclidean or spherical. Suppose one fails. Poincaré's conventionalism is the claim that either space is non-Euclidean or it is Euclidean and equipped with a force that pushes the sides of the square out of shape.[61] This is because there are mathematical invariants associated with manifolds of dimension 2 or more, such as Gaussian curvature and its Riemannian generalizations, that do not have analogues in one-dimensional spaces. Poincaré's criticisms of naive beliefs about time are, for that reason, easier to accept, because they concern what can be measured and do not directly raise the issue of what is physics and what is geometry, if indeed that distinction can be made.

5.3.3 Measuring the Infinitesimal

5.3.3.1 DU BOIS-REYMOND'S AND STOLZ'S NUMBERS

The classical theory of measurement supposes that quantities are intrinsic and can be measured by making comparisons between objects of the same kind, usually by comparing the objects with a fixed one (the unit). The outcome is a ratio. Poincaré's comments on this matter may be taken to come from a practically minded philos-

---

[61] For a full mathematical discussion of this point, and its negative implications for Reichenbach's later conventionalism, see Friedman 1983.

opher. They did not engage with a further aspect of the discussion around 1900, which concerned the kinds of numbers that results of measurements could be.[62]

An early indication that there might be unexpected things to say came from Johannes Thomae's remarks in his *Abriss* (outline).[63] This was one of the first textbooks on complex function theory, and remarks in it earned him, to quote Cantor, "the questionable fame of having infected mathematics with the cholera bacillus of infinitesimals."[64] In fact, Cantor was very jealous of his transfinite numbers and what he saw as their largely negative implications for the study of infinitesimals, and he was on this occasion simply wrong. Thomae was also the butt of many criticisms leveled at his work by his colleague at Jena, Gottlob Frege; one wonders what life was sometimes like in that small university.

Thomae was interested in how functions go to zero. The order of vanishing of the function f at the point a, where $f(a) = 0$, is the rate at which $f(x)$ tends to zero as x tends to a. It is clear from drawing graphs that the function x^2 goes to zero as x tends to zero faster than x, for example. The analysis extends in a straightforward way to rational and even irrational power of x, but Thomae noted that the natural logarithm, the function $\ln(x)$, goes very slowly to $-\infty$ as x remains positive and tends to zero, so the function $\frac{1}{\ln(x)}$ goes to zero very slowly as x tends to zero; slower, in fact, than any power of x. There are even slower functions, for example $\left(\frac{1}{\ln(x)}\right)^2$, and $\frac{1}{\ln(\ln(x))}$, and Thomae soon found he had a class of functions on his hands whose orders of vanishing could be called measures, and "these measures constitute a one-dimensional, continuous manifold (in the sense of Riemann) for the determination of which all our ordinary rational and irrational numbers do not suffice."[65]

Thomae's remarks were shortly followed by du Bois-Reymond's more extensive investigation of rates of growth of functions.[66] The idea is another very natural one. It is clear that x^3 goes to infinity faster than x^2 and slower than x^4, so although all three functions become infinite with x (whatever that might mean), there is something to say about how fast they become infinite. Du Bois-Reymond discovered there was quite a lot to say. He said that

> $f(x)$ has an infinity greater than that of $g(x)$ if $\lim_{x \to \infty} f(x)/g(x) = \infty$, and wrote $f(x) \succ g(x)$
>
> $f(x)$ has an infinity equal to that of $g(x)$ if $\lim_{x \to \infty} f(x)g(x) = a \in \mathbb{R}^+$ (the positive real numbers), and wrote $f(x) \sim g(x)$; and
>
> $f(x)$ has an infinity less than that of $g(x)$ if $\lim_{x \to \infty} f(x)/g(x) = 0$, and wrote $f(x) \prec g(x)$.

His work was not always accurate, but he went a long way to showing that something about functions was being measured, but the measurements were non-

---

[62] I am indebted to Philip Ehrlich for much of this discussion and for the papers he sent to me on the subject to communicate to *Archive for History of Exact Sciences*, now published as Ehrlich 2006. Those papers contain a much fuller discussion of these ideas in the historical and mathematical setting than can be attempted here.

[63] Thomae 1870 and subsequent editions.

[64] In Meschkowski 1965, 505.

[65] Quoted in Earlich 2006, 55.

[66] In, for example, du Bois-Reymond 1882.

standard. For example, it is reasonable to say that the function $x^{p/q}$ is infinite of order p/q, but the logarithm function $\ln x$ goes to infinity slower than any power of x and yet it does not deserve to be measured as zero. Likewise, the exponential function e^x goes to infinity faster than any power of x, but one is reluctant to say it grows too fast to be measured. In fact, du Bois-Reymond showed that given any series of functions $\phi_1 > \phi_2 > \ldots > \phi_n > \ldots$, there is a function f that has a lesser infinity: $\phi_n > f$ for all n. Borel later reworked another of du Bois-Reymond's discussions into this theorem: given $\phi_1 < \phi_2 < \ldots < \phi_n < \ldots$, there is a function f that has a greater infinity: $\phi_n < f$ for all n. This result suggests, but does not prove, that the orders of infinity of functions form a continuum different from, and richer than, the number continuum. Such was Poincaré's opinion as expressed in his 1893 paper (pp. 52–55).

Du Bois-Reymond also inspired Otto Stolz, who in 1885 produced a set of mathematical objects that could be measured, but whose measurements did not form an Archimedean continuum. He took the functions that could be obtained from positive rational powers of x and iterated exponentials and logarithms ($e^x, e^{e^x}, e^{e^{e^x}}, \ldots$, and $\ln x$, $\ln (\ln x)$, $\ln (\ln (\ln x))$). He assigned an infinity to each function f, which I shall denote here as $I(f)$, with the properties that

$I(f) > I(g)$ if and only if $f > g$,
$I(f) = I(g)$ if and only if $f \sim g$, and
$I(f) < I(g)$ if and only if $f < g$. Moreover, $I(f) + I(g) = I(f.g)$ and
$I(f) - I(g) = I(f/g)$ if $I(f) > I(g)$.

Du Bois-Reymond had offered a characterization of magnitudes in his work that seemed to hint at his perception that his orders of infinity did not allow that any $f < g$, then some multiple of f, say nf, is such that $g < nf$. Stolz's version was much more precise and was cast in the form of an axiomatization of what he called absolute magnitude.[67] These run as follows:

(i) If $A = B$, then $B = A$;
(ii) If $A > B$, then $B < A$ (and conversely)
(iii) $A = B$ or $A > B$ or $A < B$ (and precisely one is the case)
(iv) If $A = B$ and $B = C$, then $A = C$
(v) If $A = B$ and $B > C$, then $A > C$;
(vi) If $A > B$ and $B > C$, then $A > C$.

The operation "$+$" is defined so that $A + B$ is in the system whenever A and B are, and for all A, B, and C in the system it satisfies

(vii) $A + (B + C) = (A + B) + C$;
(viii) $A + B = B + A$;
(ix) $A + B = A' + B'$ if $A = A'$ and $B = B'$;'
(x) $A + B > A' + B'$ if $A > A'$ and $B = B'$;
(xi) $A + B > A$;
(xii) $A = B + X$ for some X in the system whenever $A > B$.
(xiii) For each member A of the system and each positive integer n, there is an X in the system such that $nX = A$.

---

[67] Du Bois-Reymond 1885.

It is characteristic of the modernist shift that the relations of equality and greater, or lesser, than are being defined, as are the rules for addition. An investigator of Helmholtz's generation would have taken these relations for granted and highlighted their role; here they are being introduced as new, and all their properties are spelled out. Not only did Stolz mean to exclude any properties he did not mention that do not follow as theorems from the ones he stated, he also specified some properties that less logically minded authors might have smuggled in, such as (xii).[68] It is for such reasons, as well as the overall quality of his reasoning, that Stolz is rightly hailed by Ehrlich as one of the founding fathers of the theory of non-Archimedean magnitudes.[69]

The concept of magnitude being precisely defined, Stolz could show what du Bois-Reymond had merely hinted at: the magnitudes of his class of functions form a non-Archimedean continuum. First he showed that they satisfy his axioms for an absolute magnitude. Then he noted that $I(e^x) > I(x^n) = nI(x)$ for all integers n, so although $I(x) < I(e^x)$ there is no n such that $n(Ix) > I(e^x)$. Stolz is therefore the first to show rigorously that there are non-Archimedean magnitudes, and that, moreover, they arise in a situation that is not wholly artificial.

He then attempted a proof that given assumptions about the continuity of the class of magnitudes, the magnitudes will be Archimedean, thus indicating where the new kinds of magnitude could expect to be found. The proof was defective and was criticized by Veronese and later by Hölder, and before Stolz could provide a correct proof, one was given by Bettazzi.[70]

5.3.4 Bettazzi

In 1890, the year Rodolfo Bettazzi turned twenty-nine, he began teaching at the Liceo Cavour in Turin, where he would stay until 1931.[71] His paper (Bettazzi 1890), which drew a long and glowing report from Vivanti in Darboux's *Bulletin*, was a prize-winning paper of the Lincei for 1888. In it, Bettazzi was concerned to establish a theory of magnitudes of either an Archimedean or a non-Archimedean kind, with examples in the spirit of du Bois-Reymond's work.

His approach was entirely axiomatic. It is also unavoidably technical, and readers may skip, if they wish, to the discussion of the Bettazzi-Vivanti debate below. Bet-

---

[68] Klein 1926–27, 152, praised Stolz for his logical mind.

[69] Ehrlich 2006, 9.

[70] Bettazzi 1890. For details, see also Ehrlich 2006.

[71] Bettazzi also taught at the military academy there from 1892 to 1932. In 1895, he founded *Mathesis*, an organization devoted to the proper teaching of mathematics in secondary schools. It was deeply interested in the preparation of teachers and was successful in bringing the importance of teaching to the attention of university mathematicians. Bettazzi threw himself into *Mathesis* and managed to draw in many of the luminaries of Italian mathematics. At the first congress, in 1898, Enrico d'Ovidio (the head of the Faculty of Mathematics at Turin University) presided, Giuseppe Peano and Corrado Segre were the vice-presidents. Ominously, as it would turn out, among the major items of discussion was whether to base the teaching of geometry on plane or spatial geometry (see below). Later presidents of the society included Severi, Castelnuovo, and, from 1919 to 1933, Federigo Enriques, the three most important algebraic geometers of the time. Bettazzi died in January 1941.

tazzi said that the elements of a set Γ of objects were quantities if, given any two, A and B, either $A = B$ or $A \neq B$ (but not both), where the symbols $=$ and $\neq$ are defined by the rules:

$$\text{If } A = B, \text{ then } B = A;$$
$$\text{if } A = B \text{ and } B = C, \text{ then } A = C.$$

He then considered an operation S that combines quantities into another quantity, say $S(A, B, C, \ldots) = M$, which is commutative and associative, and moreover satisfies:

$$\text{If } B = C, \text{then } S(A, B) = S(A, C), \quad \text{and if} \quad B \neq C, \text{ then } S(A, B) \neq S(A, C).$$

This allowed him to define multiples ($S(A, A), S(A, A, A)$ and so on) and differences. The difference of two quantities $S(A, B)$ and A is defined this way: $D(S(A, B), A) = B$.

He was most interested in what he called a one-dimensional class of quantities, which satisfy these five axioms with respect to an order relation $<$ and $>$:

$$\text{If } A > B \text{ and } A' = A, \text{ then } A' > B$$

$$\text{If } A > B, \text{ then } B < A$$

$$\text{If } A = B \text{ and } B > C, \text{ then } A > C$$

$$\text{If } A > B \text{ and } B > C, \text{ then } A > C$$

$$\text{If } A > B, \text{ then } S(A, C) > S(B, C).$$

It is obvious that these axioms are intended to say something about quantities such as lengths and heights. In keeping with this, Bettazzi next further narrowed his attention to what he called *proper classes*, those for which, if A and B belong to the class and $A > B$, then $D(A, B)$ also belongs to the class.

Bettazzi now showed that it was possible to divide any such class into two subclasses, Π_1 and Π_2, with the property that every element of the class belongs to exactly one subclass, and, if P_1 belongs to Π_1, then so does every P' for which $P' < P$, and if P_2 belongs to Π_2, then so does every P'' for which $P_2 < P''$. In the case when Π_1 does not have a maximum element and Π_2 does not have a minimum, one of two things can happen. Either, given any quantity A one can find P_1 in Π_1 and P_2 in Π_2 (depending on A) such that $D(P_1, P_2) < A$, or one cannot. In the former case, Bettazzi, following Dedekind, said one had a *section*. In the second case, he said there was a *jump*.

The other possibilities are what Bettazzi called *succession*: Π_1 has a maximum quantity and Π_2 a minimum; and *liaison*: either Π_1 has a maximum quantity or Π_2 a minimum quantity (but not both).

After all these definitions, Bettazzi started to prove theorems, of which this one must suffice to indicate his awareness of Cantor's point-set topology: a class with no successions and no jumps is connected.

Bettazzi now considered classes without jumps, and added the axiom of Archimedes to his list. This asserts that, given two arbitrary elements of a class Γ with no jumps, then there is a multiple of the one that exceeds the other. A class without jumps is of the first kind if the axiom of Archimedes is satisfied, and otherwise it is of the second class. Bettazzi was able to characterize the classes of the first kind, and then to turn his attention to the less well understood classes of the second kind. Among these, he noted, are Stolz's moments and du Bois-Reymond's infinities of functions.[72] Bettazzi then gave a taxonomy of classes of the second kind, indicating the different structures classes of non-Archimedean magnitudes might have.

In the second part of his paper, Bettazzi analyzed numbers as a special family of quantities. He said that two quantities A, B have the same or a different number if $A = B$ or $A \neq B$. If α and β are these new (number) concepts associated with A and B, respectively, then the number corresponding to $S(A, B)$ must be $\alpha + \beta$. Without this being a true and proper definition of a number, this is enough to establish number as a quantity. The correspondence is called metrical if $\alpha > \beta$ implies $A > B$.

As Bettazzi showed, numbers can be applied to a wide variety of different quantities. It is here that he showed that a one-dimensional directed class of the first kind, Π, can always be included in a continuous class Γ (his definition coincided essentially with Cantor's) which is in a metrical correspondence with the real numbers. This metrical correspondence is not unique, but it becomes so once an arbitrary choice of quantity corresponding to the number 1 is made. It follows that a one-dimensional Archimedean collection of magnitudes can always be mapped onto a subset of the real numbers in such a way that an arbitrary magnitude A maps to the number 1. He called this quantity A the "measure of the magnitude with respect to the chosen unit." Bettazzi also had a theorem (which we shall not discuss) about how certain kinds of non-Archimedean magnitudes could be represented.

5.3.4.1 THE BETTAZZI-VIVANTI DEBATE

If Vivanti was happy to sing Bettazzi's praises in Darboux's *Bulletin*, it was not so clear that he agreed about what had been done. Peano, the arch-rigorist of the Italian mathematical community and a leading professor at the University of Turin, was moved by the work of Bettazzi, Vivanti, and Veronese—a confused but provocative work we shall look at below—to open the pages of the journal he was in the process of founding, the *Rivista di matematica*, to a debate on the existence of infinitesimals. The debate is interesting for the light it sheds on shifting attitudes toward mathematical autonomy and the changes this had for the concept of mathematical existence.

Vivanti began.[73] He reviewed the many conflicting definitions that had been offered of infinitesimals, and singled out the idea that an infinitesimal is a quantity which, when repeated any finite number of times, still does not exceed any finite quantity. The question then becomes, he said: Do infinitesimals exist? Cantor, he noted, claimed that they do not, but he found Cantor's argument unsatisfactory. Accordingly, he then offered an argument of his own, which readers would also find substandard. This depended on a more philosophical position concerning

---

[72] Du Bois-Reymond 1882, §69.
[73] Vivanti 1891a.

the nature of mathematical objects. Usually, he said, a mathematical object is an idealization of a natural object, and the idealized properties of the object form part of the axioms or the theorems pertaining to the idealized object. Today's mathematics, however, has gone beyond this object to describe completely arbitrary objects, provided only that their properties do not contradict one another. Such entities, he suggested, should be regarded as conventional. To this philosophical perspective Vivanti added a decidedly more implausible mathematical claim that the idealized straight line had topological properties that force it to be an Archimedean magnitude.

He then considered contrary arguments, notably one with a long historical pedigree (and quite some life left in it): Is the line made up of points, or of infinitesimal pieces? Du Bois-Reymond, Veronese, and Peirce had all argued for an infinitesimalist position (as later did Hermann Weyl). Vivanti judged these arguments to be an evasion of the central issue. Like a good lawyer, he now moved to his summing up. Certainly the calculus, formerly the home of all sorts of infinitesimals, no longer needed them in any form. Nor did the concept of magnitude as represented by line segments: in that setting infinitesimals led to logical contradictions and therefore did not exist. As for the non-Archimedean magnitudes of du Bois-Reymond, Stolz, and Bettazzi, they are matters of convention.

Bettazzi replied in the same issue.[74] He dismissed the distinction between existence and matters of convention on the grounds that freedom from contradiction was the only ground for mathematical existence. Otherwise, he said, we should not study spaces of more than three dimensions or non-Euclidean geometry. Therefore, infinitesimals exist, and the question, for Bettazzi, became: In which domains in mathematics do infinitesimals exist? He too found fault with Cantor's argument, but he was willing to agree that if a line segment, "is to be useful in material practice, it must be defined with the postulate of Archimedes, which excludes the infinitesimal."[75] That said, this conclusion no more excluded segments with infinitesimals from study than did the behavior of parallels the geometry on the pseudosphere.

To this rebuttal Vivanti replied that he had perhaps not expressed himself clearly.[76] All he had meant was that it was only interesting to discuss *existence* when discussing natural phenomena. Here, somewhat Kantian considerations of how we must, inevitably, conceive of the continuum compel us, he argued, to regard its continuity as of a kind that excludes infinitesimals. The debate ended with Bettazzi's reply, which was to the effect that he and Vivanti agreed that the concept of an infinitesimal is not contradictory, but that infinitesimals are not among the kinds of segments needed in studying natural phenomena. But if the straight line is defined in a way that allows for the usual concepts of greater, smaller, sum, and difference, the choice between an Archimedean and a non-Archimedean postulate is logically free.

The use of axiomatic reasoning to clarify what is meant, the insistence that in mathematics existence means only freedom from contradiction, the lack of interest in idealizing from nature—all mark Bettazzi's position as characteristically modernist.

---

[74] Bettazzi 1891a.
[75] As quoted in Ehrlich 2006, 92.
[76] Vivanti 1891b.

5.3.5 Veronese

Guiseppe Veronese's *Fondamenti della geometria* of 1891 divided its audience more than almost any book of the period. Among those who admired it was David Hilbert, while it irritated a number of Veronese's fellow Italians, especially those around Peano. It did so by its mixture of novel mathematics, which was highly counterintuitive even as it claimed to be based on intuition, and naïve philosophizing. It was not clearly written, and it was not clear that its arguments were sound. The least of its claims was that one might speak of geometry in any number of dimensions and even have intuitive access to the geometry of four dimensions. Much more provocatively, Veronese claimed that geometry might reasonably concern itself with non-Archimedean spaces, actual infinitesimals and actual infinites.[77]

His *Fondamenti* opens with an argument that mathematics is about concepts and so is derived from the laws and operations of thought. The laws are those of identity, contradiction, and the excluded middle. The operations are those of thinking of a thing, abstracting when thinking about a thing, uniting, and comparing. What makes this activity mathematical is that the raw ingredients are the concepts of unity, plurality, more, and less. Veronese discussed these for a while, and then introduced what he called the law of continuity of thought: given a thing *A*, if it is not proved that *A* exhausts all the possibilities to be considered, then one can think of another thing, not contained in, and independent of, *A*. As Cantù notes, Cohen, Natorp, and the philosophers of the Marburg school had had a similar idea, but Veronese presented it as an intuition, which the neo-Kantians could not accept.

Veronese's principal aim was to describe, abstractly, the intuitive linear continuum. To defend his abstract ideas he argued that they corresponded with the observed continuum, a difficult position somewhat forced on him by the fact that Dedekind had made the Archimedean continuum mathematically rigorous. His style rather resembled that of Pasch, in that sundry observations are invoked to support what then become eight hypotheses, the raw ingredients for deductive reasoning. Veronese's argument appears to be that, on the one hand, there is nothing self-contradictory in the idea of the non-Archimedean infinite, and that indeed we can form a good intuitive impression of actual infinites and infinitesimals. For example, the law of continuity of thought licenses Veronese's third hypothesis, that one can always conceive of quantities larger than any finite multiple of a given quantity. On the other hand, we are also invited to discover these nonstandard objects through intuition, perhaps as a result of some education (Veronese was drawn to Klein's distinction between naive and refined intuition).

One way or another, he argued, we can construct new infinite numbers while avoiding the Archimedean axiom. These were composed of various kinds of infinity, which can be written as, for example, $\infty_1^m n \pm \infty_1^{m-1} n_1 \pm \ldots \pm n_m$.[78] Poincaré, in his report on the work of Hilbert for the Lobachevskii Prize, mistakenly assumed this was Veronese's attempt to apply Cantor's transcendental numbers to geometry. Veronese rebutted this assumption in a paper in which he showed that his new

---

[77] Our guide here is the lucid and accurate book, Cantù 1999.

[78] Veronese knew Dedekind's work but not Frege's, and like Helmholtz, he started with the concept of ordinal number.

numbers had much more natural algebraic or arithmetic properties than Cantor's, and thus began a famous controversy with Cantor. Cantor firmly denied the existence of actual infinitesimals, which he compared to a cholera bacillus; in fact they make good sense, as Veronese's pupil Tullio Levi-Civita and later Gino Fano were to show.[79] Veronese also defined continuity in the context of a non-Archimedean continuum, as Brouwer (1907, 49–50) noted without demur.[80]

In the lengthy and informative appendix to his *Fondamenti* (1891), Veronese gave a detailed discussion of the work of many of his predecessors. He disagreed strongly with Riemann that a geometric quantity was something synonymous with an *n*-tuple of real numbers. He wrote about language, where he was strikingly unsympathetic to Peano, whose *signicismo* (signism, as he called it) he found pedantic.[81] In his view (p. 606), in mathematics it is enough that the content of a proposition has the sense the author wishes to give it, and seas of words that signify nothing are not the result of an imperfection of language but a vacuity of ideas. He agreed that Peano's (1889) account of the foundations of geometry was a notable advance on Pasch's in that it imposes no restriction on the validity of his axioms, but it did this at the cost of becoming unable to say why they are mutually compatible. On the subject of higher dimensions he nodded rather less critically at Zöllner before noting that the topic originates with Grassmann's *Ausdehnungslehre (Extension Theory)*. A considerable number of authors were then reviewed, including the hostile Lotze, who in his *Logik* (1874, 217) had warned against systems of four or five dimensions as the "grimace of science," deceiving us with completely useless paradoxes.[82]

Veronese was of the opinion that his system of geometry was mixed in its premises. Some were empirical, being derived from the perception of physical facts, some semiempirical, which meant premises that had originally been empirical but were now being applied in other contexts, and some were entirely abstract. This made it a halfway house between Euclid's *Elements* and the system shortly to be preferred by his Italian contemporary, Pieri, and by Hilbert, where a consistent set of axioms creates its subject matter. Veronese nonetheless took the modernist line about mathematics: it dealt with abstract existence, which is simply that the axioms be free from contradiction. Abstract geometric space is a concept, it is not intuitive space. Rigor was needed in definitions, postulates, theorems, and proofs, which should be purely abstract, involve nothing indeterminate, and truly deductive. Unlike Pasch, but in agreement with Klein, Veronese advocated the use of figures in geometry, but with Pasch insisted the proofs be independent of the figures.

Over a decade later, in his inaugural address at Padua in 1905, Veronese took a hard line on the relations of mathematics and philosophy. If, he suggested, philosophy is intended to be the highest level of research into truth, then mathematics does so much better than philosophy "being not only the most ideal but also the most

---

[79] See Levi-Civita 1894 and Fano 1932, 435–511, which has a picture of Veronese's non-Archimedean continuum on p. 501, as Cantù points out.

[80] See Brouwer 1975, vol. 1, 13–101.

[81] As indeed it is. In his comments on Pasch in his *Principii* §5, in *Opere scelte*, vol. 2, 84, he lamented the ambiguity of natural language and gave the example of the phrase "a line through any two points," which he criticized for not specifying that the points must be distinct, and for not saying if the points are chosen successively or simultaneously. His observations are correct, but an argument for their significance is lacking.

[82] Veronese, 1891, 614 n.2.

positive of the sciences, because it is the oldest and the most precise expression of the truth."[83] It is not subject to refutation or revision, as are the sciences, nor is it merely approximate. A mathematical truth, once established logically and without error, can never be refuted. The laws of mathematics are the laws of thought, and they exist in a marvelous harmony with the external world.

These naive expressions could only be justified if the objects of mathematics were to be obliging. For Veronese, these are abstract concepts. Some correspond, more or less accurately, to real existing objects, others have an entirely abstract reality. A space in which the axiom of Archimedes is false need not have a concrete existence but it does have a formal existence.[84] So he found du Bois-Reymond's distinction between idealists and empiricists inadequate because it infringed on the liberty of the mathematician. Indeed, Veronese took every opportunity to enlarge mathematics at the expense of philosophy, as many do who have expertise in only one area. Truth and existence in mathematics were, he claimed, for the mathematician to decide, only reflection on the empirical or purely intellectual nature of mathematical ideas was for the philosopher. This separation doubtless had its advantages in warning philosophers to keep away from his new non-Archimedean geometry. They might study questions about the origin of the infinite and the infinitesimal, and reflect on the general character of human knowledge, while the mathematician developed the axiomatic form of a theory.

The odd result of Veronese's work was that it seems to have caught the attention of Hilbert, whereas the more precise but less imaginative work of the Peanians did not. It may be that non-Archimedean geometry resonated with Hilbert's own growing interest in the foundations of geometry in the 1890s; it may be that Klein, who knew Veronese, recommended his book to Hilbert. Yet one doubts if even Hilbert was persuaded by the *Fondamenti*'s strange mixture of abstract mathematics and ontological claims for the new geometry, and the defenses offered by Veronese's supporters moved the geometry firmly toward pure mathematics. On the other hand, the fact that this non-Archimedean geometry was not palpably absurd or completely abstract may have leant substance to its later study, and to a deepening sense of the novelty of geometry.

5.3.6 Hölder

Notwithstanding the papers by Stolz and the insights of Bettazzi, it is a paper by Hölder in 1901 that became definitive: his "The Axioms of Quantity and the Study of Measurement."[85] Hölder began by noting that one may speak of the axioms of arithmetic in two senses: a narrow sense to do with the natural numbers, such as Helmholtz had discussed, and a broad one, of present interest, to do with axiomatizing quantity. Hölder brushed past Helmholtz's idea (originally due to Grassmann) that the rule $a + (b + 1) = (a + b) + 1$ should be taken as an axiom of arithmetic, saying that it was better to take it as a definition. Indeed, he was not much impressed with the variety of assumptions offered concerning arithmetic, dismissing them all as

---

[83] Quoted in Cantù 1999, 237.
[84] Veronese 189 1n.2, quoted in Cantù 1999, 243.
[85] "Die Axiome der Quantität und die Lehre vom Mass" (Hölder 1901).

merely logical (and noting in a footnote that this did not mean they could be reduced to the logical formalism of philosophy).

The axioms of geometry and mechanics were another matter, however. No certain axioms were forced upon one by experience, or intuition as some would say, and he proposed to show that the general theory of measurable quantities can be based on a certain number of facts, which he proposed to call the *axioms of quantity*. These axioms were more general than those of geometry, as he proceeded to show. First, he gave his axioms:

(i) Given any two quantities, either they are equal or one is greater than the other.
(ii) To any given quantity there is a smaller one.
(iii) Any two quantities a and b have a sum, in this order, $a + b$.
(iv) $a + b$ is greater than both a and b.
(v) If $a < b$, then there is an x such that $a + x = b$ and a y such that $y + a = b$.
(vi) It is always the case that $(a + b) + c = a + (b + c)$.

Finally, and for Hölder's work most importantly:

(vii) if the collection of quantities is divided into two classes such that every quantity belongs to one of these classes and every member of the one class is smaller than every member of the other, then there is a quantity z which is greater than every member of one class and less than every member of the other. This quantity z may belong to either of the two classes.

The shrewd reader may rightly suspect that these innocuous statements are set up to realize a theorem. Note, first of all, that the addition of these quantities is not said to be commutative, otherwise Hölder would not have needed to stipulate the existence of both x and y in axiom (v).[86] Second, axiom (vii) is plainly a way of bringing Dedekind's way of completing the real numbers into the study of quantity. And third, note that the Archimedean axiom is missing. Indeed, one of Hölder's aims was precisely to show that these seven axioms imply the axiom of Archimedes, while the rational numbers are an example of a system satisfying the first six axioms and the Archimedean axiom, but not axiom (vii).

Hölder worked to this conclusion quite swiftly. He showed on the way that the x and the y in axiom (v) are unique for given a and b, and that between any two distinct quantities there is another. Then, given a quantity a and an integer n, there is a quantity b such that $nb < a$. With this done he then established the Archimedean axiom as a theorem: given quantities a and b with $a < b$, there is an integer n such that $na > b$. Next he deduced that, after all, addition is commutative. The crucial point to note is that the proof makes essential use of the Archimedean axiom. He then showed that nth parts exist, that is, given a quantity a and an integer n, there is a quantity b such that $nb = a$.

Hölder noted that the modern theory, which he attributed to Newton, differed from the classical theory of Book 5 of Euclid's *Elements*. Euclid explained a proportion between four quantities; Newton a ratio between two. Moreover, provided

---

[86] Commutative means that $a + b = b + a$.

one took care to spell out Euclid's assumptions explicitly, the classical methods were rigorous, whereas that was not true of the modern ones.

Next Hölder showed how a pair of quantities a and b gave rise to a class of ratios, which he denoted $[a:b]$, consisting of those pairs of positive integers m and n such that $na > mb$ (informally, and strictly speaking meaninglessly), the ratios m/n such that $a/b > m/n$. With this concept in place, Hölder defined the measure of a sum of quantities, developed the modern theory of proportion, and showed how to multiply quantities. He showed that given any quantity a and any real number κ there is a quantity b such that $[a:b] = \kappa$. This more or less shows that any system of quantities satisfying Hölder's seven axioms is equivalent to the real numbers. The paper concluded with an account of how the theory worked in detail for geometrical segments.

Hölder's account did two things. It established a minimal set of axioms that defined quantities that could be measured by real numbers. It also showed a number of surprising connections between one's expectations about quantities: axiom (vii) implies that the Archimedean axiom implies commutativity of addition. Accordingly, if you wanted a non-Archimedean continuum, for example, something interesting would have to be given up. But it also showed, with greater clarity than hitherto, and for the more influential German audience, that novel theories of magnitude could make perfect sense.

5.3.7 Frege

Mathematicians and physicists differ to this day in what they mean by a vector. Physicists have two kinds: one modeled on the idea of a directed length (for example, so many steps north-east), and one which is a rotation of a certain amount about a given axis. Mathematicians abstract only from the first kind, but they do not first define a vector and then a vector space—they define a vector space first and then say that a vector is an element of a vector space. Frege's approach to quantity was similar to the mathematicians' approach to vectors. He sought to define what it is for a concept to be such that its extension is what he called a "quantitative domain." A quantity is something that belongs, with other objects, to a quantitative domain.

As Dummett discusses,[87] Frege seems uncharacteristically oblivious of the sorts of quantities there might be, other than real numbers and lines. He does not discuss vectors, or angles (which have an interesting twist to their nature: is the sum of four right angles to be understood as the zero angle, or not?). As for the real numbers, Frege had to show that there is a quantitative domain, and then that it is the right type. He wished to do so in a way that produced the rational and irrational numbers simultaneously, and then to exhibit all the real numbers as ratios of quantities (with respect to an arbitrary unit).

Frege abstracted from the idea that familiar quantities (such as line segments) can be moved around on the line, ordered in respect of size, added together and subtracted (Frege allowed negative quantities). So he granted himself a set and a group of transformations on that set, and proceeded axiomatically. The upshot of his work, which was technical and precise, was that he proved theorems that are an im-

---

[87] In Dummett 1991, chapter 22.

provement on the earlier work of Hölder discussed above, of which Frege seems to have been completely unaware. In particular, Frege showed that if the ordering is dense, linear, and Archimedean, then the group must be commutative.[88] Hölder had proved a similar result but with the extra assumption that the group acts in an order-preserving way, so that if $a < b$ and g is any element of the group, then $ga < gb$. As Dummett noted, "It is an injustice that, in the literature on group theory, Frege is left unmentioned and denied credit for his discoveries."[89]

Frege then developed an insightful theory of a certain kind of quantity, albeit one he only saw fit to publish in the no-man's land of his *Grundgesetze*, where it seems no one read it. It does have, as a particular case, the theory of real numbers as a type of quantity. It differs from other theories by constructing all the real numbers at once, not by constructing the rational numbers first. But it is subject to the underlying fatal flaw of the *Grundgesetze*, and cannot be said, even by Frege, to be a success.

5.3.8 Russell: Measurement as Ordering

Perhaps the most novel and lasting reinterpretation of the task of measurement was offered by Bertrand Russell in his *Principles of Mathematics*. It occurs in part 3 of the book, on "Quantity," beginning with the chapter entitled "The Meaning of Magnitude." Here Russell pointed out that, thanks to the work of Weierstrass, Dedekind, Cantor, and others, there was now a good account of arithmetic, and that there were now also branches of mathematics that did not concern quantity at all. Measurement, moreover, was in a mess, for there were quantities that could not be measured, and some things which are not quantities can be measured.

It is one of Russell's better qualities as a philosopher that he attempted to give definitions of the key terms. A quantity, he explained, is anything that can be quantitatively equal to something else. Quantitative equality required a lengthy explanation, the conclusion of which was that when two examples of a concept may be said to be greater or lesser than one another, and the examples are being compared as to their relative magnitude, then they are magnitudes. Spatiotemporal magnitudes are quantities. Russell defined the terms "greater" and "lesser" axiomatically.

Measurement, he then argued, assigns numbers to magnitudes. Russell identified a problem: What does it mean to say that one magnitude is the double of another when both contain infinitely many parts (as when a line contains infinitely many points)? One can imagine two magnitudes, each of the same size, that add together to give a magnitude of double the size, but this does not always solve the problem. It fails to solve the problem when addition fails, and Russell gave a good example: pleasure. As he said, "The sum of two pleasures is not a new pleasure, but is merely two pleasures."[90] All one can say in such cases is that the order relations of measurement must be preserved. Magnitudes, and measurements, for Russell, need have nothing in common but that they can be ordered, and so all a measurement can be presumed to do is to assign numbers in a way that agrees with the preexisting ordering on the magnitudes being measured. Michell believes that this argument was the start

---

[88] In fact, the appeal to density is unnecessary; see Neumann, Adeleke, and Dummett 1995.
[89] In Demopolous 1995, 421.
[90] Russell 1903a, 179.

of a long and, in his view, impermissible slide into the present state of measurement in psychology, but that would take us too far.

5.3.9 Campbell on Measurement in Physics

Matters are clearer in Campbell's book of 1920, *Physics; the Elements*.[91] He picked four exemplars of the measuring process: weight, density, color, and hardness, and his key idea for measurement was that one could say of two objects that, in such-and-such a respect, one was greater than the other. Using this approach it followed immediately that color cannot be measured: one can sometimes say that this shade is redder than that one, but not, for example, that of two shades of blue one is redder than another.

Weights and densities can both be measured, but they illustrate different aspects of the measuring process. Weights can be found using a balance, and when objects are placed on the same side of a balance, their weights add together. Campbell singled out the property of weight that if a is a weight, then $a + 1 > a$, a rule he called the "First Law of Addition." This is not true of density: place two solid objects of the same density together and one does not obtain one object twice as dense. The numbers assigned to densities, Campbell said, do not obey the First Law of Addition, instead $1 + 1 = 1$. The distinction between weight and density, he explained, was that the first pertained to substance, the second to a property.

Campbell also announced a Second Law of Addition: the magnitude of a system produced by the addition of bodies depends only on the magnitude of the bodies and not on the order or method of their addition. It follows from the Second Law that the numbers assigned to quantities that obey the Second Law must obey the commutative and distributive laws of addition, and respect the usual rules for inequalities. Campbell also noted that the two laws are independent; the second does not follow from the first.

It might seem that we have returned to Helmholtz's position, perhaps amplified by the careful distinction between weight and density, but a careful comparison of Helmholtz's and Campbell's positions will clarify the fundamental nature of the shift toward a representational theory of measurement.

The key example is hardness. Mohs's original insight is that solid A is harder than solid B if a sample of A will scratch a sample of B. If A is harder than B and B is harder than C, then it turns out empirically that A is harder than C. If solid A will not scratch solid B, then either solid B will scratch solid A or it will not. In the latter case, solids A and B are said to be equally hard. Such solids are not a problem, for, if solid B is equally hard as solid A, then any solid that A scratches solid B also scratches, and any solid that scratches A also scratches B. Accordingly, any two solids A and B may be compared in respect of hardness, and either A is harder than B, or B is harder than A, or they are equally hard. So one can assign all solids a place in a linear scale of hardness. On the other hand, there is no way of assigning numbers in an additive way.

---

[91] Reissued in 1957 as *Foundations of Science*.

In the case of density, the situation is different, as Campbell discussed. The primary definition of density is that one material is denser than another if it sinks a liquid in which the other one floats. This again yields a linear ordering, and on the face of it no additive assignment of numbers can be found. But in the case of density there is what Campbell called a derived method. The ordering of material defined by the floating/sinking test agrees with another approach that defines density as the ratio of mass per unit volume. This permits densities to be measured by measuring the mass of a unit volume, and so makes density a quantity of the same (additive) sort as mass. But with this observation in place, it is clear that Campbell regarded measurement in a more general way than Helmholtz or Hölder.

In his paper (1993), Michell calls the shift from the classical to the representational theory of measurement a revolution. The revolution was complete, he says, with Campbell's *Physics; the Elements* (1920), but, he says, even members of the representational school fail to agree on a founding father. Some hint at Helmholtz, others at Hölder, and yet others at Russell. In order to adjudicate the issue, he observes that the classical theory of measurement finds the number or measurement latent in the quantity being measured, whereas the representational theory does not presume that numbers are necessary, although they may be contributed. This reflects Russell's identification of measurement with the establishing of an order relation: this magnitude A is greater than that one, B; this third object C has a magnitude greater than B's and less than A's; and so on. Whether these magnitudes are meaningfully assessed on a numerical scale is a subordinate question.

It seems likely that Campbell wrote under the influence of Russell, inasmuch as he notes in his preface that "I can make acknowledgements to my masters, Henri Poincaré and Mr Bertrand Russell; but I fear that the latter at any rate will think his pupil anything but a credit to him." The philosophical apparatus upon which Russell based his theory is lacking in Campbell's book. Rather, we may note that, writing for different audiences, Helmholtz and Campbell agreed on what the classical theory of measurement was, and that Hölder spelled out the mathematical implications of that theory in an axiomatic way, taking note of the contemporary theory of magnitudes that derived from modern geometry. Russell drew his inspiration from another part of modernist mathematics, the Cantorian theory of ordinal numbers. He and Campbell could also agree on a nonclassical theory of measurement.

The influence of Russell persuaded Campbell to distinguish the numbers that result from measurement from the numbers of the mathematicians. For Campbell, the result of a measurement would be a number (lower case "n") that might be positive or negative, whole or fractional, and which depends on a choice of unit (so might be considered to be a ratio, as Newton had suggested). It might also be amenable to arithmetic operations that derive from the operations one can perform on the substance being measured. The mathematicians' Number (upper case "N") is quite different, said Campbell (p. 305), unnecessary for the process of measurement, but possibly necessary for the use of the process. For Campbell, the nature of that use depended on the distinction he drew between a law and a theory: laws, but not theories, fail to explain or predict. A law is a regularity of measurement or the immediate result of an experiment. Campbell preferred to find the certainty that $2+2=4$ in experience rather than the pages of Russell and Whitehead's three volumes, and so he argued that number, but not Number, enters into physics. This drew the blunt if unduly credulous remark from his reviewer that "the trouble all comes, it would seem, from his having read *Principia Mathematica* and not believed it. He should have taken it on trust, unread, like the rest of us."[92] Campbell, it seems, would have preferred a philosophy of applied mathematics derived on its use in physics and entirely cut off from the consolations of logicism.

5.4 Popularizing Mathematics around 1900

5.4.1 Introduction

The years around 1900 stand out as a high point in the popularization of mathematics. A growing new audience wanted to hear about current developments, although it seems that between 1900 and 1914 the public, and the American public in particular, was ambiguous about science (see Kern 1983 and Kevles 1987). The public was happy to benefit from the motor car and the spread of the telephone, which connected New York and Chicago in 1892, New York and San Francisco in 1915, but was worried about the effect of new technology on jobs and the increased stress of modern life. The progressives wanted money spent wisely on social reform, and university presidents animated by this spirit could accept pure research because it would surely turn out to be useful. Thus Woodrow Wilson, on becoming president of Princeton in 1902, said, "Science and scholarship carry the truth forward from generation to generation and give the certain touch of knowledge to the processes of life."[93] To this end, graduate scholarships were increased, and professors became better paid. The progressives' spokesman was James McKeen Cattell, who from 1900 edited the *Popular Science Monthly* and used his influence to extol science and its uses.[94] However, the enterprise was not a lasting financial success. In 1915 the owner sold the journal, and Cattell brought out a new one, *Scientific Monthly*, which retained the popular science but dropped the social reform.[95] Reformers, led by Dewey, were no longer looking in such numbers to science.

There remained, nonetheless, an audience with more subtle requests for science, for whom scientists and mathematicians offered high-quality popularization. This audience contained teachers of mathematics, for example, and a number of academics in other disciplines, such as philosophy and psychology. To judge by the places where their books were reviewed, the audience went considerably beyond the universities and embraced many for whom learning represented an opportunity denied, or only now to be seized. A number of other professionals must also be included: lawyers and members of the Church. In America, they read intellectual magazines like *The Monist*, published in Chicago. In France they read, if they were teachers, such journals as *L'Enseignement mathématique*, founded in 1900. There was also the *Revue philosophique de la France et de l'étranger* originally edited by Théodule Ribot and first published in 1876, which covered a considerable range of topics from psychology to philosophy and reviewed books from Britain, France,

---

[92] Ritchie 1921, 214.
[93] Quoted in Kevles 1987, 70.
[94] Veblen published an article here on the "Foundations of Geometry," in vol. 68.
[95] Kevles 1987, 96.

Germany, and Italy; the *Revue de métaphysique et du morale*, founded by Xavier Léon[96] in 1893, which, as he put it in the opening foreword, was published to counter a trend toward positivistic or scientific philosophy found in the *Revue philosophique* and to direct the discussion of ideas back to philosophy properly understood;[97] and the *Revue du mois*, which was founded and edited by Émile Borel and his wife in 1906 with money from one of his prizes. In Italy, they read *Scientia*, founded by Enriques in 1910. As we will see in the next section, Americans were particularly well served by Paul Carus, but the British had to survive on much poorer fare.[98]

Rather unfairly, in 1911 the journal *The Nation* lamented the passing of the best popularization of science, epitomized by Tyndall, Huxley, and Helmholtz some forty years before. By then the authors of choice were Mach, Poincaré, and Enriques, all of them eminent, powerful scientists or mathematicians, the equal of Helmholtz and more insightful than Huxley and Tyndall. There were also a number of lesser authors who also participated, among whom might be mentioned Émile Picard, none of whose essays matched the quality of his original mathematics; the prolific if facile C. J. Keyser in the United States; and Aurel Voss in Germany. Mathematics was particularly well treated by these writers, for a number of reasons.

Mathematics had something to say under the heading of theoretical cosmology: Could there be a fourth dimension, and what does it mean for space to be curved? But more than that, mathematics was either finally sorting itself out or in a terrible and worsening mess. The mere existence of non-Euclidean geometry showed that geometry could no longer in any simple sense be true, and if it stood in need of foundations it became clear after 1903 that the only candidate, naive set theory, might buckle under the weight. The paradoxes that did the damage had the independent virtue of being intelligible without much prior study of mathematics, so the news that mathematics was under threat was plain for all to see—and doubtless exciting.

In 1897 Simon Newcomb addressed the fourth annual meeting of the American Mathematical Society on "The Philosophy of Hyperspace." Newcomb was both an entirely practical man, director of the Naval Observatory, recipient of the Huygens and Copley medals, and, in 1897–1898, the president of the American Mathematical Society. He had his blind spot—the Cantorian theory of the infinite—but he had also written original papers on non-Euclidean geometry, and one of his earliest mathematical results was that in four-dimensional space a sphere may be turned inside out without tearing. This fitted him for his subject: the fairyland of geometry.[99] It was his theme that space, logically, may have any number of dimensions, with interesting consequences. For example, if it had four, three-dimensional objects like ourselves could be moved onto their mirror images, as happens to the luckless narrator of H. G. Wells's *The Plattner Story* (1896). But there is, of course, no evidence that real

---

[96] Xavier Léon later founded the Société Française de Philosophie in 1901.

[97] For an account of these journals and the rich, overlapping worlds of French intellectual and scientific thought they display, see the forthcoming study by Christina Chimisso (2008). I am grateful to her for the chance to see the relevant pages of her book.

[98] Unless one includes the journals *Mind* and *Nature*, but these seem to be more specialist and academic than the ones cited above.

[99] See also his "The Fairyland of Geometry," *Harper's Magazine*, January 1902. There were also two books by C. H. Hinton: Hinton 1897 and Hinton 1904.

space is four-dimensional. Likewise, we may imagine three-dimensional space to be curved in such a way that the appropriate geometry is non-Euclidean, but again there is no astronomical evidence to show that it is. He did not attempt an explanation, which is a matter for some regret, because there was much confusion about how space may be curved without being in something from which it can be seen to be curved, and Newcomb understood this point well enough to help explain it.

His topic was a popular one. From 1870 to 1914, public interest grew in the idea that space might not be Euclidean. Painters were actively involved in these discussions. To quote Apollinaire again:

> The new artists have been violently attacked for their preoccupation with geometry. Yet geometrical figures are the essence of drawing. Geometry, the science of space, its dimensions and relations, has always determined the norms and rules of painting. Until now, the three dimensions of Euclid's geometry were sufficient to the restiveness felt by great artists yearning for the infinite. The new painters do not propose, any more than did their predecessors, to be geometers. But it may be said that geometry is to the plastic arts what grammar is to the art of the writer. Today, scientists no longer limit themselves to three dimensions of Euclid. The painters have been led quite naturally, one might say by intuition, to preoccupy themselves with the new possibilities of measurement which, in the language of the modern studios, are designated by the term: the fourth dimension.[100]

There was much discussion that space might be four dimensional, perhaps with time as the fourth dimension, as in H. G. Wells's *The Time Machine*, first published in 1895, and there was a successful *Scientific American* contest to solicit essays on the fourth dimension. The magazine received 245 essays from several countries. The most successful entry was awarded a prize of $500. It and three runners-up were published in 1909; they and several more were published as a book in 1910.[101]

Manning, the editor of the book, with more space at his disposal but perhaps also more awareness of what can go wrong than Newcomb, explained the idea of a geometry of four dimensions slowly and clearly. Mathematicians, he said, write down axioms that need not be true, from which to derive theorems. Four-dimensional space can be described axiomatically, and it is not to be confused with non-Euclidean geometry: it was not introduced to explain non-Euclidean geometry, but both ideas owe their recognition to the idea that geometry need not restrict itself to the study of Euclidean space. So, whether or not space is four dimensional, four-dimensional geometry may certainly be studied, and Manning, like the authors he was introducing, then proceeded to give some account of the novel features of that geometry.

Among these are generalizations of the regular solids, already studied by George Boole's daughter, Alicia Boole, who had a remarkable ability to "see" in four dimensions. Knots may be undone, and just as a square divides the plane into two regions but does not divide three-dimensional space, so in four dimensions points "inside" a closed surface such as a sphere or a cube may be moved "outside" it

---

[100] In Chipp 1968, 222–223.
[101] Manning 1910.

without ever passing through the surface. These conundra form the basis of the charming satire on Victorian orthodoxy, *Flatland: A Romance of Many Dimensions* by A. Square (Edwin A. Abbott), published in 1884 and in print ever since.[102] Less charmingly, they were invoked by those who defended the paranormal claims of the American medium Henry Slade. Slade visited Europe in 1876 but was accused of fraud, charges he escaped on a technicality. He then went to Denmark and Germany, where he was enthusiastically taken up by the Leipzig physicist Johann Zöllner, who argued that Slade really did have access to the fourth dimension.[103] Zöllner asked Slade to link two separate, unlinked wooden rings, to turn left-handed snail shells into right-handed ones, and to untie a knot whose ends were fastened to a table. Apparently he was convinced of Slade's success, and Zöllner's book *Transcendental Physics* (1878), was devoted to explaining how Slade had achieved these feats by moving into the fourth dimension. The book was a great success, its English translation ran to three editions in five years, and other physicists, Weber and Crookes, were brought in to support Zöllner's claims. Others were not so convinced. In 1881 the *Atlantic Monthly* reviewed the book in these terms: "One opens this work of Zöllner with great interest, in the expectation of something substantial and more edifying than the dreary accounts of table-tippings, and the insane conversations of great men who, entering into a Nirvana, have apparently forgotten all they learned in this world, and have nothing better to do than to move chamber furniture. Unfortunately, this hope is not realized."[104]

Zöllner and his supporters, including theosophists, had a major row with Zöllner's colleague at Leipzig, Wilhelm Wundt. Zöllner had been instrumental in bringing Wundt to Leipzig; he had just arrived in 1875 and was trying to set up a substantial psychology laboratory. Wundt performed the valuable service of debunking the American medium in his "Spiritualism as a Scientific Question," pointing out that Zöllner could only succeed when he had the opportunity to cheat. Slade's hands and the tablet on which the "spirit messages" were written down were often out of sight beneath the table, and the messages often came in English or poor German, although purportedly from native German speakers. Wundt also made the vital observation that scientists are very poor judges of psychical phenomena, because they are ill-equipped to detect deceit, and that professional magicians would do better.[105] Some

---

[102] The first to use the idea of two-dimensional beings to initiate the reader into the plausibility of four-dimensional space may well be Fechner, under the pseudonym of Dr. Mises, in his *Vier Paradoxa* (1846). In the second of these, "Der Raum hat vier Dimensionen," shadows are used to invoke a two-dimensional world, in a way that is also oddly reminiscent of Plato's cave. In an unpublished note of 1875 Fechner added that he had not known of Kant's earlier speculations in this direction, but that well-known mathematicians such as Riemann, Helmholtz, and Klein had since taken them up as had Zöllner in a profound way in conversation. See http://www.buecherquelle.com/fechner/mises/vdimens/vdimens .htm.

[103] Zöllner had already written one book, his *Principien einer electrodynamischen Theorie der Materie* (Principles of an Electrodynamic Theory of Matter), arguing that the world was four dimensional and we merely saw a three-dimensional shadow of it. Erdmann (1877, v) dismissed it as a wonderful fantasy of a bodyless and soulless world, mixed up with the scientific viewpoint of mathematical research, because it was impossible to understand the theory of matter as shadows projected from a four-dimensional world.

[104] Information from www.strangehorizons.com/2002/20020916/fourth-dimension.shtml.

[105] Fancher 1979, 155–156.

may feel that Zöllner's reply, accusing Wundt of copying his opinions from his "lord and master" Helmholtz and "the Berlin vivisectionist" Emil du Bois-Reymond makes it only too clear who was the careful thinker.

The popularity of the fourth dimension did not depend on mediums to eke out the small supply of applications. Understood as time, it had another life. When, after 1905, news of Einstein's special theory of relativity began to spread, and especially after 1909 when Minkowski so expertly promoted the new concept of space-time, another mysterious use for it emerged, and yet another literature.

5.4.2 Paul Carus

Paul Carus himself is interesting and worthy of a full-length study, as is the workings of his Open Court Press.[106] He was born in Germany in 1852. The son of a minister, he studied philosophy, philology, and science and contemplated becoming a clergyman, but some sort of religious crisis in his life turned him away from the stricter interpretations of Christianity and toward a lifelong search for religious meaning. An unhappy experience teaching at a military academy in Germany may have persuaded him to emigrate, and he came to America in 1882, settling in La Salle, Illinois. There he fell in with Edward Hegeler, another German immigrant, who made a fortune out of zinc and used some of the money to enable Carus to set up his journals, the *Open Court* and the *Monist*; Carus also married Hegeler's daughter. The journals were billed as being devoted to the "scientific study of religion" and the "religious study of science," respectively, and a brief look at them will display many of the connections that were made in this period.

The breadth of these journals was as remarkable as their success, and it was a lasting success, for many well-read books today were first printed by the Open Court Press. Carus shared with such people as Max Müller, the linguist, and Hermann Grassmann an interest in and support for studies of Eastern philosophy and religion.[107] Active involvement seems to have been sparked by what was called the Parliament of Religions, held in conjunction with the Chicago Columbian Exposition in 1893 (a seminal moment for mathematics in America, too; one wonders if Carus picked up on that as well).[108] For many, it was a two-week-long celebration of the world's religions. Ten major ones were honored by an audience of four thousand at the start of events, and seven thousand came to the closing ceremony.[109] For some it was a clash of religions and an attack on Christianity. It does seem to have promoted a sympathetic interest in non-Christian religions in the United States, and Carus himself called it the most noteworthy event of the decade, but all attempts to build a lasting organization out of the Parliament failed, much to Carus's sorrow.[110]

---

[106] I am told there is an abundance of source material.

[107] See Jackson 1968.

[108] See Joas Adiprasetya (2004), "The 1893 World Parliament of Religions," at http://people.bu.edu/wwildman/WeirdWildWeb/courses/mwt/dictionary/mwt_themes_707_worldparliamentof religions1893.htm. The mathematicians met August 21–25 in Chicago, and August 28 to September 11 in Evanston, Illinois; the Parliament met September 11–27.

[109] It was organized and led by John Henry Barrows, who described it in a two-volume book, Barrows 1893.

[110] Jackson 1968, 77.

Carus's own views are evident in the very title of his journals. The *Monist* derives from the then-fashionable philosophy of monism. Loosely understood this philosophy promoted the idea that "all is one," that there is a unity behind appearances. It is inevitably vague and for that reason compatible with a highly ecumenical view of religion that finds some truth in all religions while acknowledging the limitations of the human mind before the divine. This ecumenism, in turn, leads to the idea of an open court. But it was not a mindless tolerance; Carus sought to sweep out mystical fog by adopting a scientific approach to religion.

Among scientists, Carus had a lifelong admiration for Ernst Mach, who developed a philosophy called "neutral monism" that was to influence William James, Bertrand Carnap, and Russell.[111] It was neutral in that entities may be regarded as either objects or percepts, and mind and body are constructed out of these. The aim is to find a common ground for objects and sensations that will permit an explanation of how knowledge of the external world is possible. Carus frequently published articles by Mach in the *Monist*, and Mach often found himself in agreement with Carus's own views, writing to him that he was in full agreement with Carus's *Fundamental Problems: The Method of Philosophy as a Systematic Arrangement of Knowledge* (1889).[112] Mach was in turn favorably disposed toward most things American, and particularly valued the opportunity the *Monist* gave him to reach an American audience, in the hope that America would take the lead in overthrowing metaphysics.

Carus's monism differed from that of Ernst Haeckel, a much more divisive figure. Haeckel was a German professor of biology at Jena, known for his beautiful drawings of the radiolaria and for his energetic promotion of Darwinism in Germany.[113] But where Darwin had trod with circumspection, Haeckel marched boldly. Darwin, he argued, had shown the way reason can overcome religious superstition, notably that of the Catholic Church. It opened the door to a "monistic religion of humanity grounded on pantheism."[114] He wrote up his ideas at length in 1892 in a small book, *Der Monismus als Band zwischen Religion und Wissenschaft* (*Monism as Connecting Religion and Science*), which eventually ran to seventeen editions and formed the basis of his major work, *Die Welträthsel* (1899).[115] His fundamental view was that of panpsychism: there is a continuum of life and consciousness that runs from the smallest atoms to every living thing, varying only in intensity, on to which he grafted a evolutionary element.[116] Christianity was therefore misplaced in its theology, but, according to Haeckel, sound in its ethics. In arguing his cause, he tapped into a distrust of the Jesuits that was widespread in Protestant northern Germany, and into the uncertainties about dogma generated by German theologians.

In 1902, after many years of promoting his cause, Haeckel came out of retirement to agitate yet again against the Catholic Church, driven this time by the election of Pius X as the new pope. Haeckel's *Die Welträthsel* had proved a best-seller, and he

---

[111] In, for example, Leopold Stubenberg, "Neutral Monism," in *The Stanford Encyclopedia of Philosophy* (Spring 2005 Edition), ed. Edward N. Zalta, http://plato.stanford.edu/archives/spr2005/entries/neutral-monism/.

[112] Holton (1992, 31) records the existence of a large volume of correspondence between the two.

[113] This account follows Richards, http://home.uchicago.edu/~rjr6/articles/Articles\% 20for\% 20Jena\% 20Journal.pdf.

[114] Quoted in Richards (see fn. 113), p. 4.

[115] Translated as *The Riddle of the Universe*, 1900.

[116] See Haeckel's article in the *Monist* 2 (1892): 481–486.

converted many of his enthusiastic readers into members of a new organization he formed in 1906, the Monistenbund. Wilhelm Ostwald became its president for a time. What exactly the Monistenbund advocated was not clear. Haeckel has a reputation today for having advocated ideas later taken up by the Nazis, although the Bund dissolved itself in 1933 rather than submit to Nazi control. It had by then become somewhat pacifist and socialist in its leanings.

The Catholic Church, however, was clear that Haeckel was an enemy, and largely right that his Monistenbund was anti-Catholic, too.[117] Pius X did not mention it by name in his two encyclicals against modernism, but it is unmistakably there. The *Catholic Encyclopedia*, which was begun in 1905 and finished in 1914, commented that the Monistenbund "is openly anti-Christian, and makes active warfare against the Catholic Church. Its publications, 'Der Monist' . . . and various pamphlets. . . . are intended to be a campaign against Christian education and the union of Church and State."[118] They also saw that Paul Carus's monism was less confrontational, but they found it nonetheless wrong, of course, in leaving no room for faith as the Catholic Church construed it.

Carus's monism or panpsychism has long roots, but it had been revived most notably by Gustav Fechner.[119] Fechner's belief in psychophysical parallelism does seem to have triggered his interest in the psychology of sensation. His panpsychism appealed to Riemann, but only in unpublished writings; to William Clifford;[120] Josiah Royce (who regarded whole planets, stars, and even galaxies as alive, albeit working on a different timescale); and Lotze and Wundt. Wundt, who influenced Clifford in this respect, also advocated a theory of psychophysical parallelism.

Alongside the Monistenbund there was also the Gesellschaft für positivistische Philosophie (or Society for Positivist Philosophy) launched in 1912 to pursue a contradiction-free unitary conception of the sciences. The launch of the society was accompanied by a public appeal signed by thirty-three people, including Mach himself, Sigmund Freud, Einstein, Felix Klein, and David Hilbert, and in due course there was a journal.[121]

Carus's journals were popular. They created and cared for a public sympathetic to positivism, pragmatism, and science: C. Jackson (1968, 75, n. 8) quotes figures of 3,600 and 1,000 for subscribers to the *Open Court* and the *Monist*, respectively, in 1903 and notes that the peak was reached in 1906. He published in English not only works by Mach, but several by Poincaré; also Townsend's English translation of Hilbert's *Grundlagen der Geometrie*, Dedekind's *Essays on the Theory of Numbers*, Jourdain's edition and translation of Cantor's *Contributions to the Founding of the Theory of Transfinite Numbers*, Enriques's *Problems of Science*, Couturat's *Algebra of Logic*, George Boole's two-volume *Collected Logical Works* (his *Laws of Thought* occupies volume 2), Mach's *Space and Geometry*, Saccheri's *Euclides vindicatus*,

---

[117] They were not the only hostile critics. So was Lenin at length in his *Materialism and Empirio-Criticism*.

[118] "Monism": see http://www.newadvent.org/cathen/10483a.htm.

[119] See William Seager, and Sean Allen-Hermanson, "Panpsychism," in *The Stanford Encyclopedia of Philosophy* (Summer 2005 Edition), ed. Edward N. Zalta, http://plato.stanford.edu/archives/sum2005/entries/panpsychism/.

[120] Clifford 1878.

[121] The document is reproduced in Holton 1992, 38.

and a number of other works, as well as reprints of Halsted's translations of Lobachevskii. This does not mean that the authors subscribed to Carus's philosophy, but it does suggest that their approach to mathematics was expected to reach a favorable audience among his subscribers. It is undoubtedly an impressive roster of many of the main texts of mathematical modernism.[122]

5.4.3 Poincaré

> Not logical enough for the logician, not mathematical enough for the mathematician, not physical enough for the physicist, not psychological enough for the psychologist, nor metaphysical enough for the metaphysician, Poincaré's *Science and Hypothesis* can hardly give the satisfaction of finality to anyone; and yet it probably comes nearer to satisfying the requirements of all these classes of investigators than any single book of our acquaintance.
>
> —Wilson 1905, 187–188

Poincaré's popular essays are mostly collected in the four volumes that continue to sell more than one hundred years after most of them were written. Evidently a number of them touched a genuine issue with enough insight to merit continued reading. In this they are helped by Poincaré's limpid prose, with its almost conversational style and its seeming ability both to find the nub of the matter and to surprise. In their day, his words were contributions to the rationalist side of a debate that opposed various mystical, religious, or antiscientific currents. Poincaré was the scientist readers turned to for an expert opinion couched in language they could understand, and he was not afraid of controversy: one of his contributions was to rubbish the opinion of handwriting expert used by the prosecutors in the long-running Dreyfus affair. But he could be almost too convincing while being too casual at the same time. As we saw on §5.2.4, a cheerfully conventionalist interpretation of the motion of Foucault's pendulum, at the time a topic of intense popular interest, drew a claim from Roman Catholic apologists that the Church had not, after all, been in the wrong for forcing Galileo to withdraw his claim that the Earth truly rotated, and Poincaré had to refine his position somewhat.

Poincaré typically wrote his popular essays after having studied a topic in some depth and writing about it in research papers or lecturing on it at the Sorbonne. These lectures were often edited and published, and some of the essays are either taken from prefaces to these books or are lightly adapted. This combination of working to his own high standards for the purposes of research and then turning to general journals for popular exposition undoubtedly gives the essays their freshness and their originality. They offer his readers the opportunity—not always taken—of elucidating what he meant in an essay by turning to the much fuller academic versions.[123] The essays, too, are in a way graded. We may easily learn that Poincaré

---

[122] Among the failures one notes the opening article by Dewey on "The Present Position of Logical Theory" (*The Monist*, 1890, 1–17), which was anything but logical, not venturing beyond Kant and Hegel.

[123] See Gray 1999 and Gray 2000b.

preferred Euclidean to non-Euclidean geometry on the grounds of its simplicity by consulting his essay "Les géométries non-euclidiennes" (1891), but to discover in what precisely he thought this simplicity resided requires reading a much longer essay in the *Monist* of 1898, which is only to be found in recent reeditions of *La science et l'hypothèse*.[124]

The four books of essays—*Science et hypothèse* (1902), *La valeur de la science* (1905), *Science et méthode* (1908), and the posthumous *Derniers pensées* (1913)—cover a number of issues.[125] The nature of mathematical reasoning is one, with a particular emphasis on the principle of mathematical induction, mathematical logic, and reasoning about the infinite. Another is non-Euclidean geometry and the nature of space, where his geometrical conventionalism is set out. Relative and absolute motion and the problems posed by the motion of electrons is another, and here Poincaré is almost torn away from his faith in Newton's laws of motion. Measurement is another concern, and with it our knowledge of space and time. As many as a dozen essays are directly on various topics in physics; these are the ones closely related to the prefaces of his books of lectures. Finally, a few are more overtly philosophical, from which it emerges that Poincaré was no realist, and that he had a long-running sympathetic argument with Kantianism.[126]

If a single theme unites these essays and explains their enduring charm, it is Poincaré's concern with how we can be said to know things, and the consequent limitations on that knowledge. It is this that makes them more than popular survey articles, and has kept them from being too severely dated. As he ranges over logic, mathematics, physics, psychology, and metaphysics—not without error, as Wilson noted in his review—there were two pillars of knowledge for Poincaré: mathematical theorems and accepted scientific experiments. His novelty lies in his exploration of the space between these pillars, which calls for analysis and elucidation. Where others, and surely his readers, rushed in with confidence, Poincaré feared to tread. His message was ambiguity. Can we know the nature of space? Never, said Poincaré. There are two plausible candidates from the eight offered by mathematics (by an argument none too convincing, it should be noted). But the interpretation of mathematical terms in concrete physical objects is fraught and cannot be done except by an arbitrary act of choice. Mathematics in its own pure right was another matter. Here our knowledge was more secure, resting upon an a priori extralogical ability to conduct mathematical induction. But physics itself was much less secure, for the interpretation of an established experiment was a topic about which Poincaré was profoundly agnostic. Writing in the introduction to *Électricité et optique*, Poincaré argued that there could be no mathematical way of discriminating between one- and two-fluid theories of electricity.[127] Poincaré withheld interpretation for want of a logical way of deciding between them, and as for how a choice between conflicting explanations could be made, he contented himself with the hope that perhaps one

---

[124] The answer is that only the group of Euclidean transformations has a normal subgroup of translations!

[125] At the same time, 1901–1903, Ernst Mach had views very similar to Poincaré's on many issues. So marked was the overlap that Mach asked that some of his work be omitted from a French translation of it. See March 1906 (English translation) a reprint of his essays in *The Monist*, and his last book, Mach 1905.

[126] On Poincaré and Kant, see Folina 2006.

[127] Poincaré withheld his consent to the existence of Lorentz's ions until 1904.

day physicists would take up these questions, which were inaccessible to positive methods and had been abandoned to the metaphysicians. Quite strikingly, Poincaré wrote:

> There are other prerequisites that seem to me less reasonable. Behind the matter which impinges on our senses and which we can know from experiment, there should be another matter, the only true one in our eyes, which has no other than purely geometrical properties, and whose atoms are nothing other than purely geometrical points which obey only the laws of dynamics. By means of these colourless invisible atoms one seeks, unaware of any contradiction, to represent and consequently to approximate as closely as possible, ordinary matter.[128]

This is perhaps a plea for plausible assumptions about atoms, but in context it is surely more of a plea for making no atomistic hypotheses of any kind.

Instead of truth, Poincaré offered utility as a guide. In his address on the future of mathematics ("L'avenir des mathématiques") given to the International Congress of Mathematicians in Rome in 1908, he argued that mathematicians praise some theorems or proofs for their elegance, and prefer theorems and proofs that permit them to view a whole field at a glance, and so to say (what a lengthy calculation does not) why something is the case.[129] In exactly the same way, the laws of physics structure the facts around them. This, Poincaré reminded his audience, was nothing more than Ernst Mach's idea of the economy of thought, which depends on the clarity afforded by the discovery of general laws. According to Poincaré, theorems are organizing principles, which organize facts according to analogy and so permit their comprehension. Poincaré pursued this organizing role through his distinction between qualitative and quantitative work in mathematics, upon which he laid great weight. This was because it expedites the use of classification theorems, as, for example, the use of group theory in the study of geometry, where we have already seen that, for Poincaré, groups came before spaces.[130]

If Poincaré had a single source for his antirealist position, it was surely in his struggle to master electromagnetic theory. There he found that the experts disagreed, and indeed that he could not present a single satisfactory account of any of their theories. He may well have felt this was likely to be characteristic of any fast-moving science, when too much was unknown and obscure to make it wise to invest heavily in any interpretation. He turned this reticence into another charming feature of the popular essays: a willingness to grapple openly with conflicting interpretations. Poincaré did not so much bring to his readers the latest discoveries as tell them how difficult it was to discover, and it seems that this was what they wanted to hear. The difficulty was not the ordinary mortal difficulty in discovering a new theorem or conducting a worthwhile experiment—Poincaré made that sound only too easy—but with the very nature of discovery, and precisely because of his facility with ordinary difficulties, he was listened to when he counseled metaphysical caution.

In particular, Poincaré set limits to what mathematics can do. Mathematics, he suggested, now occupies whole new realms of unfettered thought and constructs

---

[128] Poincaré 1890, ii–iii.

[129] In Poincaré 1908, 19–43; see esp. 25.

[130] Another example, discussed in Gray 1992b, is the use of topology in studying differential equations.

logical systems that defy intuition. But it is no longer the activity that explains and holds science together by telling us what is true. It is pluralist, it contains multitudes (of theories), and it does contain contradictions. In so doing, it cuts itself off from being an a priori science. It has engaged in a process of self-criticism and analysis and emerged much enlarged. No longer can it presume to ground a single physical theory, such as Newtonian gravitation, because it produces too many.

Poincaré's attitude to the mathematical modernism he saw around him was ambiguous. On the one hand, he could not deny the merits and successes of the new mathematics, as his review of Hilbert's *Grundlagen der Geometrie* makes clear. Nor did he wish to. He merely noted that it was no longer possible to ground pure geometry in any psychological way. But he was, by inclination and indeed profession, a physicist, and he did not want to follow mathematics out of the world. So his ontological skepticism was not that of a mathematical modernist who might have denied mathematics a purpose in science. Rather, it was much broader, as unsettling to scientific realists as to any naive philosopher of mathematics.

5.4.4 Enriques

Federigo Enriques's book *Problemi della Scienza* of 1906 is a prolonged attempt to define and distinguish between facts and theories in order to analyze what constitutes reality. It is a positivist, somewhat anti-Kantian work, and it can be read as a long, if one-sided, conversation between Enriques and Poincaré, but it also illuminates issues in the philosophy of geometry that were of considerable contemporary importance and explains why Enriques's work was taken up in the United States.

The Italian geometer Federigo Enriques was the creator, with his friend Guido Castelnuovo, of the first general theory of algebraic surfaces. This work, which took twenty years, from 1894 to 1914, established him internationally as one of the leading mathematicians of his generation. In 1894, at the age of twenty-three, he had become a professor of geometry at Bologna. There he lectured on projective geometry, and he published these lectures as a book, *Lezioni di geometria proiettiva*, in 1898. David Hilbert's more famous account of geometry from an axiomatic point of view came out the next year, and Felix Klein then suggested that Enriques's book be translated into German; the translation was published in 1903, with an introduction by Klein. Klein then asked him to write the article on the principles of geometry for the *Encyklopädie der mathematischen Wissenschaften*, which he did—the essay is dated 1907.

His treatment of geometry in *Prinzipien der Geometrie* reveals several of the traits of his contemporary popularizing works. He took a deep interest in the history of geometry, so he began by describing Euclid's *Elements*. In his opinion, Euclid's definitions, postulates, and axioms differed from more modern ones by being intended to be self-evident, because they were not presumed to follow logically from the definitions, and because they were in any case incomplete, as writers like Pasch had noted. Modern presentations of geometry, Enriques observed, regarded geometric objects as belonging either to the usual intuitive space, or to physical space with which we become acquainted through experience, or to some abstract space made intelligible by abstraction or generalization. This more sophisticated approach, with its associated rise in the standards of mathematical rigor, had been initiated by

FIGURE 5.2. Federigo Enriques.

Pasch and taken up by Peano and Hilbert. It had led to questions about the arbitrariness of axioms and their independence that were connected to contemporary work on logic, and to questions about the philosophy of geometry. Geometry could be approached either empirically, through its connections to experience, or from a logical and formal standpoint, as the Italians Peano, Padoa, and Pieri had done. Enriques had strong views about the proper relation of these two approaches, which he expressed not in Klein's encyclopedia but in his popular writings.

Indeed, in 1907 Enriques was the Italian mathematician of stature who was involved in mathematics education and philosophy. His authority derived from his research in a difficult, important, but, to the general public, obscure domain; now he leant it to speculations of a quite different kind, intended for a different audience where the criteria for success, and even influence, are much more fickle. His major work in this connection is the *Problemi della scienza* of 1906, which, after some delays, was translated into English as *Problems of Science* and published in 1914. With it he joined the select group of scientists of any kind whose work reached out to the general public. He continued to do important work in algebraic geometry, and at the same time moved toward logic, education, and textbook writing, all of which broadened his involvement in Italian intellectual life and also facilitated his move toward the public platform abroad. In the years before the First World War, Enriques was an editor of the journal *Scientia* and published many articles there. A number of these were collected and reworked for publication as a book in 1912, *Scienza e razionalismo*. The topics range from the value of science, through philosophical discussions of rationalism and empiricism (including a history of rationalism from the Eleatics to Leibniz) and of rationalism and historicism (in which he grappled with Hegel), to topics on the relation of science and religion, the existence of God, and the nature of reality. Rather less grandly, he also tackled the classification of the sciences and the implications of non-Euclidean geometry for the philosophy of Kant. His work invites comparison with that of his distinguished predecessor, Henri Poincaré, with whom Enriques agreed in significant areas and with whom he also had one specific disagreement over the nature of geometry.

Erudite popularizing work, such as Poincaré and Enriques supplied, was topical. In the years before the First World War, the modern profession of mathematics was defining itself in something like its present form. In many intellectual domains, the period is marked by the first international congresses. The mathematicians met first in Zurich in 1897, then in Paris in 1900, in Heidelberg in 1904, in Rome in 1908, and Cambridge (England) in 1912, while the peripatetic philosophers went to Paris, Geneva, Heidelberg, and in 1912 to Bologna, where Enriques presided. The Italian mathematical community could see on these occasions that it ranked perhaps second, behind the mighty Germans but ahead of the French. The congresses offered images of the Italians in various ways: in Paris, for example, as logicians. Peano and his followers were prominent at both the International Congress of Philosophers and the International Congress of Mathematicians that followed it. As an eloquent spokesman for this community, Enriques benefited from the depth and range of his countrymen's work and was presumed, not altogether accurately, to be drawing on that collective wisdom.

As has been discussed at length above, one major issue for mathematicians and their audience alike was the question of non-Euclidean geometry and the nature of space. Mathematicians had come up with new geometries, and their audience wanted to know what that meant, what these other geometries are, and how we can tell which is true, even though the latter question is entirely abstract—no one was out there anxiously measuring parallaxes. There was a strong psychological aspect to this question, for psychologists had turned the investigation of the preliminaries for any kind of thought into the study of how people think. What, philosophers and psychologists alike asked, are concepts, what is it to know something? What sort of an intellectual activity is the study of geometry, regarded as the elementary appreciation of space? How can human beings to discover non-Euclidean geometry? The pioneer in this regard was the energetic, if unsystematic, Wilhelm Wundt. Enriques was enraptured by Wundt's writings, which seemed to speak directly to his philosophical interests in the 1890s. He described Wundt to his friend Castelnuovo in these terms: ". . . the most marvellous philosophical, physiological, psychological, mathematical, etc. intelligence. . . . Read the *Logik* of Wundt, at least that part about the methods of mathematics, and think that it is a physiologist who writes this, a physiologist who does not fear to scale the steep slopes of Kantian conceptions to illuminate from above the great course of all of science."[131]

Enriques's own ambitions surely shine through these words. None of his public career would have been possible had mathematics been purely technical, had psychology defined itself as the study of measurable mental processes, and had logic remained the driest kind of organized common sense. But because, as the public could easily discover, the experts had their own problems, too, and because both the public and the experts sensed that more was at stake than merely academic issues, Enriques had a platform on which to stand.

It will be recalled that the axiomatic study of geometry was something of an Italian speciality before 1899. Italian mathematicians had found unexpected proper-

---

[131] Enriques to Castelnuovo, May 4, 1896, in Bottazzini, Conte, and Gario 1996, 261.

ties of projective planes,[132] investigated a geometry in which, unlike Pasch's, the Archimedean axiom is false,[133] and come to suspect that there may be two-dimensional projective spaces in which Desargues' theorem is false.[134] They certainly did not constrain their projective geometry to the facts of everyday experience. Pieri, unlike Pasch, completely abandoned any intention of formalizing what is given in experience. Instead, as he wrote in 1895, he treated projective geometry "in a purely deductive and abstract manner, . . . , independent of any physical interpretation of the premises." Primitive terms, such as line segments, "can be given any significance whatever, provided they are in harmony with the postulates which will be successively introduced."[135] This approach to geometry was not congenial to Enriques, and in his *Problems of Science* he set out an opposing case.

Enriques's *Problems of Science* is, however, difficult both to evaluate and to present. It now inhabits that limbo of older works which are to be found on the open shelves of good university libraries, but not often consulted or argued with. However, whatever the reasons for its marginal status today, it is an extremely interesting book. It addresses a familiar conundrum, then and today: the relation between symbols and their meanings, between syntax and semantics. The philosophical problem of science it addresses is to explain how talk about mathematical and scientific concepts makes sense. Enriques was a realist. He believed it made sense to talk of objects and not merely about our ideas about objects. The way he defended and articulated such talk shows that he held with unusual vigor a viewpoint not unusual in his day, that human knowledge is acquired historically, both by the intellectual community down the ages and by each individual as he or she grows up. He observed that this process of knowledge acquisition often inverts the "logical" order, according to which (as d'Alembert had argued in the *Encyclopédie*, for example) geometry precedes mechanics because geometry is about some, but not all, of the concepts needed in mechanics. In Kant's *Critique of Pure Reason* this is elevated to the a priori status of geometrical knowledge. In *Problems in Science*, Enriques, in contrast, argued that the various sciences develop together, and geometry emerges "as a part of physics, which has attained a high degree of perfection by virtue of the simplicity, generality, and relative independence of the relations included in it" (1906/1914, 181). It is crucial to Enriques's whole approach that this process of development is, insofar as it is meaningful, necessarily and forever incomplete.

Enriques repeatedly drew a comparison between himself and Poincaré, although it does not seem that Poincaré ever thought to reply. Enriques regarded his own position as being in some sense positivist, and when he disagreed with Poincaré he found the Frenchman to be what he called transcendental. The touchstone, for Enriques, of a transcendental position was that it invoked ideas of a completed infinite process, such as abstracting away all irrelevant features (and not just more and more). A transcendental sense of space, he said, was offered when someone spoke of the inside of a ball of infinite radius, a concept he regarded as meaningless and obtained by illegitimately passing to the limit of larger and larger balls. So for example, in discussing what is meant by the geometric terms "point," "straight line,"

---

[132] Fano 1892.
[133] Veronese 1894.
[134] Peano 1894, 73.
[135] Quoted in Bottazzini 1998, 276.

and "plane," Enriques could admit that there were no such objects in reality; we do not learn what the words mean by encountering examples of them (as we do elephants or ants). They are instead abstractions: "They serve as symbols to express certain relations of position amongst bodies, relations which are stated by means of the propositions of geometry" (p. 177). Thus far, Enriques agreed with Poincaré. But that does not mean, said Enriques, that the propositions of geometry do not apply to any real fact but are merely, as Poincaré would have them, conventions by which we express physical facts. This is because these symbols find an "approximate correspondence" in the physical world "in certain objects for which they stand."

Enriques illustrated this overlap and disagreement with Poincaré with this example. The concept of a straight line is derived from several sources: the movement of solid bodies rotating about an axis, the free motion of a particle unaffected by anything else, and the path of light in a homogeneous medium. Each type of experience leads to a distinct system of geometry: the first to a metrical geometry, the third to a projective geometry. The agreement between the various experiences (each picks out the same object as a straight line) allows us to bring them together in a single geometric representation. Upon inquiry, this agreement is a symmetry in the phenomena that implies the homogeneity of space.

As discussed above, §4.1.6, Poincaré's philosophy of geometric conventionalism rests on the argument that attempts to decide if the geometry of space was Euclidean or non-Euclidean were bound to fail, because they required that straight lines be instantiated in some physical form.[136] Elsewhere Poincaré gave the example of a metal ball cooling according to a particular formula as one moves away from the centre (see §4.1.6 above).[137] With the right description of the way the disk cools, the geometry on the plate is non-Euclidean geometry, and the geodesics appear to us as arcs of circles perpendicular to the edge of the plate. However, creatures living in the plate will regard the geodesics as straight, and they would not have the same geometry as ours. But this does not mean that we can say the space is Euclidean, for, as Poincaré argued in the very next essay, one is always free to assert that the physical object that instantiates the geodesic is not, after all, straight. So Poincaré's conventionalist philosophy rests upon a supposed inability to distinguish physical from geometrical properties.

Enriques replied that in Poincaré's account we are denied any way of saying that the changes in length of rulers in the plate is due to changes in temperature, because our experience of heat is that it is a localized phenomenon to which different bodies respond differently. In discussing the Poincaré model we cannot say the things that characterize what we say about heat, so heat is not playing the role of a physical concept, but a geometrical one. The same is true of light rays, which demonstrably depart from straightness in inhomogeneous mediums. Therefore, Enriques concluded, "in this other world, geometry would be really and not merely apparently different from ours" (p. 178). To think otherwise is to make the contrast between appearance and reality into something transcendental.

A number of deep philosophical issues arise in a summary as brief as this, and in the course of *Problems of Science* Enriques touched on a number of them. One objection could be that all talk of space was meaningless, because there is nothing (no

---

[136] Poincaré 1902c, 97; English trans., 74.
[137] Poincaré 1902c, 89; English trans., 65.

object) to which the word refers. Enriques replied directly: "To Kant's thesis denying the existence of a real object corresponding to the word 'space,' we, together with Herbart, shall oppose the view that 'spatial relations' are real. And to the nominalism recently maintained by Poincaré, which declares that these relations do not possess a real significance absolutely independent of bodies, we oppose a more precise estimate of the sense in which geometry is a part of physics" (p. 174). This statement derived from his sense that there is "an actual physical significance belonging to the *spatial relations or to the positions of bodies*, whose totality may well be denoted by the word 'space.'" These spatial relations therefore, he concluded, give us knowledge of reality.

Enriques reminded his readers with several quotations from *Science et hypothèse* that Poincaré had stopped at the experience of the mutual relation of bodies—thus being a nominalist in Enriques's sense of the term, that being the antithesis of realism. Enriques believed that to talk about space was not to pass to talk of some (novel) object known transcendentally, but to talk of real relations between real bodies. Claims about space were claims about all the real (and presumably possible) relations between bodies, and, as we have seen, in discussing these claims Enriques argued that homogeneity was a property of space. Enriques went on that many knowledge claims about space reduce to claims about measurements. He gave two examples, of which the isosceles theorem was the first (p. 182). Following Klein's lectures, Enriques argued that the real meaning of the isosceles theorem is that if the side lengths differ by less than a given amount, ε, then the base angles differ by less than an amount depending on this ε in a specified way.[138] In this way the theorems of geometry can be turned into statements about bodies. The second example was a discussion of how claims about the angle sums of non-Euclidean triangles could be tested astronomically and shown to depend, within stated limits of accuracy, upon rather infeasibly large regions of space (p. 192).

However, geometry was not merely a collection of ideas about figures in space. It was, almost notoriously, not just one but two theories, two geometries of space (Euclidean and non-Euclidean). Enriques's opinion of axioms in geometry and the hypothetico-deductive side of the subject was possibly unexpected. After quoting Sartorius's remark (1856, 81) that "Gauss regarded geometry as a logical structure, only in case the theory of parallels were conceded as an axiom," he immediately disagreed with it. In his opinion, none of the postulates of geometry had the character of a logical axiom, and all of the definitions of the fundamental entities of geometry were logically defective and instead made assumptions about reality. In Enriques's opinion, geometry is not a matter of writing down some axioms (plausible or not, but in any case mutually consistent) and then reasoning entirely logically. It is concealed talk about physically possible systems, although these need not even remotely be likely to be correct, and may certainly be false, for example the Clifford-Klein space forms. According to Enriques, different systems of postulates, forming various geometries, "express different physical hypotheses" (p. 197). Enriques also noted that quite different axiom systems may describe the same geometry (qua relations of position of bodies) and gave the example of the non-Archimedean geometry of Veronese and Hilbert, which, to an infinite degree of approximation, gave theorems identical with the usual Euclidean geometry (p. 198). But, because the process of constructing

---

[138] See Klein 1901.

geometric concepts had, by 1906, gone far beyond any close link with sense data, Enriques felt compelled to explain how strikingly abstract concepts can have a certain objectivity, which he did by asserting that these abstract concepts give "a possible representation of reality" (p. 191). He also admitted that points in one space may be objects in another, for example circles in the plane may be regarded as points in a space of all coplanar circles. But this recognition of a very significant development in the mathematics of the preceding fifty years did not affect his philosophy of geometry.

Enriques's views on the nature of knowledge derived from his emphasis on the process of the acquisition of knowledge. He held that what is known is subject to a continual process of revision, so that what may be "known" at one time may be found to be false later on. Unlike Karl Popper later, he did not regard this fallibility as the characteristic feature of scientific knowledge, and indeed the example he had in mind was the exactitude of geometry, but he did write: "There is no reason however why it may not be disproved if untrue" (p. 184). His fallibilism derived from his already noted preference for meaningful discourse; Enriques was no formalist, unlike a number of his Italian contemporaries or even Hilbert. Empirical verification may be definitive if it comes up with a negative result (a counterexample), but is merely probable when it comes up with a positive (confirmatory) result (p. 155). He distinguished between explicit and implicit hypotheses in a theory (the explicit hypotheses are those particular to the theory in question, the implicit hypotheses are those needed to connect the theory to the object of study, and may well be the explicit hypotheses of logically prior theories) and wrote:

> The progress of science is a process of successive approximations, in which new and more precise, more probable and more extended inductions result from partially verified deductions, and from those contradictions that correct the implicit hypotheses. In this process certain primary and general concepts, such as those of geometry and mechanics, give us some guiding principles that are but slightly variable if not absolutely fixed. Therefore we should turn our attention to these concepts in order to explain their actual value and their psychological origin [p. 166].

Concepts, he observed come in two kinds: those appropriate to a certain physical reality, and those that are not tied down in that way (p. 117). Enriques regarded the second kind as psychological. He regarded logic as operating in a meaning-independent way, on psychologised concepts. This is not the place to discuss Enriques's theory of how the brain works, or to venture a comparison with his ideas and those of modern cognitive science. Let us merely note that, in keeping with his fallibilism, Enriques practiced what has been called "meaning finitism": the idea that the meaning of a term of concept is established upon only finitely many instances and is therefore necessarily vague (see his discussion of how one learns the meaning of legal terms, pp. 113 and 119). Infinite classes are defined by considering the conceivable objects falling under a finite number of headings (p. 129).

The mathematical community has evolved sophisticated ways of reading Enriques's work in algebraic geometry, much of which is in any case either correct or easy (these days) to put right. The same is not true of his writing as a philosopher or popularizer. He held a popular contemporary position, according to which knowledge is inseparable from the means of knowing, and logic from psychology, which at times can be a little frightening in its implications for mathematics. This dialogue has become a dead language, and with it Enriques's most original contribution lapses into the seemingly archaic. To restore him would require a public eager for issues and willing to approach them in something like his way.

In the period from 1900 to 1914, Enriques's allegiance was to Helmholtz, as he made clear in several places. Helmholtz is praised among all scientific men for offering "the clearest insight into the office which epistemology ought to fulfill in the service of science" (p. 48), specifically for saying that all sorts of scientific questions lead to epistemological problems. Enriques therefore set himself the task of deriving a positive theory of knowledge, freed of philosophic controversies, a task that he admitted required the work of the entire scientific community. He filled his *Problems of Science* with a historical and sociological positivism, in opposition to Kantianism, in metaphysics, in physics, and in biology. This version of positivism had an objective and a subjective element. In the hard sciences (not Enriques's term), the subjective element reflects the way of representing the facts, and still more of acquiring them; in the social setting the subjective aspect becomes part of the facts to be explained. The task of scientific or positive epistemology becomes that of fixing standards that correspond to our conception of objective reality so as not to be led into errors of the senses (p. 46). Progress in mental interpretation moves concepts of reality from crude to scientific facts, but how? What is objective, what is subjective, and what is arbitrary?

Enriques focused on the problems of logic and the growth of concepts, specifically those of geometry and mechanics. Logic, he wrote, might represent the ideal method of scientific construction, but positive epistemology points out its actual method (p. 47). For logic might include the methods of proof, but positive epistemology also discusses the method of discovery. It emerges in the course of a seventy-two-page discussion of the problems of logic that, whenever possible, Enriques sought to subsume logic within psychology. Psychologism was a central and dominating premise in the *Problems of Science*, and occupied many of its pages.[139]

These interests of Enriques were long standing. In the appendix to his book on projective geometry, Enriques (1898) had written:

> We have tried to show how projective geometry refers to intuitive concepts, psychologically well defined, and for that reason we have never missed an opportunity to show the agreement between deduction and intuition. On the other hand, however, we have warned that all deductions are based only on those propositions immediately inferred from intuition, which are stated as postulates. From this point of view geometry looks like a logical organism, in which the elementary concepts of "point," "line" and "plane" (and those defined through these) are simply elements of some primitive logical relations (postulates) and of other logical relations that are then deduced (the theorems). The intuitive content of these concepts is totally irrelevant. From this observation originates a very important principle that informs all of modern geometry: the principle of replaceability of geometrical elements. Let us consider some concepts, defined in whatever way, that are conventionally identified with the names of "point," "line," and "plane." Let us assume that they verify the logical relations enounced by the postulates of projective geometry. All the theorems of such a geometry will still be meaningful and valid when we want to no longer consider them not as expressing intuitive relations between "points," "lines" and "planes," but instead as relations between the given concepts,

---

[139] See the earlier discussion in §4.1.4.

which are conventionally given those names.[140] In other words, *projective geometry can be considered as an abstract science, and it can therefore be given interpretations different from the intuitive one, by stating that its elements (points, lines, planes) are concepts determined in whatever way satisfy the logical relations expressed by the postulates.* A first corollary of this general principle is the law of duality of space [pp. 376–77].

This shows quite clearly how Enriques held what might be called, with the example of Niels Bohr's dictum on quantum mechanics in our minds, a complementarity principle about geometry. In Enriques's view, projective geometry was simultaneously a matter of logic and an intuitive discipline. In the 1890s and all the way to the publication of the English edition of *Problems of Science*, Enriques wished to argue that projective geometry, however recondite, was not ultimately or merely a formal system of rules, but was grounded in intuition and, as he came to argue, in the fundamentals of human psychology.

Enriques's views were naturally opposed within the Italian community, most notably by Giovanni Vailati.[141] Vailati was loyal to Peano and the group of mathematicians and logicians who could most accurately claim to represent the logical point of view within mathematics. He wrote his essays in Peano's ideographical style, thus further acknowledging the debt he was happy to admit he owed Pieri. Vailati saw geometry as a purely logical subject concerned with postulates and deductions. The truth of mathematical statements and the existence of objects conforming to them was not the concern of the mathematician. Underpinning this disagreement was a difference of opinion about the scope, and indeed the nature, of logic. Vailati took the strict formal view, whereas Enriques advocated a complicated ongoing relation between logic and psychology. It was, however the psychologistic aspect that drew American philosophers to Enriques's book and helps account for its American success.

The *Problems of Science* was well received abroad. It was reviewed in *Mind* (17, 1908) where it was hailed as "probably the most comprehensive study that has appeared in recent years on the concepts on which modern science is built." The reviewer regarded the grasp of modern science, traditional philosophy, and psychology as such as is "rarely found united in one mind." It was translated into German, and in two parts into French. The publication of the first part, on logic, was the occasion for H. M. Sheffer at Harvard to call for an English translation (*Philosophical Review* 19 [1910]: 462–63). The translator of the second, geometrical part, introduced it as offering a middle way between Mach and Poincaré.

As Sheffer may well have known, the English translation was by then underway. It was made by Katherine Royce, the wife of Josiah Royce, and he was by then the senior professor of philosophy at Harvard and had taught Sheffer. Royce had met Enriques at the International Congress of Philosophers in Heidelberg in 1908 and been impressed by him.[142] Royce wrote to J. M. Cattell that Enriques's book "has the advantage over Poincaré's of going deeper into modern logical problems," and that "as the book of a modern geometer and a notable representative of the great Italian school of logic, it

---

[140] This passage is also largely quoted in Avellone et al. 2002. The quotation from Enriques 1898, 376–77 will be found on p. 394. My translation differs slightly from the one in Avellone et al. 2002.

[141] As discussed in Avellone et al. 2002.

[142] Paul Carus, Royce, and Catherine Ladd Franklin had been involved in setting up the International Congress of Philosophers; see Kennedy 1980.

would occupy a novel place in the literature."[143] Paul Carus of the Open Court agreed to publish the translation by Katherine Royce, but various administrative difficulties prevented Royce from finishing his wife's translation (as he had agreed to do) before the end of 1913. By then, however, Royce stated in his Introductory Note to the book, it might be useful in combating the recent rise of anti-intellectualism which he feared would prefer easy, dramatic answers to patient, critical thought.

Royce argued that although Enriques's reputation was founded on his treatise on projective geometry and his essay on the foundations of geometry, as a philosopher he said much that pragmatists could accept. This was all the more surprising because the Italian book had been published in 1906, while the vogue for pragmatism had not begun until 1908 in Heidelberg. Indeed, Royce continued, Enriques's form of pragmatism was largely original, manysided, and judicious; the book took a diverse yet synthesizing approach, and it should be welcomed above all as a treatise on methodology.

Enriques, it is clear, was given credit for a point of view that is broader than his own. He was taken to represent a school to which he barely belonged was and assimilated into traditions to which he did not belong. Enriques's interests in logic had nothing to do with the severe formalism of Peano and everything to do with psychology. But both were issues in the popular perception of mathematics, and Enriques was assumed to speak for both. Royce, for example, saw no problems in harmonizing Enriques's ideas with any others he (Royce) happens to support, and others were equally imaginative, or simply careless, in their reading. When a reviewer wrote in 1908 that "reality means, according to Enriques, the correspondence of sensations with expectation; a reality existing by itself, independent of our experience, is simply an absurdity," this projects Kantian ideas into an anti-Kantian book. Enriques indeed wrote, " 'The correspondence of the sensations with the expectation' always constitutes the true character of reality."[144] But he immediately went on: "Reality as we think of it would not cease *to exist in itself*, even if all communication between our minds and the external world were broken off" (p. 56, italics in original). Enriques then explained that the term "in itself" referred to our inability to modify our sensations as we wish, thus avoiding the position he called transcendental, which in his view lead to a skeptical idealism about an unknowable phantasm before our eyes. Ultimately, Enriques defined the real as an "*invariant in the correspondence between volition and sensation*" (p. 65, italics in original). He did not dismiss externally existing reality as absurd, but his reviewer had no trouble in summarizing him that way because such philosophical views were typical of the idealist philosophy then current.

5.5 Writing the History of Mathematics

5.5.1 History and Historians in Germany

People turn to the reading and writing of their history for many reasons apart from its intrinsic interest: for argumentative and polemical reasons, to make sense of the

---

[143] Royce 1976, letter of October 15, 1908.
[144] Enriques 1906/1914, 56.

world around them, to proclaim continuity and tradition, especially when it leads directly to them. No professional qualification is needed, and in the case of the history of mathematics many if not all of the major authors were trained as mathematicians and may never have acquired a formal training in historical research. Not that, at a time of rapid change, one should insist on such details: it is the quality of the work that matters.

By the end of the nineteenth century, the study of the history of mathematics was a significant part of German mathematical life. This derived in part from the central role academic history played in shaping the identity of the emerging German nation. Leopold von Ranke, who has been called "the father of the objective writing of history" for his pioneering work on the authoritative criticism of sources, was professor of history at the University of Berlin from 1834 until his retirement in 1871. He was made a freeman of the city of Berlin in 1885, which says something about the status of the historian at the time. He was succeeded by Heinrich von Treitschke, whose unfinished work, in five volumes, on German history 1800–1848 was apparently found in almost every middle-class home, and who was made historian of the Prussian state in 1886. From 1879 his growing anti-Semitism[145] brought him into conflict with Theodor Mommsen.[146] Mommsen, the acknowledged leader of the classical historians for his work on Roman constitutional and criminal law, became the professor of ancient history at the University of Berlin in 1858, the permanent secretary of the Prussian Academy of Arts and Sciences, and sat at various times in the Prussian parliament. He was also a powerful writer, winning the Nobel Prize for Literature in 1902; the judges had found Tolstoy's views too radical for them. He strongly opposed Treitschke's anti-Semitism and published a letter in protest signed by seventy influential figures, arguing that Jews are Germans and racist hatred will come to an end sooner or later with a real respect for the distinctiveness of the Jewish culture. He was also among the 593 people who signed the petition against anti-Semitism in 1891.

Other intellectual forces with deep social roots were also at work. German theologians took up the historical analysis of the Bible in analyses that were ultimately to alarm Pope Pius X, without, however, undermining their own Protestant faith. Among these was Adolf von Harnack, twin brother of the mathematician Axel Harnack. He urged the position that "the development of Christian theology and dogma was a strictly historical process which could be understood through historical-critical method alone, without recourse to meta-historical sources of authority."[147] He wished to continue the work of the Reformation by using new historical criticism to extract the timeless ethical message of Christianity from its various dusty texts, and his book on this theme, *Das Wesen des Christentums* (English translation, *What Is Christianity?* 1903), based on his crowd-pulling lectures at the University of Berlin in 1899–1900, was both popular and influential, and was frequently reprinted. His

---

[145] Witness his remark in the *Prussian Year-Book*, which he edited for decades: "The Jews are our misfortune" ("Die Juden sind unser Unglück," *Preussische Jahrbücher*, November and December 1879, and January 1880; reprinted as a pamphlet under the title, *Ein Wort über unser Judentum*, Berlin, 1880).

[146] Information from http://www.geschichte.hu-berlin.de/ifg/galerie/texte/treitsc2e.htm.

[147] The information in this comes from Andrew Irvine, *The Boston Collaborative Encyclopedia of Modern Western Theology*, Fall 1996, http://people.bu.edu/wwildman/WeirdWildWeb/courses/mwt/dictionary/mwt_themes_680_harnack.htm.

personal charisma and strong advocacy of freedom of scholarship made him attractive, but he was also a staunch patriot who at various times was the rector of the University of Berlin, director of the Royal Library, and the first president of the Kaiser Wilhelm Foundation. He declined an invitation to be the German ambassador to the United States.

The third major force at work was the aftermath of Hegelian philosophy. A background training in philosophy was provided in the higher reaches of school and in courses at the university, and the importance of ideas was beyond question. A degree of patriotism was instilled from an early age and sharpened in many cases by a year in the army. Hegelian philosophy suggested an inevitable progress to history that fortuitously seemed to be exemplified by the rise of Germany and to be reenacted at the level of the individual by the careers of the fortunate; the mathematician Kummer was a Hegelian of this stripe. Imperial court circles were aristocratic, militaristic and, it may go without saying, philistine; but the most eminent professors, such as Helmholtz, could become accepted there and others could at least hope to be noticed. In fact, Helmholtz's second wife made their Berlin home a salon for scientific, artistic, and literary figures, and there Helmholtz shed his taciturn laboratory persona so successfully that he became a favorite with the imperial family.[148]

The situation of Jews in Germany was, of course, more complicated. It was hard for them to go all the way with Hegelianism, because Hegel, while advocating civil rights for Jews, found no place for them as Jews in the journey from Christianity to modernity and the expression of the state. There was a slow liberalization of their position, but a perceptible anti-Semitism always operated. Some coped with it by assimilating, more moved to a sense of Jewishness that was more cultural than religious. They identified strongly with the aims of the German state, served honorably in the army, and considered themselves German in every way.

German professors therefore belonged to one of the elite groups of German society, marginal with respect to some others but well placed in many ways. This gave them a high degree of seriousness in their attitudes to their work, manifest in the high degree of productivity and the extent to which they drove themselves at times to nervous collapse. The pressure to become a full and therefore tenured professor, to build one's academic empire and secure places for one's chosen students, raised the temperature at various places and at various times. One American visitor reported most bleakly on his hosts, and one supposes classicists did not have a monopoly on this sort of behavior:

> I do not yield to any one in admiration of German learning, conscientiousness, inventiveness, grasp, but the more I have seen of the arrogance, the jealousy, the hateful manoeuvring, the shameful backbiting, the hatred, awry, malice, and all uncharitableness, which a closer knowledge of the professor's life in Germany reveals, the more glad I am to live where, if such abominations exist, they do not, like the frogs of Egypt, go up and come into our houses and into our bedchambers and upon our beds and upon our ovens and into our kneeding troughs.[149]

---

[148] Watson 1978.
[149] Gildersleeve 1884, 354, quoted in Davies 1998, 270.

5.5.1.1 GERMAN HISTORIANS OF MATHEMATICS

Mathematicians who moved into the history of mathematics knew that they had to write widely and responsibly. The first German mathematician to take up the history of the subject was Hermann Hankel, who died at the age of thirty-four but whose posthumous *Zur Geschichte der Mathematik in Altherthum und Mittelalter* (or, The History of Mathematics in Ancient Times and the Middle Ages) is notable for its attempt to demote the Greeks and promote the Indian mathematicians as having made more relevant contributions to the growth of modern mathematics. He was followed by Moritz Cantor, who wrote a monumental three-volume history of mathematics (a fourth volume, written by others, brought the coverage up to 1799). This work set the standards for subsequent work in the field, but inevitably left room for histories of the modern period. Here the most significant figure was Felix Klein.[150]

He took the lead in editing the collected works of Gauss, and the later volumes of the Gauss edition carried eight extensive essays on various aspects of Gauss's work, organized by Paul Bachmann and Ludwig Schlesinger. Klein also stimulated other mathematicians to take up historical work: Engel and Stäckel wrote the standard works on Bolyai and Lobachevskii and produced a valuable collection of sources on early work on non-Euclidean geometry. Engel himself translated Lobachevskii's two early and extensive early Russian papers into German (Lobachetschefskij 1899). He also rescued Grassmann from neglect. Klein's major project, the *Encyclopädie der mathematischen Wissenschaften*, was devoted to articles surveying the current state of the art in numerous branches of mathematics, and many of these articles have a substantial historical component. Other German mathematicians outside Klein's orbit also took to the history of mathematics. To mention just one, Alfred Pringsheim wrote not only for the *Encyclopädie der mathematischen Wissenschaften* but on Euler for the Euler edition. Finally, lectures Klein gave in Göttingen during the First World War circulated among his students and were eventually edited for publication by Courant and Neugebauer as *Vorlesungen über die Entwicklung der Mathematik* (Klein 1926–27). These were naturally biased toward his own interests and his own priorities in mathematics, and cannot be read as a simple participant's record of events, but they have a freshness and vigor of their own.

Another aspect of this scholarship was the production of collected works. Some of this is no doubt due to the growth of the profession, but it is noticeable that much of it was concentrated on the recent period. After Gauss, the next person to be remembered or honored in this way was Riemann. The aim of his first editors, Dedekind and Weber, was not, of course, to write the last word in an old story but to put into circulation ideas that they knew had by no means been fully appreciated. Other mathematicians soon to be honored in this way included Dirichlet, Hesse, Kronecker, Weierstrass, and Klein himself; and often work started before the great man was dead, so that due homage could be paid. These volumes, with their biographies and commentaries, eloquently suggest that their subjects had made history with their work. Euler (whose *Opera Omnia* and correspondence is not yet finished at seventy-six volumes) only began to follow in 1903.

---

[150] This and later paragraphs draw on Dauben and Scriba 2002.

5.5.2 History and Historians in France and Italy

The situation in France was similar as regards the professional status of historians and their willingness to take stands on public issues. The underlying concerns of French life were naturally different, dominated as they were by the perpetual desire to work through the implications of the French Revolution and, after 1871, the defeat in the Franco-Prussian War. Hippolyte Taine and Ernest Renan were the leading historians of the second half of the nineteenth century in France. Taine in particular wrote his monumental but unfinished *Origines de la France contemporaine* with a view to explaining the failures of France as consequences of its revolution. Renan, who as a young man had trained to be a priest only to abandon the attempt, was another historian who studied religion from a historical rather than theological point of view.[151] Renan's eight-volume *Histoire des origines du christianisme* (History of the Origins of Christianity) published between 1863 and 1883, began with his best-known book, the *Vie de Jésus* (the life of Jesus). This was the first biography of the founder of Christianity, and it was unsympathetic to much of the biblical accounts, especially the miracles, which Renan dismissed as frauds. Its huge sales elevated him to the role of a public intellectual and started a series of ever more skeptical accounts of the Bible, until even the historical existence of Jesus came to be disputed.

Another peculiarly French source was the Napoleonic conquest of Egypt and the consequent speculations about human history disagreeing markedly with biblical chronology. French scientists had made their way to the temple at Dendera, much of which Louis XVIII paid a considerable sum of money to have brought back to France in 1822. The focus of interest was the ceiling with its giant zodiac. At the end of a long story, which involves the decipherment of hieroglyphics, it was shown that the Dendera zodiac was in fact a recent creation (about 52 BCE), but by then astronomical dating methods had been drawn back into the study of ancient history and confidence in the literal truth of the Bible was further dented.[152]

5.5.2.1 FRENCH HISTORIANS OF MATHEMATICS

In this highly charged environment, the place for history of mathematics was pressed by Michel Chasles, whose *Aperçu historique* (Historical Observations) traced the methods of geometry from Greek times to present-day (as of 1837) France—Chasles could not read German. A generation later, Jules Hoüel not only translated the major original works on non-Euclidean geometry by Bolyai and Lobachevskii into French, thus making them widely available for the first time, but he also made continual efforts to raise standards in the subject. In this he was supported by Paul Tannery, older brother of the mathematician Jules Tannery, who wrote on Greek mathematics and coedited the *Oeuvres* of Fermat. Tannery was sympathetic to the positivist philosophy of Comte, then enjoying a revival in France, and in this he was followed by Gaston Milhaud, who wrote on the Greeks and also influentially on Descartes, and by Couturat, whose contributions to the revival of Leibniz have already been discussed. In the twentieth century this tradition produced a number of historically

---

[151] On Renan's racism, see Irwin 2006, 193–94 and 295–96.
[152] See Buchwald's article, http://pr.caltech.edu/periodicals/eands/articles/LXVI4/buchwald.html.

informed philosophers of mathematics, starting with Léon Brunschvicg and Jean Cavaillès and continuing to the present day.[153]

While it would surely be too glib to "explain" the differences between the French and the German historians of mathematics with reference to two factors alone, it is noticeable that, on the one hand, German mathematicians (but not the French) wrote on the history of mathematics, they wrote on the recent period (the nineteenth century), and they belonged to the mathematical culture most strongly drawn to modernism. On the other hand, German society was buoyant and enjoying a successful nineteenth century, whereas the French were increasingly prone to self-doubts.

5.5.2.2 ITALIAN HISTORIANS OF MATHEMATICS

The Italian situation also merits attention. The first major Italian historian of mathematics was Balthasar Boncompagni, who printed the twenty volumes of his valuable journal the *Bolletino* at his own expense from 1868 to 1887. Several of the generation of Italian mathematicians that grew up after the unification of Italy were drawn into Klein's *Encyclopädie der mathematischen Wissenschaften*, among them Pincherle, Castelnuovo, and Enriques, and of these Enriques remained committed to the history of mathematics as a means of defining and promoting it. His one-time assistant, Roberto Bonola, was the author of what became the standard history of non-Euclidean geometry. A little later, Gino Loria, writing, as he put it, "as a mathematician for mathematicians"[154] produced a number of successful histories of geometry that reached well into the nineteenth century.

5.5.3 Why the History of Mathematics Was Written

The interesting comparison in the history of nineteenth-century mathematics is therefore between Germany and Italy, the two countries most identified with mathematical modernism.[155] Two questions arise: Why did they turn with such enthusiasm to the history of mathematics, and how was the writing of that history affected by what they took mathematics to be?

5.5.3.1 NON-EUCLIDEAN GEOMETRY

To take the second question first, it is clear that in each country the one topic that really caught the mathematicians' attention was non-Euclidean geometry. If Enriques was the spokesman internationally for Italian geometers, the one who most securely grasped the historical task was Bonola, whose *La geometria non-Euclidea* (1906) grew out of an essay he had written for a collection of monographs on geometry that Enriques had edited, the *Questioni riguardanti la geometria elementare*, in 1900. It was translated into English by H. S. Carslaw and published by Open Court in 1912, where it appeared with a short introduction by Enriques, the

---

[153] See Sinaceur 2006.
[154] Quoted in Bottazzini 2002, 88.
[155] I therefore pass over a number of other distinguished historians of mathematics from the period: Eneström, Heath, Nesselmann, Suter, and others.

melancholy occasion for which was Bonola's death in 1911 at the age of thirty-seven. To produce the book, Bonola relied very sensibly on the work of Friedrich Engel and Paul Stäckel. Of course, he also played up the significance of Gerolamo Saccheri, the Italian mathematician who had died in 1733 after discovering many results that now form the cornerstone of elementary non-Euclidean geometry. His work had lapsed into almost complete obscurity, and Beltrami had recently trumpeted its merits upon rediscovering it in 1889.

Engel and Stäckel had tried to find out as much as they could about the elusive figure of János Bolyai. Subsequent writers have discovered a great many minor figures omitted by them, but, with the arguable exception of Legendre, no Western mathematician has entered this pantheon. One may conjecture that the reason is the progressive interpretation Engel and Stäckel placed on all this work. Each author marked a significant advance until, while Schweikart hesitates and Taurinus looks backwards, Gauss takes the bold step into the non-Euclidean world. From such a perspective, Legendre's attempts are reactionary, and sometimes embarrassingly flawed. The commentaries Engel and Stäckel provided are interesting because, while they amount to about one-fifth of the book, much of them are biographical and bibliographical. Their thesis, if they had one, is concealed in the choice of authors.

Bonola's work, by contrast, is more of an argument. With the work of Engel and Stäckel in print, he could content himself with summaries of the original arguments and tell a historical story of his own. He added a number of protagonists, expanded on the cryptic references to Arabic writers, included Legendre, and went past Bolyai and Lobachevskii on to Riemann. In five appendixes he considered such topics as the connection between the parallel postulate and the law of the lever, the independence of projective geometry from the parallel postulate, and the impossibility of proving the parallel postulate.

Bonola's generic account of a mathematician's contribution goes like this. The mathematician's definition of a straight line and of a parallel line is given, or, if none was supplied, one is uncovered from the use of the concepts. The original argument is then presented in something close to its own terms, and the fallacy, if any, is explained. So, when discussing an argument that would seem to show that spherical geometry cannot exist, Bonola shows how the postulate of Archimedes or its consequence, the indefinite extendibility of the straight line, has been tacitly invoked. And when Legendre produces a fallacious argument using the postulate of Archimedes, Bonola also shows how it could have been avoided, the better to explain that this was not where Legendre erred.

From first to last, Bonola's account of the origin and development of non-Euclidean geometry is rooted in an analysis of axioms—their equivalence and their independence. As a result, his analytic approach is insensitive to the methods originally used, and eventually leads him to miss the significance of a crucial step in the discovery of non-Euclidean geometry: the introduction of trigonometric methods. However, an Italian geometer, and a pupil of Enriques, writing between 1900 and 1911, would naturally see geometry as organized in this way. It may be significant that Engel and Stäckel were, if anything, more in the orbit of Klein than Hilbert, and, being forty years old in 1901–2 had several years of mature work behind them as differential geometers and analysts. At all events, Bonola's work is analytic where theirs is descriptive. For Bonola, geometry is a matter of axioms, so the history is a history of axioms.

It need not have been so. The history of non-Euclidean geometry is open to other interpretations. Had it just been a question of exhibiting an axiom system for something fairly geometrical, then spherical geometry would have done. One needs, of course, to strike out two of Euclid's axioms: the parallel postulate and the indefinite extendibility of the straight line. That this was not done suggests that the ancients were not simply investigating axiom systems. It suggests what a considerable amount of other evidence also suggests: that they were investigating something—the geometry of physical space. The ongoing question was not, "Is the parallel postulate independent of the other axioms of geometry"?, but "Is the parallel postulate independent of the other axioms of geometry when giving an account of space?" This is a different enterprise from the much more overtly logical one in fashion around 1900.

5.5.3.2 A CONNECTION TO MATHEMATICAL MODERNISM?

Can we also say that mathematicians turned to the history of mathematics because of the changes in their subject that mathematical modernism was bringing about? Historians of music generally notice that as the nineteenth century proceeded, composers constructed a canon (no pun intended) with, for example, Bach and Mozart securely in place. In this they defied popular taste, because they recognized that it is a certain small number of one's predecessors who have deep lessons to teach and to whom one can continually, valuably return. Mathematicians of the period did the same, and probably for the same reasons. But it has been said of Schoenberg that he held Bach in particularly high esteem as his forerunner in the creation of a whole system of harmony.[156] Schoenberg's view was that the greatest composers were always the ones who tried to say something new, so new that it forces changes in style and technique and yet had its roots intimately in the past.

Of course I would like to argue that it was the same with modernist mathematicians, but I think this claim cannot be sustained. It was Klein, not Hilbert or Dedekind, who wrote extensively on the history of mathematics. As we have seen, a respectable number of mathematicians followed him. But once we see that writing history was an honorable thing to do and that learning from the masters is a task the best of any generation often set for themselves, there does not seem to be a profound attempt to legitimize the new mathematics by identifying it with the core element of the tradition.

If the claim can be sustained, it would be with the renewed attention paid to Greek mathematics. It is true that the Danish scholar Johan Heiberg worked in the years around 1900 to produce the now standard editions of Euclid's *Elements* and many other works, and that Tannery in France, Heath in England, and other scholars followed him in these endeavors. Active mathematicians took note of their work, but this was not too surprising at a time when many professors would have learned Latin and Greek at school. What is missing is a figure who would urge the merits of the Greeks as axiomatizers and foundational scholars the better to legitimate modern work of the same kind. Such a reach into the past could have occasioned a favorable comparison, as modern axiomatics succeeded in creating objects while classical axiomatic work merely codified. But a few passing remarks comparing Dedekind

---

[156] See Stefan 1999 and Burkholder 1999.

with Eudoxus on the subject of the real numbers does not make recourse to history a significant aspect of mathematical modernism. In the absence of new evidence, it seems that history of mathematics was written for many reasons, but legitimating modern mathematics was not one of them.

On the other hand, it might be that a claim in the reverse direction can be sustained. The history of Greek mathematics is dominated by three great names: Euclid, Archimedes, and Apollonius. All were given new, scholarly editions of their work around 1900, notably by Heiberg. No one disputes their greatness or the impact their work would have, but there is an issue with the intensity of the darkness into which all other surviving sources were plunged. With the arguable exception of the Greek astronomers Hipparchus, Aristarchus, and Ptolemy, all other Greek mathematicians were marginalized. Some of these have only fragmentary texts to support their names, whereas several whole works of the canonical trio have survived. Some are later. But all are different, more inclined to the stating and solving of problems (Diophantus) or the applications of mathematics (Heron) or to writing commentaries (Pappus) or just not very good (Iambilichus). The net effect of the historical scholarship was to present an image of Greek mathematicians as concerned with rigor, with mathematics for its own sake, or, if applied, work of originality and depth. This was, of course, exactly the image of mathematics that Heiberg and his colleagues would pick up from contemporary mathematicians (English and Italian mathematicians would also have been brought up on Euclid's *Elements* at school). A hundred years later, it is possible to see that the historians' criteria were colored by a view of mathematics that was of its time, one that glides over the messy, mundane, but more typical work. It also fits to a Victorian view that found "our" history in the history of the classical past, but what was analytic and critical in the work of Mommsen or Renan and their followers became, as is too often the case in the history of mathematics and science, statue building.

6

MATHEMATICS, LANGUAGE, AND PSYCHOLOGY

6.1 Languages Natural and Artificial

It was the painter and aspiring but unsuccessful poet Edgar Degas who complained to his friend Stéphane Mallarmé that he had many ideas for poems but could not write the poems, to which Mallarmé replied, "Poems, my dear friend, are made of words, not ideas." The same is true, and to the same extent, of a mathematical paper, as every contemporary mathematician knew. But in what language should mathematicians write, and is, or could there be, a language best suited to their purposes?

6.1.1 National Languages in Mathematics around 1900

The international mathematical arena in 1900 was an arena of competing nationalisms, chiefly those of Europe, the United States, and Japan.[1] Mathematicians in Germany spoke and wrote German, and this included a number of visiting Americans, some of whom wrote not only their doctoral theses in this language but went on to write books in it long after their return to America.[2] The French wrote in French, just as the English wrote in English, and only occasional forays were made into another tongue. Because French and German journals dominated the field, Scandinavians, Russians, Czechs, and Poles had to decide if they would attempt French or German, or write in their native tongue with ease although for a restricted if valuable audience. Sometimes this was a political act.[3] Most mathematicians were, or presented themselves as, literary monoglots who seldom wrote outside their own linguistic boundary. Hilbert was typical, when he published in the *Bulletin of the London*

---

[1] This section is adapted from Gray 2002.
[2] Osgood 1907, also Haskell 1890 and Dickson.
[3] See Duda 1996.

Mathematical Society, writing in German,[4] and it is significant that Felix Klein, the most ardent advocate of Germany's leading role in the mathematical world, chose to publish in English when addressing an American or English audience.[5]

The work of many, therefore, had to be translated. Joseph Liouville and then Guillaume Jules Hoüel led the way in France. The American mathematical community translated many works, while the British did much less. Hilbert's twenty-three problems, presented in 1900 to the Paris International Congress of Mathematicians, were rapidly published in the *Bulletin of the American Mathematical Society* (and also in a French translation in the *Comptes rendus* of the congress).[6] The English translations of Poincaré's famous books of essays are equally divided between English and American editions. The American Mathematical Society was very active, not only translations but also summaries, reports, digests and commentaries on European work fill the pages of the *Bulletin of the American Mathematical Society* until at least 1914. Paul Carus's journal *The Monist* was also an important source.

Unsurprisingly, there were four languages for international use in mathematics around 1900: French and German because of the indigenous strength of the mathematics profession in those countries, English because of the strength of the British Empire and the growing power of the United States, and Italian, because there were so many good Italian mathematicians. Even this looked like too many to some. Schröder, in his address to the first International Congress of Mathematicians in Zurich in 1897, said there had been talk of making English the official language of the congress because it was neutral between French and German, but whatever opportunity that might have been was lost when only ten English speakers attended and many more British and American mathematicians went to the British Association for the Advancement of Science meeting held concurrently in Toronto instead.[7]

What there was not, in 1900, was an acceptable *lingua franca*. Rather, there were several rival *linguae francae*. There were, in particular, vigorous groups who aimed to create or adopt one or another international language for mathematical and even all scientific work. They were opposed, usually quietly, by speakers of any hegemonic language who wished their language to prevail. Many educated people thought that there was a small number of mainstream languages and everyone should learn one, or two, or three. This would bring cultural advantages, there were many educational opportunities, and so, such people said, there was no reason to argue with the status quo. Some, such as the linguist Michel Bréal, one of the inventors of semantics, even advocated that the entire world should learn either French or English.[8]

Still, there was a sizable minority advocating that there should be a new language, created specifically for the purpose of enhancing international communication. Some wanted to return to the idea of a universal means of communication, animated by the profusion of tongues spoken on the international stage. Others, with loftier goals, hoped that an international language would help bring about a truly international

---

[4] Hilbert 1902.
[5] Klein, 1894.
[6] Hilbert 1900.
[7] Schröder 1898, 44.
[8] Bréal 1901.

world, in which national rivalries and the dangers of war would diminish. All such people had to decide what form the new language should take

The language issue is decidedly difficult. English had abolished gender centuries before, and dispensed almost completely with declensions in nouns. Its verbs are largely simplified German verbs—but, as every nonnative speaker and many an English or American child complains, English spelling is far from phonetic. French, on the other hand, is more nearly phonetic, but it is laden with grammatical issues. Latin is a much more inflected language, and that has the virtue of allowing a very flexible word order, but this makes it very difficult to learn. This profusion of possibilities arose for reasons to do more with the nature of language and could not be held in check by school grammarians. English, for example, drew an extensive part of its vocabulary from French and Latin, and many verbs from German; these words have spellings that reflect their origins. Artificial attempts to override these linguistic facts confront more complicated problems than are apparent at first sight.

The issue of pronunciation is important. The melancholy example of Church Latin showed that a common tongue can be spoken with large, sometimes almost insuperable, regional variations. If a new international language was to enable scientific papers to be read aloud, understood, and the speaker questioned, all in the new language, then a universally intelligible spoken language—a pasilalia—is what is wanted. This is much more difficult than the production of a universal international written language or pasigraphy.

6.1.2 An International Language

Moves to bring about an international language for science were underway by 1900, although the six months of international conferences in Paris during 1900 passed without international languages being the main item of agenda for any.[9] Couturat was one of the organizers of the International Congress of Philosophy, which met in Paris from August 2 to 7, 1900, and he invited Frege to join the organizing committee. Frege agreed, and a correspondence ensued in which Couturat broached the topic of an international auxiliary language. To judge by Couturat's reply, Frege could support a scientific, commercial, and utilitarian language, but worried about the creation of neologisms.[10] The International Congress of Mathematicians, which met immediately after the philosophers, heard a paper by Charles Méray of France on the merits of Esperanto.[11] He observed that a previous international language, Volapük, had died, and then argued that the real difficulties even highly educated professors had understanding one another at the congress surely suggested that Zamenhof's simpler creation, Esperanto, had some claim on everyone's attention.[12]

The paper led to a discussion of a resolution proposed by Leau to adopt a universal scientific and commercial language to be realized by the international acade-

---

[9] There were congresses of Volapük, and international languages were mentioned in various congresses.

[10] Frege's replies are apparently lost; see Frege 1980.

[11] Méray 1902. The paper was actually read by L. Leau.

[12] Carnap was in favor of Esperanto, Ladd-Franklin of Ido.

mies. Couturat, Léopold Leau, Charles Laisant, and Alessandro Padoa were among those in favor; Schröder and the Russian Aleksandr Vasil'ev against.[13] In the end, Vasil'ev's counter-proposal, that "the academies and learned societies of all countries study the proper means for remedying the evils that arise from the increasing variety of languages used in the scientific literature" was carried.[14]

Volapük (its name means "world-speak") had been the creation of a German Roman Catholic priest, Father Schleyer, who presented it to the world in 1880. It was modeled on existing languages, making it what Couturat and Leau, in their major book (of over six hundred pages) on international languages, classified as an a posteriori language.[15] Its grammar was regular and pronunciation standardized. But despite attempts to choose a vocabulary that Europeans would find familiar, it looked strange, because the grammatical forms give it a markedly Slavonic feel and its vowel sounds were nothing like English. By 1889 its adherents claimed it was spoken by anything up to 200,000 people; yet in 1890 it was dying, torn apart by the failure of its creator to let its speakers modify it.

Esperanto was the creation of Louis Lazarus Zamenhof, a Russian Jew from Bialystock in Russia.[16] He presented it to the world in 1887 as "La Lingvo Internacia de la Doktoro Esperanto," the international language of Doctor Hopeful, and his pseudonym happily became attached to the language. As its title indicates, Zamenhof hoped his language would help bring about the reconciliation of races and nations only too obviously hostile to each other in his native part of the world. It agrees with Volapük in having standardized declensions and conjugations (Zamenhof was inspired in this respect by English, but did not go all the way), but its written form is more recognizable and more euphonious, being close to French and Spanish. It began to prosper in 1898 when the French took it up, and when in 1901 the Delegation for the Adoption of an International Language was organized, the Touring Club of France (membership over 100,000) gave Esperanto their support, and Hachette began to publish Esperantist literature. Large international conferences followed, and it seems that converts found it easy enough to learn, although Esperanto was easier to write and read than to speak. A large indigenous literature was created, to which were added translations of many great works of literature, and some mathematical works.

As is often the way, success brought its own problems. Among the many adherents to the cause were those who sought reforms, naturally fueled by the great moral claims made for the language. A major difficulty was provoked by the awkwardnesses of the original creation, and Zamenhof's original choices, of which the use of "-j" for plurals and "-n" for the accusative were perhaps the most unfortunate. It was not sufficiently a posteriori, and so it left room for alternatives. But if the rigidity of the leaders of Volapük had destroyed it, did not the lack of control over Esperanto also threaten to break it apart? Yet others sought a solution in other languages:

---

[13] Vasil'ev was a Russian mathematician in Kasan, and he had been influential in establishing the Lobachevskii Prize, first awarded in 1897 to Sophus Lie. He went on to write a book on relativity theory, which was translated into English and published with a preface by Bertrand Russell (Vasiliev 1924). His philosophy of mathematics is described in Vucinich 1999.

[14] See Kennedy 1980, 98.

[15] Couturat and Leau 1903.

[16] Bialystock is today part of Poland.

simplified Latin, or a language called Idiom Neutral, first presented in 1903 and significantly revised in 1907.

Guiseppe Peano advocated his own creation, Latino sine Flexione.[17] He based it on a rigorous attempt to identify the basic vocabulary of all European languages and gave it a simplified grammar that resembled Latin but sounded like Italian. He promoted it in his mathematical journals and in others of a more general character, and after 1908, through the formerly Volapükist Akademi de Lingu Universal, renamed the Academia Pro Interlingua. As an artificial language, Latino sine Flexione had a lot going for it. Romance speakers could become fluent quite easily, and the rest of the international professoriate had almost certainly learned Latin for some years at school. The use of essentially English or French word order came with the disappearance of the complicated word endings in Latin, and spelling was reasonably phonetic. Even the choice of name was astute: calling it hybrid Latin without inflections avoided the problems of nationalism that would have arisen if Peano called it "standard Italian."[18]

Whatever its intrinsic merits, Latino sine Flexione or Interlingua (as it also became called) could not compete successfully against Esperanto until that popular tongue fell into schism. Briefly, in 1907 a special committee was established to decide which international language should henceforth be actively promoted.[19] Peano, the distinguished chemist Wilhelm Ostwald, and the professional linguist Otto Jespersen were brought in; Zamenhof sent the Marquis de Beaufront to speak for Esperanto. The committee meetings took place at the Collège de France from October 15 to 24, 1907, and Couturat and Leau produced a new book, *Les nouvelles langues internationales* for the occasion. Unfortunately, the decision of the committee was ambiguous. Some thought that it endorsed Esperanto, or at least a simplification proposed by, of all people, the Marquis de Beaufront, and called Ido. Others thought that it found no international language acceptable and proposed to start again with Ido. Esperantists, Zamenhof among them, could not agree because there deliberately was no official body capable of deciding changes in Esperanto. Peano was among those who denied that the committee could legislate as it wished. The new language, Ido, was surely an improvement on Esperanto, particularly in its plurals and in its limited use of the "-n" ending. But whether, as further polished by Couturat, it was significantly better was an issue thrown forever into the shade of the acrimonious dispute between the Esperantists and the Idoists. An international language can surely only succeed when it is unique in its claim on people's attention; a pacifist cause only when its leaders are not themselves at war. The quality of the arguments of Couturat, Jespersen, Ostwald, and others for Ido and against Esperanto paled by comparison with the disagreeable sectarian nature of their rhetoric.

Sympathetic mathematicians were left looking for a purified mathematical language that would both obviate error and speak to all mathematicians equally. Schröder, in his address to the Zurich International Congress of Mathematicians of 1897 already mentioned, had come out strongly in favor of a pasigraphy almost

---

[17] Peano 1903. See the thorough account in Kennedy 1980.

[18] Compare the fate of C. K. Ogden's Standard English, a systematic attempt to define a basic vocabulary and a basic set of grammatical forms, which only succeeded within the bounds of the British Empire.

[19] See Guérard 1922, 145–160.

entirely of his own devising.[20] To create an international scientific language—and as he saw it, nothing could be of greater significance—one had to isolate the necessary minimum of primitive notions or categories that express all the necessary notions, using purely logical operations that can in their turn be handled using only the laws of ordinary logic. All this was to be carried out with easy signs and simple unambiguous symbols used with absolute consistency or rigor. This, he went on, put the whole enterprise on a much higher plane than the merely linguistic aims of the Volapükists. His reasons are not clear, but he seems to have thought that the Volapükists aimed at supplementing natural languages, whereas his pasigraphy would supplant them for scientific purposes.

Schröder agreed that his pasigraphy could never be spoken. He thought it was possible at all because he regarded mathematics as that branch of logic which deals with number. That being the case, he argued, Peirce's Logic of Relatives had shown that five (and perhaps only four) notions were necessary. These were equality, set-theoretic intersection, negation, an operation he called conversion, which inverts a relation (turning "parent of" into "child of," in his example), and relation in general. He then gave his reasons for preferring Peirce's system (with his own slight modifications) to that of Peano and his school, which despite many laudable achievements in his view lacked the crucial fifth category of relation in general. Then he demonstrated some of the things he believed his pasigraphy could establish before indicating that the main problem facing the project was the development of a flexible calculus. And, perhaps still smarting from Frege's criticisms of his own work, he added a caustic dismissal of Frege for having taken immense pains to perform what had already been done much better, and so delivered a stillborn child. But Schröder died in 1901, and Peirce was in no position to promote such a cause with any chance of success. Peano was left to advocate linguistic internationalism in mathematics unopposed.

Starting in 1903, with the fifth volume of his *Formulario*, Peano wrote mathematics in his pasigraphy and explained it in Interlingua. His pasigraphic language, rudely dismissed by critics as Peanian, was a purely written language with a stunted vocabulary and grammar, in which all of the mathematical relations were expressed by symbols. This written script, which filled up the pages of several Italian journals, notably Peano's journal *Rivista di mathematica* and his *Formulario*, was intended to be free of all the obscurities of a natural language. It is not clear whether mathematicians were meant to think in it, to translate their papers into it, or simply ensure that their prose was translatable into it, and significantly, papers published in this language were often accompanied by translation into plain Italian. But on any interpretation, it is clear that Peano and his followers distrusted natural languages and sought to replace them. Indeed, it is this distrust, rather than any of Peano's remedies, that marks him as tied in to larger contemporary concerns.

Although Peano's attempt to provide a pasigraphy for mathematics, and with it an internationally comprehensible script, failed, in their day they were part of a variety of ways to think through what it means for mathematics to be a language. And while the connections between mathematics and logic in the period have been much remarked upon, the connections between mathematics and language have been discussed much less often.

[20] See Peckhaus 1991.

6.1.3 Mathematics as a Language

There has always been a superficial, commonplace connection between mathematics and language. Mathematics has generally been esteemed for the quality of its reasoning and the apparently compelling quality of mathematical arguments, even when they lead to implausible results. The precision of mathematics has been attributed either to a precision of thought or of the language in which that thought is expressed, or to various combinations of the two. For Leibniz, Lambert, and others, the reliability of mathematical manipulation with its symbols was the key. For the writers of the *Encyclopédie*, language and thought advanced together in every sphere, including mathematics. Charles Bossut opened the Discours Préliminaire to the *Encyclopédie, méthodique* by saying that mathematics was a chaining together of principles, arguments, and conclusions that was always accompanied by the qualities of being certain and evident. As Hankins has noted, "Abbé Condillac claimed that algebra was the best language because it had the best symbols. There was no ambiguity in their meaning, and the grammar of this 'language' was such that conclusions followed absolutely rigorously from premises."[21]

But if the hope was that the language of mathematics would stand as a paradigm for reasoning in other spheres, it is nonetheless hard to find the commonplace view that mathematics is a language spelled out anywhere carefully and at length. It is clear that mathematics has many features in common with a natural language: it involves signs and symbols that are to be manipulated according to certain rules. But it is seldom clear where mathematics stops and natural language takes over. And—a major stopping point for any linguist—it is certainly not clear that mathematics is a spoken language, whereas all natural languages are primarily spoken and only secondarily written.

More plausibly, and with at least as much weight of tradition behind it, when, in the late nineteenth century, great claims were made for logic, notably the claim that mathematics was but a branch of it, analogous claims were made for logic as a language. Suitably enriched by talk of relations, which syllogistic logic had lacked, it too was now seen as a purified language, and as a properly clarified logical language seen as the appropriate basis for philosophy. Much of the philosophy of language became a debate about the boundaries between logic and language. Indeed, the question of the relation between logic and language was one with its roots deep in nineteenth-century German linguistics.

6.1.4 An Ideal Language

What should an ideal language do? Recent discussions of logic and language by van Heijenoort, elaborated by the Hintikkas and Kusch, have proposed an interesting distinction between language as a universal language, and language as a calculus.[22] Prominent upholders of linguistic universalism are Frege and Russell and, of language as an instrument of calculation, Peirce and Husserl. It was in this spirit in 1914

---

[21] Hankins 1985, 109.
[22] Van Heijenoort 1967b, elaborated in Jaakko Hintikka and Merrill Hintikka 1986, and also in Kusch 1989.

that Frege wrote about indefinable terms that "we have to be able to count on a meeting of minds, on others' guessing what we have in mind."[23] A typical Russellian utterance of this type is "The meaning of the fundamental terms cannot be defined, but only suggested. If the suggestion does not evoke in the reader the right idea, nothing can be done."[24] The ideal type of the first view carries the deliberate limitation that none of us can escape our universe of discourse, and the relationship of language to the world is ineffable. We cannot step outside of language to talk about it and so are reduced to hints; a meta-language for talking about a language borders on the absurd. Typically, the world is all that there is. On the second view, there is no fixed universe of discourse, language is merely properly formulated sets of symbols, and interpretation in terms of possible worlds is allowed. Meta-languages are entirely sensible.

In practice, exponents of the universalist view find semantics difficult if not impossible to explain. They cannot escape language and look at the world directly, so they find the relationship between language and the world is obscure. The correspondence theory of truth is unattractive to them, because objects in the world cannot be known "in themselves," and so a certain kind of Kantianism becomes attractive. When they fail to resolve semantic disputes concerning different languages, they become inclined toward linguistic relativism. Exponents of the second view take more or less the opposite position on all these issues. When it comes to discussing formalism, universalists and calculators again disagree. Universalists can welcome it, because syntax is the part of language that is left to them when semantics becomes inaccessible. They deny, however, that a formal system is open to many interpretations. Calculators urge interpretability, and do not wish to be caught in the thickets of mere syntax.

One way to catch such questions is to ask: If an ideal language is possible, would its users speak the truth inevitably, or would they merely be unfailingly grammatical? Some seventeenth-century writers sought to distill or to create from scratch a language in which each word had a single meaning and reasoning would be automatic (this would make it an a priori language in the terminology of Couturat and Leau). This forced them to decide what were the primitive notions and to arrange them in classificatory schemes. Words (either written or spoken) were then constructed to reflect the formerly latent, now explicit order of nature. When Mersenne wrote to Descartes about such a scheme, Descartes replied that success would depend

> on the true philosophy; for it is impossible otherwise to number and order the thoughts of men, or just to distinguish in them those which are clear and simple, which to my mind is the greatest secret in acquiring true science [or wisdom]. [But if that were done] I would then dare to hope for a universal language which was very easy to learn, speak and write. Most of all, such a language would assist our judgement by representing matters so clearly that it would be almost impossible to go wrong.

He even held, he said, that such a language was possible, and would make peasants better judges of the truth than philosophers are now, but he did not hope to see it, because it would change the world into a terrestrial paradise "and this happens only in fiction."[25]

[23] See Frege 1914, 224, in Frege 1983, 219–70; trans. in Frege 1979, 203–50.
[24] Russell 1899.
[25] Descartes to Mersenne, November 20, 1629, quoted in Gaukroger 1997, 201.

Leibniz pushed the project with remarkable energy.[26] As Eco characterizes him, Leibniz was pluralist and ecumenical in his views and hoped that a universal language would bring about world peace. But he was not an Esperantist *avant la lettre*, because he placed more hope in science than in linguistic reform. He also hoped that even if a truly general universal language for all purposes was beyond reach, a universal language of science was achievable.

What Leibniz identified as crucial to his linguistic project was a system of deductive rules that could be applied to uninterpreted symbols. Ideally, as he put it, in such a system "mental error is exactly equivalent to a mistake in calculation."[27] But the symbols were not always meaningless: they usually had interpretations built in, to prevent a valid argument about monkeys being applied to yield incorrect conclusions about humankind. Leibniz also allowed that the enumeration of primitive notions had to be open-ended, as human knowledge was forever imperfect. Accordingly, his system gave more weight to syntax than semantics.

A century later, the *encyclopédistes* tried to put an end to the construction of a priori languages. They argued that thought and language advance together, each influencing the other, and so there was no system of pure thought that could be articulated by the creation of a perfect language. People were trapped in historically changing languages, which could be improved but not escaped. This view became the orthodoxy, and as we have seen, the exponents of subsequent artificial languages took the a posteriori route of modifying existing languages. Improvement is not ruled out; perfection is.

6.1.5 Nineteenth-Century Linguistics

One of the great successes of the nineteenth century was linguistics. Much of this work has no implications for mathematical practice at all, but early ideas about philosophical grammar do. Philosophical grammarians aimed at providing a general grammar valid for all languages, perhaps with the supposition of a universal semantic structure. All languages were supposed to share a common logical structure. Wilhelm von Humboldt differed from this opinion, although he insisted that the great variety of languages did not contradict the idea that there was only one "language," and said that possession of language was what made us human. For him, language was an "intellectual instinct of reason" that spread and diversified in different human societies.[28] This raises the question of what, if anything, is the difference between logic and language.

For Humboldt, language, and any particular language, are intermediaries between thought and the world, and different languages give different views of the world that are not simply intertranslatable. And, as he showed in the case of Javanese, specific grammatical features of particular languages could be understood as arising from general ways in which, say, verbs operate and are modified. Humboldt took linguistics to be about explaining the regularities of a given language not merely codifying it (as in schoolbook grammar) or simply listing it (as in botanical accounts)

---

[26] See Eco 1995, 209–268.
[27] See Leibniz 1875, 203.
[28] Quoted in Davies 1998, 109.

and showed that language had a life and rules of its own. He therefore distinguished sharply between language and logic and claimed that language was prior, as did Heymann Steinthal in the next generation. This position rejects the tenet of philosophical grammar that all languages share a common underlying structure. On this view, grammatical categories—such as word and sentence—do not map tidily onto logical categories such as concept and judgment. This placed the study of human language ability within psychology rather than logic.

The next generation (that of the neo-grammarians) was different again, and by the end of the century more linguistic ground was shifting. There was certainly no consensus. There was, at one extreme, the position of Max Müller, a German Sanskrit scholar and from 1868 professor of comparative philology at Oxford, who argued that thought is impossible without language and wrote that "language is the true organ of the mind. We think with our words as we see with our eyes."[29] Müller looked to historical linguistics to solve the problems of philosophy: "If we fully understood the whole growth of every word, philosophy would have and could have no secrets. It would cease to exist."[30] Elsewhere in his *Science of Thought* of 1887, Müller observed that the grammatical and logical categories agree in Greek, although not in Hebrew or Chinese; Bréal in his *Sémantique* agreed. This resembles the Oxford Assyriologist Archibald Sayce's view that Aristotle's philosophy would have been very different if Aristotle had been Mexican—the argument being that Aristotle's logical categories closely resembled those of Greek grammar.[31]

Some points of agreement among linguists around 1900 can be cited. "Snow" is a noun not because it stands for a thing, but because of the places it can appear in sentences and its effect on other words in sentences, "because it can appear as the subject of a proposition, can form a plural by adding "-s."[32] On the other hand, there were functional, grammatical, and psychological analyses of what a sentence is, and the validity of these approaches was much contested, for example between Anton Marty, a pupil of Franz Brentano's, and Wilhelm Wundt.[33]

6.1.6 Semantics

But if then, as now, linguistics seems more advanced in the study of syntax than semantics, we should note that the kind of language mathematics was taken to be could only be seen by a linguist as a syntactically impoverished one. The correspondence between logic and grammar was often held to be a distant and imperfect one, and mathematics, as a language, was grammatically thin. There was to be more overlap in the field of semantics. The linguistic approach to semantics is surprisingly recent. Ideas about syntax go back a long way, as far as the ancient Greeks, who used a very similar word. But the word "*sémantique*" was created in 1883 by Michel Bréal to denote a *science des significations* or science of meaning, and his influential book

---

[29] Müller 1887, 541.

[30] In Davies 1998, 300.

[31] Littlewood (1986, 130) notes that after some contact with the Chinese language Russell was horrified to discover that the language of *Principia Mathematica* was an Indo-European one.

[32] Sweet 1875–76, 487, quoted in Davies 1998, 308.

[33] See Knobloch 1984 and Knobloch 1988.

Essai de sémantique was published (by Hachette) in 1897. He sought to emphasize the relationship between form and meaning in language, and make linguistics a human science. He stressed the importance of speech acts, of intention and communication.

This chimed in with the interests of Victoria, Lady Welby. She was, among other things, a correspondent of C. S. Peirce in his later years, the author of *What Is Meaning?* (1903) and of the article on significs in the eleventh edition of the *Encyclopaedia Britannica*.[34] "Significs" was her name for the study of meaning. She argued against the idea that meanings were universal and for their contingent cultural character. Language, in her view, is evolving (in a Darwinian sense) and used instrumentally. She distinguished between sense, meaning, and significance (reference, intention, and moral aspect). Peirce found her work exciting, but it mattered less in the long run than his own. His work, however, exerted little influence in its day, as all Peirce scholars lament, and hers was taken up by L.E.J. Brouwer.

The connection was made by the Dutch mathematician Gerrit Mannoury. Mannoury, the founder of the significs movement in the Netherlands, argued that mathematics displayed a close relationship between thinking and speaking, but not a completely accurate or consistent one. It was inevitably inadequate to the task, because it attempts to capture infinite continuous multitudes with a finite use of symbols. Mannoury called such a usage a formalization, and he allowed for successive formalizations that improve on their predecessors and allow interpersonal understanding.

Mannoury reviewed Brouwer's thesis on the foundations of mathematics in 1907, and made this response to Brouwer's claim that even by building language systems the formalists cannot ensure the reliability of the mathematical properties:

> No, Brouwer, the logicists do not ensure the reliability of the "mathematical properties," but no more will you, with your continuity-intuition ensure it, simply for the reason that it does not exist. Mathematics is human make and human devise [*sic*], containing no other truth than in relation to human language, purpose and society. "Free yourself completely from all conventions" and you will come to the conviction: "there is no unalterable truth and no unalterable measure for truth, there is no absolute unit, no absolute space and no absolute time, there is no certain knowledge." [*Wiskunde* in Dutch, meaning science.][35]

Van Dantzig, Mannoury's obituarist, points out that it was only the next year, 1908, that Brouwer first denied the law of excluded middle, and the development of his intuitionism dates from 1917. But even in 1905, in his *Levin, Kunst en Mystiek (Life, Art, and Mysticism)*, Brouwer displayed a remarkable distrust of language: "Living in the intellect carries the impossibility to communicate."[36] It is the source of confusion, because it is the slave of the delusion that there is an external reality.[37] Brouwer's distrust extended beyond language and logic to novel ideas of proof. For Brouwer a proof is an infinite mental construction that cannot be described in any language. Writing to his supervisor in 1905, he distinguished sharply between mathe-

---

[34] Welby 1903.
[35] From Georg van Dantzig's (1957, 14) obituary of Mannoury.
[36] Van Dalen 1991, 33.
[37] Thiel 1991, 32.

matical argumentation and logical argumentation, without, unfortunately, defining either of them precisely.

This distrust of language was characteristic of the significs movement as it developed in the Netherlands and was initially focused on the tendentious and dishonest language used by participants in the First World War. In a manifesto signed by Brouwer, Mannoury, and others in 1918, they deplored the lack of "a satisfactory store of words of well considered spiritual value, at least in our western languages" and proposed "to coin words of spiritual value for the languages of western nations" while removing false words. In 1919 the same group called for a new basic vocabulary and a recognition that language is always inadequate to represent any part of reality, and that meanings are defined exclusively by reference to effects (intended or presumed). From these and later writings it becomes clear that Brouwer viewed language as built up in levels: words, words in simple relationships, sentences whose meaning depends on the way words are connected, well-regulated language such as scientific language, and finally symbolic language such as mathematical logic and pure mathematics written in pasigraphic form. (Note that this is not a top-down hierarchy of values; if anything, the reverse.)

We have, therefore, two views of what an ideal mathematical language could be. One distrusts the usual mathematical language from a burdensome concern with meaning, and seeks correct definitions (as did Frege) or veridical insight (as did Brouwer). The other view distrusts mathematical language and seeks to sharpen mathematical arguments with an effective reliable syntax. Hilbert, as we saw, axiomatized geometry with much less regard for reality and experience than Pasch had. His analysis of axiom systems required him to be open to different models of his axiom systems and so to a number of semantical interpretations. In this he differed from Frege and also, by being more evenhanded in his attitude to the various models, from the Italians who were active in axiomatizing geometry at the same time.

6.1.7 Hilbert and Semantics

But if Hilbert showed some awareness of the semantic issues involved in axiomatic geometry, we saw above that his initial insights were far from always being sound. Loose talk about "true" and "proved" is what distinguishes Hilbert (and Husserl) from Frege. It comes about because they are no longer linguistic universalists, concerned to think about language as the conveyor of thoughts about the world, expressed in as logical a manner as possible. They are on their way to becoming linguistic calculators (of different kinds, in the end) open to many interpretations of structures whose syntax was all they could vouch for. Ultimately, for Frege, ontological questions reduce to determining what there is in the world, although the whole thrust of his program was to show that a remarkable amount of that knowledge could be derived logically from very little.[38] For Hilbert, although he cannot have known around 1900 what difficulties he was letting himself in for, multiple semantical interpretations of a formal system were the very essence of the

---

[38] Compare Hallett's comment that for Hilbert the truth of the axioms was reference independent, whereas for Frege, of course, sense determines reference, in Hallett 1994, 163.

mathematical enterprise, and ontological questions in mathematics were not to be solved by referring to the world.[39]

After the First World War, Hilbert's project became one of saving mathematics, not as a family of empty languages (pure syntax), nor as a body of truths about the world, but as something more complicated. Mathematics had to be rescued from Brouwerian intuition, which for a time caught the sympathies of no less a person than Hermann Weyl. Otherwise, it would not be able to deliver the richness of analysis and the vast store of applications in science. It also had to deliver familiar truths, in particular about the integers. The perspective opened up by taking a linguistic approach is that these are in part semantic questions, and so they pushed Hilbert away from semantic arbitrariness and toward semantic flexibility or multiplicity.

In unpublished lectures of 1905 Hilbert had called "the a priori of philosophers" the capability to think things and to denote them through simple signs.[40] Mental ability is reflected in the ability to represent objects and thoughts by signs and to manipulate these signs. In 1918 he introduced his paper "*Axiomatisches Denken*" (or, "axiomatic thinking") with a painfully topical analogy between the proper relations between nations and those between the sciences, and argued that axiomatics helped bring out the fundamental unity of mathematics. By 1922 Hilbert was distinguishing between meaningless signs and meaningful signs, and introducing signs of various kinds (standing, for example, for variables or formulas).[41] Certain formulas would be taken as axioms, and mathematics in the strict sense would be identified with the stock of provable formulas. Statements with content would belong to a new discipline of meta-mathematics, which should provide proofs of the consistency of sets of axioms. The aim, for Hilbert, was to create mathematics at the level of syntax, and in his paper of 1923 he wrote of a meta-mathematics that had a semantic aspect: the proofs of the mathematics under investigation. This is Hilbert's famous "proof theory," his attempt, as Bernays put it, to transfer the foundations of mathematics from the domain of epistemology to that of an appropriate mathematics.[42]

In the early 1920s Hilbert advocated the simultaneous strengthening of logic and mathematics. Two papers by Hallett show in stimulating detail how attending to semantics and syntax shed light on the progress of Hilbert's thought.[43] Hilbert said of his enterprise that "calling on mathematical methods for the investigation of the logical language is not artificial, but fully appropriate and even inevitable. . . . It is self-evident that, when we exclude the accidental features in the derivation of words, then a form of mathematical sign language arises."[44] And further that it will be possible "to frame the rules of grammar in such a surveyable way that logical inference can be carried though automatically by calculation according to simple, determined rules."[45]

---

[39] For a comparison of Hilbert and Frege along the lines suggested here, see Demopolous 1991.
[40] Quoted in Hallett 1994, 179.
[41] Hilbert 1922b and Hilbert 1923.
[42] Bernays 1922, 19.
[43] Hallett 1994 and Hallett 1995.
[44] Hilbert 1922a, 130, in Hallett 1995, 180–181.
[45] Hilbert 1922a, 79, in Hallett 1995, 181.

The delicate point in all of it is, as is well known, Hilbert's idea of a finitary argument, which is discussed elsewhere (see §7.1.2). But we may note that Hilbert's introduction of meta-mathematics as a language for analyzing mathematics surpasses anything in, for example, Schröder's work. Semantical freedom is required, but not at the expense of dismissing semantic considerations from mathematics altogether. Once Hilbert began to think that philosophical questions about the nature of mathematics could be answered by reworking them as mathematical questions about mathematical language, the debate naturally moved beyond the simplicities of pasigraphy. The real difficulties, as so often, turn out to be semantic and not syntactic. They have to be confronted not only by thinking how we speak but, however clumsily, in thinking how we speak about things and what we can speak about.

Questions about the nature of thought and the proper objects of thought are notoriously difficult. They are as much linguistic as logical, they concern the relationship between valid expressions and truths, and they require clarity about syntax and semantics. Clarity was slow in coming, but one important source for it around 1900 was the strong currents running toward the creation of ideal languages. These currents flowed out of concerns for the nature of science, about communication in science, and about the need for improved communication in general. They spilled over into analyses of the language(s) of mathematics and into the very consideration of mathematics as a language, at levels from the naive to the technical and from the disinterested to the politically charged. Different attitudes to language, such as linguistic calculationism and linguistic universalism, have implications for syntactic and semantic practice and point up distinctions between Hilbert and people like Frege and Russell. When the need for new foundations of mathematics was felt most keenly, in the decades around 1900, questions about the language of mathematics and the existence of mathematical objects were formulated as questions about logic, mathematics, and language in ways that echoed the debates about ideal grammars and vocabularies for novel languages. By reflecting on those debates one can see the confusions and the strengths of positions taken by Hilbert and those whom he opposed.

Hilbert's views were interestingly not as ruthlessly syntactical as those of the young American logician C. I. Lewis. In chapter 6 of his *A Survey of Symbolic Logic* (1918) (dropped from the later reedition), Lewis surveyed various forms of logistic: Peano's, that of Russell and Whitehead, and that of Royce. He elaborated an entirely syntactical view of mathematics, defined as follows: "A mathematical system is any set of strings of recognisable marks in which some of the strings are taken initially and the remainder derived from these by operations performed according to rules which are independent of any meaning assigned to the marks" (p. 355). He called this "mathematics without meaning." He then observed sensibly that if Russell is right, "the mathematician has given over the metaphysics of space and the infinite only to be plunged into the metaphysics of classes and functions" (p. 356). Russell's system rested not on rules but "in discussions and assumptions about the conceptual content of the mathematical system. In fact, the rules of operation are contained in the explanations of the *meaning* of the notation" (p. 358). This, in Lewis's opinion, risked a certain confusion of form and content, whereas "*the operations of any abstract and really rigorous mathematical*

system are capable of formulation without any reference to truth or meanings"
(p. 358).

6.2 Mathematical Modernism and Psychology

6.2.1 Poincaré

> There is in nature a tactile space. I might say even a manual space. . . . This is the space
> that fascinated me so much. Because that is what early Cubist painting was about, a
> research into space.
>
> —John Golding, "Braque's Studio"

Poincaré's philosophy of geometry was rooted in his analysis of how we experience
the world around us and construct our sense of geometrical space. In a set of related
essays collected in *La Science et l'hypothèse* and in *Le valeur de la science*, he argued
first that we experience space in a number of distinct ways. There is what he called
pure visual space, which is what we see. It is certainly not geometrical space: it is two
dimensional and inhomogeneous, because the retina is inhomogeneous. If we notice
the effort of accommodation of our eyes, which is a muscular activity, we construct
a three dimensional space (called, by Poincaré, complete visual space). Tactile space
is also three dimensional, but muscular space is different again. This is the space we
describe by noticing what muscular changes bring about what changes in our experience of the outside world, and it has as many dimensions as are needed to describe
the dispositions of our various muscles. This plurality of spaces shows that geometric space is not imposed on us in some Kantian fashion, and indeed that geometric
space is none of the spaces we directly experience.[46] How then, asked Poincaré, do
we come to create it?

He answered this question by invoking the concept of motion. He invoked a class
of objects external to us that could be called rigid bodies. They are external because
they invoke involuntary nonmuscular sensations, and they are rigid because their
different views and various motions can be generated in us by making suitable
corresponding motions. The sensations we experience seeing a wine glass come toward us could, in principle, be generated by our moving toward the glass, but the
sensations generated by the slight rolling of the wine in the glass could not be: the
glass is a rigid body, but the wine is not. We can now study the displacements of
rigid bodies, and when we do we construct a geometry. But this process does not at
all force us to construct Euclidean geometry. It can equally well construct non-
Euclidean geometry—and here Poincaré gave his example of the cooled metal ball—
or even, he explained, a space of, for example, four dimensions. But it cannot construct some of the spaces Hilbert described.

---

[46] Poincaré was also arguing against Russell's view that real projective geometry was imposed a
priori, a debate I have decided not to follow. See the discussion in Nabonnand 2000. It is a disappointing
debate because of Russell's poor grasp of projective geometry, which Poincaré has little difficulty showing
cannot be a candidate for the a priori form of intuition that Russell wanted it to be.

Poincaré next headed off the speculation that geometry had become an experimental science. Rather, it was the study of the possible movements of idealized objects, and, more precisely, it was the study of the possible groups these movements make up. It is the concept of a group that Poincaré claimed, perhaps a shade optimistically, preexisted latently in our minds. The role of experience is to guide us to the selection of the appropriate group among those that had recently been specified on mathematical grounds by Sophus Lie. This mental capacity had evolved, he stated explicitly, through natural selection, which had adapted our minds to select the most advantageous or convenient geometry.[47] Accordingly, geometry, said Poincaré, was not true, it was advantageous. As he put it in the next essay, Euclidean geometry is only a sort of linguistic convention, and the facts of mechanics can equally well be explained in a non-Euclidean setting.[48]

Poincaré returned to the topic of geometry in essays published in *Le valeur de la science*. In "La notion de l'espace" he indicated how different geometries could be imposed on what he called an "amorphous continuum." In other words, he took a more primitive notion of physical space as being some sort of topological space and indicated that there were distinct ways in which it could be given the structure of a metrical geometry. That done, he asked if this topological space was not imposed upon us: "We shall have enlarged the prison in which our sensibility is enclosed, but it will remain a prison." And, after a long argument, his answer once again was "No," because we can imagine other mental constructions, and because physical space is something we learn about through acquaintance, specifically via our sense of muscular space.

The principal source of Poincaré's ideas is surely the work of Théodule Ribot, although Poincaré may also have gone directly to the work of Helmholtz.[49] Ribot was the leading psychologist in France at the time. His theories included the ideas that there is a tactile and a visual space, and that motion, our own as well as that of objects, is essential to our construction of "space." He is also credited with being among the first to espouse evolutionary thinking in France, notably in his *L'hérédité psychologique* of 1873, an approach that Poincaré also shared. In his *La psychologie allemande contemporaine*,[50] Ribot argued that the subject grew out of Herbart's and Fechner's disagreements with Kant, a process consummated in Ribot's day by Wundt, and the chapter on the psychology of spatial knowledge would influence Poincaré.

Ribot noted that Müller was the first to investigate how a sense of space is elaborated through both vision and touch and briefly reviewed the nativist theories of Müller, Stumpf, and Hering and the empirical or genetic theories of Helmholtz and Wundt. In order to compare them, Ribot suggested that the senses of sight and touch create different spaces, and that the empiricists emphasize the role of the tactile, noting that the congenitally blind do not have the same sense of space that sighted people have. The nativists maintained, in various ways, that space perception is built into the nature of the retina, different parts of which do different things in accordance with the theory of local signs that goes back to Lotze. The empiricists

[47] *La science et l'hypothèse*, 107; this part was originally published as a letter to M. Mouret in the *Revue générales des sciences* 3, 74–75, and added to later editions of *La science et l'hypothèse* and does not appear in the English translation.

[48] Poincaré 1902d, 114; English trans., 90.

[49] Paul Tannery would be a possible intermediary; see Nye 1979.

[50] A book that had run to thirteen editions by 1898.

emphasized the role of motion in the construction of space. As Ribot put it: "What is the peculiar object of sight? This very simple question sums up the debate. If we reply: Color, we are empiricists. If we reply: Colored extension, we cast our lot with the nativists."[51] Ribot proceeded to side with Helmholtz and Wundt.

Théodule Ribot was in fact a pioneering figure in the development of psychology in France.[52] He eventually became a professor of experimental psychology, first at the Sorbonne and then at the Collège de France, after overcoming considerable resistance to the subject, caused, it has been suggested, by the negative influence of spiritualist philosophy.[53] Historians of psychology complain that his *La psychologie allemande contemporaine* appears to have given psychology one of its founding myths. Wozniak, in his essay on Ribot in his *Classics in Psychology, 1855–1914: Historical Essays*, disputes Ribot's pedigree for the subject, but, he says, the story was repeated by Boring and passed into folklore.[54] Be that as it may, Ribot's book had an almost immediate impact both in France and America. The emerging generation of psychologists had their own connections to the leading figures in the discipline, but Ribot gave it a philosophical aspect that would prove congenial to many in other fields. What may have done scant justice to a variety of more empirical investigators proved to be an effective way of conferring status to a novel form of enquiry.

In 1908 Poincaré talked to the Société de Psychologie in Paris about the psychology of discovery in mathematics. The lecture was later published as the third essay in his volume *Science et méthode* (1909). The ostensible circumstance was the recent relative failure of the newly founded journal *L'enseignement mathématique* to get satisfactory answers to their questionnaire about mathematicians' working methods. It emerged that not many of the best mathematicians had bothered to reply, and Poincaré is often said to have stepped in to give a speech and thereby help out in some way. Perhaps, he did, but it is also the case that Poincaré had a lifelong interest in how the mind acquires knowledge, and he was quite familiar with contemporary thinking on that question. He also offered some reasonably clear ideas about how the mathematical mind might work, ideas that were entirely psychological rather than philosophical.

In his lecture, he raised some interesting questions: How can people make mistakes in mathematics, when a healthy mind is incapable of making an error in logic? Is mathematics not simply a matter of following rules? He answered for himself that he was not capable of making the simplest calculation without error, and that almost everyone, forced to recall the results of many long and complicated arguments, will find they have faulty memories. He himself, for example, could only play chess badly because he would notice a threat, consider various alternatives, and then forget the threat when choosing his move. What sustained him in mathematics, he suggested, and let him down in chess was a good sense of the general march of an argument. So much so that he often had the feeling about good mathematical ideas he had been taught, that he had invented them himself. It was this feeling or intuition of math-

[51] Ribot 1886, 126–127.
[52] He was another of Carus's authors; his *The Psychology of Attention* was published by the Open Court Publishing Company in 1890.
[53] See Nicolas and Charvillat 2001.
[54] Boring 1950.

ematical order that he presumed some people had and others did not, and which enabled those who had it to be creative mathematicians.

But anyone, he noted, could put bits of mathematics together in new ways. The skill was in finding useful ways: "To invent," he proclaimed, "is to discriminate and to choose." Or rather, that is the work of the unconscious. This he proceeded to illustrate by drawing on his own experiences. As noted above, his conscious activity in 1880 was devoted to showing that some functions could not exist, but one night his unconscious prompted the opposite thought and "I had only to write up the results, which just took me a few hours."[55] There then followed a quite deliberate piece of work in which the new functions were investigated by means of an analogy with existing functions of a similar kind. Then his unconscious spoke again as he boarded the bus at Coutances, this time to reinterpret a web of triangles in a new, and richer, context.

It seemed to Poincaré that the unconscious played a crucial role in his work. Ideas came to him in bed, he said, when he was in a semihypnagogic state. They were not mechanically analyzed, the unconscious did not work like a machine; it exercised a selective role much as the conscious did. Here Poincaré referred to Boutroux's recent work on William James.[56] Perhaps ideas, like the atoms of Epicurus, had hooks, and were stirred up after a period of conscious work and fastened themselves together to form productive combinations. In any case, he observed, the results of unconscious work never gave the results of long calculations ready made (*tout fait*) although it may give a verification. The unconscious provides points of departure for calculations that must be made consciously, but operates by chance.

6.2.2 Intuition and Psychology in a German Setting

A bitter institutional struggle between philosophy and psychology was fought out in many a German philosophy faculty in the years between 1860 and 1914. The intellectual heart of the conflict was over the proper relation of logic to psychology. Was logic a normative discipline, holding up standards of correct reasoning, or was it a disposition of the mind? Was it normative, or natural? To take quotes from Kusch's (1995) sociological study *Psychologism*, which is an invaluable guide to the whole topic, Mill saw logic: "So far as it is a science at all, [as] a part, or branch, of Psychology . . . differing [from it] as an Art differs from a Science," and Theodor Lipps called logic "the physics of thought." Philosophers came to see this as a dreadful mistake and to this day proclaim that Frege, or, on some versions, Husserl saved the day by banishing psychologism, although, as Kusch points out, other philosophers have begun to claim that antipsychologism has problems of its own.

Kusch has no trouble showing that psychologism became a term of abuse, and no trouble showing that what the charge meant was soon very obscure. The historian's problems start to mount up when he shows that the many schools of philosophy and psychology cheerfully leveled the charge at all and sundry, so that one scholar's opponent of psychologism was another's advocate. Plainly, too much was at stake for the matter to ever have yielded to a few quick definitions. The intellectual problems ran too deep, the emerging science of psychology was moving too fast, the

---

[55] 1909, 50.
[56] Boutroux 1908.

professional turf rapidly became too contested. There was even a petition in 1913 signed by leading philosophers demanding that no new philosophy professorships be granted to experimental psychologists.[57] The controversy disappeared with the First World War.

6.2.3 Helmholtz on Knowledge and Visual Perception

Helmholtz investigated in considerable detail how visual sensations lead to a representation of the external world. Depth is learned from the motion of the eye muscles, direction of a luminous object from the eye is learned by comparing where the image of the object is on the retina with how the eye must be moved so as to look at the object directly. The local sign (the retinal element) is associated with the spatial meaning (the necessary muscular movement) and the result is the appearance of the visual object. He differed from other investigators such as Steinbuch, Lotze, and Wundt by insisting on the importance of felt location (direct physical contact), which allowed for a continuing role for experience in detecting position, but he agreed that spatial perception depended on knowledge of the orientation and position of the eye and the retinal stimulation.

What sort of knowledge is perception in Helmholtz's scheme of things? It is unconscious, said Helmholtz, but it is an inference in much the way that we infer from "All men are mortal" and "Socrates is a man" that Socrates too is mortal. As Hatfield discusses (1990, pp. 201 ff.) Helmholtz believed that the reason we agree to the syllogism is that its opening claim is acceptable shorthand for an empirical statement about very many men being mortal. The syllogism is not, on this view, deductive but inductive, but nonetheless valid. Similarly, said Helmholtz, the association between sensation and object is a piece of inductively assured knowledge. Indeed, our representation of a three-dimensional object was nothing more than the possible representations of visual images that can be obtained from it (including cross-sectional cuts). He went on to push the idea that the construction of representations is a (pictorial, nonverbal) inference into a criticism of the Kantian idea that intuition was different from thought (which he defined as concept-mediated cognition, to quote Hatfield, p. 203). Rather, intuition should be resolved into thought, and thought into associative inferences. One might say that for Helmholtz, and also on this point for Wundt, perception was a kind of thought, and so our knowledge of appearances was of a kind with propositional knowledge.

In short, for Helmholtz our perception of space is learned and is a reliable form of knowledge. The grounds for this knowledge are not that it obtained from a direct look at reality, rather it depends on the operation of causal laws in the unconscious, the existence of which might have to be taken on faith by the scientist. Whereas Kant had regarded the whole of space perception as a necessary preliminary for any experience of *space*, Helmholtz did not so much disagree as explain how we came by this ability starting from helpless infancy. Whether this merely fleshed out a process or sidestepped the whole Kantian approach is a matter of argument. Helmholtz put it this way: "That the character of our perceptions is

---

conditioned just as much by our senses as by the external things is of the greatest importance. . . . What the physiology of the senses has demonstrated experimentally in more recent times, Kant earlier tried to do . . . for the ideas of the human mind in general."[58] Nonetheless, on one point the disagreement was patent: Helmholtz could easily accommodate the perception of different geometries, Kant could not. Whatever might be said about space perception at a philosophical level, on Helmholtz's account our knowledge that space was Euclidean (if knowledge it be) was not a priori.

The extent to which Helmholtz's ideas count as psychological takes us straight into the question of what psychology was taken to be. Insofar as it was not just a branch of metaphysics, as it was with Herbart, or essentially unanalyzed, as it was with German linguists of the first half of the nineteenth century, it was an experimental subject. Helmholtz's own career shows us that this experimental aspect has its roots in physiology and what was called psychophysics, and also that his work decisively opened up the field.

The pioneer here was Fechner.[59] His psychophysical researches began in 1850 when he took up the ideas of his friend E. H. Weber, the older brother of the physicist Wilhelm Weber. Fechner discovered by careful experiment that over a range of weights, peoples' assessment of difference is directly proportional to the size of the weights. In particular, the so-called just noticeable difference between two weights was always about 1/30th of the size of the weights involved. He published his findings in his *Elemente der Psychophysik* (or, *Elements of Psychophysics*) (1860), which "gave an enormous boost to psychology's status as a potentially mathematical and experimental *science*. No longer would psychology be regarded as the merely metaphysical discipline described by Kant."[60]

In 1858, after already achieving more than most of his contemporaries could in a lifetime, Helmholtz became professor of physiology at Heidelberg, a chair he occupied until 1871. There he turned his attention not only to hydrodynamics and electrodynamics, but to the theory of color vision and the perception of sound, as well as the axioms of geometry. He produced his celebrated masterpiece *On the Sensations of Tone* in 1863, and honors and international fame flooded in. In 1870, the year he turned fifty, "the Chair in Physics at the University of Berlin became vacant and he was asked to set his own conditions of acceptance. He asked for and received a salary of 4,000 thalers (a huge sum for that day), a promise of a new institute of physics, its directorship and living quarters in that institute."[61] Helmholtz was to call his laboratory a "temple of physics." In 1887 he added the directorship of the new Physics-Technical Institute in Charlottenburg (Berlin) to his other duties.

6.2.4 Wilhelm Wundt

Helmholtz's way of creating science through experiment and observation, and leading it on to explicit philosophical conclusions, was eagerly adopted by his

---

[58] Quoted in Fancher 1979, 125.
[59] See Heidelberger 2003.
[60] Fancher 1979, 139; emphasis in original.
[61] Watson 1978.

FIGURE 6.1. Wilhelm Wundt.

one-time assistant, the prolific Wilhelm Wundt, who is rightly regarded as one of the founders of the modern science of psychology.[62] Wundt was at his most influential when he was a professor of philosophy in Leipzig, a chair he had been called to in 1875, although he had taken only one formal philosophy course in his life. Not only did he do important research, he set up a laboratory and led a school where he trained students; Watson quotes a figure of 186 completed doctorates by 1919, 70 on philosophical topics and 116 on psychological problems.[63] He was also a popular lecturer: a figure of 630 students and visitors is quoted for his afternoon lectures in 1912, when he was eighty.[64] Like Titchener and Cattell, many of his doctoral students were Americans who went on to important careers in American psychology, and they took the laboratory-based vision of psychology back with them. Wundt also wrote with abandon on a wide range of topics, producing the standard texts in the field. He also created a journal, the *Philosophische Studien*, where his students could publish their results which ran from 1881 to 1903; he had hoped to called it *Psychologische Studien*, but that title had been taken by a magazine devoted to mysticism. Wundt was eventually able to create a journal with that title in 1906; it ran until 1918.

[62] "There can be little doubt that Wilhelm Wundt (1832–1920) was the most important figure in the institutionalization of psychology" (Kusch 1995, 128).
[63] Watson 1978, chap. 11.
[64] Blumenthal 1985b, 43, quoted in Kusch 1995, 129.

Wundt's years with Helmholtz, 1858–1864, do not seem to have been happy ones. "Relations with the taciturn Helmholtz were non-existent," according to Titchener.[65] Wundt published work on space perception that made it look as if he had not read Helmholtz's work on the same topic, for example, and he eventually left under obscure circumstances. But while he was there he did the first work measuring how fast the brain responds to stimuli. He set up a pendulum clock to chime exactly as the pendulum bob passes the vertical, but found that he saw the bob only after the chime. He attributed the time difference, about one-eighth of a second, to the time it took the brain to attend to the two stimuli separately. Whereas Helmholtz had measured the speed of signals in the peripheral nerves, Wundt was now identifying and measuring processes in the brain itself. His success convinced him of several things: that there was an experimental way into psychology, that introspection should be granted only a circumscribed role; and that psychology should not be studied independently of physiology. Quite deliberately he called his first major book *Grundzüge der physiologischen Psychologie* (Fundamentals of Physiological Psychology, 1874), announcing that it was intended to "mark out a new domain of science." In this he succeeded; the book ran to six editions and grew to three volumes.

In his introduction to Wundt's work in 1999, Wozniak noted the continuities with Wundt's previous work.[66] He commented that Wundt wanted a "science that has as its subject matter the points of contact between internal and external life. . . . Physiology and psychology each by itself can easily evade this question, but physiological psychology cannot sidestep it."[67]

Wundt had views on the vexed question of the relationship between psychology and logic itself, which he set out at length in his *Logik*.[68] He distinguished between the psychology of thought, which could produce descriptive laws, and logic, whose laws were general and binding, and deplored the frequent failure to confuse the two. Only the logical laws were normative. The mistake of regarding psychological laws as essentially logical had led linguists to attempt to reduce the study of grammar to logic; the contrary mistake, regarding logical laws as essentially psychological in nature, had led others to attempt to base logic on grammar.[69] That this is a mistake, Wundt observed, can be seen from the fact that there is no general grammar composed of the sum of several individual aspects that all languages have in common. What is common to all spoken thought is not at all their grammatical form, which is manifested in infinitely many ways, but their logical laws.[70]

---

[65] Quoted in Watson 1978.

[66] Wozniak 1999, *Classics in Psychology*.

[67] Wozniak footnotes this as follows: "English translation is taken from Wundt, W. (1980). Selected texts from writings of Wilhelm Wundt. Translated with commentary notes by S. Diamond. In R.W. Rieber (Ed.). *Wilhelm Wundt and the Making of a Scientific Psychology*. New York: Plenum, pp. 155–77, p. 157."

[68] First edition 1883, third edition 1911.

[69] Wundt 1883, 84–85.

[70] Wundt was consistent on this. He wrote on language at length starting in the 1890s; see, for example, his *Völkerpsychologie* (1900). Royce (1902) reviewed the book in these terms: "Wundt's book has the merit of emphasizing the close and primary relation of language to the expression of the feelings and to the life of the will. In consequence, Wundt very decidedly sets limits to the tendency either to regard the grammatical categories as essentially logical ones, or to use the psychology of language too exclusively as a means for interpreting the psychology of the thinking process. For this very reason his book rather encourages one to look elsewhere for auxiliaries in comprehending the psychology of the intellectual life."

Wundt, it should be observed, had a lifelong habit of writing on any topic that seemed to him to border on psychology, including most of philosophy and, of course, logic. Given the chance, he would produce a work in several volumes and revise it through successive editions, and his treatment of logic is precisely of this kind. In it he took the opportunity to display what he had learned not only of what experts in logic were saying, but, observing that because of its binding character logic behaves like mathematics and like mathematics logic has a technical side, he strayed well beyond the conventional boundaries of logic to tell his readers about contemporary geometry and mathematical analysis as well. He then went on to survey the rational character of thought in each separate branch of science (physics, chemistry, and biology).

As so often with literature of this kind, its author seems to have read widely and rapidly in his library—and Leipzig was a major university center—and talked to a few colleagues. Wundt took his information on arithmetic, for example, from two books by Hankel, his *Geschichte* and his *Complexen Zahlensysteme* (1867), and Moritz Cantor's *Vorlesungen über Geschichte der Mathematik*. His treatment of algebra was standard fare. On geometry, he rattled through projective geometry, noting also the work of Grassmann and explicitly endorsing the idea that the number concept now embraced position and direction. His treatment of analysis was more remarkable. From Paul du Bois-Reymond's *Allgemeine Functionentheorie*, published only the year before, in 1882, he noted the work of Weierstrass on continuous, nowhere-differentiable functions, then, switching to the Riemannian side of the street, he went into quite some detail on complex or analytic functions, and on the exponential and periodic functions. All of this came with a hefty dose of history: Newton and Leibniz's alternative formulations of the calculus and the subsequent ideas of Euler and Lagrange. Moritz Cantor, in his review, rightly noted that the *Logik* is not a book by a mathematician for mathematicians, but by a philosopher for philosophers—and indeed for those philosophers who have come to psychology via physiology and logic via mathematics. The mathematician will notice numerous small errors, he said, but should still be able to rejoice at how much mathematics a philosopher has been able to pick up.[71]

Wundt's analysis of the mathematical concept of space may serve as an example of his way of bringing seemingly disparate fields together. He opened with a quick historical account of various philosophical views, culminating in Kant's, which he noted had been subject to several attacks. Of these, he felt that the view that we could, after all, regard space as an immediate objective reality had recently gained unexpected support from transcendental mathematical speculation. For, it had been discovered that space (as given in experience) can be regarded as a special form of a far more general concept, so the a priori intuition of space ceases to be part of our mental organization, and arises from the accidental barrier experience imposes on our representations. Wundt added that this viewpoint agrees with the endeavors of psychology.

To explain all this, he first turned to Kant's views, and various recent criticisms of them. Recent researches in geometry had promoted the need for a more general genus of concepts of which space can appear as a species, and from which, with the in-

---

[71] Moritz Cantor, *Hist.-Lit-Abtheilung*, 196–198.

troduction of special conditions, the fundamental properties of space can be developed analytically. But, said Wundt, only one space is given to us in intuition. What some writers, such as Helmholtz, called spaces are not available in intuition and are not space; it would be better to follow the lead of Klein and Riemann and call them manifolds. We can represent space, and only space; even the plane, and certainly the pseudosphere and the *n*-manifolds of Riemann, are products of abstraction. Our fundamental intuitions are what enable us to talk about space at all, so Lobachevskii's investigation of angle sums of triangles is inadmissible. Furthermore, if light didn't travel in (Euclidean) straight lines, this would be, as Lotze has already noted, a physical fact, a departure to be explained by the laws of light, never as a geometric fact. Indeed, all mathematical speculation lacks epistemological significance, and cannot touch questions of the origin and objective significance of spatial intuition. Wundt here footnoted his opposition not only to Helmholtz's conclusions, but also the hopes of other authors such as Erdmann, and of Harnack in his review of Erdmann.

So much for Kant. Next, Wundt produced his own objection: mathematics breaches Kant's philosophy of space because space is a concept, contrary to Kant's opinion. Kant had argued that we cannot have many spaces, only parts of one space. But we have the genus space (of some arbitrary number of dimensions) within which our space is a species. Wundt had already argued that the essence of a concept lies in its relations to other concepts, and space is so simply related to other concepts that a complete definition of it can be given. An intuition, without a conceptual element, could not be a definition. So space was not a form of intuition, but a concept. The definition, for what it is worth, followed: essentially (and to oversimplify) space is a continuous, infinite quantity, congruent in itself, in which points are determined by three directions.[72]

This passage illuminates what is so irritating about Wundt. He first defended Kant against Helmholtz by insisting that only one space is given in intuition, a position Helmholtz had attempted to refute. But then it emerged that space is not a form of intuition anyway, but a concept. This surely requires that Helmholtz's critique be reexamined, but Wundt did not do so. Quite often one has the feeling that Wundt is engaged in a conversation rather than a systematic inquiry, and overall consistency should not be looked for.

Wundt next turned to treat the origin of our sense of space as a question in psychology (agreeing that it had also to be influenced by epistemology and metaphysics). He reviewed various accounts, allocating Herbart credit for seeing the necessity of a psychological construction of space, but noting that Herbart's was a metaphysical theory. He then set out the stall for his own position, the theory of complex local signs. This drew on work he had done in the early 1860s, prior, that is, to Helmholtz's.[73] Unconscious inference proceeds by induction, which formulates laws according to experience. First, sensations are associated through their constant conjunction. These associations are then synthesized into an elaborate whole. This further enables inference from analogy. It is synthesis that creates our sense of space

---

[72] This definition may not achieve much, and it was severely criticized by Voss (1913) on these grounds (see p. 92), but note that Wundt did define the key terms in it.

[73] See Hatfield 1990, 198–199, which this account follows.

by bringing together muscular feelings, including touch, with local signs obtained in the eye.

Wundt's analyses display the overlapping constellations of ideas of his day, and his eagerness to shape them, in their breadth, but not their depth. It is striking, nonetheless, that someone with his background and training should want to reach so far into mathematics and philosophy. His ambition and energy are not the whole story. He also spoke successfully to a wide audience who were equally willing to see physiology and psychology alongside logic and mathematics.

6.2.5 Cognitive Foundations of Mathematics

Two authors at the end of the nineteenth century went much farther in their analyses of mathematical thinking than any others, and their work is vividly indicative of how profound was the questioning into the nature of mathematics. They did not offer logical, or even philosophical, accounts of mathematics or of the nature of number (although number was a central concern for both of them). Rather, they offered accounts of what thinking must be if it is to be capable of arithmetic. Their accounts are based on introspection, they are a priori and not at all experimentally based, and they might usefully be seen as akin to theoretical cognitive science. Of these, the one by Lipps is more like a philosopher's work, while that by Santerre reads like a mathematician's. But both show a profound influence of the mathematical style of writing. The one by Lipps, as befits someone in Wundt's circle, also shows a remarkable awareness of the history of mathematics and philosophy, and it seems best to begin this account of the work of two much less well known writers with a further look at Wundt's own views as they stood at the time.

6.2.5.1 WUNDT

Wundt wrote his essay "On Induction in Mathematics" for the first volume of his new journal, *Philosophische Studien*. It was published in 1883, after the first edition of his *Logik*. It begins by noting that mathematicians and philosophers agree that mathematics is our most complete example of a deductive science because of the certainty of its principles and the force of its proof-methods. There follows a lengthy historical review of analysis and synthesis as methods in mathematics, from their use in Euclid's *Elements* in support of deductive methods, through their transformations by Descartes, Newton, and Leibniz. The effect of these was to introduce algebraic analysis as a discovery method, or at least a new method of proof, in mathematics. Later, novel geometric methods (the use of motion, and of projection and section) brought in the modern genetic method, which allows induction and supports synthesis. Wundt now proceeded to argue for induction as a prime source of mathematical knowledge.

Wundt then turned to consider the origin of mathematical principles. He found the old terms "realist" and "nominalist" were still acceptable, if modified in the light of history. For the realist, intuition grounds the validity of mathematics on its objects. For the nominalist, mathematical knowledge rests on methods or principles (which may be pure, derive from experience, or be arbitrary creations of the human mind). In any case, proofs are compelling because the ideas in them have a constant significance, and one has the seeming paradox that mathematics is exact precisely because of the subjective, hypothetical character of its ideas. Leibniz's philosophy of mathematics left the connection between objects and ideas so mysterious, said Wundt, that Kant had aimed to reconnect them by reducing the infinitude of mathematical ideas to the pure intuitions of space and time. However, the a priori status of the forms of intuition was not proved by Kant, so one had to conclude that his philosophy of mathematics was infirm (p. 111).

Wundt then looked at what he called the entirely imaginary nature of modern mathematics, a subject whose assumptions are so far removed from immediate intuition that they look like hypotheses, even arbitrary hypotheses, so that one can say, in agreement with Hobbes and Grassmann, that it often seems that mathematics has no axioms, only definitions. This led Wundt to Mill's attempt to derive all of mathematics—exactly as in science—from intuition in the pre-Kantian sense of the term (generalization from experience). Wundt noted that this view was much trashed, and indeed it was wrong that induction in mathematics is exactly like induction in physics. For, citing Baumann, science proceeds with the aim of generalizing nature's laws until agreement is reached with experience, but in mathematics, experience is inferior (there are no true objects).[74] Induction in mathematics proceeds from representations of objects; objects are merely an aid.

Wundt believed that one reason induction was not perceived in mathematics was that generalization was confused with abstraction. They are easily distinguished, he said: intuition traces a path from the specific to the general, whereas abstraction is a process of elimination. Mathematics, for Wundt, involved both intuition and abstraction. To make matters worse, nominalists mistakenly see abstraction as a uniform process, which is the same in mathematics as in any area of human experience. Realists, however, seem to have gone so far that they have mislaid the logical foundations of mathematics. To put matters right, one could not proceed by logic alone if logic is isolated from the individual intuitions and theorems it presumes. But history helps, even though it does not have the answers. Wundt therefore sketched a case for traces of the process of intuition surviving in the axioms and definitions of mathematics, and their immediate consequences, drawing as always on the historical work of Moritz Cantor and Hermann Hankel. As for abstraction, this was a process readily apparent in the definition of a point, and visible with a little more effort in the definitions of a straight line, a circle, or any geometric figure.

Kant, he now argued, had put the subjective element of mathematical idea building ahead of the objective (the transcendental conditions) because he had denied the conceptual nature of pure intuition and postulated an immediate constructive activity of pure imaginative force. However, one could see pure intuition as an abstraction from empirical intuition, the character of mathematical abstraction being that all objective elements of a representation are removed so that only the pure form of thought activity (cognition) remains as the mathematical way of building ideas (*Begriffsgebilde*). Indeed, said Wundt, the essence of the mathematical a priori is this reduction to the formal element. Wundt then concluded his paper with an argument that mathematics, but not science, had available a concept of exact analogy, by which the great generality of abstraction was obtained. Exact analogy works because

---

[74] Baumann 1868.

mathematics is only concerned with ideas, and it licenses making general geometrical statements on the basis of figures.[75] It similarly licenses the passage from one dimension to a higher one. It comes, moreover, in two forms. The first applies to the most fundamental theorems and makes precise the general validity of certain theorems originally obtained by intuition. The second extends certain ideas or operations from their original domain to their limit, and is the basis for the most abstract speculation.

6.2.5.2 LIPPS

Gottlob Friedrich Lipps, the younger brother of the slightly less forgotten Theodor Lipps, was a student of Wilhelm Wundt who published his "Untersuchungen über die Grundlagen der Mathematik" (Researches into the Foundations of Mathematics) in five installments in Wundt's journal between 1894 and 1898. He then taught at Leipzig for many years before going to Switzerland in 1911.[76] The five papers, which are essentially his doctoral dissertation, give an interesting insight into the thinking in Wundt's circles in the 1890s. Wundt's paper of 1883 had attempted to vindicate intuition and abstraction as the processes by which mathematical knowledge is acquired, and he had only hinted at how they are then handled by cognition. Lipps now argued that it was cognition itself which generated mathematical knowledge, and even the concept of number.

For Lipps, philosophy was not a single science but rather, like Wundt's *Wissenschaftslehre*, it was charged with unifying all scientific knowledge. The task of the philosopher was to make clear what ideas actually are, how methods are to be developed from ideas, and, in the present context, what is the basis for the meaning of mathematical ideas. He also shared Wundt's belief in the historical nature of mathematical knowledge and in the circumscribed role for logic in elucidating mathematical concepts. He also had his enthusiasm for reaching from the elementary to the most advanced topics in mathematics itself.

Lipps began by briefly sketching the modernist rewriting of mathematics (although he did not use the term "modernist"). Mathematical objects were increasingly strange—geometrical figures routinely escaped to infinity, the number concept now included complex number systems, there were continuous functions that were not differentiable—and because geometry may be studied using algebra and analysis, geometers were now freed from the fetters of intuition, the foundations of geometry could be reorganized in unintuitive ways, and numbers in turn were emancipated from geometry.

Mathematicians had, of course, produced precise definitions and were sustained by the fact that their ideas arose in response to real need and operated as a harmonious whole. The foundations of the subject were therefore surely sound, but only a philosopher, he said, could do justice to the historical development of the subject. A critical analysis can be brought to bear when concepts change and intuitions clash. Here he cited the disagreement between Kronecker and Cantor, Dedekind and

---

[75] The great difficulty with the parallel postulate, he said, derived from the intuition that (given two lines crossing a third at equal angles) we can move the crossing line parallel to itself without altering the angles.

[76] Piaget studied there briefly with him in 1918.

Weierstrass over the nature of number. Faced with such contradictions, one must (p. 163) either take Mephistopheles' advice: "It's best to hear only one, and swear by the Master's word" or have the suppleness to hold different positions simultaneously, as du Bois-Reymond did in his *Allgemeine Functionentheorie*.

Lipps argued that there were a number of contradictory views flourishing in mathematics so an epistemological approach, which would treat mathematics as nominalist or realist, is not to be pursued. Instead, mathematics is to be grounded in the simple activity of thought. This will lead to an examination of the different kinds of mathematical object, and is in line with Wundt's idea (*Logik*, vol. 2, p. 75) that the task of mathematics is "to subject the thinkable images of pure intuition, along with deducible consequences of pure intuition and all their properties and relationships, to an exhaustive examination."

After a lengthy historical overview, taking in Leibniz, Kant, and Herbart, Lipps gave an analysis of the content of consciousness that divided it into a logical and a spatiotemporal ordering. The logical ordering is based on the apperception of content and its immediate separation into the same and the different. The spatiotemporal ordering is based on the bounding of individual contents of awareness within the whole. Lipps sought to ground the idea of number in the logical ordering, which he felt was more fundamental than the way number arises in the spatiotemporal ordering. He pointed to the role of synthesis, the bringing together of contents of consciousness, and which he regarded, one might say, as logical glue, or perhaps a catalyst. It brings together a series of contents, and so the series-form can be regarded as the subjective form of thought. Accordingly, Lipps embarked on a long investigation of series-forms. He argued that they are completely independent of the objects they bring together because they are merely the rules by which this bringing together is done. So this bringing together of apperceived objects in consciousness is an elementary event, just as the apperception of a single object of consciousness is. Objects are brought together in virtue of their common signs. Successive terms in the series stand in the relation of ground to consequent, one to the next. This account of cognition ended with the introduction of the concept of a normal series (*Normalreihe*), upon which Lipps placed most emphasis.

Normal series are series forms whose subjective form corresponds to an objective representation and whose terms are simply the carriers of the signs of the series form. Lipps found that a normal series must have a first element, it must be capable of indefinite extension, it is homogeneous in the sense that only its first term is distinguished, and each term must generate all its successors. It must also be memorable, and Lipps gave a lengthy technical analysis of normal series to show how, at least in principle, this could be done From this he deduced (p. 279): "The normal series is now nothing other than the number series." As confirmation of this striking result, he showed how the numeral systems of various civilizations reflected this identity. Then he showed how the operations of arithmetic can be obtained from the number series. Subtraction is a problem, and it emerges that for Lipps zero is only an improper number.[77]

In the fifth and final installment, published in 1898, Lipps sketched what he took to be the kinds of cognitive acts needed for the construction of generalized numbers,

---

[77] All of this account recalls Wundt 1883.

starting with negative numbers and reaching beyond the complex numbers to the hypercomplex numbers. It is striking that the author saw his task as going from the elementary use of numbers all the way to the contemporary study of hyper-complex numbers. There is indeed a real question as to how our ability to count small collections enables us to do advanced mathematics, and no reason to suppose that the same mechanisms operate in the two domains. It is also striking how mathematical Lipps himself became. The diagrams of chains and the use of functional notation in this part of the paper all suggest a context for this in what might be called the prehistory of cognitive science. They underline the importance that should be attached to this approach to questions about the nature of, as it might be, mathematical knowledge, sitting alongside the more technical discussions of the mathematicians and the more overtly ontological and epistemological analyses of the philosophers.

In this final section, Lipps symbolized an act of cognition using functional notation, so $\alpha a = a_1$, where α is the activity of thought (the act of cognition), a is its ground, and a_1 its consequent. If α_0 is a pure act of apperception, he wrote $\alpha_0 a = a$. The formula $\alpha a = a_1$, he said, represents the coming together of a and a_1 in their objective logical order. Acts of cognition can follow one another, so from $\alpha a = a_1$ and $\beta a_1 = b_1$ one may deduce $\beta \alpha a = b_1$. Sequences of acts of cognition form chains, which Lipps indicated graphically (he may well also have Dedekind's use of chains in formulating the number concept in mind). Acts may be iterated, when α applies to a_1, yielding $\alpha \alpha a = a_2$, and if iterated indefinitely either go on forever or close up $(\alpha \ldots \alpha a = a)$. For example, the number series may be written as $\alpha 1 = 2$, $\alpha 2 = 3$, and so on. They may even be axiomatized, said Lipps, writing 0 for an absent object and also for an act of cognition not enacted, by these rules:

If $a = 0$ or $\alpha = 0$ then $a_1 = 0$;
If neither a nor $\alpha = 0$, then a_1 is not 0;
If $a_1 = 0$, then either $a = 0$ or $\alpha = 0$.

It then only remained for Lipps to spell out in this formalism the mathematical operations involved in defining, as it might be, complex numbers, for him to have given cognitive foundations for mathematics, and this he did.

Lipps's ideas never became an orthodoxy, nor were they given a decent burial. The official lines in the history of philosophy pass them by. Husserl never responded to them, Frege never found them worthy of attack. That is not entirely because of their limited merit, rather it is because they go in a different direction, one that still lacks the full attention of historians. Lipps believed that a full account of mathematics, one that explained how it was possible, had to show how it could change. This, he believed, a strictly logical account could not do. Instead, he sought the answer is the nature of thought and thought processes. The result is oddly Kantian. It is as if Kant's transcendental aesthetic has been incarnated as an embodied thinking machine, which, precisely and solely because it can apperceive, form ideas and manipulate them—in short, think—necessarily comes to the number concept. One may not find the argument convincing, but a convincing argument along these lines would do much to weaken the feeling that Kant's synthetic a priori intuitions of space and time were psychological in the sense that philosophers deplore.

SANTERRE

Of all the accounts I have found in which psychology and arithmetic are taken seriously, the most remarkable is by a Frenchman about whom I know almost nothing else except his book. S. Santerre published his *Psychologie du nombre* in 1907, and it seems from the only review of it I have been able to find that he died before 1911.[78] It has a preface by no less a figure than Pierre Janet, who rightly notes that Santerre took an idealist approach to mathematics, that Santerre plainly locates mathematics as a species of language, and that number, for Santerre, is nothing more than a language permitting people to express properties of sets of distinct perceptions. In the endless contest between rationalism and empiricism, Santerre, he noted, offered a third alternative, which is to replace axioms by simple psychological observations. Helmholtz had already argued that the ability to preserve the order of presentation of things was the basis of our ability to count. Santerre said, more precisely, that the fundamental property is our ability to enunciate a series of words. Janet noted that the book did not read like a psychological treatise. Rather, its conclusions follow deductively from some assumptions about mental processing. This, he remarked, helps to ensure that mathematics, on Santerre's account, retains its logical force.

Even Janet's preface cannot have prepared its readers for the highly mathematical manner in which the book proceeds. Mathematicians often joke privately that the modern textbook style, with its explicit definitions of every key term and its patient enunciation of a series of theorems and their attendant proofs, can be tedious. Books of this kind sacrifice excitement, even motivation, for pedagogic clarity, and the style is slightly dismissively labeled "definition-theorem-proof." Santerre's book is exactly that. It opens with a string of definitions, of which the first is of a fact or phenomenon of consciousness—any fact that modifies our consciousness. They are accompanied by statements of principle, the first of which is that the memory of each fact of consciousness persists invariably. What we can consciously do by bringing facts of consciousness back into our awareness, associating them with others, is all itemized in this fashion.

The fundamental relation between facts of consciousness is that a person regards some as coming before others. This is a subjective time-ordering that Santerre symbolized thus $A << B$. It was impossible that two facts of consciousness could occur in each order, so one cannot have both $A << B$ and $B << A$. So there were theorems in this subject, of which the first is that $P << A$ and $A << Q$ implies $P << Q$. Most importantly, this ordering of facts of consciousness is paralleled exactly by the ordering of words and of physical phenomena. Moreover, in each case, there is a good sense of identity: if A and B are facts of consciousness, each summons up facts of consciousness earlier than it. It may be that A and B are indistinguishable, but they summon up distinct collections of earlier facts. In this case, Santerre said, the facts of consciousness were identical, and he wrote $A \equiv B$. This relation is reflexive, symmetric, transitive (in symbols $A \equiv A$, $A \equiv B$ implies $B \equiv A$ and $A \equiv B$, and $B \equiv C$ implies $A \equiv C$).

Santerre went on to define separating elements of a system of facts of consciousness (or of words), consecutive elements, and initial and final elements, and to

---

[78] G. Milhaud, in *Revue philosophique* 71 1911: 206–8.

prove theorems about them. Then he defined and studied numerous relations between systems. Next he came to the sum of two systems of words, which he defined
as a sequence of words separated by the word "plus." He proved that this sum is
associative and commutative: $A + (B + C) + (A + B) + C$, and $A + B = B + A$. He
investigated what happens when an element of one sequence is associated with an
element of another, and what happens when one tries to pair off the elements of one
sequence with another. He called such associations a correspondence, and he developed a whole theory of them, adequate to define the difference of two systems. His
result was if a system B is inferior to a system A, then there is a system D such that
$B + D = A$.[79]

The abstract psychological considerations were now over. Their purpose was to
suggest that, carefully formulated, the mental ability to handle subjectively time-
ordered sequences of facts of consciousness, of words, and of physical events was
going to provide a satisfactory basis for our ability to do arithmetic. It remained to
define number and the arithmetical operations, and this Santerre did in the second
part of his book.

A sequence of numbers is simply a sequence of distinct words, each associated
with a distinct sign, which can be made to correspond completely with the elements
of any other system, going from the first to any arbitrary element without omitting
any intermediate terms. As any linguist would agree, the choice of words is entirely
arbitrary. The existence of such a sequence followed, Santerre said, from his earlier
theorems. He now showed how to define such remaining arithmetic properties as
the equality of two numbers, the difference, product, and quotient of any two numbers, and he briefly indicated how numbers relate to magnitudes. He also showed
how his concept of number agreed with the informal concept of number in general
use. Finally, he acknowledged Hilbert's "magisterial" geometric calculus, Méray's
presentation of fractions and algebraic quantities, and Tannery's treatise on arithmetic.

Milhaud's review located it among the attempts to give a sound axiomatic foundation for numbers, but disputed, despite Janet's endorsement, that this was a matter
for psychology. He noted that those attracted to logistics specifically disdained
psychology, and even those who gave axiomatic treatments of this or that branch of
mathematics were happy to stop at that point. He suggested that the right question to
ask was the logical one: is a transitive, nonsymmetric relation adequate to generate
the integers? For this one should consult, as Santerre had not been able to, the works
of Russell and the Italian school. Other than that, Milhaud held out the hope that the
details of Santerre's book might interest the psychologists, and he thanked the author
for drawing attention to these abstract and delicate problems.

These comments rather missed the point. Quite clearly, the book aimed not to
explain what numbers are and arithmetic is, so much as to explain how we may come
to know them. We must be equipped with certain cognitive abilities. But the
mathematical nature of the argument that Santerre deployed to this purpose cannot
be underestimated. It is definition-theorem-proof from beginning to end. As Janet
observed, this surely does not look like psychology, but nor does it look like philosophy. The discursive manner of the philosopher, on display in Lipps's papers,

[79] To speak briefly but loosely where Santerre spoke precisely but at length, "inferior" here means that
the system B runs out before system A, so this is the case where subtraction readily makes sense.

where alternatives are considered if only to be rejected and misleading associations are winnowed out before definitions are settled upon, is entirely absent here. There is nothing but definitions, assumptions about conscious behavior and memory that are precisely stated, and theorems, which are clearly labeled as such. Not since Spinoza has a broadly philosophical inquiry been written in such a mathematical way, and with such mathematical curiosity. Definitions are given, but not argued for, much as a mathematician would define matters in mechanics (a *smooth*, *uniform* slope, a ball *rolls* without *sliding*, etc.). The intended response is agreement, not further discussion, and to that end the individual steps in an argument and the individual pieces of a definition are made as small as possible. It may well not be psychology, on the definitions of its day or ours, but it is something like an account of what any machine would have to be able to do in order to carry out arithmetic, and it might be fruitful to see it as a piece of cognitive science.[80]

---

[80] In this way it is very different from Husserl's *Philosophie der Arithmetik*, which, even though it opened its author to the charge of psychologizing, is much more like a work of philosophy than is Santerre's.

7

AFTER THE WAR

7.1 The Foundations of Mathematics

7.1.1 Introduction and Overview

The origins of modernism in mathematics, and its eventual acceptance by large sections of the mathematical profession, have now been traced in numerous disciplines of mathematics as well as in the philosophy of mathematics and the relations of those subjects with logic. At least another book could be written describing developments after the First World War, but this is not the place. Accordingly, this chapter takes those topics in the foundations of mathematics that simply cannot be left in midair and traces their implications for the subject as a whole in the interwar period.[1]

It would also take a book to describe and analyze the changes caused by the war itself. It must be enough here to note the obvious: some 40 percent of young French mathematicians were killed or wounded in the war, along with comparable numbers of scientists, a blow from which it took France more than a decade to recover. By then, mathematics had moved ahead fast in Germany, which, like Britain and perhaps Italy, had done a better job of using its intellectuals safely. The next generation of French mathematicians, notably those who formed the Bourbaki group in 1930, looked to Germany for inspiration and consciously brought German structural mathematics back to France. There, in their hands, it flourished precisely at the time it was being dismantled by the Nazis in Germany itself. After the war, British mathematics took hesitant steps into a modern algebra, while modern analysis flourished there as it did in the Soviet Union and, for example, in Poland. But the heartland of the modernist movement in mathematics was undoubtedly Germany, more specifically Göttingen and the schools that grew up after the war that subscribed to its view of mathematics.

---

[1] Those readers meeting friends at the Vienna station will of course share a taxi with Alberto Coffa and consult his stimulating work (Coffa 1991) and the literature it generated. I have ventured some remarks about Schlick and Carnap below, but mostly this chapter returns us to Göttingen and points beyond.

Other discontinuities were caused by the war and the straitened economic circumstances that followed. Links to philosophy, issues of language, and similar broad cultural questions were severed, or rather they became the exclusive concern of the mathematicians themselves. This did not render them any less intense. Indeed, as will be described below, in the 1920s the German mathematics profession was riven by what is called the "foundational crisis" ("Grundlagen Krise"), a sometimes bitter dispute about whether mathematics could be regarded as rational. Undoubtedly part of the intensity was caused by the collapse of German society at the end of the war and the slow, hesitant recovery that never restored German professors to the high status they had enjoyed before 1914. The mere passage of years, with its inevitable process of handover from one generation to the next, took place in this much altered world.

This chapter pursues the interrelation between three themes: mathematics, logic, and human thought. It begins with the gradual clarification of logic and its relations with set theory and then with mathematics. There was certainly no feeling that Russell and Whitehead had succeeded and that mathematics had finally been reduced to logic, but there was a widespread feeling that something like their way of formalizing mathematics while heading off the paradoxes was the right place to start any study of mathematics and logic. Accordingly, many writers, Hilbert prominently among them, were willing to use some sort of theory of types (and some axiom of reduction to make mathematics work).

It was also possible to deepen the investigation of logic itself. For example, Zermelo's otherwise impressive axiomatization of set theory relied on an undefined term, "definite." Plainly, this word meant some sensible restriction on the kind of expressions one was allowed to write—but what, exactly? In making this restriction precise, numerous writers, most notably the Norwegian Thoralf Skolem, came to separate out a core of logic (later called first-order logic) that could be taken to be logic per se, and various extensions (such as second-order logic). Roughly speaking (more precision will be given below), first-order logic is obviously acceptable but is not strong enough to give you all and only the mathematics you expect. Second-order logic will give you what you want but is so much more powerful that it cannot stand as a truly simple logical basis for mathematics. Indeed, Quine regarded it as essentially equivalent to set theory itself.

Zermelo's axioms for set theory turned out to be inadequate for mathematics, and they too were refined, most memorably by Abraham Fraenkel. The resulting system of axioms, the Zermelo-Fraenkel axioms (often called ZF), are still not proven to be consistent, but no one is worried any longer. They are adequate for large amounts of the mathematics used, for example, in science. Some of mathematics, and certain of the higher reaches of set theory, require other axioms that we have already discussed: the axiom of choice and the continuum hypothesis. It was to turn out that these axioms are independent of ZF. In each case, if ZF is consistent, so is ZF with one of these axioms added, and so is ZF with the negation of one of these axioms added. But there is no widely accepted argument to this day that says that, for example, the axiom of choice is true.[2]

---

[2] The independence of the continuum hypothesis from the axioms of Zermelo-Frankel set theory (ZF) was established in two stages: in 1939 Gödel established that the theory formed from ZF with the continuum hypothesis as an extra axiom is consistent (Gödel 1939). In 1964 Paul Cohen established that the theory formed from ZF with the negation of the continuum hypothesis as an extra axiom is also consistent (Cohen 1964; see also his 1960).

Skolem's work, and still more that of Gödel, highlights another distinction that came into focus only in the postwar period: the distinction between syntax and semantics. A study of what can be written in mathematics, and on what can be (formally, logically, validly, . . .) deduced from what has been written emphasizes syntax. But one can also ask, of a mathematical system defined by some axioms, if the axioms are consistent because the axioms are satisfied by some set of objects. When that is the case, the objects are said to form a model for the axiom system, the axioms are said to be true in this model, and the emphasis is on the semantics of the theory (what the axioms "are about"). With clarity on the issue of syntax and semantics came the realization that the existence of (essentially unique) models for this or that system of axioms depended intimately on the logic being used.

It might be argued that all these debates were victories for mathematical logic, a technical subject akin to but perhaps even more scrupulous than mathematics, over old-style philosophy of mathematics. The central philosophical debate throughout the 1920s about the nature of mathematics grew out of the foundational crisis.[3] It was initiated and advanced by the views of Brouwer, who was supported for a time by Hermann Weyl. Brouwer saw very strict limits on what the human mind can say about infinite sets, and his philosophy of intuitionism imposed grave limits on what can be done in mathematics, so much so that eventually Weyl abandoned it in order to pursue his mathematical analysis of quantum mechanics. But the philosophy alarmed Hilbert, whose proof theory and later philosophy of finitism has been said to be an attempt to outflank Brouwer by conceding as much as possible of the philosophy while securing all of classical mathematics. For a mixture of reasons, to be discussed below, something like Hilbert's views eventually triumphed, and a widely held view today is that it was all a storm in a teacup. After all, such people say, intuitionism and classical logic differ in only one axiom, and whenever it can be found an intuitionist proof yields more information than a classical one. Such was the view of Heyting, Brouwer's most able follower in mathematical logic, but I shall argue below that this represents a trivialization of Brouwer's views (as indeed Heyting came eventually to concede).[4]

If Hilbert's aim had been to ground mathematics securely in some formally impeccable system, be it logic of whatever kind or some mixture of logic and set theory, then in 1930 that dream was forever shattered by Kurt Gödel. Gödel showed, if one may speak imprecisely, that any fragment of mathematics large enough to contain arithmetic (addition and multiplication of natural numbers) is either inconsistent or incomplete. That is to say, if it is not self-contradictory then there will be true statements that cannot be proved within the system. Any such statement can of course be taken on as an axiom, but then there will be another. Hilbert's hopes of showing that all mathematical statements could be decided, if only in principle, thus collapsed. Much more precision is needed if Gödel's result is to be seen to be the theorem it is, that is, a proved result, and with that precision it would seem as if there is just a chance that Hilbert's program for mathematics and logic could still be made to work.

---

[3] Described in detail and from different points of view in Hesseling 2003, Mehrtens 1990, and Volkert 1986.

[4] The wise also know that category theory, currently the best way to establish mathematics without establishing set theory first, leads more naturally to intuitionistic logic than classical logic, which is interesting but says nothing about the limitations of the human mind.

Indeed, Gödel explicitly held out that thin hope. But no one has found a way to do it, and most probably no one today expects that a way will be found. The situation is that ordinary logic does not do enough, and no other generally acceptable starting point has been found.[5]

The second main topic of this chapter is the mechanization of thought. With the work of Charles Babbage on programmable computers, the age-old fascination with artificial life and machines that can think started to sound more and more plausible. By the same token, ideas about how humans actually think became more and more informed by ideas about how logical calculation gets done. Boole's book on an algebra for logic was, after all, grandly entitled *The Laws of Thought*. Faced with the complexity of Boole's rules, Stanley Jevons in England and Allan Marquand in the United States made machines to carry out his deductions, and Venn diagrams were speedily invented for the same purpose. Logical deduction at a level hitherto exemplified by Sherlock Holmes became mechanized.

What about mathematics—could at least some parts of it be automated? Among David Hilbert's Paris problems in 1900 was one calling for an explicit method for telling if polynomial equations have integer solutions.[6] By the 1920s, as part of his proof theory, Hilbert listed three specific targets for a proper understanding of the nature of mathematical deduction. One by one, starting with Gödel's incompleteness theorem, they were all shown to be unattainable. The third and last, called the decision problem, which called for a proof that every mathematical problem has a solution, was shown to be unsolvable by Alan Turing, and for this he introduced a most eloquent illustration of what it is to follow a mathematical procedure—the Turing machine, as it is nowadays called. The Turing machine is a universal computer. It can be programmed to do in a finite amount of time anything any kind of digital computer can do (although it takes geological periods of time to do anything worthwhile—never anywhere else does finite time seem so long). More to the point, Turing was able to show that there are certain things even Turing machines cannot do, and so certain problems are forever undecidable. Whether or not there are limits on what humans can think (we may not be Turing machines, after all), there are limits on what conventional computers can do.[7] Turing's work fitted in very well with the work of two American logicians, Alonzo Church and Emil Post, who had more formal ideas about what could be computed, and the result is a semi-philosophical view about the limits of mathematics known as Church's thesis.

Turing, meanwhile, went on to propose a test for determining if a machine was displaying intelligence, the so-called Turing test. It is often misrepresented in popular accounts, so it is worth discussing here as part of the debate about machine versus human intelligence. Somewhat earlier, McCulloch and Pitts had come up with a

---

[5] It has now become particularly easy and pleasurable to follow these developments. In addition to the books by Moore and Ferreirós and the source book by van Heijenoort already mentioned, there is the second volume of the source book edited by Ewald (1996), the source book with extensive commentaries edited by Mancosu (1998), the history of intuitionism and its reception by Hesseling, *Gnomes in the Fog* (2003), and Giaquinto's book *The Search for Certainty* (2002), which is a lucid yet trenchant guide to the technical and philosophical issues.

[6] Hilbert allowed that a problem might have an affirmative solution, a negative solution, or be proved to be insoluble. However, this Paris task was eventually shown to be impossible by Julia Robinson and Yuri Matjasevich in the 1970s.

[7] See the more precise discussion below.

theory of how neurons might operate in the thinking human mind. Their theory of neural networks influenced the other major mathematician involved in the birth of the computer, John von Neumann, and was later shown to have some affinity to the behavior of real neurons by Donald Hebb, a distinguished neuroanatomist.

I turn finally to take stock. First, I address the ways in which the philosophy of mathematics, or, more accurately, the default philosophy of the working mathematician has become Platonist. I consider what this might mean, what else it might be or have been, and the sense in which it is the natural philosophy for modernist mathematics. Two interesting positions here are Schlick's perceptive criticism of Platonism in 1918, and Bernays' typically sensitive account of what it is. It seems that his paper of 1935 may be the first to coin the term "mathematical Platonism." In this context it is also interesting to see how Gödel's avowed Platonism contrasts with Carnap's philosophy of linguistic frameworks.

Two closing sections round out the account. One is an assessment of the degree to which modernism "won" in mathematics—not wholly, but largely, I believe. The second is the voice, again, of Plato's ghost, calling out that all was not resolved. The modernist revision of mathematics accomplished much, but did it leave answers, or only deeper questions?

7.1.2 Hilbert and Proof Theory

As early as 1900, and certainly by 1904, Hilbert knew that a major problem lay in the path of his attempt to capture the essential reason mathematics was correct. It was one thing to demonstrate that this or that family of geometrical axioms could be satisfied by a suitable set of objects when these objects were points, lines, planes, and so forth and they were defined by coordinates and equations, in short, in terms of numbers. The consistency of an axiom system for a geometry simply reduced to the consistency of arithmetic. But what guaranteed that consistency? On what grounds could axioms for arithmetic be defended? If the answer was set theory then there were two problems. First, set theory was possessed by paradoxes and could not be taken as unproblematic. Second, even if set theory were sorted out, at some stage and very likely soon, it would be necessary to say that some foundations were absolute, or else one would be doomed to chase through an infinite regress. But what could those foundations be?

In his *Axiomatisches Denken* (or, *Axiomatic Thinking*) of 1918, Hilbert observed that the theory of the integers and the theory of sets must both be axiomatized. To do this, nothing other than logic could be invoked, and therefore "it appears necessary to axiomatise logic itself and to prove that number theory and set theory are only parts of logic." This project was, he said, inseparable from several others. The consistency of these theories belonged with several other problems with an "epistemological tint," such as the solvability, in principle, of every mathematical question; the subsequent checkabiltity of the results of a mathematical investigation; the relation between content and formalism in mathematics and logic; and the decidability of a mathematics question in a finite number of operations. Axiomatizing logic required clearing up these questions and the interconnections. Hilbert's position was therefore close to logicism, but also vastly more ambitious.

In 1922, in his "*Neubegründung der Mathematik*" (or, "New grounding of Mathematics"), Hilbert began to sketch out what became known as his "proof theory," a remarkable attempt to subject proof to a mathematical and logical analysis. The new approach was a consideration of certain "extra-logical discrete objects which exist intuitively as immediate experience before all thought." These objects are signs, such as repetitions of a single stroke. For Hilbert, the objects of number theory were the signs themselves. In a deliberate echo of the Gospel according to St. John, Hilbert wrote: "The solid philosophical attitude that I think is required for the grounding of pure mathematics—as well as for all scientific thought, understanding, and communication—is this: '*in the beginning was the sign*' " (p. 163). In this philosophical tangle—can numbers really just be marks on paper?—note the emphasis on communication. At the very least, one can object to Hilbert's ideas on the grounds that they are not about mathematics but the way it is written, which is something else.

Hilbert then introduced other signs in an obvious way. Starting from the initial claim that 1 is a number, he introduced the sign "+" and said that $1 + 1$ is a number, as is $1 + 1 + 1$ and so on, and these may be abbreviated by the signs 2, 3, and so on. The sign "=" is one of those that serves to make assertions, thus $2 + 3 = 3 + 2$ is the assertion that $2 + 3$ and $3 + 2$ are abbreviations for the same number sign, to wit $1 + 1 + 1 + 1 + 1$. The sign ">" as in $3 > 2$, is the assertion that the number $1 + 1 + 1$ extends beyond the sign $1 + 1$. In this way, one can write a considerable amount of elementary arithmetic without any axioms, but one cannot write down or make assertions about infinitely many objects. To do that, one had to formalize the entire theory of mathematics, and its content had to be imitated by formalisms. This formal edifice is to be made up of axioms, formulas, and proofs, and it is with them that contentual thought is done.

Hilbert suggested that there would be signs of various kinds: signs for individual objects, such as those mentioned above and $\rightarrow$ for logical implication; signs for variables, variable functions, and variable formulas; and signs to aid communication (such as a single symbol to mean a formula). The idea, although Hilbert did not put it in these terms, is that anything expressed like this would count as mathematics, and if we found enough of it on another planet we could infer that it was mathematics and hope to understand what it said. In particular, certain arrangements of signs on the page would be proofs, and the aim of proof theory was to study operations with proofs themselves. So a proof is a perfectly concrete thing, a collection of marks on paper. To show that the enterprise was feasible, Hilbert then showed that a certain system of five axioms was provably consistent, and that a system of sixteen axioms both generated arithmetic and could be shown (he claimed without proof) to be consistent. Here he was not in fact correct, but that was not fatal to the enterprise. The possibility that mathematics was an intellectual activity which could be expressed in a way that was provably consistent was now very much alive.

Hilbert's idea was that a valid mathematical argument could be displayed as an arrangement of symbols on paper. Certain expressions, called formulas, would each occupy a line and they would follow each other according to certain rules (the axioms), which could be written down in advance. The rules would be things that we could give meanings to, but the way they followed each other would be mechanically checkable. Each step would be an exemplification of an axiom. The axioms would generate a set of initial statements or formulas and generate the rest of the formulas from these. Hilbert did not use this metaphor, but you could imagine taking an axiom system and simply listing all its initial statements, all the statements that were derivable in a single step from these, then those in two steps, and so on. You could put together a conclusion drawn at one stage (a derived formula, if you prefer) with one drawn at another, all according to the axioms. Now, the most fundamental requirement of an axiom system is that it be consistent. An inconsistent axiom system can be recognized, said Hilbert, by its ability to generate a nonsense, such as $0 \neq 0$. So, to see if an axiom system is consistent, it is possible to argue as follows. First, characterize in some way all the formulas that can be validly produced. For example, the simple axiom system mentioned above cannot produce formulas with more than two occurrences of the $\rightarrow$ symbol. Second, show that the formula $0 \neq 0$ can only be generated by an argument that produces an impossible formula. For the simple axiom system, Hilbert showed that the formula $0 \neq 0$ can only be generated from a formula involving three occurrences of the $\rightarrow$ symbol; such formulas cannot arise, so the system is consistent.

Put this way, Hilbert's proof theory is simple and attractive. It was deliberately designed to have one particular virtue: it should enable mathematicians to reason correctly about all manner of infinite sets, and thus to stay in the paradise Cantor had created for them. In Hilbert's words: "No-one can drive us from the Paradise that Cantor has created for us."[8] As we shall see shortly, Hilbert was anxious to head off the challenge of those in the next generation who sought to lead the way out of Eden. Hilbert sought to do this by requiring that although there might be axioms that were controversial when made about infinite sets (such as the law of the excluded middle), the arguments about these objects were expressed in a finite number of steps, each one explicitly checkable. For this reason, Hilbert's position is much more accurately summed up by the term "finitism" than by "formalism."

Hilbert took the challenge posed by the problems of the infinite very seriously, writing in 1926 that the final explanation of the *essence of the infinite* (his italics) reaches far beyond the domain of special scientific interests and rather "has become necessary for the honour of human understanding."[9] To legitimate his approach to the infinite, Hilbert made repeated use of an analogy with ideal elements in other branches of mathematics. He mentioned specifically the ideals that Dedekind introduced into algebraic number theory (for example, the ideal that divides both 2 and $1 + \sqrt{-5}$) and the introduction of points at infinity in projective geometry. In neither case can one point to an object in a straightforward way: the ideal number and the usual numbers are different, as are ordinary points and points at infinity. But one can work with them in an entirely rigorous way. Just so, the formulas one writes stand for infinitely many statements (for example, the axiom that asserts the commutativity of the integers, $a + b = b + a$), but we can work with them. Just so, one can write down the axiom that asserts the validity of the law of the excluded middle and work with it in finite, mechanically checkable arguments. In the 1920s, but not in his work before the First World War, Hilbert also introduced a distinction that was to prove lasting. He distinguished between the purely formal aspect, which he called "proper mathematics," from the contentual side, which he called "meta-mathematics." Metamathematics was to protect mathematics from unnecessary prohibitions and the paradoxes and to provide consistency proofs. Here only finite arithmetical considerations enter (such as the enumeration of the formulas in an argument), and these

[8] Hilbert 1926, 274.
[9] See Hilbert 1926, 164, reprinted in Hilbert 1930, 265.

are entirely safe. These statements, insofar as they are mathematical at all, are self-evidently sound and do not require a rigorous theory of arithmetic before they can be used. So the accusation, leveled by Poincaré, among others, that the foundational enterprise was vitiated by a vicious circle was refuted.

It remained to write down an axiom system that would generate set theory and number theory, and that could be shown to be consistent. If such a system contained the law of the excluded middle as an axiom, then use of the law would be secured, and mathematicians could remain in paradise. As things were to turn out, the best efforts of Hilbert and his assistants Bernays and Ackermann were to fail in this regard, but only in ways that could bring no solace to Hilbert's rivals, either.

A number of issues connected with Hilbert's program are worth commenting on even if it proved unsuccessful in its day.[10] His finitism is based on the idea that the mind has immediate access to a finite collection of marks on paper. But this cannot be quite right (mathematics cannot be constrained by the size of a piece of paper, for example), and rather what is surely meant is that these marks are accessible as intuitions in something like a Kantian sense, as Mancosu has argued.[11] For Hilbert and Bernays, finitary intuitions are a new, hitherto unnoticed, source of a priori knowledge. Unfortunately, it is hard to be precise about what Hilbert meant by finitary and so to be sure what arguments are legitimate from that point of view. On the face of it, statements such as "for all n, $1 + n = n + 1$" is not finitarily meaningful because they cannot be checked on all numbers, so something must be meant that gets around this trivial objection.

Another criticism is the one raised in the mid-1920s by Hermann Weyl: all that finitism could do would be to show that mathematics is consistent. This does nothing to establish that it is true (compare the situation in geometry). For that, some account must be given of how mathematical statements are meaningful. One could, for example, try to show that formal mathematics cannot prove the validity of any statement known to be wrong to some highly reliable audience. Consistent mathematics would then be about many familiar things and some new ones. Once you go down this road, the only natural, highly reliable audience you come to is the scientific community. Weyl's much more ambitious idea was that mathematics as a whole was in some way confirmed by its profound uses in physics, a subject that was becoming unthinkable without it.

But before Weyl could get to these positive contributions to Hilbert's program, he first had to deny it outright.

7.1.3 Brouwer and Weyl

In the 1920s the opposition to Hilbert could not have come from sources more painful to him than the ones it did: L.E.J. Brouwer and Hermann Weyl. Brouwer had already established himself as the leading topologist, the dominant figure in this central and growing branch of mathematics, a man of the highest standards of

---

[10] This account follows Richard Zach, "Hilbert's Program," in *The Stanford Encyclopedia of Philosophy* (Fall 2003 edition), ed. Edward N. Zalta, http://plato.stanford.edu/archives/fall2003/entries/hilbert-program/.

[11] His argument for this shift in Hilbert's thinking is given in Mancosu 1998.

mathematical rigor. Hilbert had made sure he was among the active editors of the *Mathematische Annalen,* the cherished Göttingen journal that saw itself as the leading journal in the field. Weyl was Hilbert's most obvious successor, the gifted mathematician who had passed through Göttingen and stood out even among such talent, who had been the first to devise the right mathematics to capture Einstein's general theory of relativity and was shortly to do much the same for the new science of quantum mechanics. It can only have caused Hilbert anguish to see that the two men upon whom he placed so much hope for the future were willing to abandon so much of the mathematics that had been recently created.

Their opposition was in some ways similar, in some ways different. Both drew it from their considerable immersion in philosophy, and if Brouwer went further, and seemed oblivious to the harm he was doing to the patient mathematics by the rigor of his remedies that was because he, unlike Weyl, had no sympathy with applied mathematics and physics. Brouwer was the first to raise his criticisms, and Weyl gravitated toward him, partly because he found Brouwer to be a fine, principled man, and Weyl set considerable store by such things. Husserl described him as a "completely original, radically honest, real and very modern man."[12] Einstein, however, saw something else in Brouwer, finding him "an involuntary advocate of Lombroso's theory of the close relationship between genius and madness."[13]

Brouwer, from his earliest days as a mathematician, had entertained the gravest doubts about what the human mind can know about the infinite. He also inclined to solipsism, finding logic subordinate to mathematics and language to both. Whatever his reasons for calling his philosophy of mathematics intuitionism, it was rooted in the idea that mathematics is a languageless private activity of thought. All his distrust can be usefully concentrated on one inference in mathematics: the nonconstructive existence proof. Most mathematicians, and Hilbert above all, liked to argue that given an infinite set X and some property P that might apply to elements of X, then there is an element x of X such that the property P does hold for x if it can be shown that the claim that there is no such x leads to a contradiction. The argument is entirely unproblematic when X is a finite set, because in principle one can check P element by element. Brouwer denied that the argument was valid when X was an infinite set. He required that there be a construction that produces at least one such x.

For example, if X is an infinite set of real numbers, one wants to know if there is a least upper bound for the set, that is, a number u such that every x in X is less than u and every other number u' that is greater than every number in X is also greater than u. Very often in mathematics one argues as follows: there is a number, v say, which is greater than every number in X, and therefore there is a least upper bound u as required. It may be difficult, even impossible, to find this least upper bound exactly. Brouwer would have none of this, and only allowed that a least upper bound exists if there is a procedure for finding it to any arbitrary degree of approximation. It should be said that this forced him into some contortions, because some of the best of his prewar theorems relied on such indirect reasoning.[14] However, these contortions are not actually fatal to his position.

---

[12] Quoted in Hesseling 2003, 87.

[13] Hesseling 2003, 87, n. 395.

[14] To give one example, the famous Brouwer fixed point theorem says that any map from the disk to itself maps at least one point to itself, but the usual argument gives no way of finding such a point.

Brouwer's position can also be compared to the verdict of not proven in Scottish law. This is intermediate between guilty and innocent and means that while the accused may walk free, this is not so much because he or she is innocent as because the prosecution has not proved its case. In Brouwer's opinion, if you wanted certainty in mathematics you had to make a sharp distinction between what you could exhibit to the mind and what you must otherwise find meaningless. To say that something exists in mathematics without a construction being given for it is as incomprehensible, because contradictory, as saying guilty but not proven. This was to become a position that provoked as much dissent as the Scottish verdict does among those with, one might say, a naive realist view of life.

But one should not think that Brouwer's views were otherwise in agreement with Hilbert's, except over the law of the excluded middle. In fact, the implications of this difference went deep. Brouwer had a different concept of a set, which derived from his very precise idea of what it was to say some element belonged to a set. The set of all such-and-suches did not make sense for Brouwer without some way of determining precisely if each candidate element did or did not belong to it. This was another position he pushed to an extreme. Consider, for example, a rule that purports to define a decimal number, such as this: the nth decimal place is a 1 if and only if the decimal expansion of π contains a run of n 9's in succession, otherwise it is a 5. Hilbert would have said that this rule defines a number, but we do not know what it is. Brouwer would have denied that this rule defined anything, because there is no way of determining the number with arbitrary precision, and indeed in one of his papers he argued that not every real number has a decimal expansion. It was likewise impossible for Brouwer to accept very large sets, because they could only be "defined" by rules that flouted his criteria for being clear to the mind.

Brouwer shared with Hilbert a deep identification with the modernist view that it is the responsibility of the mathematician, and the mathematician alone, to ground mathematics with certainty, and that this can be done only by a profound analysis of mathematics itself. Thereafter, they disagreed. For Hilbert, the way forward was into a reformulation of mathematics and logic, and he was open to both syntactical and semantic linguistic analyses. He was a free spirit, optimistic that every problem can be solved, he saw no limitations on the human mind when it came to mathematics, he saw axiomatic mathematics playing a steadily greater role in the physical sciences, and, of course, he saw no real problems in applying the law of the excluded middle. Brouwer, whose intuitionism derived from the mystical position he had adopted as a young man, saw only limitations on the human mind, rejected logic and language in favor of inner and supposedly veridical insight, did not care for the connections between mathematics and science, and had abandoned the law of the excluded middle.

It is possible, as we shall see below, to separate the mathematical and philosophical parts of Brouwer's position, but in holding them so firmly together Brouwer deliberately departed from the modernist orthodoxy that had emerged over the previous few decades. Equally obviously, it was not a return to the simple-minded ways of the past. It was a response to the evident failure to ground mathematics in naive abstractionism, but it turned to the integrity of immediate mental experience. Mystical and solipsistic it might be, but it was a response to the anxiety apparent in the contemporary mathematical world. It can best be seen as akin to Heidegger's later sense of "being in the world," and in that sense the ensuing debate between the supporters of Hilbert and the supporters of Brouwer can be seen as the mathematical equivalent of the parting of the ways between Kantian and existentialist philosophers that happened at about the same time.

In 1921 Weyl changed his position of the foundations of mathematics and came close to agreeing with Brouwer. Indeed, he thought he was in complete agreement, but Brouwer felt that Weyl still had quite some way to go. The topic that provoked Weyl's change of mind was the continuum, or set of all real numbers, upon which he had written a celebrated little booklet called *Das Kontinuum* in 1918. It is worth digressing to look at this little book.[15]

Weyl was once again concerned about the paradoxes of set theory; this time, Grelling's paradox (see §4.7.3). He argued that part of the way to avoid them was to consider only meaningful sentences, and not the larger class of grammatically correct ones; here we see further tentative explorations of the semantic-syntactic distinction. This forced Weyl to explain what a meaningful statement is, and here he started by admitting some primitive intuitive concepts, such as a point in space or a number and the elementary relations between them. This put him nearer to Pasch than Hilbert on the topic of geometry, and, more importantly, very close to Poincaré on the topic of arithmetic because, like Poincaré, he took the natural numbers as given and not as objects to be constructed out of logic or set theory. Complete mathematical induction was, for Weyl, grounded in intuition. To describe the mathematical relations he needed to get to mathematical analysis, Weyl returned to his approach in 1910 but was now much more explicit that he was dealing with a formal language.

Weyl could now construct sets in a meaningful way. He adhered to a predicativist position, again much as Poincaré had advocated, started with primitive objects, and then used his system of relations to build more elaborate objects. These objects naturally have levels that reflect the way they have been constructed, and Weyl used these levels much as Russell used types to avoid the paradoxes. The risk then is that other mathematical arguments, such as the existence of a least upper bound, are imperiled because objects have the wrong levels, but Weyl dismissed Russell's axiom of reducibility (which offers a way of forgetting about the levels) as artificial and unworkable. Weyl clearly hoped to get away with just working at ground level, but here he got into trouble, because Richard's paradox can still be posed within Weyl's framework, and his prediliction for ground level cannot be made into a prescription for it.

Mathematical analysis requires a theory of real numbers, or something like it. Weyl could construct the rational numbers out of the natural numbers, and the real numbers out of the rational numbers via the theory of Dedekind cuts. But what he could prove about the real numbers was limited by his apparatus. He noted, correctly, that the least upper bound axiom for sets is false in his theory, essentially because the definition of the least upper bound is impredicative. But he also noted that the least upper bound axiom for sequences is valid in his system. These axioms are not equivalent: the one for sets implies the one for sequences (because from a sequence one can extract a set), but the converse requires the axiom of choice, and Weyl was never comfortable with the axiom of choice. Weyl therefore explored what one can do on the sequence approach, and convinced himself that one could con-

---

[15] Here our guides are Feferman 1998 and Feferman 2000.

struct a large part of the theory functions and so recover enough mathematics to do the exact sciences.

Weyl's analysis was elegant and profound. There was and is no need to start mathematics with set theory. It is entirely reasonable to accept as primitive intuitions ideas like number and point that everyone else works with quite happily. The idea of avoiding impredicative definitions in order to keep out the logical antinomies was well supported at the time. There were imperfections in the account (Richard's paradox still lurked), but further boldness might yet carry the day.[16] However, Weyl abandoned the attempt, and it was not taken up again until the 1950s. What irked him was that he had analyzed the continuum as if it were made up of isolated points. In 1921, after reading Brouwer's papers, he came to prefer to see the continuum not as already made up but as something endlessly coming into being. For this, Brouwer's view of a real number as a sequence of intervals was much more congenial.

As early as his inaugural address at the University of Amsterdam in 1912, Brouwer had argued that the "basal intuition of mathematics"[17] was that of the falling apart of moments of life into qualitatively different parts later reunited while remaining separated by time, and he referred to it as "the intuition of the bare two-oneness" (p. 85). So the intuitionist regarded Kant's apriority of space as a lost cause, but adhered, said Brouwer, to the apriority of time. From this intuition of two-oneness the intuitionist creates the finite ordinal numbers and the smallest infinite number, but there one must stop.

Weyl was attracted to this heady mixture of mathematics and philosophy and wrote:

> So I now abandon my own attempt and join Brouwer. I tried to find solid ground in the impending dissolution of the State of analysis (which is in preparation, even though it is still only recognised by few) without forsaking the order upon which it was founded, by carrying out its fundamental principle purely and honestly. And I believe I was successful—as far as this is possible. For *this order is in itself untenable*, as I have now convinced myself, and Brouwer—that is the revolution! (Weyl, 1921, 98–99).

The year 1921 was a difficult one in German history. After the Russian Revolution of 1917, there had been an attempt at a socialist revolution in Germany in 1918, which had been quite brutally put down. The German Communist Party was established as a mass party in 1920, and the Comintern was urging world revolution. The harsh implications of defeat and the terms of the Versailles Treaty left most middle-class Germans in difficult financial straits, the old certainties of Germany's position in the world were broken. Though the allegedly inevitable dissolution of the state and the coming revolution might have struck some readers as attractive, it certainly struck others as a most unpleasant prospect, Hilbert among them. He disparaged Weyl's "revolution" as a repressive dictatorship, seeking to throw overboard everything that made its authors uneasy, and as only a repetition of an attempt at a coup that had been tried before by Kronecker and failed. All it did was threaten to dismember and mutilate mathematics. All the more interesting, then, that Hilbert's and Weyl's actual political views were not that far apart: Weyl had republican

---

[16] As Feferman 1998 has shown.
[17] See Brouwer 1912, 85, in Dresden's translation.

sympathies and Hilbert was attracted to Leonard Nelson's non-Marxist left-wing breakaway from the German Socialist Party.[18]

The deliberate political undertones to these contributions to the debate on intuitionism hint at the passions it was to generate. They were fueled by a real sense that grave mathematical mistakes would be made if the nature of mathematics was misunderstood. Here, Brouwer and Weyl agreed for different reasons. For Brouwer, the primary activity was mathematical thought. If this process was unable to deal with certain kinds of infinite sets, as he felt it to be, then there was nothing more to be done. It was like picking up heavy weights: with practice and ingenuity, one might find ways of doing some things that had seemed impossible, but other activities were simply beyond human reach. In this limited sense, his philosophy is Kantian, inasmuch as Kant had imposed an unalterable structure on the human mind so that knowledge becomes possible at all. Weyl, however, was much more philosophically driven, much more inclined to locate ultimate questions in a continuing philosophical discourse. In these years he was inclined to the views of Fichte and had some sympathy with the increasingly phenomenological views of Husserl. So he believed that the basic elements of experience are the experiences of consciousness as each person has them. His more overtly idealistic philosophy, and the weight he attached to such considerations, brought him into close agreement with Brouwer's more subjective and private view, in particular when it came to the topic of human limitations.

Slowly after 1921, more rapidly after 1924, intuitionism moved to a position of importance in debates about mathematics. It became the major current in the so-called foundational crisis that dominated German mathematical life in the 1920s.[19] Exacerbated by the dire situation Germans found themselves in after the war, the foundational crisis was an exercise in self-doubt and heart searching that held in question many of the major advances of modern mathematics. What were mathematicians talking about when they said something existed? What evidence should they produce in support of their claims? How does the content of the mathematics relate to the formalism in which it is expressed? What are the abstract, infinite sets mathematicians reason about? And, when they do so, what logical principles apply, for example, the law of the excluded middle?

The crisis was given its name, and its salience, by Weyl's paper of 1921 in which he converted to the Brouwerian revolution. When Hilbert replied to it he did so in a series of five public lectures at the University of Copenhagen and then at the newly founded and mathematically strong University of Hamburg. The Copenhagen lectures were the occasion for Hilbert to receive an honorary degree, while the Hamburg lectures produced a very lively debate.[20] Perhaps for that reason, Hilbert chose the new journal of the Hamburg Mathematics Seminar to publish his paper, the "New Grounding" that was described above. In the published paper, his reaction to the intuitionism of Brouwer and Weyl was fierce. He argued that all mathematical statements should be either true or false, there was no room for half-truths; but that in pursuit of this goal, Brouwer and Weyl had chosen a false path. Indeed, he said, the alleged inconsistency of real analysis was nowhere to be seen. Few subjects had been as thoroughly studied as real analysis,

---

[18] I thank Colin McLarty for this information.
[19] It is a major focus of Mehrtens 1990. See also Volkert 1986.
[20] Information from Hesseling 2003, 136.

and not even the shadow of an inconsistency has appeared. If Weyl here sees an "inner instability of the foundations on which the empire is constructed," and if he worries about "the impending dissolution of the commonwealth of analysis," then he is seeing ghosts. . . .

To sum up, I should like to say: if one speaks of a mathematical crisis, in any case one may not speak, as Weyl does, of a new crisis. He has artificially imported the vicious circle into analysis.

Hilbert's observations about the recent history of mathematics are usually rather shallow and intended to serve a purpose, and these are no different, but they are not off the mark. It was indeed true that talk of a crisis in mathematics was not new but went back at least as far as work on the foundations of set theory, which had been publicly questioned by Poincaré, Perron, and others and had surfaced in 1905 in the debates about Zermelo's axiomatization of set theory. The shifting nature of the celebrated paradoxes of set theory also testifies to an ongoing disquiet worthy of being called a crisis. But it is also true that there was not a single genuine contradiction in the body of mathematics. None of the bizarre results discovered before the war were antinomies, they were not in contradiction with other results. Whatever one thought of the foundations of set theory, some strange principle or invisible hand seemed to keep problems in the foundations of mathematics from generating fatal problems in the higher reaches of the subject. Of perplexity there was plenty, but that was the very nature of mathematics, and there was no reason to suppose that new ideas would not be found to keep the subject moving forward.

What was new was the suggestion that there was a vicious circle in the foundations and that mathematicians were deceiving themselves if they thought otherwise. There was a famous exchange between the distinguished Hungarian analyst Polya and Weyl after Weyl had given three lectures in Lausanne in 1919 that sheds an interesting light on this.[21] The lectures formed the basis of Weyl's paper of 1921, and Polya, a colleague of Weyl's at the ETH in Zurich, asked him what he meant by the requirement that mathematical statements be not only true but meaningful. Weyl replied that this was a matter of honesty, to which Polya replied that it was a mistake to mix philosophy and science. Weyl replied: "What Polya calls sentiment and rhetoric, I call insight and truth; what he calls science I call letter pedantry. Polya's defence of set theory . . . is mysticism. Separating mathematics as formal from spiritual life kills it, turns it into a shell."

We are in the presence of the true believer who, in the intensity of his feelings, does not see disagreement as a technical matter but as arising from a terrible moral failing—a fatal separation of mathematics from *spiritual* life—and leading to the replacement of truth with rhetoric and science with letter pedantry. Mere pragmatism usually has no effect on such people, at least in the short term. In the absence of a self-contradiction in analysis, the most Weyl could produce was a series of claims that the foundations were incoherent.[22] In the 1921 paper, Weyl referred to math-

[21] Information from Hesseling 2003, 128–129.

[22] Weyl had had an earlier run-in with Polya in 1918, before he converted to intuitionism. There he had bet, among other things, that within twenty years mathematicians would come to see that there could be infinite sets that do not have a countably infinite subset, a bet he eventually admitted some years after it had expired that he had lost, "by 49 to 51 per cent." This prompts the observations that figures in straw polls carry a health warning: these numbers may reflect the choice of the electorate.

ematics as a monstrous paper economy, devoid of real value (which resides, in this analogy, in the singulars, such as food). This was published just as the extraordinary inflation of the German mark was beginning.

In 1924 reactions to the issue of intuitionism and the foundations of mathematics started to accumulate, and almost immediately there was a rapprochement between the two sides. Thus Abraham Fraenkel, whose work on axiomatic set theory will be considered in the next section, indicated in his book that he saw no reason to suppose intuitionism was a challenge to logic.[23] More importantly, in 1924 and again in 1925, Hermann Weyl moved to the center ground to suggest that mathematics could be done either as intuitionism suggested, or classically. He may already have begun to worry that mathematics à la Brouwer could not provide the necessary tools for physics, and in particular for quantum mechanics. Weyl suggested that the disagreements between Hilbert and Brouwer may be smaller than had been recognized. Both men, he suggested, saw the same limitations on intuitive thought and recognized that transfinite methods in mathematics provide no justification for saying that the transfinite expressions in mathematics are contentual truths.[24] Hilbert went on to give a purely formal account of what he saw as lying beyond the contentual; Brouwer refused. As Weyl put it in 1925: "With Brouwer, mathematics gains the highest intuitive clarity; his doctrine is idealism in mathematics thought through to the end. But, full of pain, the mathematician sees the greatest part of his towering theories dissolve in fog."[25] The difference between Brouwer and Hilbert, as Weyl now saw it, was that one was prepared to offer formal account that explained the consistency of mathematics, while the other was concerned with the much smaller part that could be said to be true.

Weyl's concern that Brouwerian mathematics might be wholly inadequate for the study of quantum mechanics was not merely a pragmatic criticism. Weyl was also shifting his views about what the successful practice of science entailed for the philosophy of mathematics.[26] Furthermore, as Epple has valuably pointed out,[27] Brouwer's own position was losing adherents as he spelled out its implications for mathematical analysis. His philosophy of set theory retained its attractions, and several spoke in its favor, but, significantly, his attempts to work out a truly constructive analysis did not win support.

Brouwer's constructive reinterpretation of the nonconstructive existence arguments in analysis hinge on what is sometimes called his "Fan theorem." Fortunately, we need not understand what this is to follow Epple's argument. He notes that these ideas of Brouwer were first published in 1924, by which time Weyl was already turning away. Neither Skolem nor Fraenkel, who were sympathetic to intuitionism, included these ideas in their later accounts of Brouwer's work. Only Menger, it seems, came close to the Fan theorem, and even he fell short. So Epple concludes that this was because these distinguished mathematicians and logicians found it difficult. They were, moreover, right to do so, Epple argues, because much later the logician

---

[23] See Fraenkel 1923. Fraenkel had become a professor at Marburg in 1922, where he remained until 1928. He then moved to Kiel, and in 1929 he emigrated and taught at the University of Jerusalem until 1959. He died in Jerusalem in 1965.

[24] Weyl 1924, 448.

[25] See Mancosu 1998, 136.

[26] See Scholz 2006 and the references there.

[27] Epple 2000.

Kleene showed that Brouwer, without clearly realizing it himself, was postulating a new formulation of intuitionistic mathematics and the introduction of a new proof scheme when he discussed the Fan theorem.[28] Promoting a new proof scheme in mathematics without realizing it was not a consequence of existing positions is seldom likely to succeed.

What prevents us from dismissing Hilbert's formalist account as a mere game, no more interesting than chess, and allows mathematics to remain a serious cultural concern, is the possibility of attaching some sense to Hilbert's game with formulas. Here, Weyl saw only one possibility: physics. The new theoretical physics offered a completely different kind of knowledge from the usual knowledge that expresses what is given in intuition. It is the theory as a whole that confronts experience and permits one to represent the transcendent, or so Weyl put it. By this he may be taken to have meant that the whole apparatus of theoretical physics permits us to speak of novel and elusive processes (in a language of extraordinary mathematics that would otherwise be a prime candidate for neglect) and yet plausibly claim that our descriptions make good sense. The process of symbolic construction, the construction, so to say, of theory by the manipulation of symbols, does not yield the kind of knowledge that we get from immediate experience, but it is not wholly arbitrary, either. The first task of mathematics is to make sure that this process of symbolic construction is self-consistent. Weyl suggested the kind of knowledge this process produces might be called belief, akin to "the belief in the reality of the own I and that of others, or belief in the reality of the external world, or belief in the reality of God." Its organ is not the eye but creativity, and it is "undeniable that there is a theoretical need . . . with a creative urge directed upon the symbolic representation of the transcendent, which demands to be satisfied."[29]

In 1927 Abraham Fraenkel gave a series of ten lectures to the Kant Society, an important gathering of German philosophers.[30] One of these lectures was on intuitionism, and there he gave as its central feature the sharp distinction between constructions and pure existence statements. But despite his own measured assessment of intuitionist and finitistic mathematics, he held out little hope of a reconciliation between the two schools because their opposition was essentially dogmatic. In this, he was both prescient and well informed, for in a paper of 1925 Hilbert had, if anything, moved toward an even stronger emphasis on mathematics as the activity that produces formulas according to rules. In 1927 Hilbert returned to Hamburg with a further defense of his position, anchored now in some increasingly sophisticated new mathematics. Weyl was in the audience, and his conciliatory remarks are interesting, for by now he was back in a harmonious relation with Hilbert, driven there by his growing feeling that Brouwer's mathematics would not give mathematicians and physicists what they need.

Weyl noted that there had been quite some sympathy for Brouwer's ideas, despite what Hilbert saw as their destructive implications, and this was because Brouwer talked about mathematics the way mathematicians always had until recently: mathematical statements were meaningful. It was Hilbert who had changed the nature of the enterprise, for his proof theory sought to save classical mathematics by

---

[28] For the details, see Epple 2000, 165, and Kleene 1956.
[29] Weyl 1925–27, quoted in Mancosu 1998, 140.
[30] Later published as Fraenkel 1927.

turning mathematics into bodies of consistent statements. For himself, Weyl found no epistemological gap between Brouwer and Hilbert. He even went so far as to claim that an intuitionist would accept proofs using ideal elements (Brouwer never agreed), and that if, as he said now seemed likely, Hilbert's views overcame intuitionism, it would be a decisive defeat for pure phenomenology.

It is sometimes claimed, usually by those strongly drawn to intuitionism, that Hilbert's finitism is a large concession to intuitionism in that it agrees that new, special rules must be used when reasoning about infinite sets. And it is surely likely that Hilbert intended it as something of a compromise designed to bring waverers over to his side (did he hope to lure Weyl in this way?). But it is also true that Hilbert knew very well that something had to be done to ensure that conclusions about infinite sets were always properly drawn, and he never contemplated restricting the scope of mathematics, as Brouwer always did. He never wavered in his belief that consistency was all that was needed to guarantee mathematical existence, and, if anything, finitism is a sharpening of that position, because all questions of meanings behind the signs are now excluded or consigned to meta-mathematics.

There are many other interesting aspects of the intuitionist debate in the 1920s that cannot be explored here but must be briefly noted. One concerns how the claim of constructibility can be made precise: Is there a useful distinction between constructible and constructed? Another concerns the intuitionist idea of negation. If there is a third category, distinct from the true and the false, is it to be thought of as the unproved, or the absurd, or in yet some other way? These and other aspects can be pursued in Hesseling's book *Gnomes in the Fog*; here we may proceed to the final confrontation between Hilbert and Brouwer in 1928.

In 1928 Brouwer gave two lectures in Vienna at Menger's invitation. They are well remembered today, perhaps because Wittgenstein and Gödel were in the audience, but they seem to have been a popular success. Also in 1928, the International Congress of Mathematicians in Bologna was the first since the First World War to invite German mathematicians. This had taken quite some negotiations behind the scenes, but Brouwer, a Dutchman, found the terms insulting and tried to organize a boycott. He had some success, bringing on board Bieberbach, who later tried to make his career as a Nazi mathematician. This was the last straw for Hilbert. He was recovering from his pernicious anemia, which might well have killed him in 1925 had it not been for the timely discovery of a cure, and he was more aware than ever that the issue of the future editors of *Mathematische Annalen* still had to be settled. Hilbert, whose motto always had been that there is no "ignorabimus" (we shall not know) in mathematics, whose greatest of many claims to fame was as a solver of deep problems, could not endure the idea that the journal (his journal) would pass to his opponent, the radical exponent of the idea that there are many things even in mathematics that we cannot know. He cast around for a way to dismiss Brouwer as an editor, and eventually wrote to the board, Brouwer included, to say that he was dismissed because of the fundamental incompatibility of their views.

Hilbert did not have the power to take such a decision on his own, and he had to seek the agreement of the other editors. Einstein found the whole business rather ridiculous, a "frog and mouse" battle, he called it.[31] Carathéodory also did not agree.

---

[31] The allusion is to a Homeric fable.

But Brouwer made the mistake of offering to accept only if Hilbert's doctor would sign a letter saying that Hilbert was of unsound mind. What someone once so admired by Weyl could have hoped to gain by such an insulting suggestion is not clear, but it cost him the support of the fourth and final editor, Blumenthal. This whole situation had to be resolved, and it was done so by formally dissolving and reconstituting the editorial board. Carathéodory and Einstein refused to join it.

Brouwer was devastated, and retired from public intellectual life. He was not the first or the last pure figure of relentless will to crumble at the dirtiness and hurly-burly of the world, but his departure took the drama and the philosophical knife out of the debate. What remained was for the various protagonists to find that there had been nothing much to disagree about after all, and that is what they did. The new spirit of consensus was most visible at the second Conference on the Epistemology of the Exact Sciences, held in Königsberg in September 1930, partly organized by supporters of the Vienna circle. The conference papers and the discussion that followed were then published in the circle's journal *Erkenntnis*, edited by Hans Reichenbach and Rudolf Carnap. The organizers had deliberately sought out younger speakers to present papers: von Neumann spoke on formalism, Carnap on logicism, and Arend Heyting (a former student of Brouwer's) on intuitionism. All three agreed that their form of mathematics was constructive: Hilbertian formalism because its proofs are finite, logicism because it constructs the real numbers out of logical ideas, intuitionism because a proof is a construction (or it is not a proof at all). As for the existence of mathematical objects, Heyting was indifferent. The only properties of an object that we could speak about were those we could claim to know, and transcendental claims to knowledge were, for him, inadmissible. Nor, however, were they asserted by finitism, either. Von Neumann spoke of the old arguments now resulting in more and more unambiguous problems for investigation that were not matters of taste.[32]

This amicable, and arguably productive, disagreement was partly possible because Heyting, the leading intuitionist after Brouwer's retirement, had made it possible to see formal ways of resolving the issue by turning it from a contentious question of philosophy to a simple matter of logic. In 1928 he had won a competition of the Dutch Mathematical Society with an essay on the theme of formalizing Brouwer's set theory. News of it spread before it was published, Bernays, for example, heard of it at the Bologna conference that year, but it briefly fell foul of the frog and mouse battle when Brouwer, with Heyting's consent, withdrew it from the *Mathematische Annalen*. It was finally published in a Berlin journal.[33]

Heyting opened the paper with a piece of Brouwerian orthodoxy: a defense of mathematics as an activity of thought to which language was secondary. He then spelled out in formal terms the rules of reasoning used in intuitionism. This was not the first time it had been done. The brilliant young Russian mathematician Kolmogorov had given one such treatment in 1925, but it did not cover all the logical rules. But this time, and for whatever reason, most of Heyting's readers engaged much less with the rhetoric than the symbolism. What they took away from the paper was clear statements of the rules of deduction that apply to both classical and intuitionistic logic, of the rules that apply only in intuitionistic logic, and of the classically valid

---

[32] For a review of the three "isms" and later developments, see Detlefsen 1997 and Mehlberg 2002.
[33] Heyting 1930.

deductions that cannot be made in the intuitionistic setting. His account was endorsed by Kolmogorov when he passed through Göttingen early in 1931. The immediate and palpable consequence of Heyting's work was that it was now possible to behave as an impeccable Hilbertian finitist while honoring Brouwer by using the intuitionists' rules. This seems to have been what happened, somewhat to Heyting's chagrin, as he later observed, remarking of his papers that they had "diverted attention from the underlying ideas to the formal system itself."[34]

7.1.4 Axiomatic Set Theory

The axiomatization of set theory proposed by Zermelo in 1908 was inadequate in two distinct ways. One was that it relied on an imprecise concept, that of a set being "definite" (see above, §4.7.6), which Poincaré was not the only one to find fault with. At best it pointed the way to making set theory rigorous while keeping out the paradoxes, but it could not stand close examination. The other objection was more technical. In his urge to keep out candidates for sets that were somehow too large, Zermelo had excluded objects that ought to be sets. For example, if given a set X we write $P(X)$ for the set of all subsets of X, then starting with $\mathbb{N}$, the set of natural numbers, the collection $\{\mathbb{N}, P(\mathbb{N}), P(P(\mathbb{N})), \ldots\}$ is not a set.

It fell to Abraham Fraenkel to revise Zermelo's axiomatization in ways that later generations found acceptable. Hermann Weyl had already taken up the problem of the indefinite nature of the "definite" concept. His final idea, in 1917, was to formulate the insight that a property was "definite" if it could be deduced from expressions of the form $x \in y$ or $x = y$ by finitely many uses of the logical rules for negation, conjunction, disjunction, existential quantification, and substitution of a constant for a variable. The problem was that he wanted to say what "finitely many" meant without previously defining the natural numbers—after all, the natural numbers are to be constructed out of sets, so the rules for set theory surely have to be specified first—and this made his theory very complicated. So much so that he decided to reverse the order and construct set theory out of the natural numbers, a policy that soon pushed him toward intuitionism.

Fraenkel was not influenced by Weyl. He took to set theory on the simple grounds that he believed it would form the foundations of mathematics and yet not many people knew it, so it was easy to become a specialist. In 1921 he wrote to Zermelo with a question about the object $\{\mathbb{N}, P(\mathbb{N}), P(P(\mathbb{N})), \ldots\}$, pointing out that it does not seem to follow from Zermelo's axioms that it is a set, but if it is not then one could not prove that the ordinal $\aleph_\omega$ exists (because it is not provably a set). Fraenkel had a poor opinion of logic and preferred to revise the axioms for sets, proposing in 1922 what he called an "axiom of replacement" that allowed a collection to be a set if it is cooked up from the empty set—any set X on Zermelo's definition—and the natural numbers by finitely many uses of set-theoretic union, taking power sets, and forming unordered pairs. This rag-bag definition, however, not only failed to allow $\{\mathbb{N}, P(\mathbb{N}), P(P(\mathbb{N})), \ldots\}$ to be a set, but von Neumann showed in 1928 that in fact this formulation of replacement was a theorem in Zermelo's set theory. Von Neumann,

---

[34] Heyting 1978, 15.

therefore, called for, and provided, a stronger axiom of replacement. This paper also gave a clear criterion for when a collection of objects is too large to be a set.[35]

The work of Fraenkel and von Neumann, welcomed by Hilbert (1929, 228) "as an awakening, as a luminous aurora," brought Zermelo back to the subject. In 1929 he gave an axiomatization of the contentious "definite" property that avoided the notion of natural number, and in 1930 he shifted his position only slightly to give a system of seven axioms for set theory. Three were taken almost verbatim from his original presentation of 1908, one derived from a paper of Fraenkel's, one from his own paper of 1929, one ruled out infinite descending sequences of elements $(X_1 \ni X_2 \ni X_3 \ni \ldots)$, and one was a modified form of Fraenkel's axiom of replacement. He dropped the axiom of infinity on the grounds that it did not belong to the general theory of sets, and he dropped the axiom of choice.

With these proposals, all the crucial ingredients for one of the standard versions of modern mathematics were in place except one, but that one was among the most important, and the most radical for the modernization of mathematics. For one might say that Zermelo's improved axiomatization was just that—an improvement. The decisive intervention came from elsewhere.

For most of the 1920s the Norwegian mathematician Thoralf Skolem had been thinking about logic. The question that most detained him was quantifiers, the very topic that had inspired the move away from commonsense logic and into its study by Boole, Schröder, and others. His attention was caught by a result of the Polish logician Leopold Löwenheim from 1915, which he reworked in 1920. Löwenheim's result is almost paradoxical. It says that if a theory allows quantification only over individuals but not over sets, then any sentence that is true over every finite domain but not over some infinite domain is in fact false over some denumerable domain. Skolem turned this argument into the theorem that such a sentence (or even countably many such sentences taken together) that has a model has a countable model.[36]

All this may seem technical, as indeed the proofs are, but Skolem was quick to draw a highly nontrivial conclusion: although Zermelo's theory created uncountably infinite sets, it could be satisfied by a countable set. Thus the idea that a given set was countable or uncountable depended on what model of set theory one was using—a disturbing suggestion known ever since as Skolem's paradox.

Skolem was the first to be clear that when talking about what one can logically do, one had to say if quantification took place only over individuals, or if quantifying over sets of individuals was allowed. The first case is today called first-order logic, the second case second-order logic.[37] First-order logic meets all the requirements one naturally asks of logic.[38] Second-order logic can strike one as rather artificial, and a

---

[35] There is an unresolved puzzle about the publication of von Neumann's essay as late as 1928, because it was preceded by a more elementary exposition of its themes in 1925. This was written at the suggestion of Fraenkel, who had gotten to know von Neumann in 1922 or 1923 when he was sent a paper of von Neumann's to review. The paper eventually became von Neumann's doctoral dissertation and the 1928 paper. Fraenkel records that perhaps he did not understand everything, "but enough to see that it was an outstanding work and to recognize *ex ungue leonem* [the lion by his paw]." Quoted in Moore 1982, 264.

[36] May the experts forgive me for my imprecision—it would serve no purpose here to put it right— and follow the accounts in Ferreirós 2001 and Moore 1982, upon which this account is based.

[37] There are higher-order logics, too, allowing one to quantify over sets of sets of . . . .

[38] For a historical account of how first-order logic became a core system of modern logic, and some philosophical criticisms of this outcome, see Ferreirós 2001.

source of mischief. But: theories constructed just using first-order logic turn out to be highly counterintuitive, as Skolem's paradox shows, and models of these theories are far from being unique as a result, so the theories are not categorical. They do not define their objects. Theories constructed with second-order logic are typically (but not always) categorical; they define their objects (up to some harmless concept of isomorphism). If you think you have produced a clean, tidy set theory using second-order logic, your work is open to the objection that perhaps second-order logic is too hospitable and the real difficulties you thought you were resolving in mathematics, and in particular in set theory, have simply relocated to your system of logic.

Skolem naturally pushed for the default logic used to construct any version of set theory to be first-order logic. This meant that such a set theory could not be categorical, a conclusion that Zermelo opposed. His formulation of set theory was essentially second-order, for that reason, and the exclusion of infinite descending sequences was made to make it even more likely that the resulting set theory was categorical. In the end, even this was seen to fail, and to this day logicians and set theorists study different set theories (they can be told apart by various assumptions involving very large cardinals) and worry which one ought to be true if mathematics is meaningfully about sets. What most mathematics, and most of the mathematics used in science, requires is the axioms Zermelo proposed in 1930, together with the axiom of choice and an honest recognition that the underlying logic is second-order. Perhaps rather selectively, this theory is known to all mathematicians as Zermelo-Fraenkel with Choice (ZFC) in honor of the contributions of both men.[39] Of all the fruits of modernism in mathematics, it is probably fair to say that the least known one, even to mathematicians, who naturally incline to Zermelo-Fraenkel, is Skolem's fundamental distinction between levels of logic.

Skolem drew a further consequence from the axiomatization of set theory, which marks a further unexpected consummation of the modernist thrust, its successes and failures. As he put it in his address of 1922: "When founded in such an axiomatic way, set theory cannot remain a privileged logical theory; it is then placed on the same level as other axiomatic theories."[40] As Ferreirós comments: "The basic relationship of membership ceases to have a fixed meaning, it can be interpreted at will in models of axiomatic set theory."[41] Axiomatization may solve certain problems, but it cannot privilege certain interpretations (models, or meanings) over others, even those that led to Skolem's paradox showing that even set theory is relative.

Skolem's paradox has a possibly unexpected bearing on the mathematical definition of the natural numbers. These were most conveniently presented, following Dedekind, in the form of Peano's five axioms (see above, §3.4.4). It might seem that nothing more need be said, that these axioms capture exactly what we believe and expect of the natural numbers. But this is not the case. For a start, they are not good enough to permit mathematicians to formalize statements about numbers so precisely that arguments about them can be followed mechanically. As the mathematician and logician Paul Cohen observed lucidly in 1966:

---

[39] There is an apple to this pc, the theory due to Bernays, Gödel, and von Neumann, which cannot be discussed here.

[40] Skolem 1922 in van Heijenoort 1967a.

[41] Ferreirós 1999, 362.

If we examine Peano's axioms for the integers, we find that they are not capable of being transcribed in a form acceptable to a computing machine. This is because the crucial axiom of induction speaks about "sets" of integers but the axioms do not give rules for forming sets nor other basic properties of sets. . . . When we do construct a formal system corresponding to Peano's axioms we shall find that the result cannot quite live up to all our expectations. This difficulty is associated with any attempt at formalization.[42]

Cohen then showed that the vague notion of a set could be replaced in a variety of ways with a precise, if infinite, list of axioms for the integers. Indeed, one could do so in several ways that captured almost the same idea. The resulting system is called *Peano arithmetic*, and it is a first-order theory. As Cohen noted, it follows from Gödel's famous incompleteness theorem that these formalizations could not be shown to form a consistent system except by methods more powerful than the system itself.

Nor do the Peano axioms capture all that we can mean by the term "natural number." The problem is the same as described by Cohen: the nonstandard or unexpected models of set theory. The structure of the natural numbers cannot be characterized uniquely in a first-order theory. Indeed, even the notion of finiteness turns out to be relative in Skolem's sense. However, if one retreats to the comfort of a second-order theory such as ZF, it is true that the finite cardinals do play the role of the natural numbers. Certainly, one can prove that for every natural number n there is a corresponding cardinal. But is every finite cardinal of the form of a natural number? ZF must remain silent, because the concept of natural number is prior to the concepts of ZF and cannot be interpreted within it. More food for the philosophers (compare §7.3.1).

7.1.5 Gödel

It remains for us to return to Königsberg in 1930 and the remarkable denouement of all the struggles to show that the conclusions of mathematics are true. In 1930 Kurt Gödel was a twenty-four-year old Austrian logician. The mathematician Hans Hahn had introduced him to the Vienna circle philosophers around Moritz Schlick, who were developing the philosophy of logical positivism, which was hostile to most forms of metaphysics and sought to make philosophy a science. This was not to Gödel's taste, but Carnap was a prominent member of the circle, and he may have encouraged Gödel to study logic. In 1929 Gödel took the propositional calculus as presented by Russell and Whitehead—and more or less as presented by Hilbert and Ackermann in their book of 1928—and showed that if a sentence in this calculus is such that its negation cannot be interpreted as a true statement, then the sentence can be proved.

This may sound rather technical, but it becomes much more interesting if it is teased apart. The claim is that a certain theory makes statements, and you are to take any one of these statements and consider if its negation can be said to be true. What does this mean? For a statement to be true, there is an interpretation of the axioms of the theory and the terms in the statement which simply makes the statement true. It might be that the axioms defining the theory can always be interpreted as laying down the rules for plane geometry, and the statement can be interpreted to say that two distinct points determine a line. Let us agree that there is no geometry in which

---

[42] Cohen 1966, 3–4.

FIGURE 7.1. Kurt Gödel. Courtesy of Princeton University Library
from the Kurt Gödel Papers on deposit from the Institute
for Advanced Study.

the negation of this statement (which would say that two points do not determine a line) is true. Then, says Gödel's theorem, the statement has a proof. That is to say, the formal statement has a proof derived solely from the axioms of the theory according to some agreed logical rules of deduction. Knowledge that the statement is always true implies that there is a proof.

Now suppose instead that some other statement could be interpreted to mean that two distinct lines determine a point. There is a geometry, indeed plane Euclidean geometry, in which the negation of this statement is true: two lines do not determine a point if they are parallel. So this formal statement does not have a proof.

Gödel's theorem said about the propositional calculus that there was an intimate connection between being true on all interpretations and being provable. Being true is a matter of interpretation, about the existence of models, about semantics. Being provable is a matter of juggling with symbols, writing down deductions, in short syntax. Gödel's theorem was one of the first to clarify and explore the relationship between the semantics and the syntax of a mathematical theory. It said that in this case every statement that was true under every semantical interpretation was also syntactically provable. The theory is therefore said to be complete.

Hilbert wanted every true statement in mathematics to be provable from some agreed axiomatized starting point. On September 7, 1930, the day before Hilbert was scheduled to speak at the Königsberg conference and to be made an honorary citizen of his native city, Gödel gave a paper there in which he proved that no mathematical theory rich enough to include arithmetic was complete.[43] More precisely, Gödel showed that if P_k is any consistent axiom system containing an axiom system, P, adequate to produce elementary arithmetic and perhaps some other axioms, then the statement that expresses the consistency of P_k in the symbolism of P cannot be proved in P_k.[44] In the published version he then explicitly remarked that "I wish to note expressly that [this theorem does] not contradict Hilbert's formalistic viewpoint. For this viewpoint presupposes only the existence of a consistency proof in which nothing but finitary means of proof is used, and it is conceivable that there exist finitary proofs that cannot be expressed in the formalism of P.[45]

Gödel's acute comment has not born fruit. No one has given Hilbert's vision a lease of life this way.[46] Von Neumann grasped the implications very quickly, and for this reason was reluctant to see the three-way discussion he had had with Carnap and Heyting published, feeling that Gödel's incompleteness theorem made it all rather irrelevant. Bernays, too, was soon convinced (Hilbert's own reaction is much harder to determine). Gradually, the feeling that there will always be true statements in mathematics that are not provable has come to feel like a liberation for mathematics from the iron claw of mechanical proof, but the existence of a gap between true and provable in mathematics has been the end of almost every nineteenth-century philosophy of mathematics.[47]

Perhaps the simplest messages to take away to any philosophical discussion to which Gödel's work might apply are these.[48]

1. Gödel's first incompleteness theorem proves that a large consistent system can make statements that cannot be proved or disproved within that system—but they may be trivially provable in a larger theory.
2. Gödel's second incompleteness theorem shows that large consistent systems cannot be proved to be consistent within the system itself—but this says

[43] The stipulation is necessary, for Mojżesz (Moses) Presburger had shown in 1928 that a formal system capable of handling addition but not multiplication is provably consistent. To quote from Feferman 2006, 435, n. 1, "Presburger's work was carried out as an 'exercise' in a seminar at the University of Warsaw run by Alfred Tarski." See Zygmunt 1991 for a short biography of Presburger, and Presburger 1991 for an English translation of his work.

[44] See Gödel, "On Formally Undecidable Propositions," in Gödel 1986.

[45] See Gödel 1986, 195.

[46] The matter is subtle and unresolved, not least because Hilbert never precisely formulated what "finitary" means. For a recent survey of the issue, see Richard Zach, "Hilbert's Program," in The Stanford Encyclopedia of Philosophy (Fall 2003 Edition), ed. Edward N. Zalta, http://plato.stanford.edu/archives/fall2003/entries/hilbert-program/.

[47] Among the many good recent guides to Gödel's ideas are the essays by Davis, Floyd and Kanamori, Sigmund, Feferman, Franzén, Dawson, and Kennedy in the Notices of the American Mathematical Society 53.4 (April 2006) and Franzén 2005.

[48] I take these from Franzén 2005 and the Raatikainen 2007 review.

nothing about their consistency as seen from another system and nothing at all about absolute consistency independent of any system whatever.

3. Gödel's incompleteness theorems apply to consistent systems, they do not apply to systems not known to be consistent.

4. The systems ZF and ZFC are large, indeed more than sufficient to derive all of ordinary mathematics. So there are arithmetical truths not provable within ZF or ZFC, but curiously no "interesting" arithmetical truth has been shown to be of such a kind, only artificial examples constructed for the purpose of being unprovable.

5. So mathematics (for the moment defined as the system ZF) cannot be shown to be complete except by systems presumably even harder to accept than ZF, so we cannot force anyone who doubts ZF to accept it—but we also have no compelling reason to doubt it.

6. Finally, Gödel's theorems cannot tell us that our attempts to graft a formal system onto our analysis of other philosophical questions, be they in physics, brain behavior, or theology, will necessarily fail to be complete, only that if the formal system is large then it is incomplete.

7.1.5.1 CODA

Even this account barely hints at the sophistication of logic by the 1930s. Improvements in Russell's type theory, notably simple type theory, have been bypassed altogether. It has not considered the remarkable work of the Polish school of logicians, from Lukasiewicz's three-valued logic to Tarski's theory of truth in formal languages. Nor is there space to describe Gerhard Gentzen's proof that arithmetic is consistent, which he published in 1935 and which does not contradict Gödel's theorem because it uses much stronger methods (a necessary minimum of transfinite induction). Gödel's incompleteness theorem did not put an end to logical and philosophical investigations into the foundations of mathematics. But it is enough for the purposes of this book to note that the movement of mathematical modernism transformed our ideas of logic out of all recognition. From organized commonsense, logic became a many-layered thing; issues of syntax and semantics were clarified and novelties on both sides began to flourish; rigor, truth, and meaning became more precise as concepts. And it must be said that mathematical modernism made the grounds for accepting mathematics contentious, but it provided no resting point. None of the prominent schools that sprang up to explain mathematics succeeded; all remain as noble fragments.

7.2 Mathematics and the Mechanization of Thought

7.2.1 Can Computers Think?

The idea of machines that can think has a long history. By the time Boole proposed his *Laws of Thought*, with the implied promise that thought could be mechanized, there was already a very detailed plan for a computer. This was Charles Babbage's analytical engine, a vast and vastly ambitious machine that stretched precision engineering to its limits, and about which Ada Lovelace rhapsodized in these terms: "There is no finite line of demarcation which limits the powers of the Analytical Engine. These powers are co-extensive with our knowledge of the laws of analysis itself," although she did later note that "it is desirable to guard against the possibility of exaggerated ideas that might arise as to the powers of the Analytical Engine."[49]

If the implication of her remarks was that the machine could think anything that we can think, she was to be strikingly wrong, for notoriously the machine proved impossible to make. But there is nonetheless something impressive about the reconstructed piece of it now in the Science Museum, London, and had it succeeded it would have done something valuable. It would have printed out tables of values of functions defined by specific rules, to a specified level of accuracy, with no mistakes in the calculation and none in the printing, either. The more puzzling question is whether this would or should count as thinking at all.

Consider just one of the questions this, or a similar machine, could answer: find $\sin(1°)$. It is unlikely that anyone not trained in mathematics would know how to do this, and many a mathematics student today would stumble their way toward the answer. To be sure, that is partly because many branches of mathematics are not concerned with this or similar questions, and all users of mathematics simply ask their nearest computer for the answer without even wondering how the machine on their desk actually finds it. But it would be a nontrivial job to teach the relevant mathematics, and successful students might take a few minutes to find the answer under exam conditions, just as they would with any question requiring thought. On the other hand, once the theory is worked out, the calculation is entirely routine. The hard work of thinking goes into finding routines that are quick and work uniformly for a large range of inputs. The actual computation is so little evidence of thought that exactly this task was delegated by the mathematician de Prony during the French Revolution to teams of unemployed hairdressers.[50]

This example illustrates what years of experience have demonstrated working with computers: there is a difference between working very fast and accurately according to a set of rules, and the way people think. At the start of the twenty-first century it seems that how humans think and how machines operate are very different indeed. That said, we may indeed be years away from building machines that can genuinely do philosophy, but if mathematics is, as Hilbert suggested, nothing more than the production of formulas according to rules, are we far away from producing machines that would do mathematics? One way in to this problem is to return to Borel's argument in his "La logique et l'intuition en mathématiques' of 1907 (see §4.2.1 above). There he observed that a machine set simply to produce provable formulas might well grind out all the algebraic formulas it can, but it cannot focus attention on the interesting ones. Indeed, not only might the formulas relevant to a particular mathematical problem be generated as numbers 344, 1027, 90112, and so on at random, but they may not come out in anything resembling the "right" order. In which case, one could never be sure one had them all. Or it might be that the steps needed to generate many interesting results were themselves so numerous that one would have to wait many lifetimes for them to arrive. Once one looks for shortcuts,

[49] Lovelace 1843.
[50] Wig manufacture had gone out of fashion, for obvious reasons.

directed at producing all and only the worthwhile formulas, one has presumably left this kind of mechanical work behind.

Boole's *Laws of Thought* was aimed at a different kind of mental activity, one more akin to human thinking. Fictional detectives excel at deductions of the kind: it cannot be the butler because he is right-handed and the murderer was left-handed, and if it was Jane then John would have not met her in London at 6 P.M. that evening, and therefore. . . . If there was a complete list of all eventualities, then, as Sherlock Holmes said, "When you have eliminated the impossible, whatever remains, however improbable, must be the truth."[51] Computer-aided statistical searches generate opportunities for sophisticated versions of this form of reasoning. Oddly enough, humans are not very good at it, and in order to speed the process up, Jevons invented a mechanical contraption, sometimes called his logical piano. This did enable people to tackle questions of this kind and those involving four terms faster than ever before, but it was clumsy and disparaged by Venn, who invented the Venn diagrams to solve the problem.[52]

More interesting for present purposes was the machine invented by Allan Marquand, who was among the students Peirce had arranged to contribute to the *Studies*. His father, Henry Gurdon Marquand, was a New York banker and one of the founders and a chief benefactor of the Metropolitan Museum of Art, and Allan Marquand was eventually to find his metier as a professor of fine arts at Princeton University, where he taught for forty years. As a young man he had been both an athlete and a scholar, with a degree from Princeton. He then studied theology for three years at the Princeton Seminary and at the Union Seminary in New York, before going to the University of Berlin for a year and then on to Johns Hopkins, where he received his Ph.D. in philosophy.

Like Jevons, Marquand worked on logic diagrams, and like Jevons found them cumbersome and set to work to automate logical deduction. While at Johns Hopkins, Marquand (1883) published two essays in the *Studies*, one displaying his immersion in classical scholarship, "The Logic of the Epicureans," and the other the work that interests us here: "A Machine for Producing Syllogistic Variations." This generates all eight forms in which a given syllogism can be written, provided it is put in the correct form. It was followed by an essay entitled "A New Logical Machine" in 1885, when Marquand was thirty-two, in which the sixteen possible combinations of four terms could be analyzed.[53] Peirce praised it in the first volume of the *American Journal of Psychology* in 1887 as "a vastly more clear-headed contrivance than that of Mr Jevons" because it used Mitchell's presentation of the syllogism throughout.[54] This seems reasonable. Baldwin, in the article he wrote in the *Dictionary of Philosophy and Psychology* on logic machines, observed that "for problems of ten terms Venn would require a new diagram of complicated form, and 1,024 keys to operate the instrument. Jevons for a ten-term machine would require 10,240 letters for his combinations, and a key-board with forty-four keys. Marquand's machine for ten terms needs only 124 letters and twenty-two keys."

[51] See Doyle 1891.
[52] See Edwards 2004.
[53] See Gardner 1983, chap. 6.
[54] Quoted in Gardner 1983, 109.

In his book Gardner tells us that his inquiries about this machine led to a search for it in Princeton University and it was eventually discovered in the stacks of the university library. Today it is preserved and on display in the Firestone Library.

Baldwin also wrote that "in 1882 Marquand constructed from an ordinary hotel annunciator another machine in which all the combinations are visible at the outset, and the inconsistent combinations are concealed from view as the premises are impressed upon the keys. He also had designs made by means of which the same operations could be accomplished by means of electro-magnets." The logician Alonzo Church found the circuit design for this machine in Marquand's papers in 1953—the first known design of this kind.[55]

7.2.2 Hilbert

In the 1920s, when Hilbert was developing his proof theory, he was not simply concerned to show that mathematics was correct. He wanted other proof-theoretic matters settled at the same time. In his "On the Infinite" (Hilbert 1926), he called for a proof that every problem can be solved. Not being a constructivist, he did not consider that proof theory would provide a general method for solving every mathematical problem; rather, he hoped that proof theory could show that the assumption that every problem has a solution is consistent. In 1928 at the International Congress in Bologna he spelled out three problems: Was mathematics complete—can every properly formulated statement be proved or disproved? Is it consistent—is it impossible to prove a false statement or derive a contradiction? And is mathematics decidable? This initially meant: given a correctly formulated sentence, can it be decided if the sentence is true?[56] Once Gödel's theorem separated the concepts of true and proved, this problem became: Given a correctly formulated sentence, can it be decided if the sentence is provable? In this form it became known as the "decision problem."

Gödel's incompleteness theorem showed that the known formalized systems of mathematics are either inconsistent or incomplete, and clearly incompleteness is to be preferred. The decision problem remained open until 1936, when it too was shown to have a negative answer. The solution was the independent work of three logicians: Alonzo Church, Alan Turing, and Emil Post (whose contribution cannot be discussed here). Church's approach is more abstract, but it will help to highlight where exactly in Turing's work the human comes in, and so to bring out the implications for human thought, an ongoing interest of Turing's. Indeed, it is worth stressing that ideas about what is computable proceeded in this period from what an idealized human being might do, given arbitrarily many pencils and sheets of paper and an arbitrarily long life.[57]

Church introduced what he called the "lambda calculus" and the concept of a lambda-definable function.[58] Independently, Gödel and the French logician Her-

---

[55] Gardner 1983, 112, tells us that the design is reproduced in *Science* 118: 281.
[56] Hilbert 1929.
[57] And, let us hope, a lot of time off.
[58] See Gandy 1995.

brand introduced the concept of a recursive function, and Church and Kleene showed that these concepts pick out the same functions. To say that a function is recursive is to say something entirely precise (what, more precisely, will be indicated below). Church was asked by Gödel what the relationship was between effectively calculable and recursive functions, where "effectively calculable" is a vague term that means something like "a function that can occur in mathematics." Church's answer was that any function that is effectively calculable by a machine of finite size is recursive. What is known as Church's thesis is therefore the philosophical thesis that the concept of "effectively calculable" is caught exactly by the mathematically precise term recursive. Post regarded it as "a fundamental discovery in the limitations of the mathematicizing power of Homo Sapiens."[59]

Alan Turing's idea was much more down to earth. He imagined a machine capable of carrying out computations unaided. It would be fed an arbitrarily long tape divided into squares. At each stage it would read the information on the square in front of it. If the square was blank, the machine could write a symbol in the square; if the square had a symbol, the machine could erase the symbol in that square. In either case, the machine would then move one place (left or right) on the tape or else stay put. What determined what the machine would do was called its configuration, and the machine could go from one configuration to another, so a typical action of the machine would be something like: "Write a 1 in this square, go one step to the left, enter configuration 6." The configurations (of which there are a finite number) can be thought of as its program or rule book. So there would be a machine for each task: adding numbers, for example, or multiplying them, or looking up the result in a table of already computed numbers. On the view that any mathematical activity can be broken down into finitely many simple activities of this kind, there would be a machine for every mathematical activity. So Hilbert's question became: Given a problem, is there a machine that solves it?

To prove that there was not, Turing focused on one particular task: computing numbers, such as the decimal expansion of π. Certainly there is a machine that will produce arbitrary many numbers in the sequence 1, 4, 1, 5, 9. . . . Turing considered what he called the computable numbers, the set of all numbers, each of which has the property that it can be written down to arbitrary levels of accuracy. It includes all rational numbers because they have repeating decimals, and it includes all the numbers specified by a suitable rule. For convenience, consider just those computable numbers that lie between 0 and 1. Suppose these numbers form a countable set, and enumerate them. Ask the machine that computes the first computable number to compute the first decimal place of that number. Ask the machine that computes the second computable number to compute the second decimal place of that number. And so on: ask the machine that computes the nth computable number to compute the nth decimal place of that number. Now read down the numbers thus generated and change them according to a rule, such as: replace each number which is not a 5 by a 5 and replace 5 by a 3. This gives you a number that is not produced by any machine—for if it were, it would have been produced by some machine, say the nth, but it differs from the number produced by the nth machine in the nth decimal place. Therefore, the computable numbers form an uncountable set.

---

[59] Quoted in Gandy 1995, 77.

But this is impossible, because the set of machines is countable. Each machine is specified by a finite set of configurations it can have, and each configuration is specified by a finite set of instructions, each written in a finite set of words. Each machine, therefore, is specified by a finite set of words. They can therefore be enumerated by writing down the rule book for each machine and assembling the rule books in alphabetical order.[60] So the set of machines is countable, and since each machine computes a computable number, the set of computable numbers is countable.

So the idea of computable numbers led Turing to a contradiction, and either it must in some way be incoherent or else the use of these machines to produce an uncomputable number must be flawed. The only way out, he realized, was that, for some number n, the computation of the nth decimal place of the nth number might fail. To be precise, the nth machine might take indefinitely long to produce the nth number from the one before. Imagine that each machine is to be checked before it goes off to do its bit. It has to be checked mechanically, of course, if the aim is to mechanize all mathematics, so what Turing saw was that there could be no checking machine that could check that every machine intended to produce a computable number in fact did so. So Turing had a mathematical problem that could not be solved by a machine, and with it the negative answer to Hilbert's decision problem.

The precise way a machine might fail gave a negative resolution to another of Hilbert's problems, the halting problem. This asks for certainty that a given Turing machine with given input will eventually halt after an explicit number of steps (the number will depend on the machine and the input). Turing showed that there could be no such certainty.

To complete the story, what Turing also proved, when he heard of Church and Kleene's work, was that every recursive function can be expressed as a Turing machine and vice versa. Thus we have what has been called the Church-Turing thesis: that every effectively calculable function can be computed by a Turing machine.

To sum up, the work of Gödel, Church, and Turing shows that any large system of mathematics (if consistent) is incomplete—there will always be true statements that are unprovable within the system—and it is undecidable: there will always be problems that cannot be shown to be solvable within the system.[61] This would seem to say that mathematics exceeds the limits that Hilbert presumed for it, it is not to be identified with the range of mechanically produced formulas, and it is therefore, one might say, creative. It also turns out that no one since the momentous years of the 1930s has produced a function that could be called effectively computable and which falsifies the Church-Turing thesis.

This does not, however, mean that every kind of computer is limited in the way that Turing machines are, or that the human mind is so limited. Only if every kind of "computation," however broadly defined, were mechanizable would there be a convincing a priori case for the human mind being modeled on a Turing machine. Such a claim has yet to be established. As for the first claim, it is simply false. There

---

[60] Allowing for translations, this might produce: Every marriage . . . ≺ It is a truth universally . . . ≺ London, Michaelmas term . . . , and, deliberate parodies and revised editions aside, all books ever written in English might be listed in this way with entries of perhaps no more than twenty letters.

[61] Here large means, for example, capable of formalizing both addition and multiplication. There are provably complete consistent systems, such as the propositional and predicate calculi.

are machines that are not Turing machines. Typically, they are infinite in size (they have infinitely long rule books), and they can handle infinite amounts of information at once. Turing ruled them out because they offended his sense of the calculable. He argued that the human eye, for example, can only take in a finite amount of information at a time, which is how he argued his way down to squares each containing one piece of information. Infinite machines are not enumerable, but one might also argue that they cannot be built precisely because they are infinite.[62] But it is an open question as to whether a finite machine obeying the laws of physics and constructible in finite time can be built that would not be a Turing machine, and quantum computers might well be such machines.[63] Nor can the seemingly unconstructible quality of the non-Turing machines weigh too heavily on the general problem, because even a Turing machine is implausible: it requires arbitrary amounts of tape, arbitrary amounts of time to run, and is assumed never to break down. This is quite a step from a human, in Turing's words, "working in a disciplined but unintelligent manner."

One class of computers that might not be Turing machines deserves brief mention: analogue computers. Some experts think that there may well be reliable physical processes that could yield computation methods that cannot be modeled by Turing machines.

It is also worth noting that Turing himself wrote about non-Turing machines in his 1936 paper. He called them oracles, or *O*-machines. One is, to oversimplify, a Turing machine with input in the form of some noncomputable number, supplied, naturally enough, by an oracle. Indeed, as Copeland shows, because a Turing machine accepts no input once it is working (everything must be supplied on the initial input) a pair of Turing machines, one accepting input from the other, can be made into a non-Turing machine.

7.2.3 Turing and the Turing Test

Turing, as already noted, was at least as interested in how humans think as in machines, and after the Second World War he took up the theme explicitly in a paper he published in *Mind* in 1950. The Turing test, as he stated it, is so different from what it is often said to be that the opening of the paper is worth quoting in full.

> I PROPOSE to consider the question, "Can machines think?" This should begin with definitions of the meaning of the terms "machine" and "think." The definitions might be framed so as to reflect so far as possible the normal use of the words, but this attitude is dangerous. If the meaning of the words "machine" and "think" are to be found by

---

[62] Consider this simple example of a perpetual motion machine: take a line and fix a unit of length. At the points 1, 2, 3, . . . units away from the origin, put a small unit weight. When the system is released from rest, it will travel down the line away from the origin, because every weight has more points beyond it than behind it. Similar obstacles might affect the construction of non-Turing machines: the rule book could never be written down completely, and so on.

[63] See the discussion in B. Jack Copeland, "The Church-Turing Thesis," in *The Stanford Encyclopedia of Philosophy* (Fall 2002 Edition), ed. Edward N. Zalta, http://plato.stanford.edu/archives/fall2002/entries/church-turing/. For a survey, see the published version of B.J. Copeland and R. Sylvan. "Beyond the Universal Turing Machine," *Australasian Journal of Philosophy* 77 (1999): 46–66.

examining how they are commonly used it is difficult to escape the conclusion that the meaning and the answer to the question, "Can machines think?" is to be sought in a statistical survey such as a Gallup poll. But this is absurd. Instead of attempting such a definition I shall replace the question by another, which is closely related to it and is expressed in relatively unambiguous words. The new form of the problem can be described in terms of a game which we call the "imitation game." It is played with three people, a man (A), a woman (B), and an interrogator (C) who may be of either sex. The interrogator stays in a room apart from the other two. The object of the game for the interrogator is to determine which of the other two is the man and which is the woman. He knows them by labels X and Y, and at the end of the game he says either "X is A and Y is B" or "X is B and Y is A." The interrogator is allowed to put questions to A and B thus: C: Will X please tell me the length of his or her hair? Now suppose X is actually A, then A must answer. It is A's object in the game to try and cause C to make the wrong identification. His answer might therefore be "My hair is shingled, and the longest strands, are about nine inches long." In order that tones of voice may not help the interrogator the answers should be written, or better still, typewritten. The ideal arrangement is to have a teleprinter communicating between the two rooms. Alternatively the question and answers can be repeated by an intermediary. The object of the game for the third player (B) is to help the interrogator. The best strategy for her is probably to give truthful answers. She can add such things as "I am the woman, don't listen to him!" to her answers, but it will avail nothing as the man can make similar remarks. We now ask the question, "What will happen when a machine takes the part of A in this game?" Will the interrogator decide wrongly as often when the game is played like this as he does when the game is played between a man and a woman? These questions replace our original, "Can machines think?"

So there are three participants, and the interrogator (recall that Turing had spent the war decoding Enigma messages) must decide who of the other two is a man or a woman. If the interrogator does no better when a machine takes the place of the man, then Turing suggested that the machine should be considered intelligent. After a lengthy discussion of what a machine can be taken to be, and a construction, which excluded such ideas as calling a person a machine and inclined toward defining a machine as a digital computer, Turing ventured his own opinion on what might happen: "I believe that in about fifty years time it will be possible to programme computers with a storage capacity of about 10^9 to make them play the imitation game so well that an average interrogator will not have more than 70 per cent chance of making the right identification after five minutes of questioning." In this, he seems to have been wrong. It is harder than he thought to make a machine talk coherent sense about complicated real world issues. Turing gave the example of the interrogator asking A to write a sonnet and A replying "Count me out on this one. I never could write poetry." Which is fair enough, but similar declarations of modesty on too many issues would surely excite suspicion.

Turing dealt, in this paper, with various objections to the idea that the test was a fair one. He characterized what he called the mathematical objection this way. It is known (from the work of Church and Turing discussed above) that there are certain things that a digital computer with an infinite capacity cannot do. There will be some questions to which it will either give a wrong answer or fail to give an answer at all however much time is allowed for a reply. This mathematical result is said to show a disability of machines to which the human intellect is not subject. Turing was not impressed with this evidence of mechanical inadequacy:

> We too often give wrong answers to questions ourselves to be justified in being very pleased at such evidence of fallibility on the part of the machines. Further, our superiority can only be felt on such an occasion in relation to the one machine over which we have scored our petty triumph. There would be no question of triumphing simultaneously over all machines. In short, then, there might be men cleverer than any given machine, but then again there might be other machines cleverer again, and so on.

Turing then briefly considered the vexing and profound topic of consciousness, which is once again at the forefront of research, and argued that it was distinct from the question of thinking.

He then turned to one of Lady Lovelace's opinions about the Analytical Engine: that it was incapable of originality. "The Analytical Engine has no pretensions to originate anything. It can do *whatever we know how to order it* to perform" (her italics). To this, Turing noted that this did not rule out being able to construct electronic equipment that would "think for itself," as certain recent developments had seemed to suggest. He gave reasons to believe that the information store of *Encyclopaedia Britannica* would fit well into a computer the size of a child's brain, and the rest of the creation of a thinking adult brain could be done by education. His account of education says more than one might expect about the rigors of an English childhood, but gave only the vaguest of accounts of how a computer capable of original thought could be taught to do so. Most likely he did not know by then that such ideas had already been animating the other powerful mathematician whose mind had turned during the war to making computers: John von Neumann.

7.2.4 Von Neumann and Neural Networks

The Psalmist (Psalm 8, verse 4) asked: "What is man, that thou art mindful of him? and the son of man, that thou visitest him?" The Quaker Warren Sturgis McCulloch tells us in his autobiography that in 1917, while a student at Haverford College in Pennsylvania and inspired by Russell's definition of number, he hoped one day to answer the questions, "What is a number, that a man may know it, and a man, that he may know a number?"[64] He spent the 1930s studying monkey and chimpanzee brains at the Yale University Medical School. At a seminar at Yale he was reminded of the work of Russell and Whitehead and asked the speaker, F. B. Fitch, to work up the symbolic formulation of the theory of neural nets. What had caught McCulloch's attention was that a logical proposition is either true or false, and a neuron either fires or it does not. Neurons are linked in the brain in what are called neural nets, the firing of some determines the firing of others, much in the way that propositions are linked and the truth of some depends on the truth of others.

In 1941 McCulloch moved to the University of Illinois Medical School. There he befriended Walter Pitts, an eighteen-year-old runaway from Detroit who had studied logic with Carnap—it seems that Bertrand Russell may have introduced them[65]—

---

[64] On McCulloch and Pitts, see Abraham 2002 and Heims 1991 and the literature cited there in.
[65] See Abraham 2002, 12–13, quoting McCorduck 1979, 73–74, n. 1.

when only fifteen and mathematical biology with Rashevsky. Together they outlined a theory of how neural nets might model complex propositions. Their idealized neuron has some excitatory and some inhibitory inputs, a threshold voltage, and some branching output to other neurons, and to that extent it is a plausible simplification of real neurons. They showed that a large class of mental functions could be carried out by such nets.

In 1942 Pitts went to MIT to work with Norbert Wiener, who regarded Pitts as "without question the strongest young scientist I have ever met."[66] Wiener also met frequently with McCulloch and was also in contact with John von Neumann. Von Neumann quickly saw that the Pitts-McCulloch model of neural nets was a good way to think about the design of general-purpose computers.

In their published paper of 1943 McCulloch and Pitts showed how to join up neurons so that, for example, neuron C will fire if and only if neurons A and B fire, or so that neuron C will fire if and only if either neuron A or neuron B fires, and so on through the logical connectives.[67] This enabled them to write wiring diagrams and to describe in formulas nets of neurons that could carry out any logical operation. They even claimed that they had "proved, in substance, the equivalence of all general Turing machines—man-made or begotten"—although mathematical logicians disputed this claim and physiologists rightly found their model oversimplified—and they happily noted (p. 121) that now the human mind no longer "goes more ghostly than a ghost." Von Neumann, however, seems to have been attracted by the unity of the biological and man-made worlds they proposed, and took their work as the crucial insight for constructing a programmable computer.[68] As Aspray remarks of von Neumann, "He made their theory of neural nets and Turing's machines the two pillars of his theory of automata."[69]

The biological and medical communities were for a long time much less enthusiastic about the work of McCulloch and Pitts than von Neumann was. Matters began to change with the work of the Canadian neurophysiologist Donald Hebb. Neurons in the brain occur in bunches, communicating by chemical signals across the small gaps between them known as synaptic gaps. In his book *The Organization of Behavior: A Neuropsychological Theory*, Hebb attempted to analyze human behavior directly in terms of the functioning of the brain and the nervous system. He offered what he called a "general theory of behavior that attempts to bridge the gap between neurophysiology and psychology." He focused on the way neurons may change as a way to explain learning and memory. In a much-quoted passage, he wrote: "When an axon of cell A is near enough to excite B and repeatedly or persistently takes part in firing it, some growth process or metabolic change takes place in one or both cells such that A's efficiency, as one of the cells firing B, is increased."[70] He argued that these cell assemblies thus formed could constitute a memory or

---

[66] Heims 1991, 40.

[67] McCulloch and Pitts 1943.

[68] See, for example, von Neumann 1951.

[69] Aspray 1990, 181. Aspray's book gives the reader a great deal of insight into the last few years of von Neumann's life and work. Gandy 1995, 87, points out that Turing also used the idea of McCulloch-Pitts logic gates in designing computers.

[70] Hebb 1949.

thought-process. Such behaviour of what became known as the Hebb synapse was first demonstrated in the 1970s.

It turn, Hebb's observations became the basis of a great deal of research into computers that could learn. The idea was that neural networks could be described that would tackle a certain problem and could be told that they had succeeded or failed, done well or badly. They would then modify themselves, strengthen or weaken various connections, and try again, learning in this way by trial and error until they were confirmed in a successful configuration. The ups, downs, and ups of this idea are beyond this book to describe.

7.3 The Rise of Mathematical Platonism

> To be a mathematician is to be an out-and-out Platonist
> —David Mumford, in Parikh 1991, xvii

7.3.1 Working Platonists?

A widely quoted aphorism circulates among mathematicians and philosophers of mathematics these days, to the effect that mathematicians are Platonists on working days but formalists on Sundays. In other words, while they are doing mathematics they believe in the existence of mind-independent mathematical objects whose properties are responsible for the truth of their theorems, but if forced to defend mathematics in philosophical terms they would resort, perhaps insincerely, to Hilbert-style formalism. Although the real situation is much more complicated, the aphorism contains a certain amount of truth. Mathematicians do speak of their most securely established theorems as true (not merely as proved). It is as true for them that 3 is a prime number as that there are no solutions to the equation $x^n + y^n = z^n$ when x, y, z, and n are integers greater than 2. Probably most mathematicians believe that these truths derive from the properties of objects (manifolds, schemes, spaces of functions, . . .) that their carefully constructed theories so accurately describe. Some even suggest that it is real insight into these objects that allows the best of them to be truly innovative and to discover deep properties previously unknown.

A fine example of a mathematical Platonist is Alain Connes, a Fields medallist in 1982. His fascinating discussions with Jean-Pierre Changeux introduce a number of important themes and contain this declaration of his faith: "There exists, independently of the human mind, a raw and immutable mathematical reality." Challenged, he admits that this reality does not exist in space-time, and indeed is "far more stable than physical reality for not being located in space-time." The mathematician perceives it via a special sense. Changeux disagrees. As a molecular neurobiologist, he prefers a materialist viewpoint, in which mathematics is a matter of syntax and semantics, of language, and of the workings of the human brain. Connes regards all that as tools with which we deepen our understanding of mathematical reality. He thinks of himself as a realist (a term he prefers to Platonist) about mathematical objects—but then, we could note, Plato was a realist too. He urges that there is no irreconcilable conflict between materialism and realism, to which Changeux objects that, on Connes' scheme of things, it is very hard to see how mathematical knowledge can be acquired at all.[71]

Mathematicians also set a high store by formal proofs, and not only on Sundays. For example, at the start of the twentieth century, Edmund Landau would not accept a theorem until the proof has been published and scrutinized. Others to this day are more willing to trust the insights of selected colleagues, and even their own.

To a considerable extent, many, even most, mathematicians are Platonists, but there is more to be said. Many mathematicians are untroubled by any reflections of this kind. They are not philosophers, and they are largely unaware of any other philosophy of mathematics, so their default Platonism is hardly well defended. Were this kind of philosophizing to matter more to mathematicians, it might be that some would espouse other philosophies, at least on weekdays.[72] In fact, when mathematicians do reflect on the nature of their subject—when, without being so labeled, they become philosophers of mathematics—they have concerns that conventional philosophy of mathematics has largely ignored since modernism succeeded. They ask about the fruitfulness of their methods and for the "right" way to think about a problem, and they are concerned about the purity of their methods and what counts as an explanation.

Mathematicians resort to philosophy in part because the best and most widely accepted formulation of mathematics leaves room for it. The axioms of set theory, say the Zermelo-Fraenkel axioms (ZF) or Zermelo-Fraenkel with choice (ZFC), do not resolve philosophical questions for several reasons: there is no proof that ZF is consistent (not that anyone seriously doubts it); ZF gives a great deal more than elementary mathematics and not everybody is comfortable with this superabundance; there is room for discussion of the ontological question about the existence of the sets constructed in ZF or ZFC; there is room for an epistemological discussion; and finally there is also room for discussion about how, and how well, the concepts generated in ZF match what they are meant to describe or capture. Much of the confidence in the de facto consistency of ZF resides in the overwhelming charm of naive set theory. Ontological problems arise over the nature of models in axiomatic mathematics: If the models are entirely mathematical, is there not a risk of a vicious circle? If the vicious circle is avoided, where and what are the ultimate undefined terms? If nonmathematical models, whether physical or naive, are admitted, is the truth claim of mathematics ultimately no better than the truth claims of these models (contrary to most mathematicians' cherished beliefs)? Epistemological issues arise, for example, in the step from denumerable to nondenumerably infinite sets, which some (Baire, for example) found unacceptable (see §4.7.5 above).

7.3.2 Schlick's Anti-Platonism

An interesting early perspective on mathematical Platonism and its errors was provided by Moritz Schlick (1918) in his *Allgemeine Erkenntnistheorie*. Early on in this book (§5), Schlick addressed the question of what is knowledge of concepts. A concept, he said, is completely precise, and here he gave the example of modern

---

[71] Changeux and Connes 1995; the quotations are on pp. 26, 28.
[72] An example is the nominalism discussed below.

mathematics. Mathematical definitions, he pointed out in §7, are implicit definitions. Mathematicians had dispensed with appeals to any kind of intuition and had offered instead specifications of exactly what could be said of a concept by saying that the primitive concepts are defined by the axioms they satisfy. Because, as Schlick put it, the intellectual labor of science consists of inferring, a concept is nothing more than that concerning which certain judgments can be expressed. The beginner chafes at the idea that the concepts thus defined are devoid of actual content, but any set of objects satisfying the axioms is as good as any other. So axiomatic geometry is not truly a science of space, because the concepts denoted "point" and "line" have no association or connection with reality at all. Its merit is rather that it determines its concepts completely and so enables us to attain strict precision of thought.

Because a concept is completely precise, it cannot be a mental image, or a real mental structure of any sort. We operate with them through their defining characteristics, we ask ourselves if an object has the properties specified in a definition. So a concept plays the role of a sign for those objects that satisfy the relevant definitions. They are represented in thought by, for example, images, and insofar as they are handled in a nonintuitive way such thinking consists of conscious acts. Stumpf, Husserl, and Külpe, he said, deserve great credit for bringing out the role of acts in psychology; one such "act" or function is thinking of a concept, being directed toward it. It is thus the conceptual function that is real, not the concept itself.

Schlick had this to say about the nature of concepts:

Concepts are not real. They are neither real structures in the consciousness of the thinker nor are they, as medieval "realism" held, some kind of actual thing within the real object that is designated by means of them. Strictly speaking, concepts do not exist at all. What does exist is a conceptual function. And this function, depending on the circumstances, can be performed on the one hand by images or various mental acts and on the other by names or written signs. . . .

The view adopted above—that concepts do not actually exist, that to talk of concepts is simply to use a kind of shorthand, that in reality there are only conceptual functions— has encountered widespread opposition. It has been argued that entire sciences exist, such as mathematics and pure logic, whose subject matter consists exclusively of concepts and their relationships. Thus it seems that we cannot deny existence to concepts without being led to absurdities of the sort Lorenz Oken expressed so well when he said "Mathematics is based on Nothingness (Nichts) and hence arises from Nothingness." For this reason, we generally prefer to say: concepts do exist, they have a kind of being just as sense objects, for example, do; but it is an ideal being rather than a real one. Granted that the concepts of triangle, of the number five, of the syllogism, and the like have no real existence. Yet they are not just nothing, since we can make various valid statements about them. Therefore we must ascribe to them a kind of being, and this we call ideal being to distinguish it from real being. There is no objection to this form of expression, of course, so long as the question is purely one of terminology. But such talk of ideal objects leads all too easily to unclear and erroneous views, views that point in the direction of the Platonic metaphysics on which these linguistic formulations lean. Almost without noticing it, we come to counterpose to the real world another world independent of it, a world of ideal being, a realm of ideas . . . in short, a timeless world of concepts.

Schlick now argued that from this seemingly innocent use of language spread entanglements that only obscured what they were meant to clarify.

He returned to this theme (in §18) when he said that clarity about the distinction between mental activities and logical relationships had only been achieved in the feud against psychologism. There was still, however, a tendency to think of a domain of ideas that "exists" independently of the real world. "But there is scarcely a Platonizing philosopher . . . who has not been led by the doctrine to entertain views that make it quite impossible to understand the true relationship of the two realms [the conceptual and the real]. These views have the same consequence in this respect as the Platonic myth, which enthrones the Ideas as real beings . . . eternally remote from our world and inaccessible to any of our senses." To get epistemic access to these remote objects, Schlick said that "idealists of the Platonist tendency . . . resort to the same expedient to which philosophers have not infrequently had recourse in similar cases: if a proposition that is close to their hearts is not correct when words are taken in their usual sense, they construct a new sense for these words. In this way, of course, it is always possible to maintain the old proposition; but now it means something different."

He quoted Husserl, who wrote about grasping, experiencing, being aware of, in connection with this ideal being, but in an entirely different sense than in connection with an empirical or individually separate being.[73] The only sense Schlick could make of this was that what actually exists are the conceptual functions, and we do actually experience an intention toward concepts. But this, he said, was no help with understanding the nature of knowledge, because we still needed to know how intentional experience relates to ideal structures. Nor was Schlick impressed with the new science of phenomenology, but we shall see that others were to be more sympathetic.

Schlick's way out was to observe that one can reason quite precisely about concepts because we can distinguish among the definitions they are associated with. Intuitive ideas are certainly vague, but they can be distinguished from one another with absolute certainty in precisely the way that exact truths can be established by inexact experiences, and so imprecise ideas can perform fully the task of concepts.[74] We need not follow him here, but can appreciate his analysis of the Platonic temptation.

7.3.3 Bernays's Formulation

The aphorism suggests that mathematicians deal with philosophical questions about mathematics uneasily, and this is further manifest in the ill-defined term "Platonism" with its vague, unanalyzed connotations of Plato's theory of forms. In fact, mathematical Platonism, however it is defined, is a twentieth-century phenomenon and arose from the triumphs of mathematical modernism.

---

[73] Schlick gave this reference: Husserl, *Logische Untersuchungen*, vol. I, p. 128.

[74] Schlick gave the delightful example "it is not simply inaccurate but utterly nonsensical to say that a person has 2.002 ears" (p. 146).

The term "mathematical Platonism" seems to have been coined by Bernays in 1935.[75] There he defined it as a tendency to see mathematical objects as "cut off from all links with the reflecting subject." He went on that the value of this philosophy was that it furnishes "models of abstract imagination" and forms representations "which extrapolate from certain regions of experience and intuition." He illustrated what he meant by a ladder of existence claims, starting from the set of the first n integers and maps from this set to itself, to the generalization that considers the set of all integers, rising to the set of all subsets of the set of all integers, and then to the real numbers as usually defined, the axiom of choice, and finally the Cantorian hierarchy of sets. He noted that talk of the existence of the first of these objects (say, the set of the first twenty-seven integers) is entirely unproblematic, and that as and when it becomes problematic the discussion is enriched with talk about how the objects can be specified, culminating in rules for forming sets. This form of Platonism, he observed, has proved itself very successful in algebra and topology, and he concluded that "platonism reigns today in mathematics" (p. 5).

Bernays then noted that this form of Platonism had been criticized. What he called absolute Platonism—by which he meant the Platonism corresponding to naive set theory—was ruled out by the famous antinomies of naive set theory, leaving what he called a quasi-combinatorial form. Some mathematicians had, however, recoiled completely from the whole endeavor, most notably Brouwer, and Bernays located the force of Brouwer's position in its insistence that the thinking subject be given a central role in the philosophy of mathematics. A variety of modern philosophical positions have adopted the latter position, which has a respectable antecedent in the later work of Husserl, and it is surely a powerful alternative to the view that would purge philosophy ruthlessly of any "psychological" component, but is fundamentally different from any form of Platonism and will not be considered further here. Bernays himself saw Brouwerian intuitionism as the polar opposite of mathematical Platonism and spent the rest of his paper trying to outline middle ways between these two extremes. He found one—naturally enough, given the course of his own work—in the underlying ideas of Hilbert's proof theory.

Bernays's definition of mathematical Platonism locates it as being concerned with mentally accessible but mind-independent objects. These objects exist, but it would seem, although he does not say so, that they do not exist as physical objects in space-time, to adopt a term from present-day philosophers of mathematics. It is the properties of these objects that mathematicians describe when they produce theorems and announce conjectures. Note that Bernays did not explicitly say that these theorems are true because they are correct descriptions of the objects. He allowed himself to say that some statements about mathematics are true, but in this essay he did not say that the conclusions of mathematics were true—he noted only that intuitionists lay claim to speak exclusively of truth. It would be a fair interpretation of what he wrote, however, that he would have agreed that the number 3 is prime, and he may well have allowed one to say that it is true that 3 is prime, but truth is a complicated subject and we shall have to return to this point.

---

[75] Platonism in Mathematics, Bernays Project, text no. 13, trans. C. D. Parsons.

7.3.4 Platonism, Nominalism, and Fictionalism

Philosophy of mathematics conducted in the analytic tradition since Bernays wrote has a lot to say about mathematical Platonism, and this helps to clarify the position into which it is suggested many mathematicians have drifted.[76] It begins by dismissing the alternatives—mentalism and psychologism—very simply. These positions imply that a mathematical object (such as the number 5 or a straight line) might be a mental state, but my mental states cannot be yours, the argument goes, so not only are we not talking about the same thing when we each speak of the number 5, but mathematically incorrect statements may actually be true of mental states. Plainly this is no way to conduct a philosophy of mathematics, so mentalism and psychologism are dismissed. This seems entirely reasonable if the aim is to conduct philosophy in a mind-independent way, and if we accept these arguments then modern philosophy offers two intriguingly similar and intriguingly antithetical alternatives: Platonism and nominalism.[77]

Nominalism is the view that there are no abstract objects. Nominalists agree that there are occasions when one sees 3 balls, even 3 red balls, they agree that this person is a good man, but they regard it as illegitimate to infer an abstract concept behind these physical objects, such as an abstract 3, or threeness, red or redness, good or goodness. An obvious problem for nominalists is that the statement that "3 is prime" is false for them, because there is no such thing as 3 and contemporary analytic philosophy agrees that statements about nonexistent objects must be false. This makes it difficult for nominalists to distinguish between the correct claim that 3 is prime and the incorrect claim that 4 is prime, given that for them both claims are false. To deal with this problem, a variant of nominalism called fictionalism has been developed that has affinities with the later philosophy of Carnap.

Mathematical Platonists and nominalists agree, however, that philosophy of mathematics is concerned with abstract objects lying outside space-time, with what they may be said to be and with how we may know them. Platonists have to show that we can meaningfully say what they are and that we really can have access to them. Nominalists have either to stop the Platonist project in its tracks by showing that there are no suitable abstract objects, or else to show that there might be philosophical sense in the idea of abstract objects but that we cannot know them.

---

[76] This treatment follows Mark Balaguer, "Platonism in Metaphysics," in *The Stanford Encyclopedia of Philosophy* (Summer 2004 Edition), ed. Edward N. Zalta, http://plato.stanford.edu/archives/sum2004/entries/platonism/.

[77] But not just two. Among other alternatives is Feferman's revival of Weyl's predicativist theory, outlined in *Das Kontinuum* but then allowed to lapse until the 1950s. See, for example, Feferman 1998, chapter 14. The aim of Feferman's work is to give logical foundations for all of the mathematics needed in physics, and in Feferman 2006 he claims to have "verified to a considerable extent that all of currently scientifically applicable mathematics can be formalized in a system that is proof-theoretically no stronger than PA." PA, for Peano Arithmetic, is a formal system considerably weaker than ZF. My point here is much simpler: mathematicians cannot claim that Platonism must be the right philosophy for them because there is no other philosophy of mathematics—plainly there are several, and nominalism is one that is close to formalist thinking.

The strongest objection to nominalism is known as the Quine-Putnam indispensability argument.[78] They argued that if the best results of science are true, and quarks exist in atoms in much the way that elephants exist in Africa, and if mathematics is indispensable to science, then it is grievously inconsistent to be a nominalist in mathematics but not in science. So, being realists about science, they suggested that nominalism in mathematics was untenable.[79]

This argument cannot, however, establish Platonism by default, because Platonism, too, has its problems. It was famously challenged by Benacerraf to meet the challenge of contemporary epistemology, which only allowed one to have knowledge of objects one can causally interact with those, in other words, which are in space-time.[80] Since it was widely supposed that Platonists did claim direct mental access to objects outside space-time by a kind of perception and, since this ability was so little shared as to seem mysterious, mathematical Platonism is often felt to be problematic for that reason.[81]

Bernays's ladder offers one way into the world of abstract objects, which might also be called the tone of voice trick. The set of integers from 1 to 27 is something about which even children can speak reliably and correctly. It contains just these squares, for example: 1, 4, 9, 16, and 25. It seems downright odd to say that there is no such set because there are no such integers. But if such talk about these 27 integers is meaningful, say because they exist, then it doesn't seem much of a stretch to speak of a map from this set to itself, such as the map that associates to each number the sum of its prime divisors (here include 1 but not the number itself, so to 5 we associate 1 and to 6 we associate 6, for example). We could even list all these maps (there are 27 choices for each number in the set, making a total of 27^{27} in all, which is a number with 39 digits, but finite all the same). The set of all integers does not seem at all unreasonable to most people, likewise the set of maps from this set to itself, and if we have a set we surely have its subsets, and so the idea that 5 exists moves quietly into the idea that all sorts of large sets exist. If the sweet reasonableness of this argument, the commonsense tone of voice with which it is uttered, seduces you, then the natural thing to do is to look for arguments that support its conclusions. Taking that route leads some to Platonism, and some nominalists to fictionalism.

First, the nominalist objects that we begged the essential question right at the start. Just as we must know what red means before we agree that we see 3 red balls before us, so we must know what 3 is, otherwise we could not distinguish this collection of 3 balls from that collection of 4. But explaining what "red" means comes down to some statements in physics, some brute facts about nature. Similarly, the abstract concept "3" will dissolve into other brute facts.

The Platonist replies that true statements about people carry the implication that the people exist, so from the evident truth of the claim that "3 is prime" we can legitimately deduce that "3" exists. The bold response of the fictionalist nominalist is

[78] See Colyvan, Mark, "Indispensability Arguments in the Philosophy of Mathematics," in *The Stanford Encyclopedia of Philosophy* (Fall 2004 Edition), ed. Edward N. Zalta, http://plato.stanford.edu/archives/fall2004/entries/mathphil-indis/.

[79] Many philosophers think that the obvious move, arguing that mathematics is not essential to science, will not work. See Balaguer 1998.

[80] Benacerraf 1973 in Hart 1996. As we have seen, Changeux raises this point, too.

[81] See the comments in Tait 1986, reprinted in Hart 1996.

to reply that the statement "3 is prime" is not true. It is not true, because there is no such thing as "3," just as the statement that "Sherlock Holmes is uncannily observant" is false because there is no such person as Sherlock Holmes. However, the fictionalist continues, just as we can distinguish usefully between the claim that "Sherlock Holmes is uncannily observant" and the claim that "Sherlock Holmes is a woman" by saying that the former claim is true in the world of certain Conan Doyle novels and the latter one is not, so we can say that "3 is prime" is true, and "4 is prime" is false in the story of mathematics. What "3 is prime" means, for the fictionalist, is that "3 is prime" follows from the axioms of arithmetic.

7.3.5 Carnap's Linguistic Frameworks

Carnap's later philosophy invoked the idea of there being various linguistic frameworks.[82] A framework is constructed within which people speak about certain types of entities (solid objects, quarks, numbers, . . .). A framework may be logical or factual. There are rules for deciding on the existence of entities within a framework; questions such as, "Is there an elephant in the room?" or "Is there a prime number which is also a square?" can be answered in principle. Carnap called these "internal questions," and distinguished them sharply from external questions, which he said only philosophers asked. These concern the existence or reality of the system of entities as a whole. Such questions, he felt, had never been answered satisfactorily because they had never been posed correctly. So within the linguistic framework that governs talk about numbers, it is logically true that no prime number is a square. But the external question, "Do numbers exist?" was, said Carnap, an improper pseudo-question. One can talk about numbers without making any claim about their reality (still less about "reality"), and the criteria for doing so, and for accepting any linguistic framework, were, for Carnap, a pragmatic matter, determined by such factors as efficiency, fruitfulness, and simplicity. He drew the following conclusion:

> Thus it is clear that the acceptance of a linguistic framework must not be regarded as implying a metaphysical doctrine concerning the reality of the entities in question. It seems to me due to a neglect of this important distinction that some contemporary nominalists label the admission of variables of abstract types as "Platonism." This is, to say the least, an extremely misleading terminology. It leads to the absurd consequence, that the position of everybody who accepts the language of physics with its real number variables (as a language of communication, not merely as a calculus) would be called Platonistic, even if he is a strict empiricist who rejects Platonic metaphysics.[83]

Carnap would have had us navigate between frameworks also on pragmatic grounds. Here a new twist to an old observation of Enriques is relevant. He objected to some purported definitions on the ground that they were not definitions at all. It was impossible, in his view, to define objects in the physical world. They can only be described. One could, however, define an abstract object (such as, in mathematics, a group or a ring). Consider now how mathematics is drawn out of ordinary life. When abstract mathematics is created and purports to offer definitions of already

---

[82] Carnap 1950.
[83] Carnap 1950, 27.

well-known objects, a process of matching must be carried out. This process checks that the formally defined object has enough of the properties of the (idealized) familiar object and no unwelcome properties, and when it is concluded we can say we have mathematically formulated what the familiar object is. Enriques was wrong: one can define, for example, a straight line, or a pair of parallel lines, but whoever does so must still admit that the formally defined object is different from the "real thing."

7.3.6 Challenges to Philosophy

Hilbert, in his controversy with Frege, argued that all it means for a mathematical object to exist is that it is described (one might say implicitly defined) by a consistent set of axioms. It is the least possible retreat from what Bernays called "absolute Platonism" and observed was refuted by the antinomies. Philosophers object to this position because it is close to saying that mathematical objects exist in a special sense of the word, incompatible with commonly accepted synonyms (you shouldn't argue for claims such as "There are flying saucers, but they don't exist"). But if talk of abstract objects were to turn out to be essential for science, and abstract objects, on an appropriate definition, do not exist in space-time, one supposes that philosophers might be willing on those grounds alone to agree that abstract objects exist.

As for the epistemological argument against Platonism, one Platonist response is to accept it, and to say that we do not have direct knowledge of these objects, but we do have indirect knowledge. That is, so much of mathematics has empirical interpretations in science, and these interpretations are so thoroughly confirmed, that we should accept the mathematics as true. Gödel, for example, argued against Carnap that the situation in mathematics is exactly analogous to the situation in science: existence assumptions can be vindicated when they lead to verifiable conclusions as indubitable and immediate as "3 is prime." Some controversy attends those parts of mathematics that are not confirmable in this way and to those parts that are not yet confirmed, say, because they have only recently been established. If emphasis is now placed on the coherence of mathematics, then another argument suggests itself, which is that vast swathes of mathematics are simply proved: starting from simple and indisputable premises, the rest of the enterprise follows of necessity.

Alternatively, one might object to Benacerraf that his challenge is simply unfair: Why should there not be at least two forms of knowledge, one appropriate to medium-sized physical objects in space-time and another for the objects of mathematics? As with the term "exists," philosophers again object that the two uses of the word "know" should be recognizably similar, and since the core case concerns what it is to know facts about physical objects, it is the nature of our knowledge of abstract objects that has to conform. But this position is not transparently justified. Our knowledge of Fermat's Last Theorem is a good deal better, more robust, more likely to survive scientific advance than our knowledge of some comparably celebrated results of science. Perhaps one should argue that a philosophy of knowledge that cannot recognize mathematical knowledge as knowledge at all is a poor thing, while acknowledging that a philosophy of mathematical knowledge that merely asserts a kind of knowledge without giving any account of how it is acquired is also inadequate.

7.3.7 Alternatives to Platonism . . .

It may be, however, that the whole philosophy of Platonism is misplaced. Mathematical Platonism is primarily intended to guarantee the mind-independent nature of mathematical results, and secondarily to respect the strong feeling of mathematicians that they are dealing with objects, which is why their hands are forced.

Confronted with words from the vocabulary of mind-dependent philosophies, in which reality is somehow constructed, or people have only signs and their (private) meanings, or in which meaning is created, mathematicians insist passionately that they have no choice in the matter. Could Sherlock Holmes be a woman, or Moriarty in disguise? The answer is "yes" if you write well enough, or if Conan Doyle turns out to have left such a story for posterity, or in a number of other ways. But Fermat's Last Theorem cannot turn out to be false, even if you are a veritable Shakespeare of mathematics, and—mathematicians would say—that is because there is something fundamentally wrong with any fictionalist account of mathematics. This is for the fictionalists to rebut.

Mathematicians, in various ways and presumably to different extents, do report that they are dealing with objects, which is why their hands are forced. The standard answer is to say that this sense is purely psychological, and therefore none of the business of a philosopher. It might, more interestingly, be in some way or another loose talk with an important kernel. Mathematics (more precisely, pure mathematics) involves axioms, rules of deduction, and concepts defined by axioms. What is involved in going from talk about concepts defined through a system of axioms to talk about some objects defined or created by those axioms?

As was discussed in §7.1.2, in 1925 Hilbert answered that question with a philosophy of mathematics that had preoccupied him for some time, and his answer is close to fictionalism. His finitary approach to mathematics is in part the founding statement of what might be called the syntactic view, namely that in many of the most advanced areas of mathematics mathematicians follow the rules of whatever axiomatized system they are working with. If this is done correctly (let the system be assumed to be consistent), then the conclusions reached after any finite number of steps are proved, and that is all anyone can ask for. Of course, Hilbert did not say or imply that this is *all* there is to mathematics: his presentation of mathematical activity is deliberately contrived and partial, for he would have agreed that mathematicians do many things, and only eventually do they come to sit down and write up their discoveries in the form of proper proofs. But then, as far as establishing mathematics knowledge is concerned, Hilbert said that only the line-by-line elaboration of a rigorous argument counts as proving things. His focus was on the written activity of proof. It isolated a feature of mathematicians' work that is essential—the establishing of claims—and suggested that from the standpoint of philosophy this is all there is. It plainly isn't. The whole way mathematics relates to science is ignored here, a topic on which Hilbert elsewhere had strong, well-informed views. But it is a view about mathematical epistemology and so, I suggest, its ontology.

The syntactic view sees mathematics as a rule-governed activity, and one can argue that we come to know the meaning of mathematical objects and concepts through our use of these rules. In his paper "On the Infinite" (1925), Hilbert argued that his rule-governed account of how to handle infinite sets in mathematics was acceptable because it had already been accepted in principle when mathematicians accepted Weierstrass's rule-governed explanation of mathematical analysis.

On this view, what it means for a strictly mathematical object, defined and treated within a system of axioms, to be said to exist and to have certain properties is nothing more than that the axioms are consistent and that the axioms and the rules for deduction permit a proof of the stated properties. This may well be what mathematicians take such talk to mean—they may not all be the hard-line Platonists philosophers sometimes take them to be. An analogy with furniture moving may be helpful. Sometimes the only proof that a new piano will fit in your living room is to move it there, whereas for much smaller and much larger objects, proof by measurement would have sufficed. By analogy, mathematical knowledge is often of the piano-moving kind: the mathematician learns something about the objects by carrying out a certain form of activity. Indeed, knowledge of mathematical objects could be understood as a way of talking about mathematical objects that makes the objects epistemologically accessible, because there is nothing to one's knowledge of a mathematical object or concept other than what can be done with it according to the rules that define it or otherwise bring it to mind. It is only great facility with the rules that allows mathematicians to rephrase talk about rules as talk about objects with properties.

7.3.8 . . . and Gödel's Platonism

Some remarks of Gödel, famous if not notorious for his Platonism, are helpful here and also instructive about his actual beliefs.[84] Gödel objected to Hilbert's finitary philosophy of mathematics not on the grounds that his own incompleteness theorem had shown it to be wrong, but in the hope that the certainty of mathematics could be secured by cultivating or deepening our knowledge of the abstract concepts themselves and seeking to gain insights into the solvability, and the actual methods for the solution, of all meaningful mathematical problems. This clarification of meaning, he said, could not consist simply in giving definitions, and so he turned to the phenomenology of Husserl. As he put it:

> Now in fact, there exists today the beginning of a science which claims to possess a systematic method for such a clarification of meaning, and that is the phenomenology founded by Husserl. Here clarification of meaning consists in focusing more sharply on the concepts concerned by directing our attention in a certain way, namely, onto our own acts in the use of these concepts, onto our powers in carrying out our acts, etc. But one must keep clearly in mind that this phenomenology is not a science in the same sense as the other sciences. Rather it is [in any case should be] a procedure or technique that should produce in us a new state of consciousness in which we describe in detail the basic concepts we use in our thought, or grasp other basic concepts hitherto unknown to us. I believe there is no reason at all to reject such a procedure at the outset as hopeless. Empiricists, of course, have the least reason of all to do so, for that would mean that their empiricism is, in truth, an apriorism with its sign reversed.[85]

---

[84] See Tieszen 2000.
[85] See Gödel 1995, vol. 3, 383.

This is not mathematical Platonism as one might have supposed it and as Connes described it, the contemplation of an abstract object by a facility akin to sight, although that is a comparison Gödel also made. It is directly connected to acts, our acts in the use of concepts, our ability to carry out those acts. We have the power because the axioms permit it, and some mathematicians have the power to a greater extent than others. Rather, in this passage at least, this is Platonism as an ontological twist to a philosophy of mathematics that singles out the finite sequence of acts that constitute giving a proof, each act being, on this interpretation, an application of a rule of deduction to a system of axioms (or a concatenation of these in the form of a lemma or theorem).

Gödel's target here was Carnap's philosophy of mathematics, which he considered inadequate because of its single-minded insistence on the syntactic aspect of mathematics. Gödel observed that insofar as mathematical proofs are syntactical and not about anything they cannot be adequately said to be true (because they are not true of some particular things), and he found this incoherent. He noted (p. 359) that a linguistic framework must be consistent to be any use, but argued that empirical induction was a poor way to ensure consistency, and the better way was "an intuition of the same power is needed as for discerning the truth of the mathematical axioms, at least in some interpretation" (p. 357). So "the syntactical interpretation of mathematics does not relieve one of the necessity to acknowledge certain, by no means trivial, propositions of a mathematical character as true in a non-conventional sense" (p. 358).

7.3.9 Hilbert's Garden

Perhaps one conclusion of all this is that mathematicians' Platonism, maybe quite generally and at least in Gödel's form, is a reification of abstract axiomatics, in which the objects return and are made known through the rules that govern them and the mental acts that constitute conjectures and proofs. There are several reasons why this might have come about. One is the subjective sensation many mathematicians have that they are dealing with objects, not rules. If this is not, or not only, an interesting insight into the psychology of mathematicians, then it may be a piece of naive philosophizing, or it may stem from the way mathematical intuition is rooted in fairly concrete objects. In his preface to *Geometry and the Imagination* Hilbert spoke of

> the tendency toward abstraction [which] seeks to crystallize the logical relations inherent in the maze of material that is being studied, and to correlate the material in a systematic and orderly manner. On the other hand, the tendency toward intuitive understanding fosters a more immediate grasp of the objects one studies, a live rapport with them, so to speak. . . . As to geometry, in particular, the abstract tendency has here led to the magnificent systematic theories of Algebraic Geometry, of Riemannian Geometry, and of Topology [which] make extensive use of abstract reasoning and symbolic calculation in the sense of algebra. Notwithstanding this, it is still as true today as it ever was that intuitive understanding plays a major role in geometry.[86]

[86] *Anschauliche Geometrie* (1932), given as lectures in Göttingen in 1920–21 and published in German in 1932 and in English in 1952.

And so, he concluded: "We want to take the reader on a leisurely walk, as it were, in the big garden that is geometry, so that each may pick for himself a bouquet to his liking." Access to the abstract, Hilbert was saying, was through the concrete, specific world of objects, and this surely contributed to the sense that the objects defined by a system of axioms shared in the same sort of existence that their intuitively defined forebears did.

Another motive may also have been at work. The more mathematics was defined as conceptual mathematics, the more it lost contact with ordinary quantity, the more it lost its sense of reference. Restoring objects in mathematics may have offered reassurance to mathematicians who felt themselves cut off from the classical subject they had studied in their youth. It can feel powerfully reassuring to say that mathematics is about objects lying just beneath the surface of familiar objects (as groups do), much less comforting to say that mathematics is about axiom systems.

7.4 Did Modernism "Win"?

This book has documented the rise of what can be called mathematical modernism. Even if the arguments it contains were to be found fully convincing, we can still take a step back and, looking at the whole practice of mathematics, ask if modernism "won." Did all of mathematics change? How profound and how broad were the successes of modernism, and what longer continuities were there?

The term "modernism" was borrowed from the artistic domains of literature, poetry, music, painting, and architecture. In each one, to varying degrees, modernism made it difficult if not impossible to work as if it had not occurred, and to that extent we can say that modernism "won." But the qualification "to varying degrees" must be noted. One can claim that to the extent that nonmodernist artists, writers, and so on resisted modernism, they were somehow reactionary or inferior, and this claim makes some sense when applied to painting. But it makes much less sense when applied to music, when it sounds more like a propaganda claim or an exhortation. Writers on Schoenberg wrestle with the apparent problem that whatever the merits of his new music, it remained resolutely unpopular; writers about Picasso do not have that problem. Modernism in architecture and in the novel are intermediate cases: modern architecture became an established style, modernist fiction was written in several forms, and none of these replaced the traditional modes. Which of these is closest to modernism in mathematics—where certainly there were modernists and also traditionalists who successfully resisted the call?

The systematic phase in the development of modernist mathematics is most visible in geometry, analysis (notably functional analysis and topology), algebra, and in the logic and foundations of mathematics. Emmy Noether and her students turned algebra into abstract structural algebra, more abstract and more general than anything Dedekind had considered. Topology, although it can be given almost a kindergarten feel at times, is actually a highly abstract subject in which the objects and methods are often inaccessible without axioms. This might make it seem as if modernism "won," in the sense that mathematics became entirely modern, axiomatic, structural, and entirely self-contained. However, this neat picture would be an oversimplification that has to be contested.

Mathematicians today generally believe that mathematics comes in two types, with a considerable degree of overlap. The highly axiomatized type just described, with its long structured chains of definitions and theorems to master, can seem a long way from and indifferent to applications, and it radically rewrites mathematics it retains from the past to adapt it to its purposes. This may be called the highly structured type of mathematics. The second type is highly delicate and subtle, much less axiomatized, much more oriented toward problems. It rewrites past mathematics much less when it takes it over, and it is closer to and more comfortable with applications. It may be called the densely structured type.

These two strands or types of mathematics go in and out of fashion. Highly structured mathematics was in the ascendant from before the First World War until quite some time after the Second, the reigns of the Göttingen and Bourbaki dynasties. In the last twenty years the fashion has swung the other way, and ad hoc theories crafted to particular problems are now much more in vogue. Undoubtedly this has something to do with the vagaries of funding, but there are substantial reasons why such swings are possible, even necessary. Densely structured mathematics contains the theories of ordinary and partial differential equations as well as dynamical systems—it cannot be marginalized. Indeed, we can ask if the modernization of mathematics is not an artifact of a value judgment about the nature of mathematics (rather than, say, an insight into it). Is the modernization of mathematics merely the view of a group of people who did well out it? The answer to this question is "No" precisely to the extent that one is prepared to say densely structured mathematics has come to coexist with the new, highly structured form.

A purely sociological view that this is just the view of the winners and their hangers-on is untenable, for the simple reason that both styles of mathematics can coexist in the same person at the same time. To give but two examples, the same Riemann who introduced modern differential geometry and rewrote the whole philosophy of geometry wrote the paper that contains what is commonly regarded as the deepest unsolved problem in mathematics to this day: the Riemann hypothesis.

This has to do with the distribution of the prime numbers. To a first approximation (noted originally by Gauss), the number of prime numbers less than an integer n increases with n like n/ln n. This is already hard to prove, and is known as the "prime number theorem." But the primes don't arrive steadily; there are sometimes long gaps between one and the next, sometimes prime numbers seem to come in a rush. The prime number theorem is more like a statistical law, and so the challenge is to explain these departures from strict regularity. That is the business of the Riemann hypothesis, which (in a way we need not describe)[87] is a conjectured explanation that relates the distribution of the primes to the complex numbers where a function, called today the Riemann zeta function, vanishes. In particular, the Riemann hypothesis is the claim that all these points lie on a line, called the critical line, because they have the same real coordinate: 1/2.

---

[87] See two very readable books on this topic: du Sautoy 2003 and Derbyshire 2003.

Our second example is Hilbert, for, as already noted, although he was the author of the most conspicuously modernist tract on geometry, his *Grundlagen der Geometrie*, his main publications between 1900 and 1910 were in analysis, where he vigorously developed the new field of integral equations and applied it to the theory of differential equations and the calculus of variations. Indeed, Hilbert was not regarded by his most intimate colleagues as a theorist so much as a problem solver. He is well remembered to this day for his so-called twenty-three problems (the actual number is around thirty; the Riemann hypothesis is number eight), which he posed at the International Congress of Mathematicians in Paris in 1900 and which succeeded, slowly but then vividly, in capturing the imagination of the mathematical community.

Continuities such as these can be documented indefinitely, and they show that modernism in mathematics is not a simple break. It is a matter of continuities and reformulations as much as dramatic novelties. To delineate this more precisely, let us consider some of the characteristic features of modernism.

The first element of modernism is its supposedly arbitrary character. We have seen it compared to chess; both are supposedly rule-governed activities. Cantor and Dedekind spoke of "free" rather than pure mathematics, meaning that this kind of mathematics was unconstrained. The return of the infinitesimal in geometry is instructive here. Bettazzi, Veronese, Hilbert, Dehn, Hölder, and others successfully built up quite a theory of infinitesimals, including a theory of non-Archimedean geometry. Or, to take a related example but nearer to analysis, consider the system of numbers called the *p*-adic numbers (discussed in §4.5) that was created around 1900 and shown to be a perfectly good way to fill in the gaps between the rational numbers (the *p*-adic numbers are also not Archimedean). It follows that there was nothing inevitable about the passage from the rational numbers to the real numbers that Dedekind, Cantor, and Heine had spent so much time describing and which was generally taken to spell out just what had hitherto been taken for granted. And once the real numbers are no longer inevitable, it can be asked if they are particularly natural or right. It happens that the familiar real numbers are the only way to extend the rational numbers that allows one to extend the ordering on the rational numbers. That is, the way one decides that $x < y$ when x and y are real numbers also says that $x < y$ when x and y are rational numbers. However, although one can compare the norms of two *p*-adic numbers x and y, this comparison does not agree with the ordinary comparison when x and y are rational numbers.

The modernist attitude is that these novelties exist, or are known to exist as soon as they are rigorously described, and so there is nothing natural or inevitable about the much more familiar real numbers (themselves only made rigorous in the 1870s). The deeper question is aesthetic: Should one care? The *p*-adic numbers spectacularly vindicated themselves with Hasse's work in algebraic number theory, so the verdict was that one should care. The more general concept of a geometric magnitude has not done so well, but that helps make the point: the existence question has to be formulated in a modernist way, but once it is settled there is a possible rapprochement with mathematicians of a nonmodernist persuasion that will proceed on pragmatic grounds.

Another aspect of modernism is its emphasis on the autonomy of the mathematician, in particular the mathematician's freedom to develop theories that have no apparent use. The very possibility of doing so, as Riemann saw very clearly, forces the realization that the relationship of mathematics to physics is a complicated one. Mathematicians offer a bunch of possible descriptions of whatever it might be, and necessity attaches to none of them, only greater or lesser degrees of plausibility. So there is no certainty that any particular one is correct, or indeed that among the known possibilities even one is correct. The relationship of mathematics to physics is as a result much more complicated and open. Should one say (and if so, how and why) that this mathematics is true? Correct? Useful? Valuable but merely hypothetical? Conversely, can any amount of success in physics vindicate at least some piece of mathematics?[88]

To the extent that this arbitrariness and autonomy are recognizable features of the mathematical world by, say, 1920, one could say that modernism "won." To get a measure of the more traditional side, one can look at the practice of mathematics in Britain, which remained resolutely unmodernist. Only two names are significant among the modernists here: Bertrand Russell and A. N. Whitehead. Of these, Russell was a philosopher of mathematics, not a mathematician. A. N. Whitehead was to Cambridge eyes an applied mathematician because that is where he began. As noted above, he then wrote a strange but original book called *Universal Algebra*, then books on axiomatic geometry, then, with Russell, *Principia Mathematica*, and after the First World War he too made his way to philosophy. This was at a time when Hardy and Littlewood were emerging into the front rank of mathematical analysis in Britain.

Hardy's great love was mathematical analysis, especially its application in the theory of numbers. His awakening came in 1897 when he read Camille Jordan's *Traité d'analyse*. He was a prolific mathematician, and his textbook *A Course of Mathematical Analysis* (1908) put modern analysis in the path of every mathematics undergraduate in Britain. In 1911 he teamed up with J. E. Littlewood, a collaboration that lasted thirty-five years and produced over hundred joint papers. In 1913 he answered a letter from an unknown clerk in Madras, and so discovered the remarkable genius of Ramanujan, whom he brought to Cambridge. Ramanujan died in 1920. Of the three, Ramanujan was the untutored genius with a staggering ability to find formulas describing recondite properties of numbers, Littlewood the determined, indeed ferocious, problem solver, and Hardy somewhat more of the theorist. Among his most important results was showing in 1914 that an infinitude of zeros of the Riemann zeta function lie on the critical line. All three men were recognizably pure mathematicians, almost the first of international stature in Britain, and their influence on the growth of British mathematics was profound.

Hardy was resolutely unimpressed by some aspects of the modernist credo. His view was: "I believe mathematical reality lies outside us, that our function is to discover to observe it, and that the theorems which we prove, and which we describe grandiloquently as our 'creations,' are simply our notes of our observations." (Hardy 1969, 123–124). In other words, and to take an example, the prime numbers are distributed in whatever way they are; the mathematician has no influence in the matter. For that matter, the zeros of the zeta function must lie where they do whether on the critical line or off it. Again, the mathematician has no choice in the matter. But Hardy would have made a distinction between the primes and the zeros, arguing that

---

[88] And indeed in the 1990s, well after the period addressed in this book, some physicists have begun to think of using the p-adic numbers in physics.

the whole approach to the distribution question via the zeta function is in a way arbitrary. It is the approach most likely to succeed—there really was no alternative in sight—but other approaches might be possible. However, once you have subscribed to this approach, the distribution of the zeros is again not something the mathematician can adjust.

Theorems in this subject were, for Hardy, deep and real, not mere artifacts of the theory. The whole immense and often delicate arguments needed to prove anything in this subject, Hardy insisted, came from a perfectly rigorous set of definitions and theorems, the already by then classic chain of arguments that starts with the integers, constructs the rational numbers, the real numbers, the complex numbers, and the whole apparatus of real and complex function theory. Together they make, as it were, the apparatus necessary for the observations.

There is little in Hardy's view—from its uncompromising stance, its implication that it is stating the obvious, and its lack of philosophical sophistication—that separates it from the opinion of many a mathematician from Greek times to today. Hardy and Littlewood even joked that the Greek mathematicians were not bright scholarship candidates, they were mathematicians from another college. It is generally the view of even today's pure mathematicians that they have choice about methods, but not results. It is, equally plainly, not modernist.

This solid block on the road to modernism requires us to push the concept of modernism. Chapter 5 discussed some of the ways in which interactions between mathematics and other disciplines, notably physics, were changed by the arrival of mathematical modernism. Here it will be more useful to see what we can do to illuminate some more philosophical questions:

1. The nature of mathematical objects
2. Mathematical proofs
3. The philosophy of mathematics itself

7.4.1 Objects

Mathematical modernism permits definition by axiomatization (one must, of course, check that the axioms make sense). It is hard to imagine groups, rings, and fields, function spaces and topological spaces, being studied in any other way. While sets became the paradigm object of study by philosophers, it is also worth noting that as more elaborate objects (such as groups and rings) became fundamental in mathematics, a point was reached where it was not so much the case that such-and-such kind of a ring was important because the ring of integers are a paradigm example, as that the integers were important because they were only one example of such-and-such a kind of a ring. This inversion of priorities is tied to the importance of generalization in mathematics. The capacity to generalize is very important in mathematics, because it opens the way to the discovery of new proofs, and because it can bring a quality of understanding.[89]

---

[89] For example, there are now numerous examples of zeta functions in various domains of mathematics, and for many of them mathematicians have been able to formulate and prove a Riemann hypothesis.

7.4.2 Proofs

Mathematicians have a variety of ways in which they evaluate approaches, even proofs. Outside of mathematics it can seem strange to hear a proof called the "wrong' proof" even though it delivers the required result, but there can be many things wrong about a proof. It can be said that a proof does not explain, and the result still seems mysterious—it does not seem to come from anywhere. Or, that it does not generalize, perhaps because it relies on features that are not known to pertain in a more general setting. Or, that it argues in a roundabout way, perhaps descending to a special case. Or, that it is a long calculation that, although correct, seems unmotivated and not intrinsically connected to the question at hand. Characteristically modernist proofs appeal to the structural quality of an argument: these results are true because the object studied is, say, a group, and the proof ought to follow from the fact that it is a group (and not, let us say, the integers). In one respect, Hardy's opinions were not so different. He distinguished mathematics from chess on the grounds that chess was trivial mathematics,[90] and what made for substantial mathematics in both theorems and proofs was "a high degree of *unexpectedness*, combined with *inevitability* and *economy*."[91] He disliked complications of detail and enumeration of cases. But he made no appeal to a higher court of structuralism.

7.4.3 The Philosophy of Mathematics

The philosophy of mathematics changed completely with the arrival of mathematical modernism, and not only with the evolution of the mathematical Platonism just discussed. After Hilbert's proof theory and the dramatic work of Gödel, mathematical philosophy hit a productive period in which mathematical logic developed rapidly and many outstanding questions were resolved, and in which the philosophy of mathematics was very productively taken to concern itself with issues in set theory. The focus on foundations was intense and was only reinforced by the debates around intuitionism. By the start of the twenty-first century, there were strong indications of a return among philosophers of mathematics to more aesthetic questions: rightness of proof, purity of method, fruitfulness of concepts, the nature of explanation and understanding.[92] These ideas are not exclusive to modern mathematics—Newton used similar ideas in his polemics against Descartes—but modern philosophers have noticed that these traditional philosophical questions were in fact particularly acute during the advent of mathematical modernism. The autonomy of the mathematician provoked it, of course, and the genuine arbitrariness of mathematics did so, too, but as this book has been at pains to describe, mathematics was undergoing a critical free-for-all in the years around 1900. Very few of these questions would have been asked at all had mathematics not changed in these ways, as the long separation of mathematics and philosophy throughout much of the nineteenth century demonstrates so clearly.

---

[90] Hardy 1949, pp. 88, 112.
[91] Hardy 1949, 113; italics in original.
[92] For one way in, see the essays in Ferreirós and Gray 2006.

It is worth insisting that it is the characteristically modernist features of the new mathematics that prompted all the discussion. Once modernism was established and the long reductionist arguments about mathematical objects were in place to reduce everything to set theory and questions about the philosophy of set theory, mathematical logic could take over, and Hilbert's hopes that the philosophy of mathematics would itself become mathematical were in large measure fulfilled. That, paradoxically, many of Hilbert's hopes for the subject were promptly dashed by Gödel and Turing may well have served to show how powerful the new methods were. But in the years from 1880 to 1920 the old philosophical questions about mathematics were debated as they had not been since Greek times.

To say that philosophy of mathematics was an arena where modernism triumphed is to say that modernism changed the practice of mathematics, because mathematicians were among the most eager participants in this debate. It was far from the case that mathematicians got on with good honest work while philosophers peered at them through the glass, each side holding the other in polite disdain. Mathematics did not cease to be densely structured, but it became highly structured as well.

7.5 The Work Is Done

The arrival of mathematical modernism was, like so many modernisms, unstable. Mathematicians and philosophers seeking to answer genuine and deep questions about the nature of mathematics, the grounds on which its conclusions should be accepted, and its relation to the sciences were drawn into difficult areas. They established remarkable results, but they never reached solid ground, and in the end their attempts were fractured beyond repair. The modernism project became in many ways the standard for good work, and in other ways had to be abandoned.

It should not be thought, therefore, that the stories told so briefly in this chapter, ending as they do in the 1930s and 1940s, represent a terminus. The various domains of pure mathematics were undoubtedly reshaped along abstract, axiomatic—in short, modernist—lines. They remain secure fields with their own standards and agendas even as the pendulum swings, in the 1950s, '60s, and '70s toward pure mathematics and since then toward applications. It is possible that we are entering for the first time an era in which the biological sciences will call forth substantial new mathematics, as the novel technologies have already done. It also remains the case that the professional division between mathematicians and scientists, even physicists, is securely in place; it is not at all clear what rigorous mathematics is rigorously employed in physics.

The work of Gödel and the other logicians of the 1930s did more than take Hilbert's program apart. It created a sophisticated body of theory that could take many somewhat philosophical questions, formulate them in terms of mathematical logic, and answer them. A considerable number of such questions about formal systems and their syntax and semantics have been answered in this way. This process gave rise to a new industry in the philosophy of mathematics which took the view that all mathematics was an outgrowth of set theory, and so the fundamental (read: philosophical) problems in mathematics were those in set theory, and this view has also had its successes. But the transition from foundations of mathematics to mathematical logic, and from the practice of mathematics to set theory, represents a narrowing of focus. There have been losses as well as gains.

The fragmentation shows up in many ways. Theoretical computation is now an autonomous discipline, but what it studies seems ever more distinct from the way humans work. We, it seems, are elaborate parallel processors; delicate issues of reliability and timing make computer scientists and their companies nervous about going in that direction. The work of Turing, McCulloch, and Pitts has given rise to a whole theory of automata. The work of Post (not discussed here) led in turn to the theory of formal grammars, the original field of work for Noam Chomsky before he turned to natural languages. But it would be hard to claim that the investigation of mathematics as a language has the resonance it once possessed. Tarski's work defining truth for a suitably rich class of formal languages explicitly stopped short of natural languages.

The most visible failing in the present situation concerns the question of what mathematics actually is. Most mathematicians sensibly get on with doing it, even innovating in remarkable ways. Developments originally begun in algebraic topology and algebraic geometry have produced, in category theory, a wholly new formulation of the foundations of mathematics that avoids, or subsumes, set theory. But the working image of mathematics for mathematicians is in many ways closer to the image they had of themselves in the nineteenth century, although with a twentieth-century underpinning. The view from outside is, if anything, less exciting: mathematicians have been left in the main to get on with it. It is not clear to this writer, at least, that mathematics, with its unique claims to knowledge and truth, animates philosophers interested in the nature of knowledge and truth.[93] The reason for this lack of curiosity about mathematics derives from the collapse of the Hilbert program and the retreat to efficient but highly technical providers of specialist results. There is not a vision of mathematics as an enterprise to which a large audience can relate. There is only a sense that mathematics is simultaneously very formal and somehow not entirely formal.

There may be signs with the new century that matters are changing again, but it seems best to conclude with three quotations cited below that date from the 1960s. The first is from Emil Post, writing in 1941 but not published until 1965. The final two come from an essay by Gödel that was published posthumously but dates from 1961. The first two represent a call for the return of meaning in mathematics as opposed to work-day Platonism and Sunday formalism. The final one, coming as it does from the most celebrated Platonist of the twentieth century, returns us most forcefully to the legacy of Kant. Reader, farewell!

> But perhaps the greatest service the present account could render would stem from its stressing of its final conclusion that *mathematical thinking is, and must be essentially creative*. It is to the writer's continuing amazement that ten years after Gödel's remarkable achievement current views on the nature of mathematics are thereby affected only to the point of seeing the need of many formal systems, instead of a universal one. Rather has it seemed to us inevitable that these developments will result in a reversal of

---

[93] Although the exceptions to this sweeping judgment are among the best twentieth-century philosophers.

the entire axiomatic trend of the late nineteenth and early twentieth centuries, with a return to meaning and truth. Postulational thinking will remain as but one phase of mathematical thinking.[94]

Obviously, this means that the certainty of mathematics is to be secured not by proving certain properties by a projection onto material systems—namely, the manipulation of physical symbol—but rather by cultivating (deepening) knowledge of the abstract concepts themselves which lead to the setting up of these mechanical systems, and further by seeking, according to the same procedures, to gain insights into the solvability, and the actual methods for the solution, of all meaningful mathematical problems. [This] procedure must thus consist, at least to a large extent, in a clarification of meaning that does not consist in giving definitions. Now in fact, there exists today the beginning of a science which claims to possess a systematic method for such a clarification of meaning, and that is the phenomenology founded by Husserl. Here clarification of meaning consists in focusing more sharply on the concepts concerned by directing our attention in a certain way, namely, onto our own acts in the use of these concepts, onto our powers in carrying out our acts, etc. But one must keep clearly in mind that this phenomenology is not a science in the same sense as the other sciences. Rather it is [or in any case should be] a procedure or technique that should produce in us a new state of consciousness in which we describe in detail the basic concepts we use in our thought, or grasp other basic concepts hitherto unknown to us. I believe there is no reason at all to reject such a procedure at the outset as hopeless.[95]

I would like to point out that this intuitive grasping of ever newer axioms that are logically independent from the earlier ones, which is necessary for the solvability of all problems even within a very limited domain, agrees in principle with the Kantian conception of mathematics. The relevant utterances by Kant are, it is true, incorrect if taken literally, since Kant asserts that in the derivation of geometrical theorems we always need new geometrical intuitions, and that therefore a purely logical derivation from a finite number of axioms is impossible. That is demonstrably false. However, if in this proposition we replace the term "geometrical" by "mathematical" or "set-theoretical" then it becomes a demonstrably true proposition. I believe it to be a general feature of many of Kant's assertions that literally understood they are false but in a broader sense contain deep truths. In particular, the whole phenomenological method, as I sketched it above, goes back in its [central] idea to Kant, and what Husserl did was merely that he first formulated it more precisely, made it fully conscious and actually carried it out for particular domains. Indeed, just from the terminology used by Husserl, one sees how positively he himself values his relation to Kant.

I believe that precisely because in the last analysis the Kantian philosophy rests on the idea of phenomenology, albeit in a not entirely clear way, and has just thereby introduced into our thought something completely new, and indeed characteristic of every genuine philosophy—it is precisely on that, I believe, that the enormous influence which Kant has exercised over the entire subsequent development of philosophy rests. Indeed, there is hardly any later direction that is not somehow related to Kant's ideas. On the other hand, however, just because of the lack of clarity and the literal incorrectness of many of Kant's formulations, quite divergent directions have developed out of Kant's

---

[94] Post, unpublished; see Post 1994, 345, quoted in Gandy 1995, 93.
[95] Gödel 1961, in Gödel 1995, 383.

thought—none of which, however, really did justice to the core of Kant's thought. This requirement seems to me to be met for the first time by phenomenology, which, entirely as intended by Kant, avoids both the death-defying leaps of idealism into a new metaphysics as well as the positivistic rejection of all metaphysics. But now, if the misunderstood Kant has already led to so much that is interesting in philosophy, and also indirectly in science, how much more can we expect it from Kant understood correctly?[96]

---

[96] Gödel 1961, in Gödel 1995, 385, 387.

APPENDIX

Four Theorems in Projective Geometry

The next four theorems are fundamental in nineteenth-century projective geometry. The first is known as *Pappus's theorem*, after the Hellenistic writer Pappus. It is a reworking of what Pappus wrote, and he may well have had something different in mind.

THEOREM 1 *(Pappus) In figure A.1, if A, B, C are three points on a line ℓ, and A′, B′, C′ are three points on a line ℓ′, and the lines AB′ and B′A meet at R, the lines BC′ and B′C meet at P, and the lines CA′ and A′C meet at Q, then the points P, Q, and R lie on a line m.*

(It is not true in general that the line *m* passes through the intersection point of the lines ℓ and ℓ′.)

   Desargues' theorem, discovered in the first third of the seventeenth century by the French mathematician and architect Girard Desargues, says:

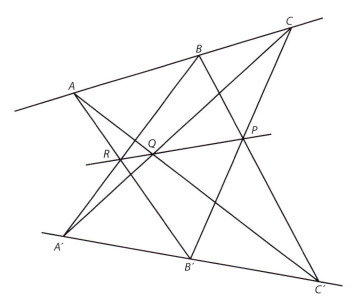

FIGURE A.1. Pappus's theorem.

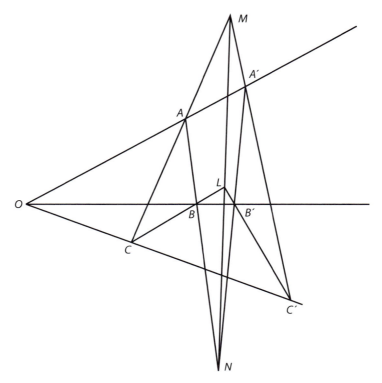

FIGURE A.2. Desargues' theorem.

THEOREM 2 *(Desargues) In figure A.2, if ABC and A'B'C' are two triangles in perspective from O (so O, A, A' lie on a line, as do O, B, B' and as do O,C,C') and if L is the point common to BC' and B'C, M is the point common to CA' and C'A, and N is the point common to AB' and A'B, then the points L, M, N lie on a line.*

THEOREM 3 *(Uniqueness of the fourth harmonic point) In figure A.3, let A, B, and C be any three distinct points on a line, and let an arbitrary point P not on this line be chosen. Draw the lines PA and PB, and draw an arbitrary line through C (other than AB) meeting AP at Q and BP at R. Draw the lines BQ and AR and let them meet at S. Draw the line PS meeting AB at D. Then the position of D is independent of all the choices made (it depends only on A, B, and C).*

The name "fourth harmonic point" arises because the position of D is determined solely by the three points A, B, and C, and because of this equality of ratios (called "harmonic" by earlier writers): $\frac{AD}{DB} = -\frac{AC}{CB}$. The minus sign arises because the length CB is measured in the direction opposite to the other lengths.

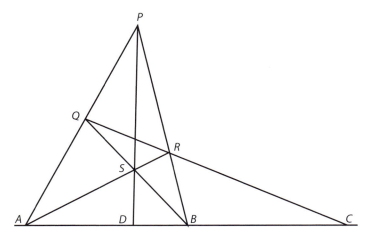

FIGURE A.3. Uniqueness of the fourth harmonic point.

THEOREM 4 *(Pascal) In figure A.4, if ABC and A'B'C' are six points on a conic section, and if P is the point common to BC' and B'C, Q is the point common to CA' and C'A, and R is the point common to AB' and A'B, then the points P, Q, R lie on a line.*

(Draw them in the order *A, B, C, C', B', A'.*)

Proofs of these results may be found in many places, for example my *Worlds Out of Nothing.*

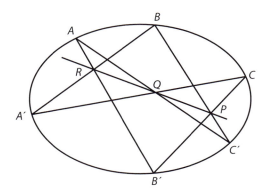

FIGURE A.4. Pascal's Theorem.

GLOSSARY

The purpose of this glossary is to give a more accurate impression of the meaning of some of the mathematical terms used in this book. When it was not possible to give precise, meaningful definitions, I have to try for intelligibility rather than precision.

Alephs The cardinality of a transfinite initial ordinal number is denoted by a symbol called an aleph (the least ordinal number of a given cardinality is called the *corresponding initial ordinal*). The axiom of choice implies that every infinite cardinal number is an aleph.

Algebraic integers An algebraic integer is a root of a monic polynomial whose coefficients are integers. (A monic polynomial is one where the coefficient of the highest power of the variable is 1.) Examples: $\sqrt{2}$, which is a root of $x^2 - 2 = 0$, and ω, a root other than 1 of $x^3 - 1 = 0$.

Archimedean, non-Archimedean axioms Given a set of elements X that can be added together and compared in size, so, for any two elements a, $b \in X$, exactly one of these three statements is true: $a < b$, $a = b$, $b < a$, the order relation $<$ is Archimedean if whenever $a < b$ there is an integer n such that $b < na$. Otherwise the relation is non-Archimedean.

Axiom of choice, denumerable axiom of choice See §4.7.1.

Axiom system: complete, categorical See §4.2.2.

Cardinal number On some approaches, the class of all sets that can be put in a 1-1 correspondence with a given set is the cardinal number of the set. This has the disadvantage that a cardinal number is not a set, but it is possible to use the axiom of choice to pick out a distinguished set from the class and make it the cardinal number of the given set.

Cesàro sum, summability A method of assigning a sum to an infinite series $\Sigma_{i=0}^{\infty} u_i$ that systematically alters the terms of the given infinite series to produce a new series. The method of alteration depends on a real parameter α and if, for a specific value of α, the new series has a conventional sum, the old series is said to be (C, α) summable. If the original series has a conventional sum it is also $(C, -1)$ summable.

Chain A term introduced by Dedekind; see §3.4.3.

Commutative law, associative law Given a set X of objects a, b, c, . . . that can be combined together to give an element of the set, the law of combination is commutative if for every a, $b \in X$, $a.b = b.a$. If, moreover, $a.(b.c) = (a.b).c$, the law of combination is associative.

Continuity of a function $f : \mathbb{R} \to \mathbb{R}$: a function $f : \mathbb{R} \to \mathbb{R}$ is continuous at a point t in $\mathbb{R}$ if and only if for every $\varepsilon > 0$ there is a $\delta > 0$ such that $|x - t| < \delta \Rightarrow |f(x) - f(t)| < \varepsilon$. The function f is continuous on a subset X of $\mathbb{R}$ if and only if it is continuous at every point of X; note that the ε may vary with $x \in X$. The function is **uniformly continuous** on a subset X of $\mathbb{R}$ if and only if for every $t \in X$ and for every $\varepsilon > 0$ there is a $\delta > 0$ such that $|x - t| < \delta \Rightarrow |f(x) - f(t)| > 0$, where the ε may depend on δ but not on t. A **sequence** u_n **converges** to a value u if and only if for every $\varepsilon > 0$ there is a $N > 0$ such that $n > N \Rightarrow |u_n - u| < \varepsilon$.

Continuum hypothesis See §3.2.2.

Convergence, uniform convergence A sequence of functions $u_n(x)$ converges pointwise to a function $u(x)$ on a subset X of $\mathbb{R}$ if and only if for every $x \in X$ the sequence $u_n(x)$ converges to $u(x)$. The sequence of functions $u_n(x)$ **converges uniformly** to a function $u(x)$ on a subset X of $\mathbb{R}$ if and only if for every $x \in X$ and every $\varepsilon > 0$ there is a $N > 0$ such that $n > N \Rightarrow |u_n - u| < \varepsilon$, and N does not vary with x.

Definit See §4.7.6.

Differentiability of a function $f: \mathbb{R} \to \mathbb{R}$. A function $f: \mathbb{R} \to \mathbb{R}$ is differentiable at a point t in $\mathbb{R}$ if and only if for every $\varepsilon > 0$ there is a $\delta > 0$ such that $|x - t| < \delta \Rightarrow |\frac{f(x) - f(t)}{x - t}| < \epsilon$.

Dimension in topology. A topological space is a set together with a distinguished collection $\mathbb{T}$ of its subsets that has the properties that arbitrary unions and arbitrary finite intersections of sets in $\mathbb{T}$ are again in $\mathbb{T}$. For example, the open intervals on the real line form a topology for the real line. Roughly speaking, the dimension of a topological space is the smallest integer n such that any collection of $n + 2$ open sets whose union is the whole space may contain a disjoint pair of open sets, but no collection of $n + 1$ open sets whose union is the whole space contains a disjoint pair of open sets. On this definition, the real line has dimension 1, as it should.

Dirichlet principle and **Dirichlet problem** See 2.2.4.

A field $\mathbb{F}$ is a set of elements satisfying three families of conditions. First, it forms a commutative group with a composition "+" called *addition*. We denote the identity element for "+" by 0. Second, the set $\mathbb{F}$ with the element 0 removed forms a commutative group with a composition "." called *multiplication*. Third, there are laws connecting the addition and the multiplication, such as (for elements a, b, c of $\mathbb{F}$): $a.(b + c) = a.b + a.c$.

Finite geometry (§4.1.1). A plane geometry is finite if it consists of a finite number of points and lines; analogous definitions hold in higher dimensions.

Fourier series, coefficients of, convergence of See §2.2.1.

Gaussian curvature A measure of how curved a surface in space, S, is at any point P on it. If the surface looks like a sphere in a small neighborhood of P, then the Gaussian curvature gives the size of the closest spherical approximation to the surface at P; and if the surface is saddle shaped in a small neighborhood of P, then the Gaussian curvature gives the size of the closest saddle-shaped approximation to the surface at P.

Geometry: Synthetic v. analytic geometry v. algebraic geometry Geometry is synthetic if the fundamental (undefined) terms are point, line, plane, etc., and there are (undefined) concepts of a point lying on a line, two lines meeting in a point, etc. Geometry is analytic or algebraic (the terms are treated as synonymous in this book) if the terms point, line, plane, etc. are defined by equations, and the incidence relations are said to hold when certain equations are satisfied.

Grelling's paradox See §4.7.3.

Group, group theory. A set X is a group when it is closed under a binary operation "." (so if a, $b \in X$ then $a.b \in X$) with the properties that

- there is an $e \in X$ such that $a.e = e.a = a$ for every $a \in X$
- for every $a \in X$ there is a $a^* \in X$ such that $a.a^* = e = a^*.a$
- for every a, b, $c \in X$, $a.(b.c) = (a.b).c$ (the combination is associative)

Harmonic function A function $f\colon \mathbb{R}^2 \to \mathbb{R}^2$ is harmonic if $\frac{\partial^2 f}{\partial x^2} + \frac{\partial^2 f}{\partial y^2} = 0$. Such a function is also said to satisfy Laplace's equation.

Hyperplane, at infinity A hyperplane is the generalization of a plane in three-dimensional space to a $(n-1)$-dimensional subspace of n-dimensional space. It is defined by an equation of degree 1. The hyperplane "at infinity" with respect to a Euclidean n-dimensional space with coordinates $(x_1, x_2, \ldots, x_n)$ is specified by passing to the corresponding projective space with homogeneous coordinates $[x_0, x_1, x_2, \ldots, x_n]$, when it is given by the equation $x_0 = 0$.

Ideal A subset A of a ring R that is closed under additions and multiplication by elements of R. In symbols, if a, $a' \in A$, then $a + a' \in A$, and if $a \in A$, $r \in R$, then $a.r \in A$. For example, in the ring of integers, the set of all multiples of 6 is an ideal.

Infinitesimal number Informally, a number is infinitesimal if it is greater than zero but less than every finite real number.

Integral (antiderivative) Informally, the integral of a function $f\colon \mathbb{R} \to \mathbb{R}$ from a to t is the (signed) area of the region between the x-axis, the vertical lines $x = a$ and $x = t$ and graph of the function. It is often regarded as a function of the upper endpoint, t.

Isotropic space, homogeneous space (§2.1.3.) A space is isotropic at a point in it if it looks the same in all directions from the point, and isotropic if it is isotropic at every point. A space is homogeneous if any two points in it have neighborhoods that are equivalent in all respects.

Jordan curve theorem A continuous closed curve in the plane that does not cross itself divides the plane into two regions, its interior and its exterior. Any two points in the same region, interior or exterior, can be joined by a curve that does not cross the given curve, no two points in different regions can be joined by a curve that does not cross the given curve. An analogous theorem about $(n-1)$-dimensional spheres in $\mathbb{R}^n$ holds in n dimensions.

Klein-Cayley metric A way of assigning a distance to every pair of points inside an arbitrarily chosen conic in the projective plane, or more generally to the points inside an arbitrarily chosen quadric in a projective space. Cayley identified a highly degenerate conic for which the corresponding notion of distance was the Euclidean one. Klein showed that when the conic is an ellipse, the corresponding distance is the non-Euclidean one.

Measure theory (see §4.4.1) is a theory of the size of sets in the plane or a higher-dimensional space, particularly well adapted to providing a theory of the integral.

Mechanics, kinematics, dynamics (Lagrangian dynamics) Mechanics is the study of the motion of particles and solid bodies. The description of the motion is given by the kinematics, its explanation in terms of forces is given by dynamics. Lagrangian dynamics is a particular formulation of a mechanical problem involving the coordinates of position and momentum of the relevant particles and bodies.

***n*-dimensional ball, sphere, space** Informally, a space has *n* dimensions when the position of a point in it depends on *n* coordinates. For example, in Euclidean three-dimensional space, a position is specified (with respect to a set of coordinate axes) by three numbers. The surface of a sphere in Euclidean three-dimensional space is a two-dimensional space. Analogously, in Euclidean *n*-dimensional space a position is specified (with respect to a set of coordinate axes) by *n* numbers. The surface of a sphere in Euclidean *n*-dimensional space is an $(n-1)$-dimensional space, and the interior of the sphere is an *n*-dimensional space.

***n*-tuple** An ugly but standard term for a string of *n* objects, typically the coordinates of a point in *n*-dimensional space, $(a_1, a_2, \ldots, a_n)$.

Notation $\mathbb{N}$ denotes the natural numbers 1, 2, 3, . . . , $\mathbb{Z}$ denotes the integers, . . . $-3, -2, -1, 0, 1, 2, 3 \ldots$, $\mathbb{Q}$ denotes the rational numbers, $\mathbb{R}$ the real numbers, and $\mathbb{C}$ the complex numbers.

***p*-adic numbers** See §4.5.

Projective geometry is the study of projective space. Projective *n*-dimensional space is the set of all homogeneous $(n+1)$-tuples $[x_0, x_1, \ldots, x_n]$, where the *n*-tuples $[x_0, x_1, \ldots, xn]$ and $[kx_0, kx_1, \ldots, kx_n]$, $k \neq 0$, represent the same point. The set of points for which, say, $x_0 \neq 0$, may all be written in the form $[1, x_1, \ldots, x_n]$, because the coordinates are homogeneous, and the corresponding points $(x_1, \ldots, x_n)$ in nonhomogeneous coordinates form a Euclidean *n*-dimensional space with the hyperplane $x_0 = 0$ "at infinity."

Projective 2-dimensional space is the set of all homogeneous triples $[x, y, z]$, where $[x, y, z]$ and $[kx, ky, kz]$, $k \neq 0$ represent the same point. Equivalently, it is the set of all lines through the origin in $\mathbb{R}^3$. Informally and originally it was the usual plane with a line at infinity added. In plane projective geometry duality is the exchange of points and lines so that collinear points go to concurrent lines and vice versa, and the dual of the dual of a point *P* is again the point *P*. The equation of a conic in plane projective geometry is a second-degree polynomial equation in the homogeneous variables *x*, *x*, *z*, thus $ax^2 + hxy + gxz + by^2 + fyz + cz^2 = 0$.

Quantifiers are used to express such statements as no *A*'s are *B*'s, or that at least one *A* is a *B*, or that all *A*'s are *B*'s. The symbol $\exists$ is used to say that there exists at least one *a* such that some condition is true. The symbol $\forall$ is used to say that for every *a* some condition is true. The symbol $\neg$ is used for negation.

Ring A set *R* is a ring if it is a commutative group under a combination law (commonly called addition and denoted "+," for which the identity element is called zero and denoted 0) and there is a second combination law (commonly called multiplication and denoted ".") with properties that if $a, b, c \in R$ then $a.b \in R$, $(a+b).c = a.c + b.c$ and $c.(a+b) = c.a + c.b$. If, moreover, for all $a, b \in R$, $a.b = b.a$, the ring is said to be commutative. If there is an element 1 such that for all $a \in R$ one has $a.1 = 1.a = a$, the ring is said to have a unit. The integers are an example of a commutative ring with a unit. A commutative ring with a unit in which it is possible to divide by every non-zero element (so for all $a \neq 0 \in R$ there is an *a'* such that $a.a' = 1$) is a field.

Schoenflies's theorem in the plane adds to the Jordan curve theorem the claim that the interior region is homeomorphic to a disk and the exterior region to a disk with a point removed. The analogous claim in higher dimensions is false—the exterior region need not be homeomorphic to an *n*-dimensional ball with a point removed.

Set-theoretic notation The union of two sets A and B is the set whose elements are elements of either A or B (in mathematics, this includes elements that belong to both A and B). The union is denoted $A \cup B$. The intersection of two sets A and B is the set whose elements are elements of both A and B. It is denoted $A \cap B$.

The three-body problem concerns the motion of three bodies in space moving under the influence of their mutual gravitational fields, and asks for the future state of motion given the initial positions and momenta of the three bodies. In this form it is intractable. Poincaré introduced and studied the restricted three-body problem, in which two bodies are allowed to have mass and to travel in circular orbits around their mutual center of gravity, while the third body is of negligible mass and moves in the common plane of the orbits of the first two. The problem now asks for the motion of the third body, which does not affect the motion of the other two, and even this can be very complicated and chaotic.

A vector space $\mathbb{V}$ is a set of elements that form a commutative group and can be multiplied by the elements of a field $\mathbb{F}$. This means that if $\mathbf{a}, \mathbf{b} \in \mathbb{V}$ and $\alpha \in \mathbb{F}$ then $\alpha\mathbf{a} \in \mathbb{V}$ and $\alpha(\mathbf{a} + \mathbf{b}) = \alpha\mathbf{a} + \alpha\mathbf{b}$. The elements of a vector space are called *vectors*. Note that one does not say what a vector is first, and then form a suitable set of vectors that make a vector space. If the field is $\mathbb{R}$, the field of real numbers, the vector space is called a *real vector space*.

ZF, ZFC Zermelo-Fraenkel set theory, or ZF, is a set of axioms for set theory (see §4.7.6 and §7.1.4). Zermelo-Fraenkel set theory with addition of the axiom of choice is commonly denoted ZFC.

BIBLIOGRAPHY

Abbott, Edwin A. 1884. *Flatland: A romance of many dimensions* by A. Square. London: Seeley.

Abraham, Tara. 2002. (Physio)logical circuits: The intellectual origins of the McCulloch-Pitts neural networks. *Journal of the History of Behavioral Science* 38.1: 3–25.

Alberts, Gerhard. 2000. The rise of mathematical modelling. *Nieuw Arch. Wiskunde* (5) 1.1: 59–67.

Anellis, Irving. 1995. Peirce rustled, Russell pierced: How Charles Peirce and Bertrand Russell viewed each other's work in logic, etc. *Modern Logic* 5.3: 270–328.

Appell, Paul. 1925. *Henri Poincaré*. Paris: Plon.

Archibald, Thomas. 1996. From attraction theory to existence proofs: The evolution of potential-theoretic methods in the study of boundary-value problems, 1860–1890. *Revue d'Histoire des Mathématiques* 2.1: 67–93.

Archibald, Thomas, and Louis Charbonneau. 1995. Mathematics in Canada before 1945: A preliminary survey. In *Canadian Mathematical Society/Société mathématique du Canada, 1945–1995. 1. Mathematics in Canada/Les mathématiques au Canada*, ed. Peter Fillmore, 1–44. Ottawa: Canadian Mathematical Society.

Aspray, William. 1990. *John von Neumann and the origins of modern computing*. Cambridge, Mass.: MIT Press.

Aspray, William, and Philip Kitcher, eds. 1988. *History and philosophy of modern mathematics*. Minnesota Studies in the Philosophy of Science, vol. 11. Minneapolis: University of Minnesota Press.

Atzema, Eisso. 1993. "The structure of systems of lines in the 19th century." Unpublished Proefschrift, University of Utrecht.

Avellone, M., A. Brigaglia, and C. Zappulla. 2002. The foundations of projective geometry in Italy from De Paolis to Pieri. *Archive for History of Exact Sciences* 56.5: 363–425.

Avigad, Jeremy. 2006. Methodology and metaphysics in the development of Dedekind's theory of ideals. In Ferreirós and Gray 2006, 159–86.

Awodey S., and E. R. Reck. 2002. Completeness and categoricity. I: Nineteenth-century axiomatics to twentieth-century metalogic. *History and Philosophy of Logic* 23.1: 1–30; Completeness and categoricity. II: Twentieth-century metalogic to twenty-first century semantics. *History and Philosophy of Logic* 23.2: 77–94.

Bacon, Francis. 1915. *The advancement of learning*. Ed. G. W. Kitchin. Dent, London, and New York: Everyman. (Based on the edition of 1605.)

Balaguer, Mark. 1998. *Platonism and anti-Platonism in mathematics*. Oxford: Oxford University Press.

Baltzer, R. 1868. *Die Elemente der Mathematik*. Leipzig: Teubner.

Banks, E. C. 2005. Kant, Herbart, and Riemann. *Kantstudien* 96: 208–34.

Barrow-Green, June. 1994. International Congresses of Mathematicians from Zurich 1897 to Cambridge 1912. *The Mathematical Intelligencer* 16.2: 38–41.

———. 1997. *Poincaré and the three-body problem*. American and London Mathematical Societies, HMath 11, Providence, R.I.

Barrows, John H., ed. 1893. *The world's parliament of religions: An illustrated and popular story of the world's first parliament of religions, held in Chicago in connection with the Columbian Exposition of 1893*. 2 vols. Chicago: Parliament Publishing Company.

Baumann, J. J. 1868. *Die Lehren von Raum, Zeit, und Mathematik in der neueren Philosophie, nach ihrem ganzen Einfluss dargestellt und beurtheilt*. Berlin.

Beaney, Michael. 1996. *Frege: Making sense*. London: Duckworth.

Belhoste, Bruno. 2003. *La formation d'une technocratie*. Paris: Belin.

Bellivier, André. 1956. *Henri Poncaré; ou la vocation souveraine*. Paris: Gallimard.

Beltrami, Eugenio. 1868a. Saggio di interpretazione della geometria non Euclidea. *Giornale di Matematiche* 6: 284–312. In *Opere matematiche*, 1:374–405. Milan: Hoepli, 1902. English translation in Stillwell 1996, 7–34.

——. 1868b. Teoria fondamentale degli spazii di curvatura costante. *Annali di Matematica pura et applicata* (2) 2: 232–55. In *Opere matematiche*, 1:406–29. Milan: Hoepli, 1902. English translation in Stillwell 1996, 41–62.

Benacerraf, Paul. 1973. Mathematical truth. *Journal of Philosophy* 70: 661–79. Rep. in Hart 1989, 14–30.

Benis Sinaceur, Hourya. 2006. From Kant to Hilbert: French philosophy of concepts in the beginning of the twentieth century. In Ferreirós and Gray 2006, 311–38.

Berman, Marshall. 1982. *All that is solid melts into air: The experience of modernity*. New York: Simon and Schuster.

Bernays, Paul. 1922. Über Hilberts Gedanken zur Grundlagen der Arithmetik. *Jahresbericht der Deutschen-Mathematiker Vereinigung* 31:10–19.

Bettazzi, Rodolfo. 1890. *Teoria delle grandezze*. Opera premiata dalla R. Accademia dei Lincei. Pisa: Spoerri.

——. 1891. Osservazioni sopra l'articolo del dott: G. Vivanti "Sull' infinitesimo attuale." *Rivista di Mat.* 1:174–82.

Bianchi, Luigi. 1898. Sugli spazi a tre dimensioni che ammettono un gruppo continuo di movimenti. *Memorie di Matematica e di Fisica della Societa Italiana delle Scienze* 11.3:267–352. In Luigi Bianchi, *Opere*, 9, 17–109. Rome: Edizione Cremonese, 1952. English translation with notes: R. T. Jantzen, On three-dimensional spaces which admit a continuous group of motions. *General Relativity and Gravitation* 33 (2001): 2157–70, 2170–253.

Birkhoff, Garrett, and M. K. Bennett. 1988. Felix Klein and his "Erlangen Program." In *History and philosophy of modern mathematics*, ed. W. Aspray and P. Kitcher, 145–76. Minnesota Studies in the Philosophy of Science, vol. 11. Minneapolis: University of Minnesota Press.

Bliss, Gilbert Ames. 1923. The reduction of singularities of plane curves by birational transformation. *Bulletin of the American Mathematical Society* 29: 161–183.

Blumenthal, Otto. 1922. David Hilbert. *Die Naturwissenschaften*, 67–72.

——. 1935. Lebensgeschichte. In Hilbert 1935, 3:388–429.

Bôcher, Maxime. 1904/5. The fundamental conceptions and methods of mathematics. *Bulletin of the American Mathematical Society* 11:115–34.

Boi, Luciano, L. Giacardi, and R. Tazzioli. 1998. *La découverte de la géométrie non-Euclidienne sur la pseudosphere*. Paris: Blanchard.

Bolyai, Farkas. 1832. *Tentamen juventutem studiosam in Elementa Matheosis purae, etc.* Maros-Vásérhely.

Bolyai, János. 1832. Appendix scientiam spatii absolute veram exhibens. In F. Bolyai 1832, trans. J. Houël, La science absolue de l'espace. *Mémoires de la Société des Sciences Physiques et Naturelles de Bordeaux* 5 (1867): 189–248; also trans. G. Battaglini, Sulla scienza della spazio assolutamente vera. *Giornale di matematiche* 6 (1868): 97–115, trans. G. B. Halsted, Science of absolute of space, appendix in Bonola 1906/1912.

Boniface, J., and Norbert Schappacher. 2001. Sur le concept de nombre en mathématique. Cours inédit de Leopold Kronecker à Berlin (1891). *Revue d'histoire des Mathématiques* 7.2: 207–276.

Bonola, Roberto 1906/1912. *La geometria non-Euclidea.* English trans., H. S. Carslaw, preface by F. Enriques. In *History of non-Euclidean geometry.* Chicago: Open Court, 1912; repr., New York: Dover, 1955.

Boole, George. 1847. *The mathematical analysis of logic.* Cambridge: Macmillan.

———. 1854. *Investigation of the laws of thought, on which are founded the mathematical theories of logic and probabilities.* London: Walton and Maberly.

Boolos, George. 1998a. Reading the *Begriffsschrift. Mind* 94: 331–344; repr. in Boolos 1998, 155–70.

———. 1998b. *Logic, logic, and logic.* Cambridge, Mass.: Harvard University Press.

Borel, Émile. 1898. *Leçons sur la théorie des fonctions.* Paris: Gauthier-Villars et Fils; 2nd ed., 1914.

———. 1907. La logique et l'intuition en mathématiques. *Revue de metaphysique* 15: 273–283.

Boring, Edwin G. 1950. *A history of experimental psychology.* 2nd ed. New York: Appleton-Century-Crofts.

Bottazzini, Umberto. 1981. *Il calcolo sublime: Storia dell'analisi matematica da Euler a Weierstrass.* Torino: Boringhieri. English trans., W. Van Egmond, *The higher calculus: A history of real and complex analysis from Euler to Weierstrass* New York: Springer, 1986.

———. 1988. Fondamenti dell'aritmetica e della geometria. In *Storia della scienza moderna e contemporanea,* ed. Paolo Rossi, 253–88. Turin: Unione Tipografico Editrice Torinese.

———. 2001. I geometri italiani e il problema dei fondamenti (1889–1899). *Bol. Unione Mat. Ital.* 4.2: 281–329.

———. 2002. Italy, in Dauben and Scriba 2002, 61–95.

———. 2003. Complex function theory, 1780–1900. In N. H. Jahnke, ed., *A history of analysis,* 213–60. American and London Mathematical Societies, HMath 24, Providence, R.I.

Bottazzini, Umberto, Alberto Conte, and Paola Gario. 1996. *Riposte armonie, lettere di Federigo Enriques a Guido Castelnuovo.* Turin: Bollati Boringhieri.

Bottazzini, Umberto, and Rosanna Tazzioli. 1995. *Naturphilosophie* and its role in Riemann's mathematics. *Revue d'Histoire des Mathématiques* 1.1: 3–38.

Boudewijnse, G.-J., D. J. Murray, and C. A. Bandomir. 1999. Herbart's mathematical psychology. *History of Psychology* 2: 163–193.

———. 2000. The fate of Herbart's mathematical psychology. *History of Psychology* 4: 107–132.

Boutroux, Etienne. 1908. Science et religion dans la philosophie contemporaine. Paris.

Bréal, Michel. 1897. *Essai de sémantique.* Paris: Hachette.

———. 1901. Le choix d'une langue internationale. *Revue de Paris* (4) 8: 229–246.

Brent, Joseph. 1993. *Charles Sanders Peirce: A life.* Bloomington: Indiana University Press.

Brouwer, Luitzen E. J. 1905. *Leven, Kunst en Mystiek.* Delft: Waltman. English trans., *Life, Art, and Mysticism,* in *Notre Dame Journal of Formal Logic* 37.3 (1996): 391–430.

———. 1912. *Intuitionisme en formalisme.* English trans., A. Dresden, *Intuitionism and formalism,* in *Bulletin of the American Mathematical Society* 20 (1913): 81–96.

———. 1975. *Collected works.* 2 vols. Amsterdam: North-Holland.

Burkholder, J. P. 1999. Schoenberg the reactionary. In *Schoenberg and his world,* ed. W. Frisch, 162–191 Princeton, N.J.: Princeton University Press.

Burrow, J. W. 2000. *The crisis of reason: European thought, 1848–1914,* New Haven, Conn.: Yale University Press.

Cahan, David, ed. 1993. *Hermann von Helmholtz and the foundations of nineteenth-century science.* Berkeley: University of California Press.

Calinon, Auguste. 1889. Les espaces géométriques. *Revue Philosophique de la France et de l'Etranger* 27: 588–595.

———. 1893. Étude sur l'indétermination géométrique de l'univers. *Revue Philosophique de la France et de l'Etranger* 36: 595–607.

Campbell, Norman. 1920. *Physics; the elements.* Cambridge: Cambridge University Press. Repr. as *Foundations of science,* New York: Dover, 1957.

Cannell, Mary. 1993. *George Green: Mathematician and physicist, 1793–1841.* London: Athlone Press.

Cantor, Georg. 1883. *Grundlagen einer allgemeinen Mannigfaltigkeitslehre.* Leipzig: Teubner.

———. 1887–88. Mittheilung zur Lehre der Transfiniten. *Zeitschrift für Philosophische Kritik* 91: 81–125 and 92: 240–265; also in Cantor 1932, 240–265.

———. 1890–91. Über eine elementare Frage der Mannigfaltigkeitslehre. *Jahresbericht der Deutschen Mathematiker-Vereinigung* 1: 75–78; also in Cantor 1932, 278–81.

———. 1895, 1897. *Beiträge zur Begründung der transfiniten Mengenlehre. Mathematische Annalen* 46: 481–512, and 49: 207–46; also in Cantor 1932, 282–354.

———. 1932. *Gesammelte Abhandlungen mathematischen und philosophischen Inhalts. Mit erläuternden Anmerkungen sowie mit Ergänzungen aus dem Briefwechsel Cantor-Dedekind herausgegeben von Ernst Zermelo. Nebst einem Lebenslauf Cantors von Adolf Fraenkel.* Berlin: Springer.

———. 1991. *Briefe.* Ed. H. Meschkowski and W. Nilson. Berlin: Springer.

Cantor, Moritz. 1894–1908. *Vorlesungen über Geschichte der Mathematik.* 4 vols. 2nd ed. Leipzig: Teubner.

Cantù, Paola. 1999. *Guiseppe Veronese e i Fondamenti della Geometria.* Milan: Unicopli.

Carnap, Rudolph. 1950. Empiricism, semantics, and ontology. *Revue Internationale de Philosophie* 4: 20–40, repr. in the supplement to *Meaning and necessity: A study in semantics and modal logic,* enlarged edition. Chicago: University of Chicago Press, 1956.

Carus, Paul. 1889. *Fundamental problems: The method of philosophy as a systematic arrangement of knowledge.* London: Longmans & Co.; repr., Chicago: Open Court Publishing, 1903.

Cassirer, Ernst. 1902. *Leibniz's System in seinen wissenschaftlichen Grundlagen.* Marburg.

———. 1907. Kant und die moderne Mathematik. *Kantstudien* 12: 1–43.

———. 1910. *Substanzbegriff und Funktionsbegriff: Untersuchungen über die Grundfragen der Erkenntniskritik.* Berlin: B. Cassirer; trans. W. C. and M. C. Swabey, *Substance and function and Einstein's theory of relativity.* Chicago: Open Court, 1923; repr., New York: Dover, 1953.

Cauchy, Augustin Louis. 1821. *Cours d'analyse. Première partie: Analyse algébrique,* ed. U. Bottazzini; rep., Bologna: Editrice Clueb, 1990.

Cayley, Arthur. 1865. Note on Lobatchewsky's imaginary geometry. *Philosophical Magazine* 29: 231–33; also in *Collected mathematical papers* 5.362: 471–72. Cambridge: Cambridge University Press.

Changeux, Jean-Pierre and Alain Connes. 1995. *Conversations on mind, matter, and mathematics.* Trans. M. B. DeBevoise. Princeton, N.J.: Princeton University Press.

Chasles, Michel. 1837. *Aperçu historique sur l'origine et le développement des méthodes en géométrie . . . suivi d'un mémoire de géométrie, etc.* In *Mémoires sur les questions proposées par l'Académie Royale des Sciences et Belles-Lettres de Bruxelles,* vol. 11. Brussels.

Chemla, Karine. 2004. Euler's work in spherical trigonometry: Contributions and applications. In *Leonhardi Euleri Opera Omnia,* ser. 4, 10:120–87. Basel: Birkhäuser.

Chimisso, Christina. 2008. *Writing the history of the mind: Science and philosophy in France, 1900–1960.* London: Ashgate.

Chipp, Herschel B. 1968 *Theories of modern art.* Berkeley: University of California Press.

Clendenning, John. 1999. *The life and thought of Josiah Royce*. Nashville: Vanderbilt University Press.

Clifford, William Kingdon. 1878. On the nature of things-in-themselves. *Mind* 3:57–67.

Coffa, J. Alberto. 1991. *The semantic tradition from Kant to Carnap: To the Vienna station*, ed. Linda Wessels. Cambridge: Cambridge University Press.

Cohen, Paul. 1966. *Set theory and the continuum hypothesis*. San Francisco: Freeman.

Coolidge, Julian Lowell. 1940. *A history of geometrical methods*. Oxford: Clarendon Press.

Corcoran, John. 1981. From categoricity to completeness. *History and Philosophy of Logic* 2:113–19.

Corry, Leo. 1996. *Modern algebra and the rise of mathematical structures*. Boston: Birkhäuser; 2nd ed., 2003.

———. 1997a. David Hilbert and the axiomatisation of physics (1894–1905). *Archive for History of Exact Sciences* 51.2:83–198.

———. 1997b. Hermann Minkowski and the postulate of relativity. *Archive for History of Exact Sciences* 51.4:273–314.

———. 1999. Hilbert and physics (1900–1915). In *The symbolic universe: Geometry and physics, 1890–1930*, ed. J. J. Gray, 145–188. Oxford: Oxford University Press.

———. 2004. *David Hilbert and the axiomatization of physics (1898–1918): From* Grundlagen der Geometrie *to* Grundlagen der Physik. Dordrecht: Kluwer Academic.

———. 2007. From *Algebra* (1895) to *Moderne Algebra* (1930): Changing conceptions of a discipline; A guided tour using *Jahrbuch über die Fortschritte der Mathematik*, in *Episodes in history of modern algebra*, ed. J. J. Gray and K. H. Parshall. HMath 32, American and London Mathematical Societies, Providence, R.I.

Courant, Richard. 1981. Reminiscences from Hilbert's Göttingen. *Mathematical Intelligencer* 3.4: 154–164.

Couturat, Louis. 1896. *De L'infini mathématique*. Paris: Alcan.

———. 1901. *La Logique de Leibniz d'après des documents inédits*. Paris.

———. 1905. *Principes des mathématiques*. Paris: Alcan.

Couturat, Louis, and Leopold Leau. 1903. *Histoire de la langue universelle*. Paris: Hachette.

Cremona, Luigi. 1868. Mémoire de géométrie pure sur les surfaces du troisième ordre. *Journal für die reine und angewandte Mathematik* 68:1–133.

Crilly, Tony. 2006. *Arthur Cayley: Mathematician laureate of the Victorian age*. Baltimore: Johns Hopkins University Press.

Crowe, Michael J. 1967 *A history of vector analysis: The evolution of the idea of a vectorial system*. Notre Dame, Ind.: University of Notre Dame Press; reprint, New York: Dover, 1985.

Dalen, Dirk van. 1991. Brouwer's dogma of languageless mathematics and its role in his writings. In *Significs, mathematics and semiotics: The significs movement in the Netherlands*, ed. E. Heijermann and H. W. Schmitz, 33–41. Münster, Germany: Nodus Publications.

———. 1995, 2005. *Mystic, geometer, and intuitionist: The life of L.E.J. Brouwer*. Vol. 1, *The dawning revolution;* vol. 2, *Hope and disillusion*. Oxford: Clarendon Press.

Dantzig, Georg van. 1957. Gerrit Mannoury's significance for mathematics and its foundations. *Nieuw Archief voor Wiskunde* (3) 5:1–18.

Darboux, Gaston. 1875. Mémoire sur les fonctions discontinues. *Annales de l'École Normale Supérieure* (2) 4:57–112.

———. 1887. *Leçons sur la théorie générale des surfaces applications géométriques du calcul infinitésmal*. 4 vols. Paris: Gauthier-Villars.

Darrigol, Olivier. 1995. Henri Poincaré's criticism of *fin de siècle* electrodynamics. *Studies in History and Philosophy of Modern Physics* 26: 1–44.

———. 2000. *Electrodynamics from Ampère to Einstein*. Oxford: Oxford University Press.

Daston, Lorraine J. 1986. The physicalist tradition in early 19th-century French geometry. *Studies in History and Philosophy of Science* 17: 269–295.

Dauben, Joseph Warren. 1979. *Georg Cantor: His mathematics and philosophy of the infinite.* Cambridge, Mass.: Harvard University Press.

Dauben, Joseph Warren, and Christoph J. Scriba, eds. 2002. *Writing the history of mathematics: Its historical development.* Science Networks—Historical Studies 27. Basel: Birkhäuser.

Davies, Anna M. 1998. *Nineteenth-century linguistics.* History of Linguistics, 4, ed. Giulio Lepschy. London: Longmans.

Dawson, John W. Jr. 1997. *Logical dilemmas: The life and work of Kurt Gödel.* Wellesley, Mass.: AK Peters.

Dedekind, Richard. 1877. Sur la théorie des nombres entiers algébriques. *Bulletin des sciences mathématiques* (2) 1: 69–92.

———. 1888. *Was sind und was sollen die Zahlen?* Braunschweig: Vieweg & Sohn. Also in R. Dedekind, *Gesammelte Mathematische Werke,* vol. 3, R. Fricke, E. Noether, and O. Ore, eds, 335–392. Braunschweig: Vieweg & Sohn, 1930. Authorized English trans. in Wooster Woodruff Beman, *Essays on the theory of numbers, I: Continuity and irrational numbers, II: The nature and meaning of number.* Chicago: Open Court, 1909.

Dedekind, Richard, and Heinrich Weber. 1882. Theorie der algebraischen Functionen einer Veränderlichen. *Journal für die reine und angewandte Mathematik* 92: 181–291. Also in R. Dedekind, *Gesammelte Mathematische Werke,* vol. 1, ed. R. Fricke, E. Noether, and O. Ore. Braunschweig: Vieweg & Sohn, 1930.

Demopolous, William. 1991. Frege, Hilbert, and the conceptual structure of model theory. *History and Philosophy of Logic* 15:211–225.

———, ed. 1995. *Frege's philosophy of mathematics.* Cambridge, Mass.: Harvard University Press.

Derbyshire, John. 2003. *Prime obsession.* Washington, D.C.: Joseph Henry Press.

Descartes, René. 1649. *Les passions de l'ame.* Paris: Chez Henry Le Gras.

Detlefsen, Michael. 1997. Philosophy of mathematics in the twentieth century. In *Routledge History of Philosophy,* ed. Stuart G. Shanker, vol. 9, 50–123. London: Taylor and Francis.

Didon, Père Henri. 1884. *Les Allemands.* Paris: Plon.

———. 1891. *Jesus Christ.* Paris: Plon.

Dirichlet, Peter Gustav Lejeune. 1829. Sur la convergence des séries trigonométriques. *Journal für die reine und angewandte Mathematik* 4: 157–169; also in *Gesammelte Werke,* 2 vols., ed. L. Fuchs and L. Kronecker, vol. 1, 117–132. Berlin, 1889 and 1897. Berlin: Preussische Akademie der Wissenschaften; rep. New York: Chelsea, 1969.

———. 1876. *Vorlesungen über die im umgekehrten Verhältniss des Quadrats der Entfernung wirkenden Kräfte,* ed. F. Grube. Leipzig: Teubner.

Dorier, J.-L. 1995. A general outline of the genesis of vector space theory. *Historia Mathematica* 22.3: 227–61.

Dow, Arthur. 1917. Modernism in art. *American Magazine of Art* 8:113–16.

Doyle, Arthur Conan. 1891. *The sign of four.* London: George Newnes.

du Bois-Reymond, Paul. 1882. *Die allgemeine Functiontheorie.* Tübingen: Laupplschen Buchhandlung.

du Sautoy, Marcus. 2003. *The music of the primes.* Oxford: Oxford University Press.

Duda, Roman. 1996. Fundamenta Mathematica and the Warsaw school of mathematics. In *L'Europe mathématique,* ed. C. Goldstein, J. J. Gray, and J. Ritter, 479–98. Paris: Éditions de la Maison des sciences de l'homme.

Dugac, Pierre. 1973. Éléments d'analyse de Karl Weierstrass. *Archive for History of Exact Sciences* 10:41–176.

———. 1990. Lettres de René Baire à Emile Borel. *Cahiers du Séminaire d'Histoire des Mathématiques* 11:33–120.

Duhem, Pierre. 1908. *La théorie physique. Son objet et sa structure.* English trans., P. P. Wiener, *The aim and structure of physical theory,* Princeton: Princeton University Press, 1954.

Dummett, Michael. 1991. *Frege: Philosophy of mathematics.* London: Duckworth. Chap. 22 reprinted in Demopolous 1995.

Dupin, Pierre Charles François. 1819. *Essai historique sur les services et les travaux scientifiques de G. Monge.* Paris.

Eco, Umberto. 1995. *The search for a perfect language.* English trans., James Fentress. London: Fontana.

Edwards, Anthony W. F. 2004. *Cogwheels of the mind: The story of Venn diagrams.* Baltimore: Johns Hopkins University Press.

Edwards, Harold M. 1977. *Fermat's last theorem.* New York: Springer.

———. 1980. The genesis of ideal theory. *Archive for History of Exact Sciences* 23:321–78.

———. 1987. Dedekind's invention of ideals. *Studies in the History of Mathematics*, ed. E. R. Phillips, 8–20. Studies in Mathematics 26, Mathematical Association of America.

———. 2005. *Essays in constructive mathematics.* New York: Springer.

Ehrlich, Paul. 2006. The rise of non-Archimedean mathematics. *Archive for History of Exact Sciences* 60.1:1–121.

Einstein, Albert. 1912a. The speed of light and the statics of the gravitational field. English trans., in *The collected papers of Albert Einstein*, vol. 4: *The Swiss years: Writings, 1912–1914*, 95–106. Princeton, N.J.: Princeton University Press.

———. 1912b. On the theory of the static gravitational field. In English trans. of *The collected papers of Albert Einstein*, vol. 4, *The Swiss years: Writings, 1912–1914*, 107–120. English trans. Anna Beck and Don Howard. Princeton, N.J.: Princeton University Press, 1996.

———. 1917. *Ueber die spezielle and die allgemeine Relativitätstheorie: Gemeinverständlich.* Braunschweig: Vieweg. English trans., *Relativity, the special and the general theory: A popular exposition.* London: Methuen, 1920, 1954; New York: Holt, 1921.

———. 1921. Geometrie und Erfahrung. *Sitzungsberichte der Königlich Preussischen Akademie der Wissenschaften* (Berlin), 123–130. English trans., Sonja Bargmann, Geometry and experience, in Einstein, *Ideas and opinions.* New York: Crown, 1982; rep. in English trans. of *The collected papers of Albert Einstein*, vol. 7: *The Berlin years: Writings, 1918–1921*, English trans., Alfred Engel and Engelbert Schucking. Princeton, N.J.: Princeton University Press, 2002.

———. 1995. *The collected papers of Albert Einstein* vol. 5, *Correspondence: The Swiss years, 1901–1914.* English trans., Anna Beck and Don Howard. Princeton, N.J.: Princeton University Press.

———. 1997. *The collected papers of Albert Einstein*, vol. 6, *The Swiss years: Correspondence, 1902–1914.* English trans., Anna Beck and Engelbert Schucking. Princeton, N.J.: Princeton University Press.

———. 1998. *The collected papers of Albert Einstein*, vol. 8A, B, *The Berlin years: Correspondence, 1914–1918.* English trans., Ann Hentschel and Klaus Hentschel. Princeton, N.J.: Princeton University Press.

Engel, Friedrich, and Paul Stäckel. 1895. *Theorie der Parallellinien von Euklid bis auf Gauss.* Leipzig: Teubner.

Enriques, Federigo. 1898. *Lezioni di geometria proiettiva.* Bologna: Zanichelli.

———, ed. 1900. *Questioni riguardanti la geometria elementare.* Bologna: Zanichelli.

———. 1906. *Problemi della scienza.* English trans., Katherine Royce, *Problems of science.* Chicago: Open Court, 1914.

———. 1907. Prinzipien der Geometrie. In *Encyklopädie der Mathematischen Wissenschaften* 3:1: 1–129.

———. 1912. *Scienza e razionalismo.* Bologna: Zanichelli.

Epple, Moritz. 1998. Topology, matter and space. I: Topological notions in 19th-century natural philosophy. *Archive for History of Exact Sciences* 52.4: 297–392.

———. 1999. *Die Entstehung der Knotentheorie.* Braunschweig: Vieweg.

———. 2000. Did Brouwer's intuitionistic analysis satisfy its own epistemological standards? In *Proof theory: History and philosophical significance*, ed. V. F. Hendricks, S. A. Pedersen, and K. F. Jørgensen, 153–78. Synthese Library, 292. Dordrecht: Kluwer.

———. 2002. From quaternions to cosmology: Spaces of constant curvature, ca. 1873–1925. In *Proceedings of the International Congress of Mathematicians, Beijing, 2002*, vol. 3, 1–3.

———. 2003. The end of the science of quantity: Foundations of analysis, 1860–1910. In *A history of analysis*, ed. N. H. Jahnke, 291–324. American and London Mathematical Societies, HMath 24, Providence, R.I.

———. 2004. An unusual career between cultural and mathematical modernism: Felix Hausdorff, 1868–1942. Paper presented to Research Practices of Jewish Scientists and Scholars in the 19th and 20th Centuries, a conference of the Leo Baeck Institute London Centre for German Jewish Studies at the University of Sussex, October 4–5, 2004.

Erdmann, Benno. 1877. *Die Axiome der Geometrie: Eine philosophische Untersuchung der Riemann-Helmholtz'schen Raumtheorie*. Leipzig.

Erdmann, Benno. 1892. *Logische Elementarlehre*. Halle a. S.: Max Niemeyer. 2nd ed, 1907.

Euler, Leonhard. 1770. *Lettres à une princesse d'Allemagne sur divers sujets de physique et de philosophie*, 3 vols. English trans., Henry Hunter, *Letters of Euler to a German princess, on different subjects in physics and philosophy*. London, 1795.

Everdell, William. 1997. *The first moderns*. Chicago: University of Chicago Press.

Ewald, William. 1996. *From Kant to Hilbert: A source book in the foundations of mathematics*. 2 vols. Oxford: Clarendon Press.

Fancher, Raymond E. 1979. *Pioneers of psychology*. New York: Norton.

Fano, Gino. 1892. Sui postulati fondamentali della geometria proiettiva. *Giornale di Matematiche* 30: 106–131.

———. 1932. Geometrie non euclidee e non archimedee. In *Enciclopedia delle matematiche elementari e complementi*, ed. L. Berzolari, G. Vivanti, and D. Gigli, vol. 2.2, art. 38: 435–511.

Fechner, Gustav T. (writing as Dr. Mises). 1846. *Vier Paradoxa*. Leipzig.

———. 1851. *Zend-Avesta, oder über die Dinge des Himmels, und des Jenseits, vom Standpunkt der Naturbetrachtung*. Leipzig.

———. 1860. *Elemente der Psychophysik*. 2 vols. Leipzig. English trans., Helmut E. Adler, *Elements of psychophysics*, ed. Davis H. Howes and Edwin G. Boring. New York: Holt, Rinehart & Winston.

Feeman, Timothy G. 2002. *Portraits of the earth*. Providence, R.I.: American Mathematical Society.

Feferman, A. B., and S. Feferman. 2004. *Alfred Tarski: Life and logic*. Cambridge: Cambridge University Press.

Feferman, Solomon. 1998. Weyl vindicated: *Das Kontinuum* seventy years later. In *In the light of logic*. Oxford: Oxford University Press.

———. 2000. The significance of Hermann Weyl's *Das Kontinuum*. In *Proof theory: History and philosophical significance*, ed. V. F. Hendricks, S. A. Pedersen, and K. F. Jørgensen, Synthese Library, 292. Dordrecht: Kluwer.

———. 2006 The impact of the incompleteness theorems on mathematics. *Notices of the American Mathematical Society* 53.4: 434–39.

Fermat, Pierre de. 1891. *Oeuvres de Fermat*. Vol. 1, ed. P. Tannery and C. Henry. Paris.

Ferraro, Giovanni. 1999. The first modern definition of the sum of a divergent series: An aspect of the rise of 20th-century mathematics. *Archive for History of Exact Sciences* 54.2: 101–35.

Ferreirós, José. 1996. Traditional logic and the early history of sets, 1854–1908. *Archive for History of Exact Sciences* 50: 5–71.

———. 1999. *Labyrinth of thought*. Basel: Birkhäuser, Science Networks.

———. 2001. The road to modern logic—an interpretation. *Bulletin of Symbolic Logic* 7.4: 441–484.

Ferreirós, José, and Jeremy J. Gray, eds. 2006. *The architecture of modern mathematics.* Oxford: Oxford University Press.

Fine, Henry B. 1905. *College algebra.* Boston: Ginn; repr., New York: Dover, 1961.

Fisch, Menachem. 1994. The emergency which has arrived: The problematic history of 19th-century British algebra—A programmatic outline. *British Journal for the History of Science* 27:247–76.

———. 1999. The making of Peacock's treatise on algebra: A case of creative indecision. *Archive for History of Exact Sciences* 54:137–79.

Fisher, Gordon. 1981. The infinite and infinitesimal quantities of du Bois-Reymond and their reception. *Archive for History of Exact Sciences* 24:101–63.

Fiske, Thomas Scott. 1988. The beginnings of the American Mathematical Society: Reminiscences of Thomas Scott Fiske. In *A century of mathematics in America*, part 1, ed. P. Duren, 13–17.

Flament, Dominique, ed. 1997. *Le nombre—Une hydre à n visages: Entre nombres complexes et vecteurs.* Paris: Maison des sciences de l'homme.

Folina, Janet. 2006. Poincaré's circularity arguments for mathematical intuition. In *Kant's scientific legacy in the nineteenth century*, ed. M. Friedman and A. Nordmann, 275–94. Cambridge, Mass.: Dibner Press, MIT.

Føllesdal. Dagfinn. 1958. *Husserl und Frege: Ein Beitrag zur Beleuchtung der Entstehung der phänomenologischen Philosophie.* Oslo: I kommisjon hos Aschehoug.

Forman, Paul. 1971. Weimar culture, causality, and quantum theory: Adaptation by German physicists and mathematicians to a hostile environment. *Historical Studies in the Physical Sciences* 3:1–115.

Fourier, Joseph. 1822. *Théorie analytique de la chaleur.* Paris.

Fraenkel, Abraham (Adolf). 1923. *Einleitung in die Mengenlehre.* 2nd ed. Berlin: Springer.

———. 1927. *Zehn Vorlesungen über die Grundlagung der Mengenlehre.* Leipzig: Teubner.

———. 1932. Das Leben Georg Cantors. In Cantor 1932, 452–83.

Franzén, Torkel. 2005. *Gödel's theorem: An incomplete guide to its use and abuse.* Wellesley, Mass.: A. K. Peters.

Frege, Gottlob. 1879. *Begriffsschrift: Eine der arithmetischen nachgebildete Formelsprache des reinen Denkens.* Halle, Germany.

———. 1884. *Die Grundlagen der Arithmetik: Eine logisch mathematische Untersuchung über den Begriff der Zahl.* Breslau. Germany. English trans., J. L. Austin, *The foundations of arithmetic: A logico-mathematical enquiry into the concept of number.* London: Basil Blackwell, 1950.

———. 1893–1903. *Grundgesetze der Arithmetik.* Jena. English trans. and ed. Montgomery Furth, *The Basic Laws of Arithmetic: Exposition of the system.* Berkeley: University of California Press, 1964.

———. 1894. Review of Husserl's *Philosophy of arithmetic. Zeitschrift für Philosophie und Kritik* 103: 3133–332. Trans., H. Kaal, *Collected papers on mathematics, logic, and philosophy*, ed. Brian McGuinness. Oxford: Blackwell.

———. 1906a. Reply to Mr Thomae's Holiday *Causerie. Jahresbericht der Deutschen Mathematiker-Vereinigung* 15: 586–90. Also in *Collected papers on mathematics, logic, and philosophy*, ed. Brian McGuinness, trans., Max Black et al., 341–45. Oxford: Blackwell, 1984.

———. 1906b. Über die Grundlagen der Geometrie, I, II, III. *Jahresbericht der Deutschen Mathematiker-Vereinigung* 15: 293–309, 377–403, 423–30. Trans. Max Black et al., On the foundations of geometry, first and second series, in *Collected papers on mathematics, logic, and philosophy*, ed. Brian McGuinness, 273–84 and 293–340. Oxford: Blackwell, 1984.

———. 1979. *Posthumous writings.* Ed. Hans Hermes, Friedrich Kambartel, Friedrich Kaulbach et al.; trans. Peter Long, Roger White, with the assistance of Raymond Hargreaves. Oxford: Blackwell.

———. 1980. *Philosophical and mathematical correspondence.* Various editors. Chicago: Chicago University Press.

———. 1983. *Nachgelassene Schriften und wissenschaftlicher Briefwechsel*, ed. Hans Hermes, Friedrich Kambartel, and Friedrich Kaulbach. Hamburg: Felix Meiner Verlag.

———. 1984. *Collected papers on mathematics, logic, and philosophy.* Ed. Brian McGuinness, trans. Max Black et al. Oxford: Blackwell.

Freudenthal, Hans. 1962. The main trends in the foundations of geometry in the 19th century. In *Logic, methodology and philosophy of science*, ed. E. Nagel, F. Suppes, and A. Tarski, 613–21. Stanford, Calif.: Stanford University Press.

Friedman, Michael. 1983. *Foundations of space-time theories.* Princeton, N.J.: Princeton University Press.

———. 1999. *Reconsidering logical positivism*, Cambridge: Cambridge University Press.

———. 2000a. Geometry, construction, and intuition in Kant and his successors. In *Between logic and intuition: Essays in honor of Charles Parsons*, ed. G. Sher and R. Tieszen. Cambridge: Cambridge University Press.

———. 2000b. *A parting of the ways: Carnap, Cassirer, and Heidegger.* Chicago: Open Court.

Fries, Jakob Friedrich. 1822. *Mathematische Naturphilosophie nach philo sophischer Methode bearbeitet.* Heidelberg.

Galison, Peter. 2003. *Einstein's clocks and Poincaré's maps: Empires of time.* New York: Norton.

Gandon, Sébastian. 2004. Russell et l'*Universal Algebra* de Whitehead: La géométrie projective entre ordre et incidence (1898–1903), *Revue d'Histoire des Mathématiques* 10.2: 187–256.

Gandy, Robin. 1995. The confluence of ideas in 1936. In *The universal Turing machine—A half-century survey*, ed. Rolf Herken, 55–111. Oxford: Oxford University Press.

Gardner, Martin. 1983. *Logic machines and diagrams.* 2nd ed. Brighton, U.K.: Harvester Press.

Gardner, Sebastian. 1999. *Kant and the* Critique of Pure Reason. London: Routledge.

Gaukroger, Stephen. 1997. *Descartes: An intellectual biography.* Oxford: Clarendon Press.

Gauss, Carl Friedrich. 1801. *Disquisitiones arithmeticae.* Brunswick. English trans., Arthur A. Clarke, *Disquisitiones Arithmeticae*, revised by William C. Waterhouse, with C. Greither and A. W. Grootendost. New York: Springer, 1986.

———. 1818. Neue Beweise und Erweiterung des Fundamentalsatzes in der Lehre von den quadratischen Resten. *Commentationes soc. reg. sc. Götting. recentiores* 4, Göttingen. Also in *Untersuchungen über höhere Arithmetik*, ed. H. Maser. 496–510. New York: Chelsea.

———. 1900. *Werke.* Vol. 8. Göttingen: Königliche Gesellschaft der Wissenschaften.

Gergonne, Joseph Diaz. 1827. Géométrie de situation. *Annales des Mathématiques*, 18, 150–152.

Giaquinto, Marcus. 2002. *The search for certainty.* Oxford University Press.

Gispert, Hélène. 1991. La Société Mathématique de France. *Cahiers d'Histoire & de Philosophie des Sciences* (n.s.) 34: 13–180.

Gispert, Hélène, and Renata Tobies. 1996. A comparative study of the French and German Mathematical Societies before 1914. In *L'Europe Mathématique*, ed. C. Goldstein, J. J. Gray, and J. Ritter, 407–430. Paris: Éditions de la Maison des sciences de l'homme.

Gödel, Kurt. 1930. Die Vollständigkeit der Axiome des logischen Funktionenkalküls. *Monatshefte für Mathematik und Physik* 37: 349–360. Trans. with a reprint of the original as The completeness of the axioms of the functional calculus, in Gödel 1995, vol. 1, 102–23.

———. 1931. Über formal unentscheidbare Sätze der *Principia mathematica* und verwandter Systeme I. Trans. with a reprint of the original as On formally undecidable propositions of *Principia mathematica* and related systems I, in Gödel 1995, vol. 1, 144–95.

———. 1995. *Collected works.* Vol. 3, numerous editors. Oxford: Oxford University Press.

———. 1986–2005. *Collected works.* 5 vols., numerous editors. Oxford: Oxford University Press.

Goldstein, Catherine, Jeremy J. Gray, and Jim Ritter, eds. 1996. *L'Europe Mathématique.* Paris: Éditions de la Maison des Sciences de l'Homme.

Goldstein, Catherine, Norbert Schappacher, and Joachim Schwermer, eds. 2007. *The shaping of arithmetic.* Berlin: Springer.

Grabiner, Judith V. 1981. *The Origins of Cauchy's rigorous calculus.* Cambridge, Mass.: MIT Press.

Grassmann, Hermann. 1844. *Die lineale Ausdehnungslehre of 1844.* English trans., Lloyd C. Kannenberg, *A new branch of mathematics: The "Ausdehnungslehre" of 1844 and other works.* Chicago: Open Court, 1995.

———. 1847. *Geometrische Analyse, geknüpft an die von Leibniz erfundene geometrische Charakteristik.* Leipzig: Weidmann.

———. 1862. *Die Ausdehnungslehre: Vollständig und in strenger Form bearbeitet.* Berlin: Enslin. English trans., Lloyd C. Kannenberg, *Extension Theory,* American and London Mathematical Societies, HMath 19, Providence, R.I., 2000.

Grattan-Guinness, Ivor. 1970. An unpublished paper by Georg Cantor. *Acta Mathematica* 124:65–107.

———. 2000. *The search for mathematical roots, 1870–1940: Logics, set theories, and the foundations of mathematics from Cantor through Russell to Gödel.* Princeton, N.J.: Princeton University Press.

Gray, Jeremy J. 1989. *Ideas of space: Euclidean, non-Euclidean, and relativistic.* 2nd ed. Oxford: Oxford University Press.

———. 1991. Did Poincaré say "Set theory is a disease"? *Mathematical Intelligencer* 13.1: 19–22.

———. 1992a. A nineteenth-century revolution in mathematical ontology. In *Revolutions in mathematics,* ed. D. Gillies, 226–248. Oxford: Oxford University Press.

———. 1992b. Poincaré, the stability of the solar system and topological ideas in dynamics. In *An investigation of difficult things: Essays on Newton and the history of the exact sciences,* ed. P. Harman and A. Shapiro, 503–24. Cambridge: Cambridge University Press.

———. 1996. Poincaré and electromagnetic theory. In *Henri Poincaré, science et philosophie,* ed. G. Heinzmann et al., 193–208. Berlin, Akademie Verlag; Paris: Blanchard.

———. 1997. Around and around: Quaternions, rotations, and Olinde Rodrigues. In *Le nombre—une hydre à n visages: Entre nombres complexes et vecteurs,* ed. D. Flamant, 89–101. Paris: Maison des sciences de l'homme.

———. 1998. The foundations of geometry and the history of geometry. *Mathematical Intelligencer* 20.2: 54–59.

———. 1999a. Mathematicians as philosophers of mathematics: Part 2. *For the Learning of Mathematics* 19.1: 28–31.

———. 1999b. Geometry—formalisms and intuitions. In *The symbolic universe: Geometry and physics, 1890–1930,* ed. J. J. Gray, 58–83. Oxford: Oxford University Press.

———. 2000a. *Linear differential equations and group theory from Riemann to Poincaré.* 2nd ed. Basel: Birkhäuser.

———. 2000b *The Hilbert challenge.* Oxford: Oxford University Press.

———. 2002 Languages for mathematics and the language of mathematics in a world of nations. In *Mathematics unbound: The evolution of an international mathematical community, 1800–1945,* ed. K. H. Parshall and A. C. Rice, 201–228. American and London Mathematical Societies, HMath 23, Providence, R.I.

———. 2003. Gauss and non-Euclidean geometry. In *Non-Euclidean geometries: János Bolyai memorial volume,* ed. A. Prékopa and E. Molnar, 61–80. New York: Springer.

———. 2004a. *János Bolyai, non-Euclidean geometry and the nature of space.* Cambridge, Mass: Burndy Library, MIT.

———. 2004b. Anxiety and abstraction in nineteenth-century mathematics. *Science in Context* 17.2: 23–48.

———. 2006a. A history of prizes in mathematics. In *The Millennium Prize Problems*, ed. J. Carlson, A. Jaffe, and A. Wiles, 3–27. Cambridge, Mass., and Providence R.I. Clay Mathematics Institute and American Mathematical Society.

———. 2006b. Poincaré between physics and philosophy. In *The Kantian legacy in nineteenth-century science*, ed. M. Friedman and A. Nordmann, 295–314. Cambridge, Mass: Dibner Press, MIT.

———. 2006c. *Worlds out of nothing: A course on the history of geometry in the 19th century.* London: Springer.

———. 2006d. Enriques: Popularising science and the problems of geometry. In *Interactions: Mathematics, physics, and philosophy, 1850–1940*, ed. V. F. Hendricks, K. F. Jorgensen, J. Lützen, and S. A. Pedersen, 135–54. Boston Studies in the Philosophy of Science 251. Dordrecht: Springer.

Green, George. 1828. An essay on the application of mathematical analysis to the theories of electricity and magnetism. Nottingham, U.K.

———. 1871. *Mathematical papers.* London.

Gregory, Frederick. 1983. Neo-Kantian foundations of geometry in the German Romantic period. *Historia Mathematica* 10.2: 184–201.

———. 2006. Extending Kant: The origins and nature of J. F. Fries's philosophy of science. In *The Kantian Legacy in Nineteenth-Century Science*, ed. Michael Friedman and Alfred Nordmann 81–100. Cambridge, Mass.: MIT Press.

Grelling, Kurt, and Leonard Nelson. 1908. Bemerkungen zu den Paradoxieen von Russell und Burali-Forti. *Abhandlungen der Fries'schen Schule*, n.s. 2, no. 3.

Griffin, Nicholas. 1991. *Russell's idealist apprenticeship.* Oxford: Clarendon Press.

Guérard, Albert L. 1922. *A short history of the international language movement.* London: Fisher Unwin.

Hadamard, Jacques. 1906. La logistique et la notion de nombre entier. *Revue générale des sciences* 17: 906–909. Also in *Oeuvres de Jacques Hadamard*, vol. 4, 2145–55. Paris: CNRS, 1968.

Haddock, Guilliermo E. 2000. To be a Fregean or to be a Husserlian: That is the question for Platonists. In *Husserl or Frege? Meaning, objectivity, and mathematics*, ed. Claire Ortiz Hill and Guillermo E. Rosado Haddock, 199–220. Chicago: Open Court.

Haeckel, Ernst. 1893. *Der Monismus als Band zwischen Religion und Wissenschaft.* Bonn.

———. 1899. *Die Welträthsel: Gemeinverständliche Studien über monistische Philosophie.* Bonn. English trans., Joseph McCabe, *The Riddle of the Universe.* Rationalist Press Association, Watts Co., 1900.

Hallett, Michael. 1984. *Cantorian set theory and limitation of size.* Oxford: Clarendon Press.

———. 1994. Hilbert's axiomatic method and the laws of thought, In *Mathematics and mind*, ed. A. George, 158–98. Oxford: Oxford University Press.

———. 1995. Hilbert and logic. *Québec Studies in the Philosophy of Science* 1:135–187.

Hallett, Michael, and Ullrich Majer, eds. 2004. *David Hilbert's lectures on the foundations of geometry, 1891–1902.* Berlin: Springer.

Hamel, Georg. 1905. Eine Basis aller Zahlen und die unstetigen Lösungen der Funktional-gleichung: $f(x + y) = f(x) + f(y)$. *Mathematische Annalen* 60: 459–462.

Hamilton, William Rowan. 1843. Quaternions. In *Mathematical papers*, vol. 3, 101–105. Cambridge: Cambridge University Press, 1967.

Hankel, Hermann. 1867. *Complexen Zahlensysteme.* Leipzig.

———. 1874. *Zur Geschichte der Mathematik in Altherthum und Mittelalter.* Leipzig.

Hankins, Thomas L. 1985 *Science and the enlightenment.* Cambridge: Cambridge University Press.

Hardy, Godfrey Harold. 1941. *A mathematician's apology*. Cambridge: Cambridge University Press.

Harkness, James, and Frank Morley. 1893. *A treatise on the theory of functions*. New York: Stechert.

Harman, Peter. 1982. *Energy, force, and matter*. Cambridge: Cambridge University Press.

Harnack, Axel. 1878. Review of Erdmann 1877 in *Vierteljahrschrift für wissenschaftliche Philosophie* 2:119–26.

Harnack, Adolf von. 1900. *Das Wesen des Christentums*. Leipzig. English trans., *What is Christianity?* London: S.P.C.K., 1902.

Hart, W. D., ed. 1996. *The philosophy of mathematics*. Oxford: Oxford University Press.

Haskell, Mellen W. 1890. Über die zu der Curve $x^3y + y^3z + z^3x = 0$ im projektiven Sinne gehörende mehrfache Überdeckung der Ebene. *American Journal of Mathematics* 13:1–52.

Hatfield, Gary. 1990. *The natural and the normative: Theories of spatial perception from Kant to Helmholtz*. Cambridge, Mass.: MIT Press.

Hausdorff, Felix. 1914. *Grundzüge der Mengenlehre*. Leipzig: Veit & Comp.

———. 1915. Bemerkung über den Inhalt von Punktmengen. *Mathematische Annalen* 75: 428–33.

———. 2002–. *Gesammelte Werke*. Berlin: Springer.

Hawkins, Benjamin S. 1977. Peirce and Russell: The history of a neglected "controversy". In Houser et al. 1997, 111–64.

Hawkins, Tom. 1975. *Lebesgue's theory of integration: Its origins and development*. 2nd ed. Providence, R.I.: American Mathematical Society.

———. 1977. Another look at Cayley and the theory of matrices. *Archives internationales d'histoire des sciences* 27. 100: 82–112.

———. 2000. *Emergence of the theory of Lie groups*. New York: Springer.

Hebb, Donald Olding. 1949. *The organization of behavior: A neuropsychological theory*. New York: John Wiley & Sons; London: Chapman & Hall.

Heelan, P. A. 1981. *Space perception and the philosophy of science*. Berkeley: University of California Press.

Heidelberger, Michael. 2003. *Nature from within: Gustav Theodor Fechner and his psychophysical worldview*. Trans., Cynthia Klohr. Pittsburgh: University of Pittsburgh Press.

Heijenoort, Jean van. 1967a. *From Frege to Gödel: A source book in mathematical logic, 1879–1931*. Cambridge, Mass.: Harvard University Press.

———. 1967b. Logic as language and logic as calculus. *Synthese* 17:324–30.

Heijermann, E., and H. W. Schmitz, eds. 1991. *Significs, mathematics and semiotics: The significs movement in the Netherlands*. Münster, Germany: Nodus Publications.

Heims, Steve J. 1991. *Constructing a social science for postwar America: The cybernetics group, 1946–1949*. Cambridge, Mass.: MIT Press.

Heine, Eduard. 1872. Die Elemente der Functionenlehre. *Journal für die reine und angewandte Mathematik* 74:172–88.

Heinzmann, Gerhard. 2001. The foundations of geometry and the concept of motion: Helmholtz and Poincaré. *Science in Context* 14.3:457–70.

Helmholtz, Hermann von. 1868. On the facts underlying geometry. *Göttingen Nachrichten* 9. Also in Helmholtz 1977, 39–57.

———. 1870. On the origin and significance of the axioms of geometry. In Helmholtz 1977, 1–38.

———. 1878. The facts in perception. Address given during the anniversary celebrations of the Friedrich Wilhelm University in Berlin, in 1878; reprinted in *Vorträge und Reden*, vol. 2, 215–47, 387–406. Also in Helmholtz 1977, 115–62.

———. 1887. Numbering and measuring from an epistemological viewpoint. In *Philosophische Aufsätze: Eduard Zeller zu seinem fünfzigjährigen Doctor-Jubiläum gewidmet*. Leipzig, 1887. Also in Helmholtz 1977, 72–102.

———. 1977. *Schriften zur Erkenntnistheorie*. Ed. P. Hertz and M. Schlick. Berlin: Springer, Berlin, 1921. Trans., M. F. Lowe, *Epistemological Writings*, ed. R. S. Cohen and Y. Elkana. North Holland: Reidel.

Henderson, Linda D. 1983. *The fourth dimension and non-Euclidean geometry in modern art*. Princeton, N.J.: Princeton University Press.

Hensel, Kurt. 1908. *Theorie der algebraischen Zahlen*. Leipzig: Teubner.

Hentschel, Klaus. 1990. *Interpretationen*. Science Networks, vol. 6. Boston and Basel: Birkhäuser.

Herbart, Johann Friedrich. 1802. *Pestalozzis Idee eines A B C der Anschauung*. Göttingen.

———. 1824–25. *Psychologie als Wissenschaft neu gegrundet auf Erfahrung, Metaphysik, und Mathematik*. 2 vols.

———. Possibility and necessity of applying mathematics in psychology. Trans. H. Haanel. In *Classics in the history of psychology*, http://psychclassics.yorku.ca/Herbart/mathpsych.htm.

Hertz, Heinrich. 1895. *Gesammelte Werke*. Ed. P. Lenard. 3 vols. Leipzig. Vol. 3 English trans., D. E. Jones and J. T. Walley, *The principles of mechanics presented in a new form*. New York: Dover, 1955.

Hesse, Hermann. 1943. *Das Glasperlenspiel*. Zurich. Trans. R. and C. Winston, *The glass bead game*. London: Jonathan Cape, 1970.

Hesseling, Dennis E. 2003. *Gnomes in the fog: The reception of Brouwer's intuitionism in the 1920s*. Science Networks Historical Studies 28. Basel: Birkhäuser.

Hessenberg, Gerhard. 1904. Über die kritische Mathematik. *Sitzungsberichte, Berlin Mathematischen Gesellschaft* 3:21–28.

———. 1905. Beweis des *Desargues*schen Satzes aus dem *Pascal*schen. *Mathematische Annalen* 61:161–72.

———. 1906. *Grundbegriffe der Mengenlehre*. Göttingen: Vandenhoeck & Ruprecht.

Heyting, A. 1930. Die formalen Regeln der intuitionistischen Logik. 3 parts. *Sitzungsberichte der Preussischen Akademie von Wissenschaften, physikalisch-mathematische Klasse, Berlin*, 42–56, 57–71, 158–169. English trans. of pt. 1 in Mancosu 1998, 311–27.

———. 1978. History of the foundation of mathematics. *Nieuw Arch. Wiskunde* (3) 26.1: 1–21.

Hilbert, David. 1897. Die Theorie der algebraischen Zahlkörper (Zahlbericht). *Jahresbericht der Deutschen Mathematiker-Vereinigung* 4: 175–546. Also in Hilbert 1935, vol. 1, 63–363. English trans., Iain Adamson, *The theory of algebraic number fields*. Berlin: Springer, 1998.

Hilbert, David. 1899. *Grundlagen der Geometrie*. Festschrift to celebrate the dedication of the Gauss-Weber memorial in Göttingen. Leipzig: Teubner. Many subsequent editions, e.g., *Foundations of geometry*, 10th English ed., trans. of the 2nd German ed. L. Unger. Chicago: Open Court, 1971.

———. 1900. Theorie der algebraischen Zahlkörper. In *Encyclopädie der mathematischen Wissenschaften* 1C4a, 675–714.

———. 1901a. Mathematische Probleme. *Archiv für Mathematik und Physik* 1:44–63, 213–37. Also in Hilbert 1935, vol. 3, 290–329. Repr., New York: Chelsea, 1965. French trans., L. Laugel, Sur les problèmes futurs des mathématiques. *In Compte rendu du deuxième Congres Internationale des Mathématiciens*, 58–114. Paris: Gauthier-Villars, 1902. English trans., Mathematical problems. *Bulletin of the American Mathematical Society* 8 (1902): 437–479; repr, Mathematical problems, in *Mathematical developments arising from Hilbert problems*. Proceedings of Symposia in Pure Mathematics, American Mathematical Society, 1976. 2 vols., part 1.

———. 1901b. Über Flächen von konstanter Gaussscher Krümmung. *Trans. American Mathematical Society* 2: 87–99. Also in Hilbert 1935, vol. 2, 437–448.

———. 1902. Über den Satz von der Gleichheit der Basiswinkel im gleichschenkligen Dreieck. *Proceedings of the London Mathematical Society* 35:50–68. Also in *Grundlagen der Geometrie*, 133–158. Leipzig: Teubner, 1930.

———. 1905. Über die Grundlagen der Logik und der Arithmetik. *Verhandlungen des dritten internationalen Mathematiker-Kongresses in Heidelberg*, 174–85. Leipzig: Teubner.

———. 1918. Axiomatisches Denken. *Mathematische Annalen* 78; 405–418. Also in Hilbert 1935, vol. 3, 146–56; repr., New York: Chelsea, 1965.

———. 1922a. *Wissen und mathematisches Denken*. Ed. William Ackermann. Mathematisches Institut, University of Göttingen.

———. 1922b. Neubegründung der Mathematik: Erste Mitteilung. *Abhandlungen aus dem Mathematischen Seminar der Hamburg Universität* 1: 157–177. Also in Hilbert 1935, vol. 3, 157–177; repr., New York: Chelsea, 1965.

———. 1923. Die logischen Grundlagen der Mathematik. *Mathematische Annalen* 88: 151–165. Also in Hilbert 1935, vol. 3, 178–91. Berlin: Teubner.

———. 1926. Über das Unendliche. *Mathematische Annalen* 95: 161–190. Also in *Grundlagen der Geometrie*, 7th German ed., part 8, 262–288.

———. 1929. Probleme der Grundlegung der Mathematik. *Mathematische Annalen* 102: 1–9. Also in *Grundlagen der Geometrie*, 7th German ed., pt. 10, 313–23. English trans., Problems of the Grounding of Mathematics, Mancosu 1998, 227–33.

———. 1932. *Anschauliche Geometrie*. Berlin: Springer. English trans., P. Nemenyi, *Geometry and the Imagination*. New York: Chelsea, 1952.

———. 1935. *Gesammelte Abhandlungen*. 3 vols. Berlin: Teubner.

Hill, Claire Ortiz. 1995. Husserl and Hilbert on completeness. In *From Dedekind to Gödel: Essays on the Development of the Foundations of Mathematics*, ed. Jaakko Hintikka, 143–63. Synthese Library 251. Dordrecht: Kluwer Academic.

Hill, Claire Ortiz, and Guillermo E. Rosado Haddock. 2000. *Husserl or Frege? Meaning, objectivity, and mathematics*. Chicago: Open Court.

Hintikka, Merrill B., and Jaakko Hintikka. 1986. *Investigating Wittgenstein*. Oxford: Basil Blackwell.

Hinton, C. H. 1897. *What is the fourth dimension?* 3rd ed. London: Swan Sonnenschein.

Hinton, C. H. 1904. *The fourth dimension*. London and New York: Sonnenschein, J. Lane.

Hirosige, T. 1976. The ether problem, the mechanistic worldview, and the origins of the theory of relativity. *Historical Studies in the Physical Sciences* 7:3–82.

Hobson, E. W. 1907. *The theory of functions of a real variable and the theory of Fourier's series*. Cambridge: Cambridge University Press.

Hochkirchen, Thomas. 1999. *Die Axiomatisierung der Wahrscheinlichkeitsrechnung und ihre Kontexte*. Göttingen: Vandenhoeck & Ruprecht.

Hölder, Otto. 1900. *Anschauung und Denken in der Geometrie*. Academic lecture, July 22, 1899. Leipzig: Teubner.

———. 1901. Die Axiome der Quantität und die Lehre vom Mass. *Leipzig Berichte* 53: 1–64.

———. 1924. *Die mathematische Methode*. Leipzig: Teubner.

Holton, Gerald. 1992. Ernst Mach and the fortunes of positivism in America. *Isis* 83.1:27–60.

Houser, Nathan, Don D. Roberts, and James Van Evra, eds. 1997. *Studies in the logic of Charles Sanders Peirce*. Bloomington: Indiana University Press.

Howard, Don. 2004. Einstein and the development of twentieth-century philosophy of science. In *Cambridge Companion to Einstein*. Forthcoming.

———. 2005. Albert Einstein as a philosopher of science. *Physics Today*, December.

Hunt, Bruce. 1991. *The Maxwellians*. Ithaca, N.Y.: Cornell University Press.

Huntington, Edward. 1902. A complete set of postulates for the theory of absolute continuous magnitude. *Trans. AMS* 3:264–279.

Husserl, Edmund. 1900–1901. *Logische Untersuchungen*. Vol. 1. Halle: Niemayer.

————. 1970. *Philosophie der Arithmetik, mit ergändzenden Texte (1890–1901)*. Ed. Lothar Eley. Husserliana 12. The Hague: Martinus Nijhoff.

————. 1994. *Early writings in the philosophy of logic and mathematics*. Trans. Dallas Willard, in *Collected Works*, vol. 5. Dordrecht: Kluwer.

Irvine, A. D. 2006. Principia Mathematica. In *The Stanford Encyclopedia of Philosophy*, ed. Edward N. Zalta, http://plato.stanford.edu/archives/fall2006/entries/principia-mathematica.

Irwin, Robert. 2006. *For lust of knowing: The orientalists and their enemies*. London: Allen Lane, Penguin.

Jackson, C. 1968. The meeting of East and West: The case of Paul Carus. *Journal of the History of Ideas* 29.1:73–92.

Jevons, William Stanley. 1864. *Pure logic; or, the logic of quality apart from quantity: With remarks on Boole's system, and on the relation of logic and mathematics*. London: Stanford.

————. 1874. *The principles of science: A treatise on logic and scientific method*. London: Macmillan.

Jordan, Camille. 1870. *Traité des substitutions et des équations algébriques*. Paris: Gauthier-Villars.

————. 1892. *Cours d'analyse de l'École Polytechnique*. 2nd ed. 3 vols. Paris: Gauthier-Villars et Fils.

Jourdain, Philip E. B. 1915. Introduction to *Georg Cantor: Contributions to the founding of the theory of transfinite numbers*. New York: Dover, 1955.

Joyce, James. 1924. *Ulysses*. Paris: Shakespeare & Co.

Jungnickel, Christa, and Russell McCormmach. 1986. *The intellectual mastery of nature—Theoretical physics form Ohm to Einstein*. 2 vols. Chicago: University of Chicago Press.

Kandinsky, Wassily. 1926. *Punkt und Linie zu Fläche: Beitrag zur Analyse der malerischen Elemente*. Munich: Albert Langen. English trans., Howard Dearstyne and Hilla Rebay, *Point and line to plane*. New York: Dover, 1979.

Kant, Immanuel. 1781. *Critik der reinen Vernunft*. Riga; 2d ed., Jena, 1787.

————. 1929. *Immanuel Kant's critique of pure reason*. Trans. Norman Kemp Smith. 2nd ed. repr., 1970.

Kargon, Robert, and Peter Achinstein, eds. 1987. *Kelvin's Baltimore lectures and modern theoretical physics: Historical and philosophical perspectives*, Cambridge, Mass.: MIT Press.

Kasner, Edward. 1904/5. The present problems of geometry. *Bulletin of the American Mathematical Society* 11:283–314.

Kempe, Alfred Bray. 1890. On the relation between the logical theory of classes and the geometrical theory of points. *Proceedings of the London Mathematical Society* 21:147–82.

Kennedy, Hubert C. 1980. *Peano: Life and works of Guiseppe Peano*. Dordrecht: Reidel.

Kern, Stephen. 1983. *The culture of time and space, 1880–1918*. Cambridge, Mass.: Harvard University Press.

Kerry, Benno. 1890. *System einer Theorie der Grenzbegriffe*. Ed. and completed G. Kohn. Leipzig and Vienna.

Kevles, Daniel. 1987. *The physicists: The history of a scientific community in modern America*. Cambridge, Mass.: Harvard University Press.

Killing, Wilhelm. 1892. Ueber die Grundlagen der Geometrie. *Journal für die reine und angewandte Mathematik* 109:121–86.

————. 1893, 1898. *Einführung in die Grundlagen der Geometrie*. 2 vols. Paderborn: F. Schöningh.

Kleene, S. 1956. Representation of events in nerve nets and finite automata, In *Automata studies*, ed. C. Shannon and J. McCarthy, 3– 42. Princeton, N.J.: Princeton University Press.

Klein, C. Felix. 1872. *Vergleichende Betrachtungen über neuere geometrische Forschungen (Erlangen Programm)*. 1st pub. Erlangen: Deichert. Also in *Gesammelte Mathematische Abhandlungen* 1.27 (1921):460–97.

———. 1890. Zur nicht-euklidischen Geometrie. *Mathematische Annalen* 37: 544–572. Also in *Gesammelte Mathematische Abhandlungen* 1.21(1921):353–83.

———. 1892–93. *Höhere Geometrie.* 2 vols. Friedrich Schilling. 2nd ed., 1907.

———. 1894. *Lectures on mathematics.* The Evanston Colloquium. New York: Macmillan.

———. 1895. Ueber Arithmetisirung der Mathematik. *Gött. Nachr. Geschäftl. Mitt.,* 82–91. Also in *Gesammelte Mathematische Abhandlungen* 2.47:232– 40. English trans., *Bulletin of the American Mathematical Society* 2 (1896):241– 49.

———. 1897. Gutachten betreffen der . . . ersten Verteilung des Lobatschewskypreises. *Nachrichten phys.-mat. Gesellschaft der Universität Kasan* (2) 8. Also in *Gesammelte Mathematische Abhandlungen* 1 (paper no. 22): 384–401.

———. 1901. *Anwendung der Differential- und Integral-Rechnung auf Geometrie, eine Revision der Prinzipien.* Autographed lecture notes. Ed. C. Müller.

———. 1921–23. *Gesammelte Mathematische Abhandlungen.* 3 vols. Berlin: Springer.

———. 1923. Autobiography. *Göttingen Mitteilungen des Universitätsbundes Göttingen* 5.1.

———. 1926–27. *Vorlesungen über die Entwicklung der Mathematik im 19. Jahrhundert.* 2 vols. Ed. R. Courant and O. Neugebauer. Berlin: Springer; rep., New York: Chelsea, 1948.

Kline, Morris. 1972. *Mathematical thought from ancient to modern times.* Oxford: Oxford University Press.

Knobloch, Clemens. 1984. Sprache und Denken bei Wundt, Paul und Marty: Ein Beitrag zur Problemgeschichte der Sprachpsychologie. *Historiographia Linguistica* 11.3:413–84.

———. 1988. *Geschichte der psychologischen Sprachauffassung in Deutschland von 1850 bis 1920.* Tübingen: Max Niemeyer Verlag.

Koenigsberger, Leo. 1906. *Hermann von Helmholtz.* Trans. Welby. Repr., New York: Dover, 1965.

Köhnke, K. C. 1986. *Entstehung und Aufstieg des Neukantianismus.* Frankfurt: Suhrkamp. English trans., R. J. Hollingdale, *The rise of Neo-Kantianism: German academic philosophy between idealism and positivism.* Cambridge: Cambridge University Press, 1991.

Kowalewski, Gerhard. 1923. Ueber Bolzanos nichtdifferenzierbare stetige Funktion. *Acta Mathematica* 44:315–19.

Korselt, Alwin. 1908. Über die Logik der Geometrie. *Jahresbericht der Deutschen Mathematiker-Vereinigung* 17:98–124.

———. 1911. Über mathematische Erkenntnis. *Jahresbericht der Deutschen Mathematiker-Vereinigung* 20:364–80.

Kronecker, Leopold. 1881. Über die Discriminante algebraischer Functionen einer Variabeln. *Journal für die reine und angewandte Mathematik* 91:301–334. In Kronecker 1895–1931, vol. 2, 193–236.

———. 1882. *Grundzüge einer arithmetischen Theorie der algebraischen Grössen. Journal für Mathematik,* 1882. Also in Kronecker 1895–1931, vol. 2, 239–387.

———. 1887. Ueber den Zahlbegriff. *Journal für die reine und angewandte Mathematik* 101:337–55. Also in *Philosophische Aufsätze: Eduard Zeller zu seinem fünfzigjährigen Doctor-Jubiläum gewidmet.* Leipzig, 1887. Also in Kronecker 1895–1931, vol. 3.1, 249–74.

———. 1895–1931. *Leopold Kroneckers Werke.* 5 vols. Ed. K. Hensel. Deutsche Akademie der Wissenschaften zu Berlin. Leipzig: Teubner.

Krüger, L., L. J. Daston, and M. Heidelberger, eds. 1989. *The Probabilistic revolution.* Cambridge, Mass.: MIT Press.

Kuclick, B. 1985. *Josiah Royce: An intellectual biography.* Indianapolis: Hackett.

Kuhn, Thomas S. 1987. *Black body theory and the quantum discontinuity, 1894–1912.* Oxford: Clarendon Press.

Kummer, E. Eduard. 1851. Mémoire sur la théorie des nombres complexes composés de racines de l'unité et de nombres entiers. *Journal de Mathématiques* 16:377–498. Also in *Collected papers,* ed. A. Weil, 363–484. Berlin: Springer, 1975.

Kusch, Martin. 1989. *Language as calculus vs. language as universal medium.* Dordrecht: Kluwer.

———. 1995. *Psychologism: A case study in the sociology of philosophical knowledge.* London: Routledge.

Laugwitz, Detlef. 2000. Comments on the paper: "Two letters by N. N. Luzin to M. Ya. Vygodskii." *American Mathematical Monthly* 107.1:64–82.

Le Roy, Édouard. 1907. *Dogme et critique.* Paris.

Lebesgue, Henri. 1902. Intégrale, longeur, aire. *Annali di matematica* (3) 7:231–359.

Lechalas, Georges. 1896. *Étude sur l'espace et le temps.* Paris: Alcan.

Lehto, Olli. 1998. *Mathematics without borders: A history of the International Mathematical Union.* Berlin: Springer.

Leibniz, Gottfried W. 1986. *Principes de la nature et de la grâce fondés en raison: Principes de la philosophie, ou, monadologie, présentés d'après des lettres inédites par André Robinet.* Paris: Presses Universitaires de France.

———. 1875. De scientia universalis seu calculo philosophico. In C. I. Gerhardt, *Die philosophischen Schriften von G. W. Leibniz*, vol. 7. Berlin: Weidmann.

———. 1904. *Neue Abhandlungen über den menschlichen Verstand.*

———. n.d. *Philosophical writings.* English trans., Mary Morris. London: Everyman's Library, no. 905.

Leitch, Alexander. 1978. *A Princeton companion.* Princeton, N.J.: Princeton University Press.

Lenin, Vladimir I. 1927. *Materialism and empirio-criticism: Critical notes concerning a reactionary philosophy.* New York: International Publishers. (Russian original, 1909.)

Lenoir, Timothy. 2006. Operationalizing Kant. In *Kant's scientific legacy in the nineteenth century*, ed. M. Friedman and A. Nordmann, 141–210. Cambridge, Mass.: Dibner Press, MIT.

Levi, Beppo. 1904. Fondamenti della metrica projettiva. *Torino Memorie* (2) 54:281–353.

———. 1907. Geometrie proiettive di congruenza e geometrie proiettive finite. *Transactions of the American Mathematical Society* 8:354–65.

Levi-Civita, Tullio. 1894. Sugli infiniti ed infinitesimali attuali quali elementi analitici. *Atti dell' Istituto Veneto delle scienze*, ser. 7, 4 (1765–1815): 1–39. Also in *Opere matematiche*, Bologna, 1954.

Lewis, Clarence Irving. 1918. *A survey of symbolic logic.* Berkeley: University of California Press.

Lie, Sophus. 1893. *Theorie der Transformationsgruppen.* Leipzig: Teubner.

Liebmann, Otto. 1865. *Kant und die Epigonen: Eine kritische Abhandlung.* Stuttgart: C. Schober. Reissued in 1912, Berlin: Reuther & Reichard.

———. 1876. *Die Analyse der Wirklichkeit: Philosophische Untersuchungen.* Strassburg: Trübner.

———. n.d. Ueber die Phenomenalität des Raumes. *Philosophisches Monatshefte* 7.8.

Lipps, Gottlob Friedrich. 1894–1898. Untersuchungen über die Grundlagen der Mathematik. *Philosophische Studien*, several vols.

Littlewood, John Edensor. 1986. *Mathematician's miscellany.* Cambridge: Cambridge University Press.

Lobachevskii, Nicolai Ivanovich. 1840. *Geometrische Untersuchungen.* Berlin; repr., Mayer & Müller, 1887. French trans. J. Houël, Etudes géométriques sur la théorie des parallèles, *Mémoires de la Société des sciences physiques et naturelles de Bordeaux* 4 (1867): 83–128. Paris: Gauthier-Villars. English trans., G. B. Halsted, Geometric researches in the theory of parallels, appendix in Bonola 1906/1912.

———. 1856. *Pangéométrie, ou précis de géométrie fondée sur une théorie générale des parallèles.* Kasan. Trans. and ed. H. Liebmann, *Pangéométrie.* Leipzig: Engelmann, 1912.

Lobachetschefskij, Nikolai Ivanovich. 1899. *Zwei geometrische Abhandlungen.* Trans. F. Engel. Leipzig: Teubner.

Lotze, Hermann. 1874. *Logik: Drei Bücher vom Denken, vom Untersuchen und vom Erkennen.* Leipzig: S. Hirzel. First edition, 1843.

Lovelace, Ada Countess of. 1843. Notes by the translator to L. F. Menabrea, Sketch of the analytical engine invented by Charles Babbage. *Taylor's Scientific Memoirs* 3.

Lützen, Jesper. 2003. The foundations of analysis in the 19th century. In *A History of Analysis*, ed. N. H. Jahnke, 155–96. HMath 24. American Mathematical Society and London Mathematical Society, Providence, R.I.

———. 2005. *Mechanistic images in geometric form.* Oxford: Oxford University Press.

Mach, Ernst. 1905. *Erkenntniss und Irrthum: Skizzen zur Psychologie der Forschung.* Leipzig: Joh. Ambr. Barth.

———. 1906. *Space and geometry in the light of physiological, psychological and physical inquiry.* English trans., T. J. McCormack. La Salle, Ill.: Open Court.

Machover, Moshe. 1996. *Set theory, logic and their limitations.* Cambridge: Cambridge University Press.

Majer, Ullrich. 1997. Husserl and Hilbert on completeness. *Synthese* 110.1:37–56.

Mancosu, Paolo, ed. 1998. *From Brouwer to Hilbert: The debate on the foundations of mathematics in the 1920s.* New York: Oxford University Press.

Manning, H. P., ed. 1910. *The fourth dimension simply explained: A collection of essays selected from those submitted in the Scientific American's prize competition.* New York: Munn and Co.

Mannoury, Gerrit. 1909. *Methodologisches und philosophisches zur Elementar-Mathematik.* Haarlem: P. Visser.

Marchisotto, Elena A. 1989. Mario Pieri: His contributions to the foundations and teaching of geometry. *Historia Mathematica* 16: 287–88.

———. 1993. Mario Pieri and his contributions to geometry and foundations of mathematics. *Historia Mathematica* 20: 285–303.

———. 1995. In the shadow of giants: The work of Mario Pieri in the foundations of mathematics. *History and Philosophy of Logic* 16: 107–19.

———. 2006. The projective geometry of Mario Pieri: A legacy of Georg Karl Christian von Staudt. *Historia Mathematica* 33.3: 277–314.

Marchisotto, Elena A., and James T. Smith. 2007. *The legacy of Mario Pieri in geometry and arithmetic.* Boston: Birkhäuser.

Marquand, Allan. 1883. A machine for producing syllogistic variations. In *Studies in Logic*, ed. C. S. Peirce, 12–15.

Maxwell, James Clerk. 1873. *Treatise on electricity and magnetism.* 2 vols. Oxford: Clarendon Press.

———. 1879. Thomson and Tait's *Natural philosophy. Nature* 20: 776–785. Also in *The scientific papers of James Clerk Maxwell.* Cambridge: Cambridge University Press, 1890: repr., New York: Dover, 1965.

Maz'ya, Vladimir, and Tatyana Shaposhnikova. 1998. *Jacques Hadamard: A universal mathematician.* American and London Mathematical Societies, History of Mathematics, HMath 14, Providence, R.I.

McCorduck, P. 1979. *Machines who think: A personal inquiry into the history and prospects of artificial intelligence.* San Francisco: W. H. Freeman.

McCulloch, Warren S., and W. Pitts. 1943. A logical calculus of the ideas immanent in nervous activity. *Bulletin of Mathematical Biophysics* 5:115–33.

McLarty, Colin. 2006. Emmy Noether's set-theoretic topology: From Dedekind to the rise of functors. In *The architecture of modern mathematics*, ed. J. Ferreirós and J. J. Gray, 211–36. Oxford: Oxford University Press.

Mehlberg, Henry. 2002. The present situation on the philosophy of mathematics. In *Philosophy of mathematics: An anthology*, ed. Dale Jacquette, 65–82. Oxford: Blackwell.

Mehrtens, Herbert. 1990. *Moderne Sprache, Mathematik: Eine Geschichte des Streits um die Grundlagen der Disziplin und des Subjekts formaler Systeme.* Frankfurt: Suhrkamp.

Meinecke, Wilhelm. 1906. Die Bedeutung der Nicht-Euklidischen Geometrie in ihrem Verhältnis zu Kants Theorie der mathematischen Erkenntnis. *Kantstudien* 11: 209–32.

Menand, Louis. 2001. *The metaphysical club.* New York: Farrar, Straus, and Giroux.

Méray, Charles. 1902. Sur la langue internationale auxiliare de M. de Zamenhof, connue sur le nom d'Esperanto. In *Compte rendu du deuxième Congress Internationale des Mathématiciens*, 429–432. Paris: Gauthier-Villars.

Merrill, Daniel D. 1997. Relations and quantification in Peirce's logic. In Houser et al. 1997, 158–72.

Meschkowski, Herbert. 1965. *Noneuclidean geometry.* Trans. A. Shenitzer. New York: Academic Press.

Meyer, Walther Franz. 1904. Kant und das Wesen des Neuen in der Mathematik. *Archiv der Mathematik und Phyisk*, ser. 3, 8:287–305.

Michell, Joel. 1993. The origins of the representational theory of measurement: Helmholtz, Hölder, and Russell. *Studies in History and Philosophy of Science* 24:185–206.

———. 1999. *Measurement in psychology: Critical history of a methodological concept.* Cambridge: Cambridge University Press.

Miller, Arthur. 1981. *Albert Einstein's special theory of relativity.* New York: Wiley.

Minkowski, Hermann. 1896 *Geometrie der Zahlen.* Leipzig: Teubner.

———. 1905. Peter Gustav Lejeune Dirichlet und seine Bedeutung für die heutige Mathematik. *Jahresbericht der Deutschen Mathematiker-Vereinigung* 14:149–63.

———. 1908a. Raum und Zeit. In *Jahresbericht der Deutschen mathematiker Vereinigung* (1908), 75–88. Also in *Gesammelte Abhandlungen*, vol. 2, 431–444. Leipzig: Teubner. English trans. in Albert Einstein et al., *The Principle of Relativity.* New York: Dover, 1952.

———. 1908b. Die Grundlagen für die elektromagnetischen Vorgänge in bewegten Körpern. *Göttingen Nachrichten*, 53–11, and in *Gesammelte Abhandlungen* 2: 352–404.

Molk, Jules. 1885. Sur une notion qui comprend celle de la divisibilité. *Acta Mathematica* 6:1–166.

Mongré, Paul. 1898. *Das Chaos in kosmischer Auslese*; rep. in Hausdorff 2004, vol. 7.

Moore, Gregory H. 1982. *Zermelo's axiom of choice: Its origins, development and influence.* New York: Springer.

———. 1995. The axiomatization of linear algebra, 1875–1940. *Historia Mathematica* 22.3:262–303.

Moran, Dermot. 2000. Introduction to Edmund Husserl, *Logical investigations.* London: Routledge.

Müller, Max. 1887. *Science of thought.* London: Longmans, Green.

Mumford, David, Caroline Series, and David Wright. 2002. *Indra's pearls: The vision of Felix Klein.* Cambridge: Cambridge University Press.

Murphey, Murray G. 1961. *The development of Peirce's philosophy.* Cambridge, Mass.: Harvard University Press; repr., Indianapolis: Hackett, 1993.

Musil, Robert. 1930–1943. *Der Mann ohne Eigenschaften.* Berlin: Ernst Rowohlt Verlag. English trans., Sophie Wilkins, *The man without qualities.* London: Picador, 1995.

Myhill, John. 1974. The undefinability of the set of natural numbers in the ramified *Principia*. In *Bertrand Russell's philosophy*, ed. G. Nakhnikian, 19–27. London: Duckworth.

Nabonnand, Philippe. 2000. La polémique entre Poincaré et Russell au sujet du statut des axiomes de la géométrie. *Revue d'Histoire des Mathématiques* 16.2:219–70.

———. 1999. The Poincaré–Mittag-Leffler relationship. *Mathematical Intelligencer* 21.2: 58–64.

Nagel, Ernest. 1939. The formation of modern concepts of formal logic in the development of geometry. *Osiris* 7:142–224.

Nelson, Leonhard. 1905/06. Bemerkungen über die nicht-euklidische Geometrie und den Ursprung der mathematischen Gewissheit. *Abhandlungen der Fries'schen Schule*, N.F., nos. 1–3:373–430.

Netto, Eugen. 1877. Review of B. Erdmann, *Die Axiome der Geometrie: Eine philosophische Untersuchung der Riemann-Helmholtz'schen Raumtheorie*, in *Jahrbuch über die Fortschritte der Mathematik*, 09.0032.01.

Neumann, Peter M., S. A. Adeleke, and M. Dummett. 1995. On a question of Frege's about right-ordered groups, and postscript. In Demopolous 1995, 405–21.

Newcomb, Simon. 1902. The fairyland of geometry. *Harper's Magazine*, January.

Newton, Sir Isaac. 1707. *Arithmetica universalis; sive, De compositione et resolutione arithmetica liber*. Cambridge.

Nicolas, S., and A. Charvillat. 2001. Introducing psychology as an academic discipline in France: Théodule Ribot and the Collège de France (1888–1901). *Journal of the History of Behavioural Science* 37.2:143–64.

Noether, Max. 1878. Review of Erdmann 1877 in *Zeitschrift für Mathematik und Physik, Hist.-Lit. Abtheilung* 32:76–84.

Norton, John D. 2004. Einstein's investigations of Galilean covariant electrodynamics prior to 1905. *Archive for History of Exact Sciences* 59.1:45–105.

Nye, Mary Jo. 1979. The Boutroux Circle and Poincaré's conventionalism. *Journal of the History of Ideas* 40:107–20.

Olesko, Kathryn M. 1991. *Physics as a calling*. Ithaca, N.Y.: Cornell University Press.

Osgood, William Fogg. 1903. A Jordan curve of positive area. *Transactions of the American Mathematical Society* 4:107–12.

———. 1907, 1924. *Lehrbuch der Funktionentheorie*. 2 vols. Leipzig: Teubner; repr., New York: Chelsea, 1965.

Pais, Abraham. 1982. *'Subtle is the Lord . . . ': The science and life of Albert Einstein*. Oxford: Oxford University Press.

Parikh, Carol. 1991. *The unreal life of Oscar Zariski*. Boston: Academic Press.

Parker, J. R. L. 2005. *Moore: Mathematician and teacher*. Washington, D.C.: Mathematics Association of America.

Parshall, Karen Hunger. 2006. *James Joseph Sylvester: Jewish mathematician in a Victorian world*. Baltimore: Johns Hopkins University Press.

Parshall, Karen Hunger, and David E. Rowe. 1994. *The emergence of the American mathematical research community, 1876–1900: J. J. Sylvester, Felix Klein, and E. H. Moore*. American Mathematical Society and London Mathematical Society, HMath 8, Providence, R.I.

Pasch, Moritz. 1882. *Vorlesungen über neuere Geometrie*. Leipzig: Teubner.

Paul, Harry. 1979. *The edge of contingency: French Catholic reaction to scientific change from Darwin to Duhem*. Gainesville: University Presses of Florida.

Peano, Giuseppe. 1888. Intégration par séries des équations différentielles linéaires. *Mathematische Annalen* 32:450–56.

———. 1889. *Arithmetices Principia*. Turin: Bocca.

———. 1890. Sur une courbe, qui remplit toute une aire plane. *Mathematische Annalen* 32:157–60.

———. 1894. Sui fondamenti della geometria. *Rivista di matematiche* 4:73.

———. 1903. De Latino sine flexione—lingua auxiliare internationale. In *Opere scelte*, vol. 2, 439–47. Rome: Edizione Cremonese, 1958.

Peckhaus, Volker. 1990. *Hilbertprogramm und kritische Philosophie*. Göttingen: Vandenhoeck & Ruprecht.

———. 1991. Ernst Schröder und die "pasigraphische Systeme" von Peano und Peirce. *Modern Logic* 1.2:174–205.

———. 1994. Benno Kerry: Beiträge zu seiner Biographie. *History and Philosophy of Logic* 15:1–8.

———. 1997 *Logik, Mathesis universalis und allgemeine Wissenschaft*, Berlin: Akademie Verlag.

———. 1999. 19th-century logic between philosophy and mathematics. *Bulletin of Symbolic Logic* 5.4:433–50.

Peckhaus, Volker, and Reinhard Kahle. 2002. Hilbert's paradox. *Historia Mathematica* 29.2:157–75.

Peirce, Charles Sanders. 1867. Upon the logic of mathematics. *Proceedings of the American Academy of Arts and Sciences* 7:402–12. Also in Peirce 1960–1966, vol. 3, 20–44.

———. 1870. Description of a notation for the logic of relatives, resulting from an amplification of the conceptions of Boole's calculus of logic. *Memoirs of the American Academy of Arts and Sciences* 9:317–78. Also in Peirce 1960–1966, vol. 3, 45–149.

———. 1880. On the algebra of logic. *American Journal of Mathematics* 3:15–57. Also in Peirce 1960–1966, vol. 3, 154–251.

———. 1881. On the logic of number. *American Journal of Mathematics* 4:85–95. Also in Peirce 1960–1966, vol. 3, 252–88.

———. 1883. *Studies in logic by members of the Johns Hopkins University*. Baltimore.

———. 1885. On the algebra of logic: A contribution to the philosophy of notation. *American Journal of Mathematics* 7:180–202. Also in Peirce 1960–1966, vol. 3, 359–403.

———. 1960–1966. *Collected papers of Charles Sanders Peirce*. Ed. Charles Hartshorne and Paul Weiss. Cambridge, Mass.: Belknap Press of Harvard University Press.

Perron, Oskar. 1907. Was sind und sollen die irationalen Zahlen? *Jahresbericht der Deutschen Mathematiker-Vereinigung* 16:142–55.

———. 1911. Über Wahrheit und Irrtum in der Mathematik. *Jahresbericht der Deutschen Mathematiker-Vereinigung* 20:196–211.

Pfungst, Oskar. 1907. *Clever Hans (The Horse of Mr. von Osten)*. English trans. in *Classics in Psychology*, ed. Robert H. Wozniak. Bristol: Thommes Press, 1911.

Pieper, H. 1980. Gegen die Schmach des Belagerungszustands. *Spectrum, Monatsbericht der AdW der DDR* 1:22–24.

Pieri, Mario. 1895, 1896a, 1896b. Sui principi che reggono la geometria di posizione. *Atti della Reale Accademia delle Scienze di Torino*. Nota 1, 30 (1894–1895): 607–41; Nota 2, 31 (1895–1896): 381–99; Nota 3, 31 (1895–1896): 457–70.

———. 1898. Principii della Geometria di Posizione, composti in sistema logico deduttivo. *Memorie della Reale Accademia delle Scienze di Torino* (2) 48:1–62.

———. 1899b. Della geometria elementare come sistema ipotetico-deduttivo: Monografia del punto e del mote. *Memorie della Reale Accademia delle Scienze di Torino* (2) 49:173–222.

———. 1900. Sur la géométrie envisagée comme une système purement logique. *International Congress of Philosophy, 1900–1903*, vol. 3, 367–404.

———. 1907. Sopra gli assiomi aritmetici. *Bolletino delle sedute della Accademia Gioenia di Scienze Naturali in Catania* (2) 1.1:26–30. Trans. in Marchisotto and Smith 2007, 308–12.

———. 1908. La geometria elementare istituita sulle nozioni di "punto" e "sfera." *Memorie di matematica e di fisica della Società Italiana delle Scienze* (3) 15:345–450. Trans. in Marchisotto and Smith 2007, 157–270.

Pierpont, James. 1899. On the arithmetisation of analysis. *Bulletin of the American Mathematical Society* 5:394–406.

Plücker, Julius. 1839. *Theorie der algebraischen Curven*. Bonn.

Plato, Jan von. 1994. *Creating modern probability*. Cambridge: Cambridge University Press.

Poincaré, Henri. 1883. Mémoire sur les groupes Kleinéens. *Acta Mathematica* 3:49–92. Also in *Oeuvres*, vol. 2, 258–99.

———. 1887. Sur les hypothèses fondamentales de la géométrie. *Bulletin de la Société Mathématique de France* 15: 203–216. Also in *Oeuvres*, vol. II (1956), 79–91.

———. 1890. *Électricité et optique.* Paris: Gauthier-Villars.

———. 1891. Les géométries non-euclidiennes. *Revue générales des sciences pures at appliquées* 2:769–774. Also in Poincaré, 1902d, 63–76, quote on 69. English trans., Non-Euclidean geometries, 1952, 35–51.

———. 1893. Le continu mathématique. *Revue de métaphysique et de morale* 1:26–34.

———. 1894. *Cours sur les oscillations électriques.* Paris: Gauthier-Villars.

———. 1898a. On the foundations of geometry. *The Monist* 9.1:1–43. Trans. T. J. McCormack. Repr. in Ewald 1996, vol. 2, 982–1011.

———. 1898b. La mesure du temps. *Revue de métaphysique et de morale* 6:1–13, in Poincaré 1905b, 35–58.

———. 1902a. La grandeur mathématique et l'expérience. In Poincaré 1902d, 47–60. English trans., Mathematical magnitude and experiment, in Poincaré 1905c, 17–34.

———. 1902b. L'espace et la géométrie. In Poincaré 1902d, 77–94. English trans., Space and geometry, in Poincaré 1905c, 51–71.

———. 1902c. L'expérience et la géométrie. In Poincaré 1902d, 95–110. English trans., Experiment and geometry, in Poincaré 1905c, 72–88.

———. 1902d. La *Science et l'hypothèse.* Paris: Flammarion; repr., ed. J. Vuillemin. English trans., W. J. Greenstreet, *Science and hypothesis.* London: Walter Scott, 1968.

———. 1903. Review of Hilbert's *Foundations of Geometry. Bulletin of the American Mathematical Society* 10:1–23. Partially repr. in *Bulletin of the American Mathematical Society*, n.s., 37.1:77–78, and in *The Way It Was*, ed. Donald G. Saari, 273–96. Providence, R.I: American Mathematical Society, 2003. Original in Darboux's *Bulletin* 26.2 (1902): 249–72.

———. 1905a. Les définitions mathématiques et l'enseignement, in Poincaré 1908, 123–51.

———. 1905b. *La valeur de la science.* Paris: Flammarion. English trans., G. B. Halsted, *The value of science.* New York: Science Press, 1907; repr., New York: Dover, 1958.

———. 1905c. Reprint of Poincaré 1902d. New York: Dover, 1952.

———. 1908. *Science et méthode.* Paris: Flammarion. English trans., F. Maitland, *Science and method,* with a preface by Bertrand Russell. New York: Scribner, 1914; repr., New York: Dover, 1952.

———. 1909. La logique de l'nfini. In Poincaré 1913, 7–31. English trans., The logic of infinity, in Poincaré 1963, 45–64.

———. 1912a. La logique de l'nfini, in Poincaré 1913, 84–96. English trans., Mathematics and logic, in Poincaré 1963, 65–74.

———. 1912b. L'espace et le temps. *Scientia* 12.25: 159–70. Also in Poincaré 1913, 97–110. English trans., Space and time, in Poincaré 1963, 15–24.

———. 1912c. Pourquoi l'espace a trois dimensions. *Revue de Métaphysique et de Morale* 20.4:483–504. English trans., Why space has three dimensions, in Poincaré 1963, 25–44.

———. 1913. *Derniéres pensées.* Paris: Flammarion.

———. 1916–1934, 1951–1956. *Oeuvres.* 11 vols. Paris: Gauthier-Villars.

———. 1963. Repr. of Poincaré 1913. English trans., J. W. Bolduc, *mathematics and* science: Last essays. New York: Dover.

———. 1997. *Three supplementary essays on the discovery of Fuchsian functions.* Edited with an introductory essay by J. J. Gray and S. A. Walter. Berlin and Blanchard, Paris: Akademie Verlag.

Poncelet J. V., 1862–64. *Applications d'analyse.* 2 vols. Paris: Mallet-Bachelier.

———. 1822. *Traité des propriétés projectives des figures.* Paris: Gauthier-Villars.

Pont, Jean-Claude. 1986. *L'aventure des paralleles.* Berne: Lang.

Post, Emil L. 1994. *Solvability, provability, definability: The Collected Works of Emil L. Post,* ed. Martin Davis. Boston: Birkhäuser.

Potter, Michael D. 2000. *Reason's nearest kin: Philosophies of arithmetic from Kant to Carnap.* Oxford: Oxford University Press.

Presburger, Mojżesz. 1991. On the completeness of a certain system of arithmetic of whole numbers in which addition occurs as the only operation. Trans. Dale Jacquette, *History and Philosophy of Logic* 12:225–33.

Prym, F. E. 1871. Zur Integration der Differentialgleichung, etc. *Journal für* Mathematik 73:340–64.

Pulte, Helmut. 1998. Jacobi's criticism of Lagrange: The changing role of mathematics in the foundations of classical mechanics. *Historia Mathematica* 25.2:154–84.

Purkert, Walter. 2002. Grundzüge der Mengenlehre: Historische Einführung. In *Felix Hausdorff, Gesammelte Werke*, numerous editors, vol. 2, 1–89. Berlin: Springer.

Pyenson, Lewis. 1985. *The young Einstein: The advent of relativity*. Bristol, U.K.: Adam Hilger.

Reed, David. 1994. *Figures of thought: Mathematics and mathematical texts*. London: Routledge.

Reichenbach, Hans. 1958. *The philosophy of space and time*. New York: Dover.

Remmert, Volker R. 1999. Mathematicians at war: Power struggles in Nazi Germany's mathematical community—Gustav Doetsch and Wilhelm Süss. *Revue d'Histoire de Mathématique* 5.1:7–59.

Remmert, Volker R. 2004. Die Deutsche Mathematiker-Vereinigung im "Dritten Reich": Fach- und Parteipolitik. *Mitt. Deutsch. Math.-Ver.* 12.4:223–45.

Ribot, Théodule. 1873. *L'Hérédité: Étude psychologique sur ses phénomènes, ses lois, ses causes, ses conséquences*. Paris.

———. 1881. *La psychologie allemande contemporaine*. 2nd ed. Paris. English trans., James Mark Baldwin, *German Psychology of To-Day*. New York: Charles Scribner's Sons, 1886.

———. 1890. *The psychology of attention*. Chicago: Open Court.

Richard, Jules. 1905. Les principes des mathématiques et le problème des ensembles. *Revue générale des sciences pures et appliquées* 16:541-42. English trans. in Heijenoort 1967a, 142–44.

Richards, Joan L. 1979. The reception of a mathematical theory: Non-Euclidean geometry in England, 1865–1883. In *Natural order: Historical studies of scientific culture*, ed. B. Barnes and S. Shapin. Beverley Hills, Calif.: Sage Publications.

———. 1986. Projective geometry and mathematical progress in mid-Victorian Britain. *Studies in the History and Philosophy of Science* 17.3:297–325.

———. 1988. *Mathematical visions: The pursuit of geometry in Victorian England*. New York: Academic Press.

Riemann, Bernhard. 1851. Grundlagen für eine allgemeine Theorie der Functionen einer veränderlichen complexen Grösse. PhD diss. University of Göttingen. Also in Riemann 1990, 35–77.

———. 1854. Ueber die Hypothesen welche der Geometrie zu Grunde liegen. *K. Ges. Wiss. Göttingen* 13:1–20, pub. 1867. Also in Riemann 1990, 304–19.

———. 1857. Theorie der Abel'schen Functionen. *Journal für Mathematik* 54:88–142. Also in Riemann 1990, 120–74.

———. 1876. *Schwere, Elektricität und Magnetismus: Nach den Vorlesungen von B. Riemann bearbeitet von K. Hattendorff*. Hannover.

———. 1990. *Gesammelte Mathematische Werke, Wissenschaftlicher, Nachlass und Nachträge, Collected Papers*, ed. R. Narasimhan. New York: Springer.

———. 2004. *Collected Papers*. Trans., Roger Baker, Charles Christenson, and Henry Orde. Heber City, Utah: Kendrick Press.

Riesz, Friedrich. 1906. Die Genesis des Raumbegriffs. *Mathematische und Naturwissenschaftliche Berichte aus Ungarn* 24: 309–353. In *Oeuvres complètes*, 2 vols. Vol.1, 111–54. Budapest: Akademai Kiado, 1960.

Ringer, Fritz K. 1990. *The decline of the German mandarins: The German academic community, 1890–1933*. Cambridge, Mass.: Harvard University Press.

Ritchie, A. D. 1921 Review of physics: The elements. *Mind* (2) 30:207–14.

Rodríguez Hernández, Laura Regina. 2006. *Friedrich Riesz' Beiträge zur Herausbildung des modernen mathematischen Konzepts abstrakter Raüme: Synthesen intellectueller Kulturen in Ungarn, Frankreich und Deutschland.* PhD diss., Johannes. Gutenberg-Universität, Mainz.

Rosen, Charles. 1996. *Arnold Schoenberg.* Chicago: University of Chicago Press.

———. 1999. Mallarmé the Magnificent. *New York Review of Books* 56.9 (May 20): 42–46.

Rowe, David E. 1983. A forgotten chapter in the history of Felix Klein's Erlanger Programm. *Historia Mathematica* 10: 448–57.

———. 1986. "Jewish mathematics" at Göttingen in the era of Felix Klein. *Isis* 77:422–49.

———. 1992. Klein, Mittag-Leffler, and the Klein-Poincaré correspondence of 1881–1882. In *Amphora, Festschrift für Hans Wussing*, 597–618. Boston: Birkhäuser.

———. 1994. The philosophical views of Klein and Hilbert. In *The intersection of history and mathematics*, 187–202. Science Networks Historical Studies 15. Basel.

———. 1997. Perspective on Hilbert. *Perspectives in Science* 5.4:533–70.

———. 2000. Episodes in the Berlin-Göttingen Rivalry. *Mathematical Intelligencer* 22.1:60–69.

Royce, Josiah. 1902. Recent logical inquiries and their psychological bearings: Address of the president before the American Psychological Association, Chicago, January, 1902. *Psychological Review* 9:105–33.

———. 1905. The relation of the principles of logic to the foundations of geometry. *Trans-American Mathematical Society* 6.3.

———. 1970. *The letters of Josiah Royce.* Edited with an introduction by John Clendenning. Chicago: University of Chicago Press.

Rüdenberg, L., and H. Zassenhaus, eds. 1973. *Hermann Minkowski—Briefe an David Hilbert.* Berlin: Springer.

Russ, Steve. 2004. *The mathematical works of Bernhard Bolzano.* Oxford: Oxford University Press.

Russell, Bertrand. 1897. *An essay on the foundations of geometry.* Cambridge: Cambridge University Press.

———. 1899 On the axioms of geometry. In *The Collected Papers of Bertrand Russell*, ed. Nicholas Griffin and Albert C. Lewis, vol. 2, 394–415. London: Hyman Unwin, 1990.

———. 1900. *A critical exposition of the philosophy of Leibniz.* Cambridge: Cambridge University Press.

———. 1901. Mathematics and the metaphysicians. Repr. in *Mysticism and logic*. London: Unwin, 1963.

———. 1903a. *Principles of mathematics.* Cambridge: Cambridge University Press.

———. 1903b. Recent work on the philosophy of Leibniz. *Mind* (2) 12.46:177–201.

Russell, Bertrand, and Alfred North Whitehead. 1910–12. *Principia Mathematica.* 3 vols. Cambridge: Cambridge University Press.

Russinoff, I. Susan. 1999. The syllogism's final solution. *Bulletin of Symbolic Logic* 5.4:451–69.

Russo, L. 1998. The definitions of fundamental geometric entities contained in Book I of Euclid's *Elements. Archive for History of Exact Sciences* 52.3:195–219.

Sagan, Hans. 1994. *Space-filling curves.* New York: Springer.

Samuel, Pierre. 1988. *Projective geometry.* New York: Springer.

Santerre, S. 1907. *Psychologie du nombre.* Paris: Octave Doin.

Sartorius, W. von Waltershausen. 1856. *Gauss zum Gedächtnis.* Leipzig: Hirzel, Repr. Martin Sandig oHG, 1965.

Sauer, Tilman. 2005. Einstein equations and Hilbert action. *Archive for History of Exact Sciences* 59.6:577–590.

Scanlon, M. 1991. Who were the American postulate theorists? *Journal of Symbolic Logic* 56:981–1002.

Scharlau, Winfried, and Hans Opolka. 1985. *From Fermat to Minkowski*. Heidelberg: Springer.

Schlick, Moritz. 1918. *Allgemeine Erkenntnistheorie*. 2nd. ed. Berlin: Springer, 1925. English trans., A. E. Blumberg, *General Theory of Knowledge*. Berlin: Springer, 1974. Repr. Chicago: Open Court, 1985.

Schoenflies, Arthur. 1902. Die Entwickelung der Lehre von den Punktmannigfaltigkeiten. *Jahresbericht der Deutschen Mathematiker-Vereinigung* 8.2:1–250.

———. 1906a. Über die logischen Paradoxien der Mengenlehre. *Jahresbericht der Deutschen Mathematiker-Vereinigung* 15:19–25.

———. 1906b. Die Beziehungen der Mengenlehre zur Geometrie und Funktionentheorie. *Jahresbericht der Deutschen Mathematiker-Vereinigung* 145:557–76.

———. 1911. Über die Stellung der Definition in der Axiomatik. *Jahresbericht der Deutschen Mathematiker-Vereinigung* 20:222–55.

Scholz, Erhard. 1982. Herbart's influence on Bernhard Riemann. *Historia Mathematica* 9.4:413–40.

———. 1989. *Symmetrie Gruppe Dualität*. Boston: Birkhäuser.

———. 2004. C. F. Gauss' Präzisionsmessungen terrestricher Dreiecke und seine Überlegungen zur empirischen Fundierung der Geometrie in den 1820er Jahren. In *Form, Zahl, Ordnung: Studien zur Wissenschaftsund Technikgeschichte*, ed. R. Seising, M. Folkerts, and U. Hashagen, 355–380. Franz Steiner Verlag.

———. 2006. Practice-related symbolism in relation to H. Weyl's mature view of mathematical knowledge. In *The architecture of modern mathematics*, ed. J. Ferreirós and J. J. Gray, Oxford: Oxford University Press.

Scott, Geoffrey. 1914. *The architecture of humanism*. London: Constable. Repr., New York: Norton, 2004.

Schröder, Ernst. 1880. Rezension von Freges Begriffsschrift. *Zeitschrift für Mathematik und Physik, Hist.-Lit. Abtheilung* 25:81–94.

———. 1890, 1891, 1895, 1905. *Algebra der Logik*. 3 vols. Repr., New York: Chelsea, 1966.

———. 1898. Über Pasigraphie, ihren gegenwärtigen Stand und die pasigraphische Bewegung in Italien. In *Verhandlungen des Ersten Internationalen Mathematiker-Kongresses vom 9. bis 11. August 1897*. Leipzig: Teubner. Trans., On pasigraphy: Its present state and the pasigraphic movement in Italy. *The Monist* 9 (1899): 44–62, Corrigenda, 320.

Schubring, Gerd. 1996. Changing cultural and epistemological views on mathematics and different institutional contexts in nineteenth-century Europe. In *L'Europe Mathématique*, ed. C. Goldstein, J. J. Gray, and J. Ritter, 361–88. Paris: Éditions de la Maison des sciences de l'homme.

Schur, Friedrich. 1899. Über den Fundamentalsatz der projectiven Geometrie. *Mathematische Annalen* 51:401–409.

Schwarzschild, Karl. 1900. Ueber das zulässige Krümmungsmass des Raumes. *Vierteljahrsschrift des Astronomische Gesellschaft* 35:337–47.

Segal, Sanford L. 2003. *Mathematicians under the Nazis*. Princeton, N.J.: Princeton University Press.

Segre, Corrado. 1884. Studio delle quadriche in uno spazio lineare a un numero qualunque di dimensioni. *Memorie della Regia Accademia delle Scienze di Torino* (2) 36:3–86.

Sieg, Wilfried. 2002. Beyond Hilberts reach? In *Reading natural philosophy—Essays in the history and philosophy of science and mathematics*, ed. D. Malament, 363–405. Evanston, Ill.: Open Court.

Sieg, Wilfried, and Dirk Schlimm. 2005. Dedekind's analysis of number: Systems and axioms. *Synthese* 147.1: 121–170.

Skolem, Thoralf. 1922. Einige Bemerkungen zur axiomatischen Begründung der Mengen-lehre. In *5. Kongress Skandinav. in Helsingfors vom 4. bis 7. Juli 1922*, 217–232. Hel-singfors: Akademische Buchhandlung. English trans, Some remarks on axiomatized set theory, in Heijenoort 1967, 290–301.

Sluga, Hans. 1980. *Gottlob Frege*. London: Routledge.

Stachel, John. 1989. The rigidly rotating disc as the "Missing Link" in the history of general relativity. In *Einstein and the history of general relativity*, ed. D. Howard and J. Stachel, 48–62. Boston: Birkhäuser.

———. 1995. History of relativity. In *Twentieth-Century Physics*, ed. L. M. Brown, A. Pais, and B. Pippard, 249–356. Bristol: Institute of Physics Publishing.

Stäckel, P. 1913. *Wolfgang und Johann Bolyai: Geometrische Untersuchungen, Leben und Schriften der beiden Bolyai*. Leipzig: Teubner.

Staudt, Carl Georg Christian von. 1847. *Geometrie der Lage*. Nürnberg.

Stefan, R. 1999. Schoenberg and Bach. In *Schoenberg and his world*, ed. W. Frisch, 126–40. Princeton, N.J.: Princeton University Press.

Stillwell, John, ed. 1996. *Sources of hyperbolic geometry*. Trans. J. Stillwell. American and London Mathematical Societies, HMath 10, Providence, R.I.

Stolz, Otto. 1885. *Vorlesungen über allgemeine Arithmetik*. Leipzig: Teubner.

Stout, George Frederick. 1930. *Studies in philosophy and psychology*. New York: Macmillan.

Strauss, David Friedrich. 1835. *Das Leben-Jesu*. Tübingen. English trans., Marian Evans (George Eliot), *The life of Jesus, critically examined*. London, 1860.

Stubhaug, Arild. 2002. *The mathematician Sophus Lie*. English trans., Richard H. Daly. Berlin: Springer.

Study, Eduard. 1914. *Die realistische Weltansicht und die Lehre vom Raume; Geometrie Anschauung und Erfarhrung*. Braunschweig.

Stumpf, Carl. 1873. *Ueber den psychologischen Ursprung der Raumvorstellung*. Leipzig: Hirzel.

Sweet, Henry. 1875–76. Words, logic and grammar. *Transactions of the Philological Society*, 470–503.

Tait, W. W. 1986. Truth and proof: The platonism of mathematics. *Synthese* 69: 341–370. Repr. in Hart 1996, 141–67.

Tannery, Jules. 1886. *Introduction à la théorie des fonctions d'une variable*. Paris: Gauthier-Villars.

Tannery, Jules, and Jules Molk. 1893–1902. Éléments de la théorie des fonctions elliptiques. 4 vols. Paris: Gauthier-Villars.

Tappenden, Jamie. 2006. The Riemannian background to Frege's philosophy. In *The Architecture of Modern Mathematics*, ed. J. Ferreirós and J. J. Gray, 97–132. Oxford: Oxford University Press.

Taton, René. 1951. *L'Oeuvre scientifique de Monge*. Paris.

Taub, Abraham. 1951. Empty space-times admitting a three-parameter group of motions. *Annals of Mathematics* 53:472–90.

Thiel, Christian. 1991. Brouwer's philosophical language research and the concept of the ortho-language in German constructivism. In *Significs, mathematics and semiotics: The significs movement in the Netherlands*, ed. E. Heijermann and H. W. Schmitz, 21–32. Münster: Nodus Publications.

Thomae, Johannes. 1870. *Abriss einer Theorie der complexen Functionen und der Theta-functionen einer Veränderlichen*. Halle.

Thomson, Sir William (Lord Kelvin), and Peter Guthrie Tait. 1879. *Treatise on natural philosophy*. 2 vols. Cambridge: Cambridge University Press.

Thurston, William. 1997. *Three-dimensional geometry and topology*. Ed. Silvio Levy. Princeton, N.J.: Princeton University Press.

Tieszen, Richard. 2000. Gödel and Quine on meaning and mathematics. In *Between logic and intuition*, ed. Gila Sher and Richard Tieszen. Cambridge: Cambridge University Press.

Tobias, Wilhelm. 1875. *Grenzen der Philosophie constatirt gegen Riemann und Helmholtz, vertheidigt gegen von Hartman und Lasker.* Berlin: Müller.

Tobies, Renate. 1981. *Felix Klein*. Biographien hervorragender Naturwissenschaftler, Techniker und Mediziner, vol. 50. Leipzig: Teubner.

Toepell, Michael-M. 1986. *Über die Entstehung von David Hilberts "Grundlagen der Geometrie."* Göttingen: Vandenhoeck & Ruprecht.

Torretti, Roberto. 1984. *Philosophy of Geometry from Riemann to Poincaré.* Dordrecht: Reidel.

Trendelenburg, Friedrich. 1840. *Logische Untersuchungen.* 3d ed., Berlin, 1870.

van der Waerden, Bartel Leendert. 1930–31. *Moderne Algebra.* Berlin: Springer.

Van Vleck, E. B. 1914. The influence of Fourier's series upon the development of mathematics. *Science,* n.s. 39. nr.995:113–24.

Vasiliev, A. V. 1924. *Space Time Motion: An historical introduction to the general theory of relativity.* London: Chatto and Windus.

Veblen, Oswald. 1904. A system of axioms for geometry. *Transactions of the American Mathematical Society* 5:343–84.

———. 1905. Theory of plane curves in non-metrical analysis situs. *Transactions of the American Mathematical Society* 6:83–98.

———. 1907. Collineations in a finite projective geometry. *Transactions of the American Mathematical Society* 8:366–68.

Veblen, Oswald, and W. H. Bussey. 1906. *Finite projective geometries. Transactions of the American Mathematical Society* 7:241–259.

Veblen, Oswald, and J. W. Young. 1910. 1917. *Projective geometry.* 2 vols. Boston: Ginn and Co.

Veronese, Giuseppe. 1891. *Fondamenti di geometrie a più dimensioni.* Padua: Tipografia del Seminario.

———. 1894. *Grundzüge der Geometrie von mehreren Dimensionen und mehreren arten gradliniger Einheiten in elementarer Form entwickelt.* Leipzig: Teubner.

Vilkko, R. 1998. The reception of Frege's Begriffsschrift. *Historia Mathematica* 25.4:412–422.

Vivanti, G. 1891a. Sull' infinitesimo attuale. *Rivista di Matematica* 1: 135–153.

———. 1891b. Ancora sull' infinitesimo attuale. *Rivista di Matematica* 1:248–255.

Voelke, Jean-Daniel. 2005. *Renaissance de la géométrie non euclidienne entre 1860 et 1900.* Berne: Peter Lang.

Volkert, Klaus Th. 1986. *Die Krise der Anschauung.* Göttingen: Vandenhoeck & Ruprecht.

von Neumann, John. 1951. The general and logical theory of automata. In *Cerebral mechanisms in behavior—the Hixon Symposium*, 1–33. New York: Wiley. Also in *Collected Works*, ed. A.H. Taub, vol. 5, 288–328. Oxford: Oxford University Press.

Voss, Aurel. 1913. *Über das Wesen der Mathematik.* Rede. Erweitert und mit Anmerkungen versehen. 2d ed. Leipzig: B. G. Teubner.

———. 1914. Über die mathematische Erkenntnis. In *Die Kultur der Gegenwart*, pt. 3, sec. 1, 3d ed., E1–148 E.

Vucinich, Alexander. 1999. Mathematics and dialectics in the Soviet Union: The Pre-Stalin period. *Historia Mathematica* 26.2: 107–24.

Walter, Scott. 1997. La vérité en géométrie: Sur le rejet mathématique de la doctrine conventionnaliste. *Philosophia Scientiæ* 2:103–35.

———. 1999. The non-Euclidean style of Minkowskian relativity. In *The symbolic universe: Geometry and physics, 1890–1930*, ed. J. J. Gray, 91–127. Oxford: Oxford University Press.

Warwick, Andrew. 2003. *Masters of theory: Cambridge and the rise of mathematical physics.* Chicago: University of Chicago Press.

Watson, R. I., Sr. 1978. *The great psychologists*. 4th ed. New York: J. B. Lippincott.

Webb, Judson Chambers. 1980. *Mechanism, mentalism and meta-mathematics: An essay on finitism*. Synthese Library, vol. 137. Dordrecht: Reidel.

Weber, Heinrich L. 1891–92. Kronecker. *Jahresbericht der Deutschen Mathematiker-Vereinigung* 2:5–23.

———. 1906. Elementare Mengenlehre. *Jahresbericht der Deutschen Mathematiker-Vereinigung* 15:173–84.

Weierstrass, Karl. 1870. Über die sogenannte Dirichlet'sche Princip. In Weierstrass 1895, vol. 2, 49–54.

———. 1872. Über continuirliche Functionen eines reellen Arguments. In Weierstrass 1895, vol. 2, 71–74.

———. 1895. *Mathematische Werke*. 2 vols. Berlin: Mayer & Müller.

Welby, Lady Victoria. 1903. *What is meaning? Studies in the development of significance*. London: Macmillan.

———. 1910. Significs. In *Encyclopaedia Britannica*, 11th ed.

Wellstein, Joseph. 1905. Grundlagen der Geometrie. In *Enzyklopädie der Elementar-Mathematik: Ein Handbuch für Lehrer und Studierende*, ed. H. Weber and J. Wellstein, vol. 2. Leipzig: Teubner.

Weyl, Hermann. 1910. Über die Definitionen der mathematischen Grundbegriffe. *Mathematisch-naturwissenschaftliche Blätter* 7:93–95, 109–113; in *Gesammelte Abhandlungen*, vol. 1, 298–304.

———. 1918. *Das Kontinuum: Kritische Untersuchungen über die Grundlagen der Analysis*. Leipzig: Viet & Co.

———. 1924. Randbemerkungen zu hauptproblemen der Mathematik. *Mathematische Zeitschrift* 20:131–50. Also in *Gesammelte Abhandlungen*, vol. 2, 433–52.

———. 1925–27. Die heutige Erkenntnislage in der Mathematik. *Symposion* 1: 1–32. Also in *Gesammelte Abhandlungen* vol. 2, 511–42. English trans., B. Müller in Mancosu 1998, 123–42.

———. 1940. *Algebraic theory of numbers*. Annals of Mathematics Studies 1. Princeton, N.J.: Princeton University Press.

Whitehead, Alfred North. 1898. *A treatise on universal algebra, with applications*. Cambridge: Cambridge University Press.

———. 1906. *The axioms of projective geometry*. Cambridge: Cambridge University Press Tract 4.

Wiechert, Emil. 1899. *Grundlagen der Elektrodynamik*. Festschrift zur Feier der Enthüllung des Gauss-Weber-Denkmals in Göttingen.

Wiles, Andrew. 1995. Modular elliptic curves and Fermat's last theorem. *Annals of Mathematics* 141:443–551.

Wiles, Andrew, and R. Taylor. 1995. Ring-theoretic properties of certain Hecke Algebras. *Annals of Mathematics* 141:553–572.

Wilson, E. B. 1906. The foundations of science, review of *Wissenschaft und Hypothese von H. Poincaré*, trans. F. and L. Lindemann. *Bulletin of the American Mathematical Society* 12.4:187–93.

Winter, Maximilien. 1911. *La méthode dans la philosophie des mathématiques*. Paris: Alcan.

Wittgenstein, Ludwig. 1958. *Philosophical investigations*. Trans. G.E.M. Anscombe, Oxford: Blackwell.

Woods, Frederick S. 1902. Space of constant curvature. *Annals of Mathematics* (2) 3:71–112.

———. 1903. Forms of non-Euclidean space. In The Boston Colloquium, *Lectures on Mathematics* 31–74.

Wozniak, R. H. 1999. *Classics in psychology, 1855–1914: Historical essays*. Bristol: Thoemmes Press.

Wundt, Wilhelm. 1874. *Grundzüge der physiologischen Psychologie*. Leipzig: Engelmann.

———. 1900. *Völkerpsychologie: Eine Untersuchung der Entwicklungsgesetze von Sprache, Mythus und Sitte*. Leipzig: Engelmann.

———.1980. Selected texts from writings of Wilhelm Wundt. Trans. S. Diamond. *Wilhelm Wundt and the making of a scientific psychology*, ed. R.W. Rieber, 155–77. New York: Plenum.

Wussing, Hans. 1969. *Die Genesis des abstrakten Gruppenbegriffes*. VEB Deutscher Verlag der Wissenschaften. English trans., A. Shenitzer with H. Grant, *The Genesis of the Abstract Group Concept*. Cambridge, Mass.: MIT Press, 1984.

Yandell, Ben. 2002. *The honors class: Hilbert's problems and their solvers*. Natick, Mass.: AK Peters.

Zermelo, Ernst. 1908. Untersuchungen über die Grundlagen der Mengenlehre. *Mathematische Annalen* 65: 261–281. English trans. in Heijenoort 1967a, 199–215.

Zöllner, J.C.F. 1876. *Principien einer electrodynamischen Theorie der Materie*. Leipzig.

Zöllner, J.C.F. 1878. *Transcendental physics: An account of experimental investigations*. From the scientific treatises of J.C.F. Zöllner. Trans. and ed. C. C. Massey.

Zygmunt, Jan. 1991. Mojżesz Presburger: Life and work. *History and Philosophy of Logic* 12: 211–23.

INDEX

For technical terms, see the glossary.